Groups and Symmetries

From Neolithic Scots to John McKay

Volume 47

CRM PROCEEDINGS & LECTURE NOTES

Centre de Recherches Mathématiques
Montréal

Groups and Symmetries

From Neolithic Scots to John McKay

John Harnad
Pavel Winternitz
Editors

The Centre de Recherches Mathématiques (CRM) of the Université de Montréal was created in 1968 to promote research in pure and applied mathematics and related disciplines. Among its activities are special theme years, summer schools, workshops, postdoctoral programs, and publishing. The CRM is supported by the Université de Montréal, the Province of Québec (FQRNT), and the Natural Sciences and Engineering Research Council of Canada. It is affiliated with the Institut des Sciences Mathématiques (ISM) of Montréal. The CRM may be reached on the Web at www.crm.math.ca.

American Mathematical Society
Providence, Rhode Island USA

The production of this volume was supported in part by the Fonds Québécois de la Recherche sur la Nature et les Technologies (FQRNT) and the Natural Sciences and Engineering Research Council of Canada (NSERC).

2000 *Mathematics Subject Classification.* Primary 20D08; Secondary 11F03.

Library of Congress Cataloging-in-Publication Data

Groups and symmetries : from Neolithic Scots to John McKay / John Harnad, Pavel Winternitz, editors. p. cm. – (CRM proceedings & lecture notes ; v. 47)
Includes bibliographical references.
ISBN 978-0-8218-4481-6 (alk. paper)
1. Finite groups—Congresses. 2. Galois theory—Congresses. 3. Symmetry (Mathematics)—Congresses. 4. Modular functions—Congresses. 5. Geometry, Algebraic—Congresses. 6. McKay, John, 1939—Congresses. I. McKay, John, 1939- II. Harnad, J. P. (John P.), 1946- III. Winternitz, Pavel.

QA174.G755 2009
512′.23—dc22 2009010376

∞ The paper used in this book is acid-free and falls within the guidelines
established to ensure permanence and durability.
This volume was submitted to the American Mathematical Society
in camera ready form by the Centre de Recherches Mathématiques.
Visit the AMS home page at `http://www.ams.org/`

10 9 8 7 6 5 4 3 2 1   14 13 12 11 10 09

For Trinh

Contents

Preface ix

Workshop Poster xi

Workshop Photos xii

List of Participants xv

Introduction and Background
John McKay 1

Symmetric Sums Associated to the Factorization of Grunsky Coefficients
Hélène Airault 3

A Monstrous Proposal
Daniel Allcock 17

Quivers and Difference Painlevé Equations
Philip Boalch 25

Families of Ramanujan Graphs and Quaternion Algebras
Denis X. Charles, Eyal Z. Goren, and Kristin E. Lauter 53

Normal Forms, K3 Surface Moduli, and Modular Parametrizations
Adrian Clingher, Charles F. Doran, Jacob Lewis, and Ursula Whitcher 81

Spontaneous Generation of Hilbert Modular Functions
Harvey Cohn 99

On a Class of Congruence Subgroups
C. J. Cummins and Noushin Sabetghadam Haghighi 105

McKay's Correspondence for Cocompact Discrete Subgroups of $\mathrm{SU}(1,1)$
Igor V. Dolgachev 111

Arithmetic Groups and the Affine E_8 Dynkin Diagram
John F. Duncan 135

The Galois Action on Character Tables
Terry Gannon 165

Hecke Operators in Equivariant Elliptic Cohomology and Generalized Moonshine
Nora Ganter 173

Sculptural Presentation of the Icosahedral Rotation Group
George W. Hart 211

Spherical Harmonics and the Icosahedron
Nigel Hitchin 215

Alternating Group and Multivariate Exponential Functions
Anatoliy Klimyk and Jiří Patera 233

Moonshine Elements in Elliptic Cohomology
Jack Morava 247

The Generalized Artin Conjecture and Arithmetic Orbifolds
M. Ram Murty and Kathleen L. Petersen 259

McKay Correspondence
Iku Nakamura 267

On Ground Fields of Arithmetic Hyperbolic Reflection Groups
Viacheslav V. Nikulin 299

Moonshine-type Functions and the CRM Correspondence
Simon P. Norton 327

Monodromy Evolving Deformations and Halphen's Equation
Yousuke Ohyama 343

Integral Solutions of Apéry-like Recurrence Equations
Don Zagier 349

Preface

The conference bearing the name of this volume, honoring the remarkable contributions of John McKay over four decades of research, took place at the CRM in Montréal during the three day period: April 27–29, 2007. There were a total of 46 invited participants, of whom eighteen gave one hour talks (one of them, Yu. Manin, gave two), and one (George Hart) gave a workshop session on polyhedral sculpture construction.

The talks covered a wide range of topics, with principal focus given to two of the areas in which John McKay has made pioneering contributions: "Monstrous Moonshine" and the "McKay Correspondence." The general areas included were: number theory, modular functions, group representations, Galois theory and algebraic geometry.

There were many who contributed to the success of this event, which had a great significance in marking the importance of John Mckay's ideas and honoring his contributions. Financial and organizational support for this conference was generously provided by Concordia University, the CRM and the CRM CICMA Laboratory. Above all, those without whose support and encouragement this event could not have taken place were:

(1) CRM Director François Lalonde, who assured the total support of the CRM, putting its facilities, resources and personnel fully behind its realization, and helping to communicate the value of decades of exceptional contributions by John McKay to mathematics.

(2) David Graham, who at the time was Dean of the Faculty of Arts and Science at Concordia University (and currently is Provost) and assured the enthusiastic and generous support of the Faculty, thus providing the recognition of John McKay's remarkable merits and contributions by his home university that he had long merited.

(3) Very substantial practical, material and moral support generously provided by:

- Graham Carr, Vice Dean, Faculty of Arts and Sciences, Concordia University
- Truong Vo Vanh, Associate Vice President for Research, Concordia University
- Louise Dandurand, Vice President for Research, Concordia University.

(4) Several others who communicated their enthusiastic support in letters and representations to Concordia University: Ram Murty (FRSC, Queens Univ.), Henri Darmon (FRSC, McGill Univ.), Jacques Hurtubise (FRSC, McGill Univ.), André Joyal (FRSC, UQÀM), as well as Université de Montréal Rector Luc Vinet.

Several mathematicians of international renown, including Yuri Manin, John Conway, Nicholas Katz, Igor Dolgachev, came as invited speakers and others, who were unable to attend, such as Don Zagier and Nigel Hitchin, have nevertheless contributed to this volume. In all, there were a total of about sixty participants, from all corners of the globe (Japan, France, United States, Canada, United Kingdom, Germany, Argentina). There was also a very stimulating and entertaining presentation, in the form of a group workshop devoted to assembling a sculpture by mathematician-sculptor extraordinaire George Hart, consisting of a work of his, entitled "People," which is based on the symmetry group of the icosahedron, in which the participants were obliged, under the gentle coaxing of the sculptor, to mobilize their knowledge of the structural properties and cyclic subgroups.

The topics covered, within the general areas listed above, included:

- Moonshine and the Monster;
- McKay correspondence;
- Finite groups;
- Hyperbolic geometry;
- Galois theory;
- Modular groups and functions.

By all accounts of those who took part, the workshop was a very great success.

John Harnad
Director, Mathematical Physics Laboratory
Centre de recherches mathématiques, and
Department of Mathematics and Statistics
Concordia University

Montréal, December 2008

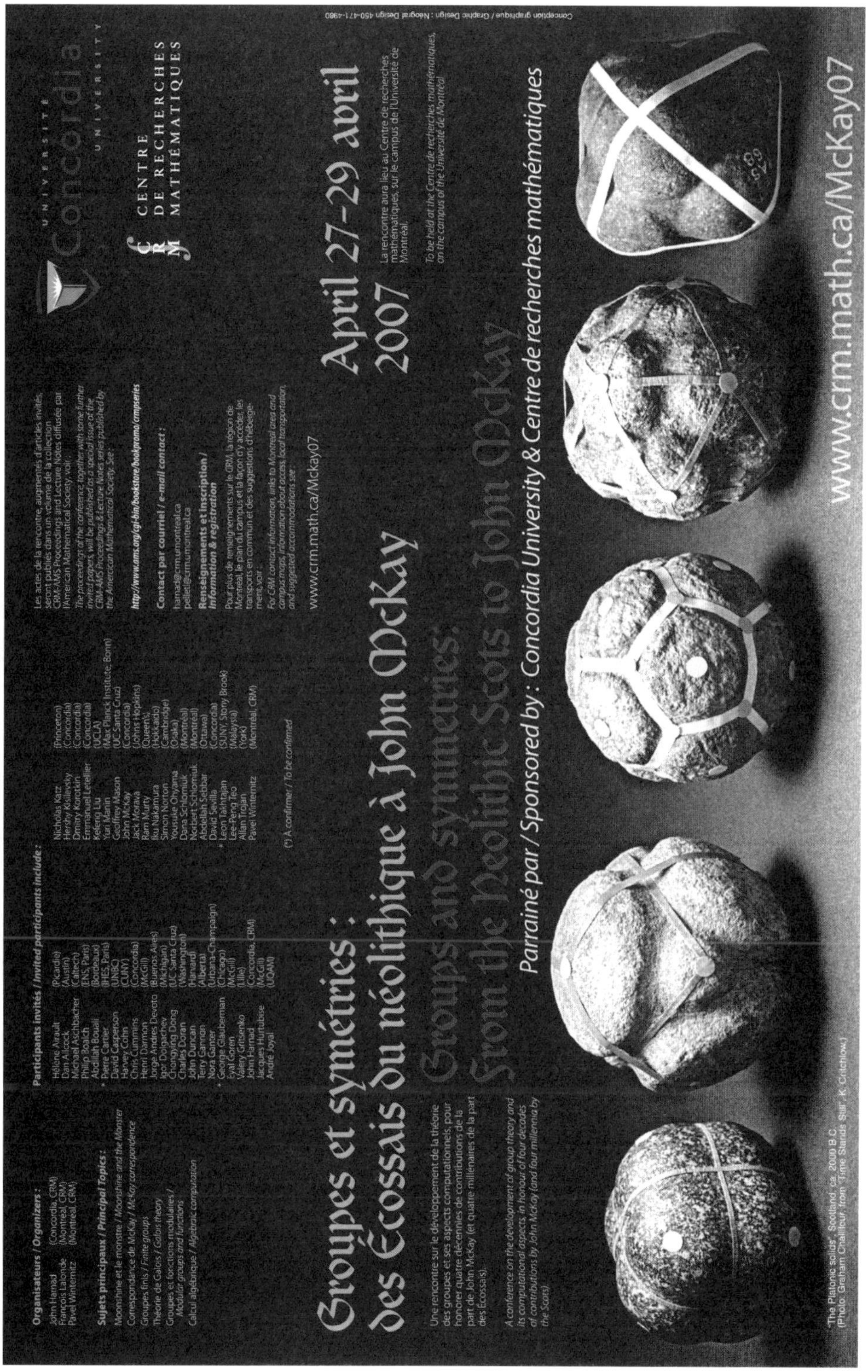
Organisateurs / Organizers :
John Harnad (Concordia, CRM)
François Lalonde (Montréal, CRM)
Pavel Winternitz (Montréal, CRM)
Sujets principaux / Principal Topics :
Moonshine et le monstre / Moonshine and the Monster
Correspondance de McKay / McKay correspondence
Groupes finis / Finite groups
Théorie de Galois / Galois theory
Groupes et fonctions modulaires / Modular groups and functions
Calcul algébrique / Algebraic computation
Participants invités / Invited participants include :
Hélène Airault (Picardie)
Dan Allcock (Austin)
Michael Aschbacher (Caltech)
Philip Boalch (ENS, Paris)
Abdlilah Bouali (Bordeaux)
* Pierre Cartier (IHES, Paris)
David Casperson (UNBC)
Harvey Cohn (CUNY)
Chris Cummins (Concordia)
Henri Darmon (McGill)
Jorge Andres Devoto (Buenos Aires)
Igor Dolgachev (Michigan)
Chongying Dong (UC Santa Cruz)
Charles Doran (Washington)
John Duncan (Harvard)
Terry Gannon (Alberta)
Nora Ganter (Urbana-Champaign)
* George Glauberman (Chicago)
Eyal Goren (McGill)
Valery Gritsenko (Lille)
John Harnad (Concordia, CRM)
Jacques Hurtubise (McGill)
André Joyal (UQAM)
Nicholas Katz (Princeton)
Hershy Kisilevsky (Concordia)
Dmitry Korotkin (Concordia)
Emmanuel Letellier (Concordia)
Kefeng Liu (UCLA)
Yuri Manin (Max Planck Institute, Bonn)
Geoffrey Mason (UC Santa Cruz)
John McKay (Concordia)
Jack Morava (Johns Hopkins)
Ram Murty (Queen's)
Iku Nakamura (Hokkaido)
Simon Norton (Cambridge)
Yousuke Ohyama (Osaka)
Dana Schlomiuk (Montréal)
Norbert Schlomiuk (Montréal)
Abdellah Sebbar (Ottawa)
David Sevilla (Concordia)
* Leon Takhtajan (SUNY, Stony Brook)
Lee-Peng Teo (Malaysia)
Allan Trojan (York)
Pavel Winternitz (Montréal, CRM)
(*) À confirmer / To be confirmed
Les actes de la rencontre, augmentés d'articles invités, seront publiés dans un volume de la collection CRM-AMS Proceedings and Lecture Notes diffusée par l'American Mathematical Society, voir :
The proceedings of the conference, together with some further invited papers, will be published as a special issue of the CRM-AMS Proceedings & Lecture Notes series published by the Amercican Mathematical Society. See :
http://www.ams.org/cgi-bin/bookstore/bookpromo/crmpseries
Contact par courriel / e-mail contact :
harnad@crm.umontreal.ca
pelleti@crm.umontreal.ca
Renseignements et inscription / Information & registration
Pour plus de renseignements sur le CRM, la région de Montréal, le plan du campus et la façon d'y accéder, les transports en commun et des suggestions d'hébergement, voir :
For CRM contact information, links to Montreal area and campus maps, information about access, local transportation, and suggested accommodations see :
www.crm.math.ca/Mckay07
UNIVERSITÉ Concordia UNIVERSITY
CENTRE DE RECHERCHES MATHÉMATIQUES
Groupes et symétries : des Écossais du néolithique à John McKay
April 27-29 avril 2007
La rencontre aura lieu au Centre de recherches mathématiques, sur le campus de l'Université de Montréal.
To be held at the Centre de recherches mathématiques, on the campus of the Université de Montréal
Une rencontre sur le développement de la théorie des groupes et ses aspects computationnels, pour honorer quatre décennies de contributions de la part de John McKay (et quatre millénaires de la part des Écossais).
A conference on the development of group theory and its computational aspects, in honour of four decades of contributions by John McKay (and four millennia by the Scots).
Groups and symmetries: From the Neolithic Scots to John McKay
Parrainé par / Sponsored by : Concordia University & Centre de recherches mathématiques
"The Platonic solids", Scotland, ca. 2000 B.C.
(Photo: Graham Challifour, from "Time Stands Still", K. Critchlow.)
www.crm.math.ca/McKay07
Conception graphique / Graphic Design : Néograf Design 450-471-4980

D. Sevilla Gonzalez

H. Airault and H. Cohn

J. Conway and V. V. Nikulin

J. Harnad

T. Gannon and C. Cummins

N. M. Katz

People Sculpture and J. Conway

Creating People Sculpture

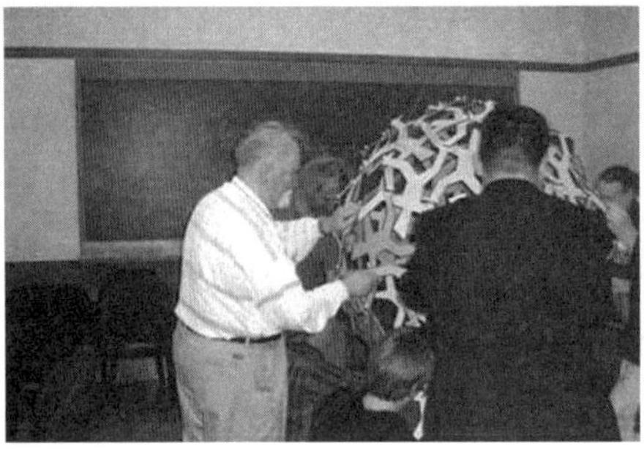

J. McKay

J. McKay

J. Hurtubise, E. Goren,
and S. P. Norton

Y. Ohyama and J. McKay

List of Participants

Hélène Airault
Université de Picardie
hairault@insset.u-picardie.fr

Syed Twareque Ali
Concordia University
stali@mathstat.concordia.ca

Daniel Allcock
University of Texas at Austin
allcock@math.utexas.edu

François Bergeron
Université du Québec à Montréal
bergeron.francois@uqam.ca

Philip Boalch
École Normale Supérieure
boalch@dma.ens.fr

David Casperson
University of Northern British Columbia
casper@unbc.ca

Harvey Cohn
Institute for Defense Analyses
bhcohn@fea.net

John Conway
Princeton University
jhorcon@yahoo.com

C. J. Cummins
Concordia University
cummins@mathstat.concordia.ca

Abdessalem Daghbouche
University of Michigan
amine1970@yahoo.com

Henri Darmon
McGill University
darmon@math.mcgill.ca

Jorge Andres Devoto
Instituto Tecnologoco de Buenos Aires
jdevoto@itba.edu.ar

Igor V. Dolgachev
University of Michigan
idolga@umich.edu

Chongying Dong
University of California, Santa Cruz
dong@math.ucsc.edu

John Duncan
Harvard University
duncan@math.harvard.edu

Abdelkrim El Basraoui
University of Ottawa
aelba026@uottawa.ca

Noam D. Elkies
Harvard University
elkies@math.harvard.edu

David Ford
Concordia University
ford@cse.concordia.ca

Terry Gannon
University of Alberta
tgannon@math.ualberta.ca

Nora Ganter
University of Illinois at Urbana-Champaign
ganter@math.uiuc.edu

Alexandru E. Ghitza
Colby College
aghitza@colby.edu

Eyal Z. Goren
McGill University
eyal.goren@mcgill.ca

Valery Gritsenko
Université de Lille 1
valery.gritsenko@math.univ-lille1.fr

Richard L. Hall
Concordia University
rhall@mathstat.concordia.ca

Ian Hambleton
McMaster University
hambleton@mcmaster.ca

Koichiro Harada
The Ohio State University
haradako@math.ohio-state.edu

John Harnad
Concordia University and CRM
harnad@crm.umontreal.ca

George Hart
Stony Brook University
george@georgehart.com

Gerald Hoehn
Kansas State University
gerald@math.ksu.edu

Jacques Hurtubise
McGill University
jacques.hurtubise@mcgill.ca

Jennifer Hyndman
University of Northern British Columbia
hyndman@unbc.ca

André Joyal
Université du Québec à Montréal
joyal@math.uqam.ca

Nicholas M. Katz
Princeton University
nmk@math.princeton.edu

Arthemy Kiselev
Université de Montréal
kiselev@dms.umontreal.ca

Hershy Kisilevsky
Concordia University
kisilev@mathstat.concordia.ca

Anatoliy Klimyk
Bogolyubov Institute for Theoretical Physics
aklimyk@bitp.kiev.ua

Dmitry Korotkin
Concordia University
korotkin@mathstat.concordia.ca

François Lalonde
Université de Montréal
lalonde@dms.umontreal.ca

Clement Lam
Concordia University
lam@cse.concordia.ca

Emmanuel Letellier
Université de Caen
emmanuel.letellier@math.unicaen.fr

Yuri Manin
Max-Planck-Institut für Mathematik
manin@mpim-bonn.mpg.de

Kishore Marathe
Brooklyn College, CUNY
kmarathe@brooklyn.cuny.edu

Alexander Masters
alexander.masters@btinternet.com

John McKay
Concordia University and CRM
mckay@encs.concordia.ca

Jack Morava
The Johns Hopkins University
jack@math.jhu.edu

M. Ram Murty
Queen's University
murty@mast.queensu.ca

Iku Nakamura
Hokkaido University
nakamura@math.sci.hokudai.ac.jp

Maryna Nesterenko
National Academy of Sciences of Ukraine
maryna@imath.kiev.ua

Viacheslav V. Nikulin
The University of Liverpool
vnikulin@liv.ac.uk

Simon P. Norton
University of Cambridge
S.Norton@dpmms.cam.ac.uk

Yousuke Ohyama
Osaka University
ohyama@math.sci.osaka-u.ac.jp

Victor Ostromoukhov
Université de Montréal
ostrom@iro.umontreal.ca

Jiri Patera
Université de Montréal
patera@dms.umontreal.ca

Noushin Sabetghadam Haghighi
Concordia University
n_sabet@mathstat.concordia.ca

Dana Schlomiuk
Université de Montréal
dasch@dms.umontreal.ca

Norbert Schlomiuk
Université de Montréal
dasch@dms.umontreal.ca

Abdellah Sebbar
University of Ottawa
asebbar@uottawa.ca

Ahmed Sebbar
Université Bordeaux 1
sebbar@math.u-bordeaux1.fr

David Sevilla Gonzalez
Concordia University
dsevilla@cs.concordia.ca

Francisco Thaine
Concordia University
ftha@alcor.concordia.ca

Allan Trojan
York University
allantrojan@hotmail.com

Pavel Winternitz
Université de Montréal
wintern@crm.umontreal.ca

Ismet Yurdusen
Université de Montréal
yurdusen@CRM.UMontreal.CA

Centre de Recherches Mathématiques
CRM Proceedings and Lecture Notes
Volume **47**, 2009

Introduction and Background

In November 1978 I initiated a correspondence with John Thompson. It contained the astonishing $196884 = 1 + 196883$ and led to some mathematics which he later called 21st century mathematics. We are in the right century now and, complemented by our knowledge of the neolithic carved stones whose photograph is here included in the conference poster, are in a good position to draw open the curtains on what must surely be some of the most remarkable mathematics ever.

From 27 to 29 April 2007, at the Centre de recherches mathématiques, Montréal, we explored where we are moving in resolving some mysteries of the Monster and its moonshine. We seek to understand the primes dividing the order of the monster; why the moonshine Riemann surfaces are genus zero, what is the reason for the bijection between the three exceptional sets of Dynkin data for E_8, E_7, and E_6 and the three sporadic groups F_1, F_2, and F_3 centralizing elements in the Monster. Ultimately it would be wonderful to resolve how many of the 26 sporadic groups can be constructed within a scheme generalizing Chevalley's for Lie groups.

The papers of Allcock, Basak, and Duncan speak to the geometry of the Monster, and contain new ideas to be pursued. Airault is following her interest in the important Faber polynomials that provide the Hecke action needed with replicable functions. Boalch explores the discrete Painlevé equations and their connection with the affine Weyl groups of type E_8, E_7, and E_6. Harvey Cohn speculates on extending our paper on the "spontaneous generation" of the elliptic modular function j to Hilbert modular functions. Cummins extends tables on the fixing groups of replicable functions to those of congruence subgroups of the modular group, of genus zero and one. Dolgachev explores my correspondence with $SU(1,1)$ replacing $SU(2)$. Doran studies lattice-polarized K3 surfaces with the lattice $H \oplus E_8 \oplus E_8$ of signature $(1,17)$. Duncan throws new light on connections between affine E_8 Dynkin data and the monster. Gannon discusses the nature of Galois action on character tables. Ganter introduces us to the intricacies of elliptic cohomology and how it relates to Norton's generalized moonshine. Goren et al. discusses Ramanujan graphs and quaternion orders.

Hart provided us with the enjoyable task of assembling a large (more than a metre wide) model having icosahedral symmetry — this no doubt to remind us of the role of this symmetry and its interpretation by Empedocles up to modern cosmological observations. Hitchin investigates a mysterious connection between the solution of Laplace's equation and icosahedral symmetry.

2000 *Mathematics Subject Classification.* 20C34, 20D08, 20C15, 20C20, 20C33, 20C34, 11F03, 20H05,57S17.

This is the final form of the paper.

A. Klimyk and J. Patera study multivariate exponential functions that are symmetric with respect to the alternating group, relating to determinants of matrices with exponential entries, which are eigenfunctions of the Laplacian. Ram Murty and Kathleen Petersen make use of the remark of Galois that $\mathrm{PSL}(2,\mathbb{Z})$ has a subgroup of index p only if $p \leq 11$ to clarify the topology of $\mathrm{PSL}(2, O_K)$ for an order O_K. Nakamura surveys his work on Hilbert schemes in connection with the 2-dimensional McKay correspondence. Nikulin writes on arithmetic hyperbolic reflection groups. These have significance in Borcherds' work. Norton provides tables and his imprimatur to the nomenclature and definitions of replicable functions. Ohyama studies monodromy evolving equations related to Halphen's quadratic equations.

Zagier writes of the differential equations that generalize those specializing to Apéry's equation that leads to Apéry's proof of the irrationality of $\zeta(3)$.

John McKay
Concordia University
Montréal, December 2008

Centre de Recherches Mathématiques
CRM Proceedings and Lecture Notes
Volume **47**, 2009

Symmetric Sums Associated to the Factorization of Grunsky Coefficients

Hélène Airault

ABSTRACT. Properties of the Grunsky coefficients and Schwarzian derivative associated with powers of a given series are examined.

Introduction

Faber polynomials, Grunsky coefficients and Schwarzian derivatives are of great significance in the context of univalent functions and conformal mapping, see [11, 22]. On the other hand, Grunsky coefficients and Faber polynomials have been introduced in modular function theory, see [14, 15, 19] and they are closely linked to replication, see [9, Section 8]. In the present work, we define extended Grunsky coefficients and we consider the Schwarzian derivative for a function $f_p(z) = zh(z)^p$ where $h(z) = 1+\sum_{n\geq 1} b_n z^n$. For the Grunsky coefficients, we take mainly the point of view of [21] and for the Schwarzian derivative of $f_p(z)$, we show how to extend to arbitrary $p \in \mathbb{C}$, the method which has been developed in [4] in the case $p = 1$. Let

$$g(z) = z + b_1 + \frac{b_2}{z} + \frac{b_3}{z^2} + \cdots .$$

The classical Grunsky coefficients are given by the expansion

$$\log \frac{g(z) - g(\zeta)}{z - \zeta} = - \sum_{\substack{n\geq 1 \\ k\geq 1}} \frac{1}{n} \beta_{n,k} z^{-k} \zeta^{-n}. \tag{1}$$

The Grunsky coefficients $\beta_{n,k}$ are homogeneous polynomials in the variables b_1, $b_2, \ldots$ An explicit expression of the coefficients of these polynomials has been given in [4]. In the following, let $p \neq 0$, $p \in \mathbb{Z}$,

$$f_p(z) = z(1 + b_1 z + b_2 z^2 + \cdots)^p. \tag{2}$$

2000 *Mathematics Subject Classification.* 17B66, 30B50.

Key words and phrases. Faber polynomials, Grunsky coefficients, Schwarzian derivative, power series.

This is the final form of the paper.

Consider the expansion

$$\log \frac{f_p(z)-f_p(\zeta)}{z-\zeta} = -p\sum_{k\geq 1}\frac{F_k}{k}(z^k+\zeta^k) - \sum_{\substack{n\geq 1\\ k\geq 1}}\frac{1}{n}\beta_{n,k}(p)z^n\zeta^k. \tag{3}$$

The Faber polynomials are the coefficients $(F_k)_{k\geq 1}$ in terms of the $(b_j)_{j\geq 1}$ and we call *extended Grunsky coefficients*, the $\beta_{n,k}(p)$ in terms of the $(b_j)_{j\geq 1}$. If $\zeta = 0$ in (3), then (3) reduces to

$$\log(1+b_1z+b_2z^2+\cdots+b_nz^n+\cdots) = -\sum_{k\geq 1}\frac{F_k}{k}z^k. \tag{4}$$

Let $g_p(z) = 1/f_p(1/z) = z(1+b_1/z+b_2/z^2+\cdots)^{-p} = z[h(1/z)]^{-p}$, then

$$\log\left(\frac{f_p(1/\zeta)-f_p(1/z)}{1/\zeta-1/z}\right) = p\log h\left(\frac{1}{z}\right) + p\log h\left(\frac{1}{\zeta}\right) + \log\left(\frac{g_p(z)-g_p(\zeta)}{z-\zeta}\right). \tag{5}$$

If $p=-1$, the expansion (3) reduces to (1). However, except in the cases $p=1$ and $p=-1$, it is not clear at that moment how to calculate the coefficients $\beta_{n,q}(p)$ in terms of polynomials in $b_1^{\mu_1}b_2^{\mu_2}\cdots$ with the method of [4, Proposition 3]. This is why we shall generalize Schur's procedure of factorization of Grunsky coefficients, see [21]. In Section 1, we define the basic polynomials and we enumerate some of their properties. In Section 2, we establish the factorization of the extended Grunsky coefficients for any p. There exist several attempts at studying more general Grunsky coefficients, see for example [12], the case of our function f_p in (2) is not covered by [12]. We give elementary properties of the $\beta_{n,k}(p)$. Section 2.2 Proposition 4 extends Schur's result ($p=-1$) [21] to the case of the Grunsky coefficients of $f_p(z)$ for any p. In Section 2.3, we do a desintegration of the Grunsky coefficients and obtain identities between finite sums, we find that

$$\psi_{n,q}(n+q) = \sum_{k=1}^{n}\frac{k^{n+q-1}(-1)^{n-k}}{(n-k)!\,(q+k)!}\times(n+q)!\times(-1)^{n+q}b_1^{n+q} \tag{6}$$

is symmetric in n and q

as it can be verified on the two following examples $\psi_{4,2}(7)$ and $\psi_{7,4}(11)$. If $n=4$, $q=2$, then

$$-\frac{6!}{3!\,3!}+\frac{6!\,2^5}{2!\,4!}-\frac{6!\,3^5}{5!}+\frac{6!\,4^5}{6!} = -\frac{6!}{5!}+\frac{6!\,2^5}{6!}.$$

If $n=7$, $q=4$, then

$$\frac{11!}{6!\,5!}-\frac{11!\,2^{10}}{5!\,6!}+\frac{11!\,3^{10}}{4!\,7!}-\frac{11!\,4^{10}}{3!\,8!}+\frac{11!\,5^{10}}{2!\,9!}-\frac{11!\,6^{10}}{10!}+\frac{11!\,7^{10}}{11!}$$

is equal to

$$-\frac{11!}{3!\,8!}+\frac{11!\,2^{10}}{2!\,9!}-\frac{11!\,3^{10}}{10!}+\frac{11!\,4^{10}}{11!}.$$

The first identity ($n=4$, $q=2$) comes from $1/(4p)\beta_{4,2}(p) = 1/(2p)\beta_{2,4}(p)$ and in fact the coefficients of s^{4p} and s^{2p} in $\big((1-s^{1+p})/(1-s)\big)^6$ are the same. Compare with Proposition 4 in Section 2. For the second identity ($n=7$, $q=4$), it results from $1/(7p)\beta_{7,4}(p) = 1/(4p)\beta_{4,7}(p)$. In that case, the coefficients of s^{4p} and s^{7p} in $\big((1-s^{1+p})/(1-s)\big)^{11}$ are the same. The evaluation of the coefficient of s^j in products as in Proposition 4 appears in many situations. A classical example is the Poincaré dice problem, [20, Chapter II, p. 51]. On the other hand, for the Schwarzian derivative of a function f

$$S_f(z) = \left(\frac{f''}{f'}\right)' - \frac{3}{2}\left(\frac{f''}{f'}\right)^2 \tag{7}$$

the methods of [4] are efficient to calculate the coefficients in the expansion of $z^2 S_{f_p}(z)$ for arbitrary p. The Schwarzian derivative is linked to the Grunsky coefficients by the identity, see [7, p. 64]

$$Sf(z) = -6\frac{\partial^2}{\partial\zeta\partial z}_{|\zeta=z} \log\frac{f(z)-f(\zeta)}{z-\zeta} \tag{8}$$

For a function $f(z) = z + b_1z^2 + b_2z^3 + \cdots$, the Schwarzian derivative $S_f(z)$ expands as

$$\begin{aligned} z^2 S_f(z) &= \sum_{n\geq 2} \mathcal{P}_n(b_1, b_2, \ldots, b_n) z^n \\ &= 6(b_2 - b_1^2)z^2 + 6(4b_3 + 4b_1^3 - 8b_1b_2)z^3 \\ &\qquad + 6(10b_4 - 20b_1b_3 - 12b_2^2 + 34b_1^2b_2 - 12b_1^4) + \cdots . \end{aligned}$$

In [4], Y. Neretin gave a method to calculate the numerical coefficients of $b_1^{\mu_1}b_2^{\mu_2}\cdots$ in $\mathcal{P}_n(b_1, b_2, \ldots, b_n)$. In the following work, we take $f_p(z) = z(1 + b_1z + b_2z^2 + \cdots)^p$ where $p \in \mathbb{C}$, $p \neq 0$ as in (2). We show that Neretin's method is convenient to obtain the coefficients of $S_{f_p}(z)$ for arbitrary $p \neq 0$ in terms of a sum of monomials in $b_1^{\mu_1}b_2^{\mu_2}\cdots$. The expressions for the coefficients given in [4] are relative to the case $p = 1$. The case $p = -1$ can also be deduced from [4] using (8). For arbitrary p, the procedure is more complicated. Let $z^2Sf_p(z) = \sum_{n\geq 2}\widetilde{\mathcal{P}}_n(b_1, b_2, \ldots, b_n)\, z^n$. In Section 3, Proposition 5, we give the expression of the coefficients of the polynomials $\widetilde{\mathcal{P}}_n$ in the expansion of $S_{f_p}(z)$ for any p. It is surprising that for $p = -1$, the coefficients in Neretin polynomials $\widetilde{\mathcal{P}}_n$ are all negative, the coefficients in extended Grunsky coefficients $\beta_{nk}(-1)$ are all positive, and when $p = -1$, none of these polynomials depend upon b_1. Remark that from (8), it holds

$$\widetilde{\mathcal{P}}_q = -6 \sum_{\substack{n\geq 1, k\geq 1\\ n+k=q}} k\beta_{n,k}(p). \tag{9}$$

In Section 4, we consider a generalization of Kirillov's vector fields [13] to the "p-case" and we relate these new vector fields to the Schwarzian derivative of $f_p(z)$. See [13; 16; 17, p. 157] when $p = 1$.

1. Basic polynomials

For $n \geq 0$, $p \in Z$, the polynomials K_n^p are defined by

$$(1 + b_1z + b_2z^2 + \cdots + b_nz^n + \cdots)^p = \sum_{n\geq 0} K_n^p z^n \tag{10}$$

The polynomial K_n^p is homogeneous of weight n in the variables $(b_1, b_2, \ldots, b_n)$, each variable b_j having weight j. Since

$$(1 + b_1 z + b_2 z^2 + \cdots + b_n z^n + \cdots)^p = \exp\left[-p \sum_{k\geq 1} \frac{F_k}{k} z^k\right]$$
$$= 1 + \sum_{m\geq 1} (-p)^m \frac{1}{m!} \left[\sum_{k\geq 1} \frac{F_k}{k} z^k\right]^m$$

we deduce easily that

$$K_n^p(b_1, b_2, \ldots, b_n) = \sum_{m=1}^{n} (-p)^m \times \frac{1}{m!} \times D_n^m\left(\frac{F_1}{1}, \frac{F_2}{2}, \ldots, \frac{F_n}{n}\right)$$

where $D_n^m(b_1, b_2, \ldots, b_n)$ is the coefficient of z^n in $(b_1 z + b_2 z^2 + \cdots + b_n z^n + \cdots)^m$. From the binomial formula,

$$(11)\quad (b_1 z + b_2 z^2 + \cdots + b_n z^n + \cdots)^m$$
$$= \sum_{s\geq 0} D_{m+s}^m z^{s+m} = \sum_{\substack{\mu_1, \mu_2, \ldots \\ \mu_1+\mu_2+\cdots=m}} \frac{m!}{\mu_1!\, \mu_2! \cdots} b_1^{\mu_1} b_2^{\mu_2} \cdots z^{\mu_1+2\mu_2+3\mu_3+\cdots}$$

and for $m \leq n$,

$$(12)\qquad D_n^m(b_1, b_2, \cdots, b_n) = \sum_{\substack{\mu_1, \mu_2, \ldots \\ \mu_1+\mu_2+\cdots=m \\ 1\mu_1+2\mu_2+\cdots=n}} \frac{m!}{\mu_1!\, \mu_2! \cdots} b_1^{\mu_1} b_2^{\mu_2} \cdots$$

$$D_1^1 = b_1, \quad D_2^1 = b_2, \quad D_2^2 = b_1^2, \quad D_3^1 = b_3, \quad D_3^2 = 2b_1b_2, \quad D_3^3 = b_1^3,$$
$$D_4^1 = b_4, \quad D_4^2 = 2b_1b_3 + b_2^2, \quad D_4^3 = 3b_1^2b_2, \quad D_4^4 = b_1^4,$$
$$D_5^1 = b_5, \quad D_5^2 = 2b_1b_4 + 2b_2b_3, \quad D_5^3 = 3b_1^2b_3 + 3b_1b_2^2, \quad D_5^4 = 4b_1^3b_2, \quad D_5^5 = b_1^5,$$
$$D_6^1 = b_6, \quad D_6^2 = 2b_1b_5 + 2b_2b_4 + b_3^2, \quad D_6^3 = 3b_1^2b_4 + 6b_1b_2b_3 + b_2^3,$$
$$D_6^4 = 4b_1^3b_3 + 6b_1^2b_2^2, \quad D_6^5 = 5b_1^4b_4, \quad D_6^6 = b_1^6,$$
$$D_7^1 = b_7, \quad D_7^2 = 2b_1b_6 + 2b_2b_5 + 2b_3b_4, \quad D_7^3 = 3b_1^2b_5 + 3b_2^2b_3 + 3b_3^2b_1 + 6b_1b_2b_4,$$
$$D_7^4 = 4b_1^3b_4 + 4b_2^3b_1 + 12b_1^2b_2b_3, \quad D_7^5 = 5b_1^4b_3 + 10b_1^3b_2^2, \quad D_7^6 = 6b_1^5b_2, \quad D_7^7 = b_1^7,$$
$$D_8^1 = b_8, \quad D_8^2 = 2b_1b_7 + 2b_2b_6 + 2b_3b_5 + b_4^2,$$
$$D_8^3 = 3b_1^2b_6 + 3b_2^2b_4 + 3b_3^2b_2 + 6b_1b_3b_4 + 6b_1b_2b_5,$$
$$D_8^4 = 4b_1^3b_5 + 6b_1^2b_3^2 + 12b_1^2b_2b_4 + 12b_1b_2^2b_3 + b_2^4, \quad D_8^5 = 5b_1^4b_4 + 20b_1^3b_2b_3 + 10b_1^2b_2^3,$$
$$D_8^6 = 6b_1^5b_3 + 15b_1^4b_2^2, \quad D_8^7 = 7b_1^6b_2, \quad D_8^8 = b_1^8,$$
$$D_9^1 = b_9, \quad D_9^2 = 2b_1b_8 + 2b_2b_7 + 2b_3b_6 + 2b_4b_5,$$
$$D_9^3 = 3b_1b_4^2 + 3b_1^2b_7 + 3b_2^2b_5 + b_3^3 + 6b_1b_3b_5 + 6b_1b_2b_6 + 6b_2b_3b_4,$$
$$D_9^4 = 4b_1^3b_6 + 4b_2^3b_1 + 12b_1^2b_3b_4 + 12b_1^2b_2b_5 + 12b_1b_2^2b_4 + 12b_1b_2b_3^2,$$
$$D_9^5 = 5b_1^4b_5 + 5b_1b_2^4 + 10b_1^3b_3^2 + 20b_1^3b_2b_4 + 30b_1^2b_2^2b_3,$$
$$D_9^6 = 6b_1^5b_4 + 30b_1^4b_2b_3 + 20b_1^3b_2^3, \quad D_9^7 = 7b_1^6b_3 + 21b_1^5b_2^2, \quad D_9^8 = 8b_1^7b_2, \quad D_9^9 = b_1^9,$$
$$D_{10}^1 = b_{10}, \quad D_{10}^2 = 2b_1b_9 + 2b_2b_8 + 2b_3b_7 + 2b_4b_6 + b_5^2,$$

$$D_{10}^3 = 3b_1^2b_8 + 3b_2^2b_6 + 3b_3^2b_4 + 3b_4^2b_2 + 6b_1b_4b_5 + 6b_1b_3b_6 + 6b_2b_3b_5 + 6b_1b_2b_7,$$
$$D_{10}^4 = 4b_1^3b_7 + 4b_2^3b_4 + 4b_3^3b_1 + 6b_2^2b_3^2 + 6b_1^2b_4^2 + 12b_1^2b_3b_5 + 12b_1b_2^2b_5 + 12b_1^2b_2b_6 + 24b_1b_2b_3b_4,$$
$$D_{10}^5 = 5b_1^4b_6 + 20b_1^3b_3b_4 + 20b_1^3b_2b_5 + 30b_1^2b_2^2b_4 + 30b_1^2b_2b_3^2 + 20b_1b_2^3b_3 + b_2^5,$$
$$D_{10}^6 = 6b_1^5b_5 + 15b_1^4b_3^2 + 15b_1^2b_2^4 + 30b_1^4b_2b_4 + 60b_1^3b_2^2b_3,$$
$$D_{10}^7 = 7b_1^6b_4 + 42b_1^5b_2b_3 + 35b_1^4b_2^3, \quad D_{10}^8 = 8b_1^7b_3 + 28b_1^6b_2^2,$$
$$D_{10}^9 = 9b_1^8b_2, \quad D_{10}^{10} = b_1^{10}.$$

In D_n^m, all the terms have m factors and the sum S of the coefficients is equal to the coefficient of z^n in $(z + z^2 + \cdots + z^n)^m$,

$$S = \mathrm{C}_{n-1}^{m-1} = \frac{(n-1)!}{(m-1)!\,(n-m)!} = \sum_{\substack{\mu_1,\mu_2,\ldots \\ \mu_1+\mu_2+\cdots=m \\ 2\mu_2+3\mu_3+\cdots=n}} \frac{m!}{\mu_1!\,\mu_2!\,\cdots}. \tag{13}$$

The polynomials D_n^m have been related to the Faber polynomials F_k in [23], we have, see [4, 23],

$$\frac{1}{n}F_n(b_1, b_2, \ldots, b_n) = \sum_{m=1}^{n} \frac{(-1)^m}{m} D_n^m(b_1, b_2, \ldots, b_n). \tag{14}$$

On the other hand,

$$K_n^p = K_n^p(b_1, b_2, \ldots, b_n) = \sum_{m=1}^{n} \mathrm{C}_m^p D_n^m(b_1, b_2, \ldots, b_n) \tag{15}$$

with $\mathrm{C}_m^p = p(p-1)\cdots(p-m+1)/m!$.

Since $z^m h(z)^m = (z+b_1z^2+\cdots)^m = \sum_{s\geq 0} D_{m+s}^m(1, b_1, b_2, \ldots)z^{s+m}$, we deduce that

$$K_n^p(b_1, b_2, \cdots, b_n) = D_{p+n}^p(1, b_1, b_2, \ldots). \tag{16}$$

It is immediate that K_n^p is a polynomial of degree n in p, thus it is defined for any complex number p. Moreover $p = 0$ is a root of K_n^p, since $for n \geq 1$, $K_n^0 = 0$. For $j \geq 1$,

$$\lim_{p\to 0} \frac{1}{p} K_j^p = -\frac{F_j}{j} \tag{17}$$

As in [2, 8], for $n \geq 1$, we put $G_n(b_1, b_2, \ldots, b_n) = K_n^{-1}(b_1, b_2, \ldots, b_n)$, then

$$\begin{aligned} &D_n^m\big(G_1(b_1), G_2(b_1, b_2), \ldots, G_n(b_1, b_2, \cdots)\big) \\ &\qquad = \sum_{j=0}^{n-m} (-1)^{j+m} \frac{(j+m-1)!}{j!\,(m-1)!} D_n^{m+j}(b_1, b_2, \ldots, b_n) \\ &D_n^m(1, b_1, \ldots, b_{n-1}) = \sum_{j=0}^{n-m} (-1)^j \frac{(m+j-1)!}{(m-1)!\,j!} D_{n-m}^j(G_1(b_1), G_2(b_1, b_2), \ldots). \end{aligned} \tag{18}$$

In [23, formula (30), p. 274], the Faber polynomials $(\phi_n(t))_{n\geq 1}$ of the function $g_1(z) = z(1+b_1/z+b_2/z^2+\ldots)^{-1}$ are in a simple way in terms of the $D_n^m(b_1, b_2, \ldots)$

and of the shifted polynomials $D_n^m(1, b_1, b_2, \dots, b_{n-1})$. We have the Taylor expansion,

$$\frac{1}{n}\phi_n(t) = -\frac{1}{n}F_n(b_1, b_2, \dots, b_n) + \sum_{k=1}^{n} \frac{t^k}{k} D_n^k(1, b_1, b_2, \dots, b_{n-1}) \tag{19}$$

The polynomials (K_n^p) are important because the coefficients of reverted series are given in terms of these polynomials, for example, let $g(z) = z + b_1 + b_2/z + \cdots + b_{n+1}/z^n + \cdots$, then

$$\frac{1}{g^{-1}(z)} = \frac{1}{z}\left[1 + \sum_{n\geq 1} \frac{1}{n+1} K_n^{n+1} \frac{1}{z^n}\right].$$

These reverted series have been related to the characters of the Monster, see [10]. The polynomial K_n^{n+1} contains only terms in $b_1^{\mu_1} b_2^{\mu_2} \cdots$ with $\sum_{j\geq 1} j\mu_j = n$. With (15), we obtain the coefficient λ of

$$\frac{1}{\prod_{j\geq 1}(\mu_j!)} b_1^{\mu_1} b_2^{\mu_2} \cdots \quad \text{with} \sum_{j\geq 1} \mu_j = m \text{ and } \sum_{j\geq 1} j\mu_j = n \text{ in } \frac{1}{n+1} K_n^{n+1}.$$

It gives

$$\lambda = \frac{m!}{n+1} C_m^{n+1} = \frac{n!}{(n+1-m)!} = \frac{(\sum_{j\geq 1} j\mu_j)!}{(1 + \sum_{j\geq 2}(j-1)\mu_j)!}.$$

2. Factorization of the Grunsky coefficients

In (3), the coefficient $\beta_{n,k}(p)$ is a polynomial of degree $n+q$ in the variable p, and a homogeneous polynomial of degree $n+q$ in the variables $(b_1, b_2, \dots, b_{n+q})$, each b_j having weight j. We shall prove that the following factorization holds

$$\begin{aligned} &\frac{1}{n}\beta_{nq}(p)(b_1, b_2, \dots, b_{n+q}) \\ &\qquad = \sum_{k=1}^{n} \frac{1}{k} K_{n-k}^{pk}(b_1, b_2, \dots, b_{n-k}) K_{q+k}^{-pk}(b_1, b_2, \dots, b_{q+k}). \end{aligned} \tag{20}$$

Similar factorizations of $\beta_{n,k}(p)$ for $p = 1$ and $p = -1$ are explicited in [5], extending the classical factorization of [21] when $p = -1$. It is immediate that $(1/n)\beta_{n,k}(p)$ is symmetric in n and k and that $\beta_{nk}(0) = 0$. With an expansion of (3) at $p = 0$, we see that

$$\lim_{p\to 0} \frac{1}{pn}\beta_{n,k}(p) = \frac{1}{k+n} F_{k+n}. \tag{21}$$

We know from the property of Faber polynomials that there exists a unique polynomial $\phi_n(z; p)$ of degree n in the variable z such that

$$\phi_n\left(\frac{1}{f_p(z)}; p\right) = \frac{1}{z^n} + \sum_{k\geq 1} \beta_{n,k}(p) z^k. \tag{22}$$

The coefficients of $\phi_n(z;p) = F_n(K_1^p - z, K_2^p, K_3^p, \dots, K_n^p)$ are easily calculated with Taylor formula and the derivatives of Faber polynomials explicited in [2, (1.6)],

$$\begin{aligned} \phi_n(z;p) &= pF_n(b_1, b_2, \dots, b_n) \\ &\qquad + nK_{n-1}^{-p}(b_1, b_2, \dots, b_{n-1})z + \frac{n}{2!} K_{n-2}^{-2p} z^2 + \cdots + z^n. \end{aligned} \tag{23}$$

2.1. Expressions of $\beta_{n,k}(p)$. We put $h(z) = 1 + b_1 z + b_2 z^2 + \cdots$. From (3), we have

$$\log(zh(z)^p - \zeta h(\zeta)^p) - \log(z - \zeta) = -p \sum_{k\geq 1} \frac{F_k}{k}(z^k + \zeta^k) - \sum_{\substack{n\geq 1\\ k\geq 1}} \frac{1}{n}\beta_{n,k}(p) z^n \zeta^k. \tag{24}$$

Proposition 1. *For $p = -1$, then $(\beta_{n,k}(-1))_{n\geq 1, k\geq 1}$ are independent of b_1.*

PROOF. We take the derivative with respect to b_1 in (3). For the left side, since $\partial h(z)/\partial b_1 = z$, it gives

$$\frac{\partial}{\partial b_1} \log(zh(z)^p - \zeta h(\zeta)^p) = \frac{z^2 p h(z)^{p-1} - \zeta^2 p h(\zeta)^{p-1}}{zh(z)^p - \zeta h(\zeta)^p}.$$

On the other hand,

$$z^2 p h(z)^{p-1} - \zeta^2 p h(\zeta)^{p-1} + z\zeta\left(\frac{h(z)^p}{h(\zeta)} - \frac{h(\zeta)^p}{h(z)}\right) = (zh(z)^p - \zeta h(\zeta)^p) \times \left(\frac{z}{h(z)} + \frac{\zeta}{h(\zeta)}\right).$$

If $p = -1$, the second term cancels. For the right side of (3), as in [8], we have $\partial(-F_k/k)/\partial b_1 = G_{k-1}$, where $(G_k)_{k\geq 0}$ are defined by $1/h(z) = \sum_{k\geq 0} G_k z^k$. Thus

$$\sum_{k\geq 1} \frac{\partial}{\partial b_1}\left(-\frac{F_k}{k}\right)(z^k + \zeta^k) = \frac{z}{h(z)} + \frac{\zeta}{h(\zeta)}.$$ □

Proposition 2. *$\beta_{nq}(p)$ is a polynomial of degree $n + q$ in the variable p, and $\beta_{nq}(0) = 0$. We have the factorization* (20).

PROOF.

$$\log \frac{f_p(z) - f_p(\zeta)}{z - \zeta} = \log f_p(z) - \log(z - \zeta) + \log\left(1 - \frac{f_p(\zeta)}{f_p(z)}\right)$$

and when $\zeta \to 0$,

$$\log\left(1 - \frac{f_p(\zeta)}{f_p(z)}\right) = -\sum_{k\geq 1} \frac{1}{k} \frac{f_p(\zeta)^k}{f_p(z)^k} = -\sum_{k\geq 1} \frac{1}{k} \frac{\zeta^k}{z^k} \sum_{j\geq 0} K_j^{pk} \zeta^j \times \sum_{r\geq 0} K_r^{-pk} z^r.$$

For $k + j = q$ and $r - k = n$, we obtain $(1/n)\beta_{n,q} z^n \zeta^q$. Thus $(1/n)\beta_{nq} = \sum_{k=1}^{q} (1/k) K_{q-k}^{pk} K_{k+n}^{-pk}$. Because of the symmetry in n, q of $(1/n)\beta_{n,q}$, the factorization can also be written as $(1/n)\beta_{nq} = \sum_{k=1}^{n} (1/k) K_{n-k}^{pk} K_{k+q}^{-pk}$. Remark that if $p = 0$, we have $K_{q+k}^{0} = 0$ and $\beta_{nq}(0) = 0$. □

Proposition 3. *For $p = -1$, then $(\beta_{n,k}(-1))_{n\geq 1, k\geq 1}$ are the classical Grunsky coefficients of expansion* (1).

PROOF. We write $g(z) = zh(1/z)$. Then

$$\log\big(g(z) - g(\zeta)\big) - \log\left(zh\left(\frac{1}{z}\right)\right) = \log\left(1 - \frac{\zeta h(1/\zeta)}{zh(1/z)}\right) = -\sum_{k\geq 1} \frac{1}{k} \frac{\zeta^k h(1/\zeta)^k}{z^k h(1/z)^k}$$
$$= -\sum_{k\geq 1} \frac{1}{k} \frac{\zeta^k}{z^k} \sum_{j\geq 0} K_j^{k} \zeta^{-j} \times \sum_{r\geq 0} K_r^{-k} z^{-r}.$$

We take $k - j = -n$ and $k + r = q$ for the coefficient of $\zeta^{-n} z^{-q}$. It gives the factorization $(1/n)\beta_{n,q} = \sum_{k=1}^{q} (1/k) K_{k+n}^{k} K_{q-k}^{-k}$. This proves Proposition 3. □

2.2. Some examples of the factorization. With (20) and K_n^p as in [5], we find $\beta_{1q}(p) = K_{q+1}^{-p}(b_1, b_2, \ldots, b_{q+1})$, $\frac{1}{2}\beta_{2q}(p) = K_1^p K_{1+q}^{-p} + \frac{1}{2}K_{2+q}^{-2p}$,

$$\frac{1}{2}\beta_{22}(p) = p(p+1)\left[\frac{(2p^2+4p+3)}{3!\,2}b_1^4-(p+1)b_1^2b_2+b_1b_3\right]-pb_4+\frac{p(2p+1)}{2}b_2^2$$

$$\frac{1}{2}\beta_{23}(p) = p(p+1)\left[-\frac{(p+2)(11p^2+17p+12)}{5!}b_1^4+\frac{(5p^2+10p+6)}{3!}b_1^3b_2-(p+1)b_1^2b_3\right]$$
$$+p(p+1)b_1b_4-\frac{p(p+1)(3p+2)}{2}b_1b_2^2+p(2p+1)b_2b_3-pb_5$$

$$\frac{1}{2}\beta_{24}(p) = p(p+1)\left[\frac{(p+2)(13p^3+51p^2+56p+30)}{5!\,3}b_1^6-\frac{(p+2)(3p^2+5p+3)}{3!}b_1^4b_2\right]$$
$$+p(p+1)\left[\frac{(5p^2+10p+6)}{3!}b_1^3b_3-(3p+2)[b_1^2b_4+b_1b_2b_3]+\frac{3(p+1)^2}{2}b_1^2b_2^2+b_1b_5\right]$$
$$+\ p(2p+1)\left[\frac{b_3^2}{2}+b_2b_4\right]-\frac{p(p+1)(2p+1)}{3}b_2^3-pb_6$$

$$\frac{1}{3}\beta_{3q}(p) = K_2^pK_{q+1}^{-p}+\frac{1}{2}K_1^{2p}K_{q+2}^{-2p}+\frac{1}{3}K_{q+3}^{-3p}$$

$$\frac{1}{3}\beta_{33}(p) = K_2^pK_4^{-p}+\frac{1}{2}K_1^{2p}K_5^{-2p}+\frac{1}{3}K_6^{-3p}$$
$$= p(p+1)^3b_1^3b_3$$
$$+\frac{p(p+1)}{2}\left[-\frac{(11p^3+31p^2+30p+12)}{6}b_1^4b_2+(5p^2+7p+3)b_1^2b_2^2-2(p+1)b_1^2b_4\right]$$
$$+p(p+1)\left[-2(2p+1)b_1b_2b_3+b_1b_5+\left(\frac{11p^4+44p^3+71p^2+54p+20}{5!}\right)b_1^6\right]$$
$$+p(2p+1)b_2b_4+\frac{p(3p+1)}{2}b_3^2-\frac{p(3p^2+3p+1)}{3!}b_2^3-pb_6$$

$$\frac{1}{3}\beta_{34}(p) = K_2^pK_5^{-p}+\frac{1}{2}K_1^{2p}K_6^{-2p}+\frac{1}{3}K_7^{-3p}$$
$$= \frac{p(p+1)(p+2)}{7!}(-302p^4-2489p^3-2130p^2-1014p-360)b_1^7+\cdots.$$

In the case of $\beta_{2q}(p)$,

(1) For $\mathbf{m} = (m_1, m_2, \ldots)$, let $b^{\mathbf{m}} = b_1^{m_1}b_2^{m_2}\cdots$, $|\mathbf{m}| = m_1 + m_2 + \cdots$, $\|\mathbf{m}\| = \sum_{j\geq 1} jm_j$. With the expression of K_n^p as in (15), it holds

$$\text{(25)}\quad \frac{1}{2p}\beta_{2q}(p) = \sum_{\substack{\mathbf{m}\\ \|\mathbf{m}\|=2+q}} \frac{(-1)^{|\mathbf{m}|}}{\prod_{j\geq 1}(m_j!)}\left[\frac{(2p+|\mathbf{m}|-1)!}{(2p)!} - \frac{(p+|\mathbf{m}|-2)!}{(p-1)!}m_1\right]b^{\mathbf{m}}$$

(2) If p is an integer and $p \geq 1$, then

$$K_n^{-p} = (-1)^p\frac{1}{(p-1)!\,(n+p)}\frac{\partial^p}{\partial b_1^p}F_{n+p},$$

then $\beta_{2q}(p) = 2K_1^p K_{q+1}^{-p} + K_{q+2}^{-2p}$ implies

$$(26)\quad \frac{1}{2p}\beta_{2q}(p) = \frac{(-1)^p}{(p-1)!(q+1+p)}\left(b_1\frac{\partial^p}{\partial b_1^p}\right)F_{q+1+p} + \frac{1}{(2p)!(q+2+2p)}\frac{\partial^{2p}}{\partial b_1^{2p}}F_{q+2+2p}.$$

The following proposition is a consequence of (20). The case $p=-1$ occurs in [21]. Both cases $p=-1$ and $p=1$ can be obtained directly without factorization as in the proof of [4, Proposition 3.3]. When $p \neq 1$ or $p \neq -1$, it is more delicate and the factorization (20) seems to be a good tool.

Proposition 4. *Let $p \neq 0$, integer, $\mathbf{n} = (n_1, n_2, \ldots, n_k, \ldots, 0, 0, \ldots)$ be a sequence of integers with only a finite number of them nonzero. Then the coefficient $\mathcal{C}(p;\mathbf{n})$ of $b_1^{n_1}b_2^{n_2}\cdots$ in $\beta_{kq}(p)/(pk)$ is zero if $\sum_{j\ge1} jn_j \neq q+k$. If $\sum_{j\ge1} jn_j = q+k$, the coefficient $\mathcal{C}(p;\mathbf{n})$ is equal to*

$$(27)\quad \mathcal{C}(p;\mathbf{n}) = (-1)^m\frac{(m-1)!}{\prod_{j\ge1}(n_j!)} \times \textit{the coefficient of } s^{kp} \textit{ in } \prod_{j\ge1}\left(\frac{1-s^{1+pj}}{1-s}\right)^{n_j} \quad \textit{where } m = \sum_{j\ge1} n_j.$$

$\mathcal{C}(p;\mathbf{n})$ is a polynomial in p of degree $k+q-1$. If $n_1 \neq 0$, then $p=-1$ is a root of $\mathcal{C}(p;\mathbf{n})$.

PROOF. We put $\|n\| = \sum_{j\ge1} j\mu_j$ and expand with the binomial formula

$$(28)\quad \prod_{j\ge1}(1-s^{1+pj})^{n_j} = \sum_{\nu=0}^{m}\sum_{\alpha=\nu}^{\|n\|}(-1)^\nu s^{p\alpha+\nu}\prod_{\nu_j}\frac{n_j!}{\nu_j!(n_j-\nu_j)!} \quad \text{where } \sum_{j\ge1}\nu_j = \nu \text{ and } \sum_{j\ge1} j\nu_j = \alpha.$$

Then we multiply by $1/(1-s)^m = \sum_{t=0}^{+\infty}(t+m-1)!/(t!(m-1)!)s^t$. For the coefficient of s^{pk}, it is enough to keep only the terms in $s^{p(k-n)+\nu}$, $n\ge1$, in $\prod_{j\ge1}(1-s^{1+pj})^{n_j}$. Thus we consider a sum of terms such that $t+p(k-n)+\nu = pk$, they have as coefficients $(pn-\nu+m-1)!/\big((pn-\nu)!(m-1)!\big)$. Then we compare with the expression of $\mathcal{C}$ obtained using the factorization (20) and (15),

$$\mathcal{C} = (-1)^m\sum_{n=1}^{k}\frac{1}{n}\times\sum_{0\le\nu_j\le n_j}(-1)^\nu\frac{(pn)!}{(pn-\nu)!\prod_{j\ge1}(\nu_j!)}\times\frac{(pn+m-\nu-1)!}{(pn-1)!\prod_{j\ge1}((n_j-\nu_j)!)}$$

where $\nu = \sum_{j\ge1}\nu_j$. □

Remark 1. Assume that $\sum_{j\ge1} jn_j = k+q$. Then the coefficients of s^{kp} and s^{qp} in $A = \prod_{j\ge1}\big((1-s^{1+pj})/(1-s)\big)^{n_j}$ are equal. This comes from a residue calculus and the change of variables $s = 1/u$ in the line integral $\int A\,\mathrm{d}s/s^{kp+1}$.

Examples. The calculation of the coefficient of b_1^{n+q} in $(1/n)\beta_{n,q}$ by the factorization (20) is exactly the same as to calculate the coefficient of s^{np} in

$\left((1-s^{1+p})/(1-s)\right)^{n+q}$ doing the multiplication of

$$(1-s^{1+p})^{n+q} = 1-(n+q)s^{1+p}+\frac{(n+q)(n+q-1)}{2}s^{2+2p}+\cdots\frac{(q+2)\cdots(q+n)}{(n-1)!}s^{(n-1)(1+p)}$$

by $1/1-s)^{n+q}$. For the coefficient of a term product of (b_j), we giver the example of $b_2b_1^5$ in $\frac{1}{3}\beta_{3,4}$. We develop $(1-s^{1+p})^5(1-s^{1+2p}) = 1-5s^{1+p}+10s^{2+2p}-s^{1+2p}+\cdots$ forgetting the powers of s which are bigger than $3p$ and we multiply by $1+\sum_{k\geq 1}(k+5)!/(k!5!)s^k$. We look at the coefficient of s^{3p}, it gives the sum of 4 polynomials in p and these are the same as those obtained from the factorization of the Grunsky coefficient $\frac{1}{3p}\beta_{3,4}$.

Remark 2.

$$\frac{1}{p}\log\left(\frac{zh(z)^p-\zeta h(\zeta)^p}{z-\zeta}\right) = \frac{z\log[h(z)]-\zeta\log[h(\zeta)]}{z-\zeta}-\frac{z\zeta}{2}\left(\frac{\log[h(z)]-\log[h(\zeta)]}{z-\zeta}\right)^2 p$$
$$+\frac{z\zeta(z+\zeta)}{3!}\left(\frac{\log[h(z)]-\log[h(\zeta)]}{z-\zeta}\right)^3 p^2$$
$$-\frac{z\zeta(z^2+\zeta^2+4z\zeta)}{4!}\left(\frac{\log[h(z)]-\log[h(\zeta)]}{z-\zeta}\right)^4 p^3+\cdots.$$

PROOF. Let $\phi(p)=\log\left((zh(z)^p-\zeta h(\zeta)^p)/(z-\zeta)\right)$,

$$\phi'(p)=\frac{z[\log h(z)]h(z)^p-\zeta[\log h(\zeta)]h(\zeta)^p}{zh(z)^p-\zeta h(\zeta)^p}$$
$$\phi''(p)=-z\zeta\left(\log h(z)-\log h(\zeta)\right)^2\frac{h(z)^p h(\zeta)^p}{(zh(z)^p-\zeta h(\zeta)^p)^2}.$$

To expand $A(p)=h(z)^p h(\zeta)^p/(zh(z)^p-\zeta h(\zeta)^p)^2$ in a series of powers of p, we put $u=\log h(z)$, $v=\log h(\zeta)$ and $\alpha=(u-v)/2$, then $A(p)=\mathrm{e}^{pu}e^{pv}/(z\mathrm{e}^{pu}-\zeta\mathrm{e}^{pv})^2 = 1/(z\mathrm{e}^{p\alpha}-\zeta\mathrm{e}^{-p\alpha})^2$, see for example [6] for the expansion of $A(p)$ in powers of p. □

2.3. The symmetric sums $\psi_{n,q}(m)$. In Proposition 4, we obtained polynomials $\mathcal{C}(p;\mathbf{n})$. In this section, we see how the coefficients of these polynomials in p can be obtained as symmetric sums. From (3), it holds

$$\frac{1}{n}\beta_{n,q}(p)=\sum_{m=1}^{q+n}\frac{(-1)^m}{m!}p^m\times\psi_{n,q}(m). \tag{29}$$

Each function $\psi_{n,q}(m)$ is symmetric in n and q. Since $\psi_{n,q}(n+q)$ is the coefficient of p^{n+q} in $\frac{1}{n}\beta_{n,q}(p)$, (6) is obtained with (27)–(29) and the factorization formula.

3. The Schwarzian derivative of $f_p(z)=z(1+b_1z+b_2z^2+\cdots+b_nz^n+\cdots)^p$

Let $h(z)=1+b_1z+b_2z^2+b_3z^3+\cdots$.

Lemma. *The Schwarzian derivative of $\tilde f(z) = z[h(z)]^p$ is given by*

$$(30)\quad z^2 S_{\tilde f}(z) = -\frac{(p+1)(p-1)}{2} \times \left[\frac{z^2h'^2}{h^2} + \frac{2}{p}\frac{zh'}{h}\right] + \frac{(p+1)(p-1)zh' + p^2z^3h''' + 3p^2z^2h''}{p(h+pzh')} - \frac{3}{2}\left[\frac{(p+1)zh' + pz^2h''}{h+pzh'}\right]^2.$$

PROOF.

$$\frac{z^2h'^2}{h^2} + \frac{2}{p}\frac{zh'}{h} = \frac{1}{p^2}\left(\frac{z^2\tilde f'^2}{\tilde f^2} - 1\right),$$

$$\text{(i)}\quad \frac{z\tilde f'}{\tilde f} = (1-p) + p\frac{zf'}{f} = \frac{h+pzh'}{h} \quad \text{and}$$

$$\frac{z^2\tilde f'^2}{\tilde f^2} = p^2\frac{z^2f'^2}{f^2} - 2p(p-1)\frac{zf'}{f} + (p-1)^2$$

$$\text{(ii)}\quad z^2S_{\tilde g}(z) = u^2S_{\tilde f}(u) \quad \text{where } \tilde g(z) = \frac{1}{\tilde f(1/z)} = z\left[h\left(\frac{1}{z}\right)\right]^{-p} \text{ and } u(z) = \frac{1}{z}$$

$$\frac{\tilde f''}{\tilde f'} = (p-1)\frac{h'}{h} + \frac{(p+1)h' + pzh''}{h+pzh'} \quad \text{and} \quad \frac{(h')^2}{h(h+pzh')} = \frac{1}{pz}\left[\frac{h'}{h} - \frac{h'}{h+pzh'}\right].$$

□

Notation. Let $\tilde f(z) = z[h(z)]^p = z + \sum_{n\ge1} K_n^p(b_1,\dots,b_n)z^{n+1}$. The Schwarzian derivative of $\tilde f$ is given by $z^2S_{\tilde f}(z) = \sum_{n\ge2}\widetilde{\mathcal P}_n(b_1,b_2,\dots,b_n)z^n$ with

$$(31)\qquad \widetilde{\mathcal P}_n(b_1,b_2,\dots,b_n) = \mathcal P_n\big(K_1^p(b_1), K_2^p(b_1,b_2),\dots,K_n^p(b_1,b_2,\dots,b_n)\big)$$

Proposition 5. *Let $m = \sum_{j\ge1}\mu_j$, $N_1 = \sum_{j\ge1} j\mu_j$, $N_2 = \sum_{j\ge1} j^2\mu_j$, then the coefficient of $b_1^{\mu_1}b_2^{\mu_2}\cdots$ in $\widetilde{\mathcal P}_n(b_1,b_2,\dots,b_n)$ is*

$$(32)\quad (-1)^m\frac{(m-1)!}{\prod_{j\ge1}(\mu_j!)} \times [\mathcal A] \quad \textit{with}$$

$$[\mathcal A] = \left[\left(\frac{N_1}{p} + \frac{1}{2}N_2 - \frac{3}{2}N_1^2\right)\prod_{j\ge1}(1+jp)^{\mu_j} + (p+1)(p-1)\left(\frac{N_1}{p} + \frac{1}{2}N_2 - \frac{1}{2}N_1^2\right)\right].$$

Moreover $[\mathcal A]$ is a polynomial in p with $p=0$ as a root. If $\mu_1 \ne 0$, then $p=-1$ is also a root of $[\mathcal A]$.

If $p=-1$, then $\widetilde{\mathcal P}_n(b_1,b_2,\dots,b_n)$ is independent of b_1, $\widetilde{\mathcal P}_2 = -6b_2$, $\widetilde{\mathcal P}_3 = -24b_3$, $\widetilde{\mathcal P}_4 = -12(b_2^2+5b_4)$,.... With (32), it is easy to verify that when $p=-1$, all the coefficients of $b_2^{\mu_2}\cdots b_j^{\mu_j}$ are negative. For general p,

$$\mathcal P_2(K_1^p,K_2^p) = 6\left(pb_2 - \frac{p(p+1)}{2}b_1^2\right)$$

$$\mathcal P_3(K_1^p,K_2^p,K_3^p) = 24\left(pb_3 - p(p+1)b_1b_2 + \frac{p(p+1)(p+2)}{3!}b_1^3\right)$$

$$\mathcal P_4(K_1^p,K_2^p,K_3^p,K_4^p) = 6p[10b_4 - 10(p+1)b_1b_3 - (5+7p)b_2^2 + (p+1)(10+7p)b_1^2b_2] - \frac{p(p+1)(11p^2+31p+30)}{2}b_1^4.$$

PROOF.

$$(p+1)(p-1)zh' + p^2z^3h''' + 3p^2z^2h'' = \sum_{n\geq 1}(p^2n^2-1)nb_nz^n$$

and

$$h + pzh' = 1 + \sum_{n\geq 1}(1+pn)b_nz^n.$$

With lemma [4, (2.8)], taking $c_n = (1+pn)b_n$ and $\alpha_j = j(pj-1)$, we find that the coefficient of $b_1^{\mu_1}b_2^{\mu_2}\cdots$ in

$$\frac{(p+1)(p-1)zh' + p^2z^3h''' + 3p^2z^2h''}{p(h+pzh')}$$

is

$$(-1)^m\frac{(m-1)!}{\prod_{j\geq 1}(\mu_j!)} \times \frac{N_1 - pN_2}{p} \times \prod_{j\geq 1}(1+pj)^{\mu_j}.$$

Thus the coefficient of $b_1^{\mu_1}b_2^{\mu_2}\cdots$ in (30) is $(-1)^m(m-1)!/\prod_{j\geq 1}(\mu_j!) \times [\mathcal{A}]$ with

$$\begin{aligned}
[\mathcal{A}] = &-\frac{(p+1)(p-1)}{2}(N_1^2 - N_2) && \text{coming from } \frac{z^2(h')^2}{h^2}\\
&+\frac{(p+1)(p-1)}{p}N_1 && \text{coming from } -\frac{zh'}{h}\\
&+\left(\frac{N_1 - pN_2}{p}\right)\prod_{j\geq 1}(1+pj)^{\mu_j} - \frac{3}{2}(N_1^2 - N_2)\prod_{j\geq 1}(1+pj)^{\mu_j}.
\end{aligned}$$

Adding, we find $[\mathcal{A}]$. It is immediate that $[\mathcal{A}]$ is a polynomial in p. To prove that $p = 0$ is a root, we expand

$$\prod_{j\geq 1}(1+jp)^{\mu_j} = 1 + N_1p + \frac{1}{2}(N_1^2 - N_2)p^2 + \cdots. \qquad \square$$

Remark 3. With the same notations as in Proposition 5, we have

$$(33)\quad \frac{\partial^2}{\partial s\partial z}_{|z=s}\prod_{j\geq 1}\left(\frac{s^{1+pj} - z^{1+pj}}{s-z}\right)^{m_j} = \frac{p^2}{6}\left[\frac{3}{2}N_1^2 - \frac{1}{2}N_2 - \frac{N_1}{p}\right]\prod_{j\geq 1}(1+jp)^{\mu_j} \times s^{pN_1-2}.$$

If $v \neq -1$,

$$(34)\quad \frac{\partial^2}{\partial s\partial z}_{|z=s}\prod_{j\geq 1}\left(\frac{s^{1+pj} + vz^{1+pj}}{s+vz}\right)^{m_j} = -p^2\frac{2v}{(1+v)^2} \times \left[\frac{N_1}{p} + \frac{N_2}{2} - \frac{N_1^2}{2}\right] \times s^{pN_1-2}.$$

Compare with the coefficients in Proposition 5. With (33) and (8), we obtain the $\beta_{n,k}(p)$ when $p = 1$ or when $p = -1$. With (34), we obtain the expansion of

$$(35)\quad \frac{\partial^2}{\partial\zeta\partial z}_{|z=\zeta}\log\frac{f(\zeta) + vf(z)}{\zeta + vz}) = \frac{v}{(1+v)^2}\left[\frac{1}{\zeta^2} - \frac{f'(\zeta)^2}{f(\zeta)^2}\right] \quad \text{if } v \neq -1.$$

Remark 4. To a function $f(z) = zh(z)$, there are many ways to associate a new function of the form $\tilde{f}(z) = z\tilde{h}(z)$ where $\tilde{h}(z) = 1+\tilde{b}_1 z+\tilde{b}_2 z^2+\cdots$. One of the simplest examples is $\tilde{f}(z) = zh(\lambda z) = 1/\lambda f(\lambda z)$. In that case, by the formula of the Schwarzian derivative of the composition of functions, it is immediate that $S_{\tilde{f}}(z) = \lambda^2 S_f(\lambda z)$. Proposition 5 shows that it is possible to calculate the Schwarzian derivative of arrangement of functions which are more complicated than a simple composition of functions. When p is a positive integer, the function $h(z)^p$ enters in the class of multiplicative vectors, see [18, Section 5], since $h(z)^p = \prod_{k=1}^{+\infty} h(\lambda_k z)$ with the sequence $(\lambda_k)_k$, $\lambda_k = 1$ for $1 \le k \le p$ and $\lambda_k = 0$ for $k \ge p$.

4. Differential operators associated to $z \to z(h(z))^p$

Let for $k \ge 1$,

$$R_k = \frac{1}{p}\left(\frac{\partial}{\partial b_k} + \sum_{j\ge 1}(pj+1)b_j \frac{\partial}{\partial b_{j+k}}\right). \tag{36}$$

On the manifold of coefficients, the vector field R_k is the image of the vector field

$$L_k = \frac{\partial}{\partial c_k} + \sum_{j\ge 0}(j+1)c_j \frac{\partial}{\partial c_{j+k}} \tag{37}$$

through the map $(b_1, b_2, \dots) \to (K_1^p(b_1), K_2^p(b_1, b_2), \dots)$. This means that if we put $c_1 = K_1^p(b_1) = pb_1$, $c_2 = K_2^p(b_1, b_2) = pb_2 + p(p-1)/2b_1^2$, $c_3 = K_3^p(b_1, b_2, b_3) = pb_3 + p(p-1)b_1b_2 + p(p-1)(p-2)/3!b_1^3, \dots$, then for a differentiable function $\phi(b_1, b_2, \dots)$, we have

(38)
$$R_k\left[\phi(pb_1, pb_2 + \frac{p(p-1)}{2}b_1^2, \dots, K_n^p(b_1, \dots), \dots)\right] = (L_k\phi)(c_1, c_2, \dots, c_n, \dots).$$

Lemma. *The operators $(R_k)_{k\ge 1}$ in (36) are determined by the condition (38).*

Proof. To obtain (38), it is enough to verify that $R_k[f(z)] = z^{1+k}f'(z)$ for $f(z) = zh(z)^p$ and this is true if R_k is given by (36). Conversely, we look for $(R_k)_{k\ge 1}$ in the form $R_k = \sum_{j\ge 1}\phi_j \partial/\partial b_j$. Let $f(z) = zh(z)^p$, we wish $R_k[f(z)] = z^{1+k}f'(z)$. It implies $R_k[h(z)] = z^k[h(z)/p + zh'(z)]$. Since $R_k[h(z)] = \sum_{j\ge 1}\phi_j z^j$, we obtain ϕ_j by identifying equal powers of z in $\sum_{j\ge 1}\phi_j z^j = z^k[1/p+\sum_{j\ge 1}(1/p+j)b_j z^j]$. □

Proposition 6. *Let $\tilde{f}(z) = z \times (1 + b_1 z + b_2 z^2 + \dots)^p$ and $z^2 S\tilde{f}(z) = \sum_{n\ge 2}\widetilde{\mathcal{P}}_n(b_1, b_2, \dots)z^n$, then*

$$R_k(\widetilde{\mathcal{P}}_n) = (k+n)\widetilde{\mathcal{P}}_{n-k} + (k^3-k)\delta_{k,n}. \tag{39}$$

This extends the identity obtained by [13; 17, p. 157] in the case $p = 1$.

Proof. Compare with [3, (A.6.8) p. 662]. From $R_k[f(z)] = z^{1+k}f'(z)$, we deduce that $\phi(z) = f''(z)/f'(z)$ satisfies $R_k[\phi(z)] = z^{1+k}\phi'(z) + (k+1)z^k\phi(z) + k(k+1)z^{k-1}$. □

Remark 5. In the same way, $R_k[z\tilde{f}'/\tilde{f}] = kz^k\, z\tilde{f}'/\tilde{f} + z^{k+1}(z\tilde{f}'/\tilde{f})'$ and we verify that $\psi(z) = (z\tilde{f}'/\tilde{f})^2$ satisfies $R_k[\psi(z)] = z^{1+k}\psi'(z) + 2kz^k\psi(z)$.

Acknowledgments. The author thanks John McKay and the organizers of the conference for her enjoyable and interesting stay in Montréal in April 2007, and Heidi Daboussi, coordinator of the CCCI (Commission des colloques et congrès internationaux, France) for financing the traveling. The result of Section 4 was shown in a short communication [1] (ICM Madrid 2006). The author thanks Heidi Daboussi and the CCCI for helping her in the participation to the ICM where she met John McKay.

References

1. H. Airault, *Partial differential equations related to generating functions*, International Congress of Mathematicians (Madrid, 2006). short communication.
2. H. Airault and A. Bouali, *Differential calculus on the Faber polynomials*, Bull. Sci. Math. **130** (2006), no. 3, 179–222.
3. H. Airault and P. Malliavin, *Unitarizing probability measures for representation of Virasoro algebra*, J. Math. Pures Appl. (9) **80** (2001), no. 6, 627–667.
4. H. Airault and Yu. A. Neretin, *On the action of Virasoro algebra on the space of univalent functions*, Bull. Sci. Math. **132** (2008), no. 1, 27–39.
5. H. Airault and J. Ren, *An algebra of differential operators and generating functions on the set of univalent functions*, Bull. Sci. Math. **126** (2002), no. 5, 343–367.
6. G. E. Andrews, R. Askey, and R. Roy, *Special functions*, Encyclopedia Math. Appl., vol. 71, Cambridge Univ. Press, Cambridge, 1999.
7. S. Bergman, *The kernel function and conformal mapping*, 2nd ed., Math. Surveys, vol. 5, Amer. Math. Soc., Providence, RI, 1970.
8. A. Bouali, *Faber polynomials, Cayley–Hamilton equation and Newton symmetric functions*, Bull. Sci. Math. **130** (2006), no. 1, 49–70.
9. J. H. Conway and S. P. Norton, *Monstrous moonshine*, Bull. London Math. Soc. **11** (1979), no. 3, 308–339.
10. D. J. Ford and J. McKay, *Monstrous Moonshine — problems arising: Tate characters* (2002).
11. H. Grunsky, *Koeffizientenbedingungen für schlicht abbildende meromorphe Funktionen*, Math. Z. **45** (1939), no. 1, 29–61.
12. J. A. Hummel, *The Grunsky coefficients of a schlicht function*, Proc. Amer. Math. Soc. **15** (1964), 142–150.
13. A. A. Kirillov, *Geometric approach to discrete series of unireps for Virasoro*, J. Math. Pures Appl. (9) **77** (1998), no. 8, 735–746.
14. J. McKay, *Finite simple groups, quantum logic, replicable functions, dispersionless flow, and Witten's manifold*, Montréal, 2004. CRM-Fields Prize lecture.
15. J. Morava, *Symmetric functions, Faber polynomials and replicable functions.* in this volume.
16. Yu. A. Neretin, *Holomorphic extensions of representations of the group of diffeomorphisms of the circle*, Math. USSR-Sb. **67** (1990), no. 1, 75–97.
17. ———, *Representations of Virasoro and affine Lie algebras*, Encyclopaedia Math. Sci., vol. 22, Springer, Berlin, 1994.
18. ———, *Structures of boson and fermion Fock spaces in the space of symmetric functions*, Acta Appl. Math. **81** (2004), no. 1-3, 233–268.
19. S. P. Norton, *More on moonshine*, Computational Group Theory (Durham, 1982) (M. D. Atkinson, ed.), Academic Press, London, 1984, pp. 185–193.
20. H. Poincaré, *Calcul des probabilités*, Gauthier-Villars, Paris, 1912.
21. I. Schur, *On Faber polynomials*, Amer. J. Math. **67** (1945), 33–41.
22. H. A. Schwarz, *Über einige Abbildungsaufgaben*, J. Reine Angew. Math. **70** (1869), 105–120.
23. P. V. Todorov, *On the Faber polynomials of the univalent functions of class* Σ, J. Math. Anal. Appl. **162** (1991), no. 1, 268–276.

Institut Supérieur des Sciiences et Technologie, Université de Picardie Jules Verne, 48 rue Raspail, 02100 Saint-Quentin, France

E-mail address: `hairault@insset.u-picardie.fr`

Centre de Recherches Mathématiques
CRM Proceedings and Lecture Notes
Volume **47**, 2009

A Monstrous Proposal

Daniel Allcock

Dedicated to Domingo Toledo and to John McKay

Abstract. We explain a conjecture relating the monster simple group to an algebraic variety that we discovered in a non-monstrous context.

The purpose of this note is to explain a conjecture I have circulated privately since 1997. On one level it is purely about group theory and complex hyperbolic geometry, but if it is true then the most natural explanation for it would be algebra-geometric: a certain ball quotient would be the moduli space for some sort of objects, the objects admitting some sort of marking related to the monster simple group. My original grounds for the conjecture were flimsy, but in his dissertation, Tathagata Basak discovered some very suggestive coincidences. The conjecture is still speculative, but now I am taking it seriously.

In brief, Conway described the bimonster $\mathrm{M} \times \mathrm{M} : 2$ as being generated by 16 involutions satisfying some braid and commutation relations, subject to the additional relation that a certain word w has order 10. On the other hand, I discovered a certain complex analytic orbifold X having a nice uniformization by complex hyperbolic 13-space B^{13}. There is nothing obviously monstrous about X, and I had no thought of the monster when constructing it. Then I noticed that its fundamental group $\pi_1^{\mathrm{orb}}(X)$ contains 16 elements satisfying exactly the same braid and commutation relations, but having order 3 rather than 2. Later, Basak discovered that they generate all of $\pi_1^{\mathrm{orb}}(X)$, that w has order 20, and that the generators can be extended to a set of 26 in exactly the same way as Conway's bimonster generators. This note formulates a conjecture that explains these coincidences by giving a uniform interpretation of both groups.

2000 *Mathematics Subject Classification.* Primary 20D08; Secondary 20F36, 20F05.
Partly supported by NSF grant DMS-0245120.
This is the final form of the paper.

1. Conjecture

In [1], we considered a certain lattice L over the Eisenstein integers $\mathcal{E} = \mathbb{Z}[\omega = \sqrt[3]{1}]$, of signature $(1, 13)$; a lattice means a free module equipped with an $\mathcal{E}$-valued Hermitian form $\langle\ |\ \rangle$. L is the most natural such lattice, in the sense that its underlying $\mathbb{Z}$-lattice is a scaled version of the unique even unimodular lattice of signature $(2, 26)$. We will describe L explicitly below, but for now it suffices to describe L as the unique $\mathcal{E}$-lattice of signature $(1, 13)$ satisfying $L = \theta L^*$, where $\theta = \omega - \overline{\omega} = \sqrt{-3}$ and the asterisk denotes the dual lattice. In particular, all inner products in L are divisible by θ. Because of this, if $r \in L$ has norm $r^2 = \langle r \,|\, r\rangle = -3$ (such an r is called a root of L), then the map

$$x \mapsto x - (1 - \omega)\frac{\langle x \,|\, r\rangle}{\langle r \,|\, r\rangle} r \tag{1}$$

is an isometry of L; it sends r to ωr and fixes $r^\perp$ pointwise. This is a complex reflection of order 3, also called a triflection, and $r^\perp$ is called its mirror.

The importance of L in [1] was that it allowed the construction of a complex hyperbolic reflection group of record dimension, namely the the subgroup Γ of $\operatorname{Aut} L$ generated by the triflections in the roots of L. Now, $\operatorname{Aut} L$ acts on the complex 13-ball B^{13} consisting of positive lines in $\mathbb{C}^{1,13} = L \otimes_{\mathcal{E}} \mathbb{C}$. I proved that Γ has finite-volume fundamental domain, and Basak has gone further [4], proving that Γ is all of $\operatorname{Aut} L$.

We define $X = \mathrm{B}^{13}/P\Gamma$, which is an algebraic variety and a complex analytic orbifold. We write $\mathcal{H}$ for the union of the mirrors of the triflections, Δ for the image of $\mathcal{H}$ in X, and G for the orbifold fundamental group $\pi_1^{\mathrm{orb}}(X - \Delta)$. By a meridian we mean an element of G represented by a small loop in $X - \Delta$ that encircles Δ once positively at a generic point of Δ, or any conjugate of such a loop.

Conjecture. *The quotient of G by the normal subgroup generated by the squares of the meridians is the bimonster, i.e., the semidirect product* $\mathrm{M} \times \mathrm{M} : 2$, *where* M *is the monster simple group and* $\mathbb{Z}/2$ *acts by exchanging the factors.*

The conjecture gives a uniform description of the bimonster and $P\Gamma$, as the deck groups of the covering spaces of X which are universal among those having 2- and 3-fold branching along Δ. A geometric way to understand it is to consider the orbifold structure along $\Delta \subseteq X$. Because of the triflections, a generic point of Δ has local group $\mathbb{Z}/3$. The idea of the conjecture is to rub out this $\mathbb{Z}/3$ and replace it by $\mathbb{Z}/2$. This should change the orbifold fundamental group from $P\Gamma$ to the bimonster. The stated form of the conjecture expresses this idea without having to make sense of the new orbifold structure at non-generic points of Δ. (This can be done, but involves a digression; see remark (6) in Section 3.)

If the conjecture holds then we should expect some reason for it to be true. One possibility that would be very pretty is for X to have a moduli interpretation. If $\mathcal{M}$ is a moduli space parameterizing some sort of algebra-geometric object, then a cover $\widetilde{\mathcal{M}}$ of $\mathcal{M}$, possibly ramified somewhere, can be thought of as a moduli space parameterizing suitably marked versions of the objects.

For example we may take $\mathcal{M}$ to be the moduli space of unordered 12-tuples in $\mathbb{CP}^1$ (say, in the sense of geometric invariant theory) and $\widetilde{\mathcal{M}}$ to be the moduli space of ordered 12-tuples. In this case the marking is the ordering, the deck group is the symmetric group S_{12}, and $\widetilde{\mathcal{M}}$ is the universal cover of $\mathcal{M}$ with 2-fold branching along the discriminant. This example is especially relevant because $\mathcal{M}$ may also be described as a 9-ball quotient; indeed B^9 is the universal cover of $\mathcal{M}$ – (one point) with 3-fold branching along the discriminant. So this situation is an exact analogue of that of the conjecture, with B^{13} replaced by B^9 and the bimonster by S_{12}. In fact, the group acting on B^9 is also the analogue of our $P\Gamma$. See [1, 10, 22] for more information.

It would be very pretty if the same phenomenon happened for X; then the objects it parameterized would admit a sort of marking, and varying the objects in a family would permute the markings by an action of the bimonster. The branched covering $B^{13} \to X$ would parameterize the same objects, but equipped with a different notion of marking.

2. Evidence

The origin of the conjecture is a coincidence of diagrams. Conway conjectured that the bimonster can be presented as the quotient of the Coxeter group with diagram Y_{555}

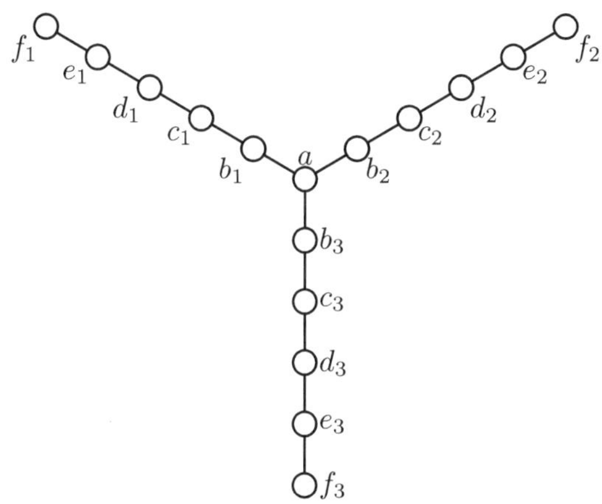

by a single extra relation (see [7]). There are several ways to write the extra relation, one of which is $(ab_1c_1ab_2c_2ab_3c_3)^{10} = 1$, called the spider relation. Ivanov [13] and Norton [18] proved his conjecture.

The diagram Y_{550} arose as Figure 5.1 of my paper [1], in a manner suggesting its extension to Y_{555}. It describes an arrangement of 11 vectors in $\mathbb{C}^{1,9}$ of norm -3 such that their triflections braid ($aba = bab$) or commute ($ab = ba$) according to whether the corresponding nodes of the figure are joined or not. The obvious generalization to Y_{555} is the arrangement of vectors in $\mathbb{C}^{1,13}$

$$\begin{array}{rl}
a = (& \;\;,\;\;,\;\;,\;\;;\;\;,\;\;,\;\;,\;\;;\;\;,\;\;,\;\;,\;\;;\;1,\;\bar{\omega}) \\
b_1 = (& \;\;,\;\;,\;\;,\;\bar{\theta};\;\;,\;\;,\;\;,\;\;;\;\;,\;\;,\;\;,\;\;;\;0,\;1) \\
c_1 = (& \;\;,\;1,\;1,\;1;\;\;,\;\;,\;\;,\;\;;\;\;,\;\;,\;\;,\;\;;\;\;,\;\;) \\
d_1 = (& \;\;,\;\theta,\;\;,\;\;;\;\;,\;\;,\;\;,\;\;;\;\;,\;\;,\;\;,\;\;;\;\;,\;\;) \\
e_1 = (& -1,\;-1,\;1,\;\;;\;\;,\;\;,\;\;,\;\;;\;\;,\;\;,\;\;,\;\;;\;\;,\;\;) \\
f_1 = (& \;\bar{\theta},\;\;,\;\;,\;\;;\;\;,\;\;,\;\;,\;\;;\;\;,\;\;,\;\;,\;\;;\;\;,\;\;)
\end{array}$$

with blanks indicating zeros and $b_2, \ldots, f_2$ and $b_3, \ldots, f_3$ got from $b_1, \ldots, f_1$ by permuting the first three blocks of coordinates. Here we are referring to the Hermitian form with inner product matrix $\operatorname{diag}[-1, \ldots, -1] \oplus \left(\begin{smallmatrix} 0 & \bar{\theta} \\ \theta & 0 \end{smallmatrix}\right)$. The triflections in these roots braid or commute according to the Y_{555} diagram, and the roots span a copy of L in the form $L = \Lambda_4^{\mathcal{E}} \oplus \Lambda_4^{\mathcal{E}} \oplus \Lambda_4^{\mathcal{E}} \oplus H$. Here, $\Lambda_4^{\mathcal{E}}$ is the E_8 root lattice regarded as a 4-dimensional $\mathcal{E}$-lattice and H is the "hyperbolic cell" $\left(\begin{smallmatrix} 0 & \bar\theta \\ \theta & 0 \end{smallmatrix}\right)$. For each $i = 1, 2, 3$, the ith summand $\Lambda_4^{\mathcal{E}}$ is spanned by $c_i, \ldots, f_i$.

The Artin group A of Y_{555} is the group with one generator for each node of the diagram, subject to the same commutation and braiding relations as in the Coxeter group. Forcing the generators to have order 2 yields the Coxeter group of Y_{555}, so Conway provides us with a surjection from A to the bimonster. On the other hand, forcing them to have order 3 gives a map from A into Γ by sending the Artin generators to the triflections in a and the $b_i, \ldots, f_i$. Basak has proven that these 16 triflections generate $\Gamma = \operatorname{Aut} L$, so both the bimonster and Γ are quotients of A, with generators of order 2 and 3 respectively. I expect that there is also a map $A \to G$, sending the Artin generators to meridians, so that $A \to P\Gamma$ is the composition $A \to G \to P\Gamma$.

This suggests that G is really the central object, and that it is 'like' A, subject to some extra relations. In the presence of these extra relations, forcing the meridians to have order 3 reduces the group to Γ. On the other hand, with luck, when the meridians are forced to have order 2, the extra relations become equivalent to the spider relation, giving the conjectured map $G \to \mathrm{M} \times \mathrm{M} : 2$. A natural sanity check is whether the spider relation is compatible with this picture. Basak considered the word $S = ab_1c_1ab_2c_2ab_3c_3$ as an element of A and computed the order of its image in Γ, which turns out to be 20. This implies that S, regarded as a element of G, has order a multiple of 20 (or infinite order), which is certainly compatible with the spider relation $S^{10} = 1$ in the bimonster. Indeed, 20 is a notably small multiple of 10. Basak also found similar compatibilities using other words.

He has also made the striking discovery that the 16 triflections may be extended to a set of 26 in exactly the same way as Conway's 16 bimonster generators. Namely, Conway observed that the map from the Coxeter group of Y_{555} to the bimonster extends to a map from the Coxeter group of a larger graph, the incidence graph of the 13 points and 13 lines of the projective plane over $\mathbb{F}_3$. The symmetries of this finite projective plane, including the dualities exchanging points and lines, extend to automorphisms of the bimonster. Basak found a set of 26 roots of L, the reflections in which commute or braid according to this graph; indeed all inner products are 0 except for $\langle p \,|\, l \rangle = \theta$ when p is the root corresponding to a point of $\mathbb{P}^2(\mathbb{F}_3)$ and l is the root corresponding to a line containing it. It is not really surprising that the 16 roots of Y_{555} fit in $\mathbb{C}^{1,13}$, but it *is* surprising that 26 vectors with specified inner products just happen to fit in a 14-dimensional space, and just happen to span a very natural lattice. The mirrors in B^{13} corresponding to the 13 points (resp. lines) of $\mathbb{P}^2(\mathbb{F}_3)$ are mutually orthogonal and meet at a single point, say P (resp. L). The midpoint of the segment joining P and L has stabilizer $\mathrm{PGL}_3(\mathbb{F}_3) : 2$ in $P\Gamma$, realizing every automorphism of $\mathbb{P}^2(\mathbb{F}_3)$ including dualities exchanging points and lines. Presumably there is a map from the Artin group of the 26-node graph to G, but I have not investigated this.

A final coincidence is that one of the maximal subgroups of the monster has structure $3^{1+12} \cdot 2 \cdot \mathrm{Suz} : 2$, where Suz denotes Suzuki's sporadic finite simple group, while $P\Gamma$ contains a subgroup K with structure $(\operatorname{Im} \mathcal{E}) \cdot \Lambda_{12}^{\mathcal{E}} : (6 \cdot \mathrm{Suz})$. Here, $\operatorname{Im} \mathcal{E} \cong \mathbb{Z}$ and $\Lambda_{12}^{\mathcal{E}}$ denote the additive groups of the imaginary part of $\mathcal{E}$ and of the complex Leech lattice. The extension defining the Heisenberg group $(\operatorname{Im} \mathcal{E}) \cdot \Lambda_{12}^{\mathcal{E}}$ is given by the imaginary part of the inner product on $\Lambda_{12}^{\mathcal{E}}$, and $6 \cdot \mathrm{Suz}$ is $\operatorname{Aut} \Lambda_{12}^{\mathcal{E}}$.

Identifying the scalar ω of $6\cdot\mathrm{Suz}$ with a generator of $\mathrm{Im}\,\mathcal{E}$ reduces K to $3^{1+12}\cdot 2\cdot\mathrm{Suz}$. There are two ways to do this, corresponding to the two generators of $\mathrm{Im}\,\mathcal{E}$; this mimics the construction of $3^{1+12}\cdot 2\cdot\mathrm{Suz}:2\subseteq M$ as a quotient of $3^{1+12}:6\cdot\mathrm{Suz}:2$ in [15, Sections 3.3 and 5.2].

K is a very natural subgroup of $P\Gamma$, namely the stabilizer of the null vector $(0;0,1)$ in the realization of L as $\Lambda_{12}^{\mathcal{E}}\oplus\left(\begin{smallmatrix}0&\bar\theta\\ \theta&0\end{smallmatrix}\right)$. Since there are no mirrors in B^{13} passing through $(0;0,1)$, K is a subgroup of G, not just a subquotient of G. Therefore we have a natural subgroup of G, which modulo a natural relation has index 2 in a maximal subgroup of the monster.

3. Remarks

We close with some remarks that seem relevant but are not in the main line of ideas.

(1) There are infinitely many triflections in Γ, but their mirrors form a locally finite arrangement in B^{13} because Γ is discrete. Γ acts transitively on roots, so all the triflections are conjugate (up to inversion) and $\Delta\subseteq X$ is irreducible. The only complex reflections in Γ or $P\Gamma$ are the triflections we have considered.

(2) There exist points of B^{13} that lie on no mirrors, yet have nontrivial stabilizer in $P\Gamma$. For example, the stabilizer $6\cdot\mathrm{Suz}$ is possible for points near the ideal point of B^{13} given by $(0;0,1)\in\Lambda_{12}^{\mathcal{E}}\oplus\left(\begin{smallmatrix}0&\bar\theta\\ \theta&0\end{smallmatrix}\right)$. A consequence is that $X-\Delta$ has nontrivial orbifold structure even though the obvious orbifold points have been removed. This phenomenon does not occur for finite complex reflection groups or real hyperbolic Coxeter groups.

(3) Conway and Pritchard [8] used real hyperbolic geometry to study the bimonster and certain other finite groups as quotients of Coxeter groups. I don't know of any connection between their real hyperbolic geometry and complex hyperbolic 13-space B^{13}.

(4) The study of complex reflection groups as quotients of Artin groups goes back to Coxeter [9]. In our language, he studied the Artin groups for the diagrams $A_{n=1,\dots,5}$ modulo the relations that the generators have order 3. In the first four cases one gets a finite complex reflection group acting on $\mathbb{C}^n$. The last case gives a group of structure $(\mathrm{Im}\,\mathcal{E})\cdot\Lambda_4^{\mathcal{E}}:\mathrm{Aut}\,\Lambda_4^{\mathcal{E}}$, which acts as a complex reflection group on the 5-ball, fixing a point on the boundary. Coxeter showed that quotienting by the central $\mathrm{Im}\,\mathcal{E}$ gives a complex reflection group acting cocompactly on $\mathbb{C}^4$.

(5) One can build up Y_{555} from smaller diagrams by beginning with three A_4's, which describe finite complex reflection groups, "affinizing" them by enlarging them to A_5's (see the previous remark), and then "hyperbolizing" the result by adjoining a single extra node, joined to each of the three affinizing nodes. This is analogous to (say) extending the E_8 Coxeter group, which acts on the 7-sphere, to E_9, which acts on Euclidean 8-space, and then to E_{10}, which acts on real hyperbolic 9-space. Essentially the same process leads to diagrams like the one in [6], which is a "hyperbolization" of $A_{11}D_7E_6$.

(6) We have avoided the issue of making sense of the alteration of orbifold structures described right after the conjecture. But this dodge is not necessary. Suppose $b\in B^{13}$, S is its stabilizer in $P\Gamma$, $R\subseteq S$ is the subgroup generated by the triflections which fix b, and U is a small ball around b. Then R is a direct product of some copies of the triflection groups associated to $A_1,\dots,A_4$ in remark (4). These act on U by the direct product of their triflection representations and

possibly a subspace where all the factors act trivially. By Chevalley's theorem, U/R is a smooth variety. Obviously, the image of $\mathcal{H}$ therein is a divisor D, and S/R acts on U/R, preserving D. Let U' be the cover of U/R which is universal among those having 2-fold ramification along D and no other ramification, and let R' be the deck group of U' over U/R. By [19], R' is the direct product of Coxeter groups associated to the same diagrams $\mathrm{A}_1, \ldots, \mathrm{A}_4$ as for R, and U' is an open set in $\mathbb{C}^{13}$. Also, R' acts on U' as the product of the Coxeter groups' standard representations, again possibly with some fixed subspace. Now, an element of S/R preserves D, so it has a lift to an automorphism of U', in fact $|R'|$ many lifts. We take S' to be the group consisting of all such lifts. Then

$$U'/S' \cong (U'/R')/(S'/R') \cong (U/R)/(S/R) \cong U/S,$$

where the first and third isomorphisms are of complex analytic orbifolds. The middle isomorphism is one of complex analytic varieties, and is a complex analytic orbifold isomorphism away from the image of $D \subseteq U/R = U'/R'$. We have equipped $U/S \subseteq \mathrm{B}^{13}/\mathrm{P}\Gamma$ with an orbifold structure in which the generic point of Δ has local group $\mathbb{Z}/2$ rather than $\mathbb{Z}/3$. It is easy to see that such a structure is unique.

(7) The truth of the conjecture would imply that the orbifold of the previous remark is the quotient of a complex manifold by an action of the bimonster. Then each component of the preimage of Δ is a smooth hypersurface fixed pointwise by an involution α in the bimonster. The centralizer of α is $\langle\alpha\rangle$ itself times a copy of the monster, so the monster acts on the fixed-point set of α. We wonder if this complex 12-manifold (or a suitable compactification of it) could serve as the 24-dimensional "monster manifold" sought by Hirzebruch et. al. [12, pp. 86–87] in their study of elliptic cohomology.

(8) We have suggested that X may be a moduli space; we do know that it contains several lower-dimensional moduli spaces. Namely, the moduli space of unordered 12-tuples in $\mathbb{C}\mathrm{P}^1$ is the quotient of B^9 by $\mathrm{PAut}\big(\Lambda_4^{\mathcal{E}} \oplus \Lambda_4^{\mathcal{E}} \oplus H\big)$, the moduli space of genus 4 curves is the quotient of B^9 by $\mathrm{PAut}\big(\Lambda_4^{\mathcal{E}} \oplus N \oplus [-3] \oplus H\big)$, where N is the orthogonal complement of a root in $\Lambda_4^{\mathcal{E}}$, and the moduli space of cubic threefolds is the quotient of B^{10} by $\mathrm{PAut}\big(\Lambda_4^{\mathcal{E}} \oplus \Lambda_4^{\mathcal{E}} \oplus [-3] \oplus H\big)$. We remark that this last lattice may be constructed from Y_{551} in the same manner as L was from Y_{555}. See [1, 2, 10, 14, 16, 22] for more information about these ball quotients.

(9) The Coxeter group of the Y_{555} diagram appears in the monodromy of the T_{666} surface singularity $x^6 + y^6 + z^6 + \lambda xyz = 0$, where λ is a nonzero constant (see [3, Section 3.8; 11]), and also in Mukai's analysis [17] of the moduli space of 12 points in $(\mathbb{P}^5)^5$. I don't know any way to fit these facts into my conjectural framework. If there is a connection then Conway and Pritchard's real-hyperbolic constructions are probably also relevant; see remark (3).

(10) Simons [20] has studied a simpler version of our situation, concerning Y_{333} and a quotient of the Coxeter group with diagram the incidence graph of the 7 lines and 7 points of $\mathbb{P}^2(\mathbb{F}_2)$, which turns out to be $\mathrm{O}_8^-(2):2$. I found 14 tetraflections (order 4 complex reflections) of B^7 that satisfy the commutation and braid relations given by this diagram. One can choose the roots to have norm -2 and inner products 0 or $1 \pm \mathrm{i}$, spanning the lattice $\Lambda_2^{\mathcal{G}} \oplus \Lambda_2^{\mathcal{G}} \oplus \Lambda_2^{\mathcal{G}} \oplus \left(\begin{smallmatrix} 0 & 1+\mathrm{i} \\ 1-\mathrm{i} & 0 \end{smallmatrix}\right)$, where $\mathcal{G}$ is the Gaussian integers $\mathbb{Z}[\mathrm{i}]$ and $\Lambda_2^{\mathcal{G}}$ is the D_4 root lattice regarded as a lattice over $\mathcal{G}$. Because these are tetraflections, it is easy to turn them into involutions: one simply reduces the lattice modulo $1 + \mathrm{i}$. Then the tetraflections act on an

8-dimensional vector space over $\mathbb{F}_2$ equipped with a quadratic form of minus type. This provides a nice perspective on Simons' group, although it does not seem to give a new proof of his theorem. Basak has developed these ideas in a quaternionic context in [5].

References

1. D. Allcock, *The Leech lattice and complex hyperbolic reflections*, Invent. Math. **140** (2000), 283–301.
2. D. Allcock, J. Carlson, and D. Toledo, *The moduli space of cubic threefolds as a ball quotient*, Mem. Amer. Math. Soc., to appear.
3. V. Arnold, S. Gusein-Zade, and A. Varchenko, *Singularities of differentiable maps*, Vol. II, Birkhäuser, 1988.
4. T. Basak, *The complex Lorentzian Leech lattice and the bimonster*, J. Algebra **309** (2007), 32–56.
5. ______, *Reflection group of the quaternionic Lorentzian Leech lattice*, J. Algebra **309** (2007), 57–68.
6. R. Borcherds, J. H. Conway, L. Queen, and N. Sloane, *A monster Lie algebra?*, Adv. Math. **53** (1984), 75–79; reprinted in J. Conway and N. Sloane (eds.), Springer, 1993.
7. J. H. Conway, S. P. Norton, and L. H. Soicher, *The bimonster, the group* Y_{555} *and the projective plane of order* 3, Computers in Algebra (M. C. Tangora, ed.), Dekker, New York, 1988, pp. 27–50.
8. J. H. Conway and A. D. Pritchard, *Hyperbolic reflexions for the bimonster and* $3Fi_{24}$, Groups, Combinatorics and Geometry (Durham, 1990) (Liebeck M. and J. Saxl, eds.), London Math. Soc. Lecture Notes, vol. 165, Cambridge Univ. Press, Cambridge, 1992, pp. 23–45.
9. H. S. M. Coxeter, *Factor groups of the braid group*, Proceedings of the Fourth Canadian Mathematical Congress (Banff, 1957), University of Toronto Press, Toronto, 1959.
10. P. Deligne and G. D. Mostow, *Monodromy of hypergeometric functions and nonlattice integral monodromy*, Publ. Math. Inst. Hautes Études Sci. **63** (1986), 5–89.
11. A. M. Gabrielov, *Bifurcation, Dynkin diagrams and the modality of isolated singularities*, Funct. Anal. Appl. **8** (1974), no. 2, 94–98.
12. F. Hirzebruch, T. Berger, and R. Jung, *Manifolds and modular forms*, Friedr. Vieweg & Sohn, Braunschweig, 1992.
13. A. A. Ivanov, *A geometric characterization of the Monster*, Groups, Combinatorics and Geometry (Durham, 1990) (Liebeck M. and J. Saxl, eds.), London Math. Soc. Lecture Notes, vol. 165, Cambridge Univ. Press, Cambridge, 1992, pp. 46–62.
14. S. Kondō, *The moduli space of curves of genus* 4 *and Deligne–Mostow's complex reflection groups*, Algebraic Geometry 2000 (Azumino (Hotaka)), Adv. Stud. Pure Math., vol. 36, Math. Soc. Japan, Tokyo, 2002, pp. 383–400.
15. S. Linton, R. Parker, P. Walsh, and R. Wilson, *Computer construction of the monster*, J. Group Theory **1** (1998), 307–337.
16. E. Looijenga and R. Swierstra, *The period map for cubic threefolds*, to appear.
17. S. Mukai, *Geometric realization of* T*-shaped root systems and counterexamples to Hilbert's fourteenth problem*, Algebraic Transformation Groups and Algebraic Varieties, Encyclopaedia Math. Sci., vol. 132, Springer, Berlin, 2004, pp. 123–129.
18. S. P. Norton, *Constructing the Monster*, Groups, Combinatorics and Geometry (Durham, 1990) (Liebeck M. and J. Saxl, eds.), London Math. Soc. Lecture Notes, vol. 165, Cambridge Univ. Press, Cambridge, 1992, pp. 63–76.
19. P. Orlik and L. Solomon, *Discriminants in the theory of reflection groups*, Nagoya Math. J. **109** (1988), 23–45.
20. C. Simons, *Deflating infinite Coxeter groups to finite groups*, Proceedings on Moonshine and Related Topics (Montréal, QC, 1999), CRM Proc. Lecture Notes, vol. 30, Amer. Math. Soc., Providence, RI, 2001, pp. 223–229.
21. J. Swift, *A modest proposal*, 1729, pamphlet.
22. W. P. Thurston, *Shapes of polyhedra and triangulations of the sphere*, The Epstein Birthday Schrift, Geom. Topol. Monogr., vol. 1, Geom. Topol. Publ., Coventry, 1998, pp. 511–549.

Note added in proof. The author has found a computer-free proof of Basak's result that Γ is all of $\operatorname{Aut} L$. This will appear in *On the* Y_{555} *complex reflection group.* Also, Basak has proven that there is a map from the 26-generator Artin group to G, which we presumed to exist in Section 2. His proof uses a computer to verify that certain surfaces in B^{13} are disjoint from the arrangement of mirrors.

Department of Mathematics, University of Texas at Austin, Austin, TX 78712, USA
E-mail address: allcock@math.utexas.edu
URL: http://www.math.utexas.edu/~allcock

Centre de Recherches Mathématiques
CRM Proceedings and Lecture Notes
Volume **47**, 2009

Quivers and Difference Painlevé Equations

Philip Boalch

Dedicated to John McKay

ABSTRACT. We will describe natural Lax pairs for the difference Painlevé equations with affine Weyl symmetry groups of types E_6, E_7 and E_8, showing that they do indeed arise as symmetries of certain Fuchsian systems of differential equations.

1. Introduction

The theory of the Painlevé differential equations is now well established and their solutions, the Painlevé transcendents, appear in numerous nonlinear problems of mathematics and physics, from Einstein metrics to Frobenius manifolds.

However in certain applications these differential equations arise from a limiting process, starting with a *discrete* nonlinear equation. For example (see [41]) the difference equation

$$f_{n+1} + f_n + f_{n-1} = \frac{an+b}{f_n} + c$$

arose in work of Brézin and Kazakov [12] and Gross and Migdal [23] in the calculation of a certain partition function of a 2D quantum gravity model, and it becomes a differential Painlevé equation in a certain limit. The classification and study of such "discrete Painlevé equations" has been an active area in recent years (see for example the reviews [22] or [41]). On one hand many such equations have been found by studying a discrete analogue of the Painlevé property, such as singularity confinement. On the other hand Sakai [40] has discovered a uniform framework to classify all second order discrete Painlevé equations in terms of rational surfaces obtained by blowing up $\mathbb{P}^2$ in nine, possibly infinitely near, points. In this framework, to each equation is associated an affine root system R embedded in the affine E_8 root system, and thus an affine Dynkin diagram. (Upon removing the extending node, the Dynkin diagrams which occur correspond to the irreducible reflection subgroups of the E_8 Weyl group.) The (extended) affine Weyl group of

2000 *Mathematics Subject Classification.* 39A10, 34M55, 53C26.
This is the final form of the paper.

the orthogonal $R^\perp$ of R is then the symmetry group of the corresponding equation. The possible affine Weyl symmetry groups are roughly as follows:

$$\begin{array}{ccccccccccccccccc} \mathrm{E}_8 & \longrightarrow & \mathrm{E}_7 & \longrightarrow & \mathrm{E}_6 & \longrightarrow & \mathrm{D}_5 & \longrightarrow & \mathrm{A}_4 & \longrightarrow & \mathrm{A}_2+\mathrm{A}_1 & \searrow & & & & & \\ & & & & & & & \searrow & & \searrow & & & \mathrm{A}_1+\mathrm{A}_1 & \searrow & & & \\ & & & & & & & & \mathrm{D}_4 & \longrightarrow & \mathrm{A}_3 & \nearrow & & & \mathrm{A}_1 & \longrightarrow & \mathrm{A}_0 \\ & & & & & & & & & & & \searrow & \mathrm{A}_2 & \nearrow & & & \end{array}$$

(See [40, p. 182] for the precise version.) There are three types of discrete Painlevé equations: difference, q-difference or elliptic and we will be concerned here solely with the (ordinary) difference equations — this rules out the cases of D_5, A_4 and $\mathrm{A}_2 + \mathrm{A}_1$.

It turns out that many of these difference Painlevé equations may be understood as discrete symmetries of Painlevé differential equations: First the differential Painlevé equation may be obtained as the equation controlling the isomonodromic deformations of a rational system of linear differential equations on $\mathbb{P}^1$ of the form

$$\frac{\mathrm{d}}{\mathrm{d}z} - A \tag{1}$$

for some matrix A of rational functions. Then the difference Painlevé equation arises as Schlesinger transformations (or contiguity relations) of (1), obtained by applying rational gauge transformation to shift the exponents at the poles by integers (see [40, pp. 204 – 205]). This only applies to the bottom two rows of Sakai's list and the corresponding differential Painlevé equations are:

$$\begin{array}{ccccccccc} & & & \nearrow & \mathrm{P}_{\mathrm{III}} & \searrow & & & \\ \mathrm{P}_{\mathrm{VI}} & \longrightarrow & \mathrm{P}_{\mathrm{V}} & & & & \mathrm{P}_{\mathrm{II}} & \longrightarrow & \mathrm{P}_{\mathrm{I}} \\ & & & \searrow & \mathrm{P}_{\mathrm{IV}} & \nearrow & & & \end{array}$$

so for example the sixth Painlevé equation P_{VI} has symmetry group affine D_4 (which may be extended to affine F_4 by including diagram automorphisms).

Thus, looking at Sakai's table, this misses the cases with the richest symmetry groups, namely we still have the cases of E_6, E_7 and E_8.

A basic question is thus to try to find linear systems (1) with the desired symmetry groups, and living in two-dimensional moduli spaces. (Explicitly this is Problem A of [41].) A key observation is that to some extent it is irrelevant that there are corresponding differential Painlevé equations — as we will see the solution to this question turns out to be amongst certain Fuchsian systems which do not admit nontrivial isomonodromic deformations (in spite of the fact they are nonrigid). Another point is that systems of higher rank appear. (Recall [27] that all the differential Painlevé equations may be obtained by considering isomonodromy deformations of *rank two* systems (1).) Indeed the ranks will be given by the dimension of the largest irreducible representation of the finite group associated to the given Dynkin diagram via the McKay correspondence, viz:

rank 3, 4, or 6 for the cases of E_6, E_7, or E_8

respectively.

Said differently finding a linear system (1) with the desired affine Weyl symmetry group (and living in a two-dimensional moduli space) amounts to finding a 'Lax pair' for the corresponding Painlevé difference equation (see [22, pp. 266 and 274]; the Schlesinger transformations themselves give the other half of the Lax pair). The fact that the system lives in a complex two-dimensional moduli space implies that, if written out explicitly in coordinates, the translation subgroup of the symmetry group constitutes a second order nonlinear difference equation.

Thus the aim of this article is to describe natural linear systems (1) with the appropriate symmetries appearing in Lax pairs for the outstanding cases of affine Weyl groups of types E_6, E_7 and E_8 in Sakai's table. The E_8 case is the most symmetric difference Painlevé equation. (The E_6 case we will present here appeared recently from a different 'Mellin transformed' viewpoint in Remark 2.14 of [2], which we are thus generalising.) The origin of the desired spaces of Fuchsian systems lies in Kronheimer's work [32] which is very closely related to the McKay correspondence. The main work to be done is to relate Kronheimer's work to Sakai's.

The layout of this article is as follows. Section 2 recalls the geometry of the Painlevé VI phase spaces as a motivating example. Section 3 then recalls the complex symplectic approach to Kronheimer's asymptotically locally Euclidean hyperKähler four manifolds, explaining how they arise as quiver varieties. Section 4 recalls how quiver varieties for star-shaped quivers parameterise logarithmic connections on the trivial vector bundle over $\mathbb{P}^1$ (i.e. Fuchsian systems). Then Section 5 describes the spaces of Fuchsian systems corresponding to quivers whose underlying graph is a star-shaped affine Dynkin diagram. Then (families of) such spaces inherit a natural action of the corresponding finite Weyl groups. Section 6 then describes this action and how it extends to a birational action of the full affine Weyl group (the translation subgroup of which will constitute the desired difference Painlevé equation). Section 7 ties things together by describing directly how to relate the spaces of Kronheimer to those of Sakai on which the nonlinear difference equation were originally defined (in essence one glues an affine line onto Kronheimer's spaces such that the affine Weyl group action becomes biregular rather than birational). Finally in the appendix we give a direct geometrical approach to the Weyl group symmetries (as well as some generalisations), addressing the philosophical question of "where do the symmetries come from?" in terms of operations on connections.

The possibility of finding the q-difference analogue (or the elliptic analogue) of the Lax pairs of this article will be studied elsewhere.

2. The Painlevé VI phase space

We will begin by describing the phase space of the sixth Painlevé differential equation, in a slightly nonstandard way. (The corresponding contiguity relations lead to the so-called dP_V equation which, as for the P_{VI} differential equation itself, has symmetry group the affine Weyl group of type D_4.)

Painlevé VI arises as the equation controlling the isomonodromic deformations of 2×2 Fuchsian systems of the form

$$\frac{d}{dz} - A; \quad A = \frac{A_1}{z} + \frac{A_2}{z-t} + \frac{A_3}{z-1} \tag{2}$$

where $t \in \mathbb{C} \setminus \{0, 1\}$ is the deformation parameter. Tensoring (2) by a logarithmic connection on a line bundle allows us to assume without loss of generality that each of the residues A_i, $i = 1, 2, 3$ is a rank one matrix (i.e. we may add a scalar matrix to each A_i). By convention we choose $\theta_i \in \mathbb{C}$ so that A_i has eigenvalues $0, \theta_i$, $i = 1, 2, 3$. Finally we choose a fourth rank one matrix A_4 with eigenvalues $0, \theta_4$ such that $\sum_1^4 A_i$ is a scalar matrix:

$$A_1 + A_2 + A_3 + A_4 = \nu.1 \tag{3}$$

with $\nu \in \mathbb{C}$, so that, upon taking the trace we see

$$\sum_1^4 \theta_i = 2\nu. \tag{4}$$

(Note that the residue at infinity of $A dz$ is $A_4 - \nu$ so in all cases θ_i is the difference, with appropriate sign, between the eigenvalues of the corresponding residue.) Let $\mathcal{O}_i \subset \mathfrak{gl}_2(\mathbb{C})$ be the adjoint orbit of rank one matrices with eigenvalues $0, \theta_i$. Each orbit may be identified with a coadjoint orbit using the trace pairing and so has a natural invariant complex symplectic structure, and the phase space we are interested in is the subspace of the product of the four orbits satisfying the constraint (3), modulo diagonal conjugation. But this is just the symplectic quotient of the product of orbits by diagonal conjugation at the value ν of the moment map, since the moment map is just the sum:

$$\mathcal{O}_1 \times \cdots \times \mathcal{O}_4 \underset{\nu}{/\!\!/} \mathrm{GL}_2(\mathbb{C}) = \{(A_1, \dots, A_4) \mid A_i \in \mathcal{O}_i, (3) \text{ holds}\} / \mathrm{GL}_2(\mathbb{C}). \tag{5}$$

Here the double slash ($/\!\!/$) denotes the complex symplectic quotient and the single slash denotes the affine GIT quotient, taking the variety associated to the ring of invariant functions. (If $\theta_i = 0$ we should take $\mathcal{O}_i$ to be *closure* of the orbit of nonzero nilpotent matrices.) These phase spaces are complex surfaces in general. Since any rank one matrix can be written as the tensor product of a vector and a linear form, we see that each orbit $\mathcal{O}_i$ is itself the symplectic quotient of a vector space, the cotangent bundle of $\mathbb{C}^2$, by the natural scaling action of $\mathbb{C}^*$, at the value θ_i of the moment map:

$$\mathcal{O}_i = T^*\mathbb{C}^2 \underset{\theta_i}{/\!\!/} \mathbb{C}^* = \{(v, \alpha) \in \mathbb{C}^2 \times (\mathbb{C}^2)^* \mid \alpha(v) = \theta_i\} / \mathbb{C}^*.$$

The map to $\mathcal{O}_i$ is simply $(v, \alpha) \mapsto v \otimes \alpha$. Combining these four $\mathbb{C}^*$ quotients together with the above $\mathrm{GL}_2(\mathbb{C})$ quotient yields a description of the phase space (5) as a symplectic quotient of a vector space $(T^*\mathbb{C}^2)^4$ by the group $(\mathbb{C}^*)^4 \times \mathrm{GL}_2(\mathbb{C})$.

Now we know *a posteriori* that the set of these phase spaces admits a birational action of the affine D_4 Weyl group (due mainly to work of Okamoto [36]), with the set of parameters θ corresponding to the complex span of the D_4 root lattice. However the description above leads to a direct path from the D_4 root system to the phase spaces, as we will recall in the following sections. This will then generalise immediately to the root systems of types E_6, E_7 and E_8.

One can see an immediate link to D_4 since the condition for the phase spaces to be nonsingular is given by $\theta_i \neq 0$ (so there are no nilpotent orbits appearing) and $\sum_1^4 \pm\theta_i \neq 0$ for any choice of signs (this implies the system cannot be conjugated into triangular form and thus that the action of $\mathrm{PGL}_2(\mathbb{C})$ is free). Now the 24 vectors $\pm\epsilon_i$, $\sum_1^4 \pm\epsilon_i/2$ make up a D_4 root system so the nonsingularity conditions are just that the vector $\theta = \sum \theta_i \epsilon_i$ should not lie on any of the reflection hyperplanes for the D_4 Weyl group. The key facts (see [32]) to see the generalisation are that for $\theta = 0$ the space (5) is the quotient $\mathbb{C}^2/\Gamma$ (where $\Gamma \subset \mathrm{SL}_2(\mathbb{C})$ is the finite group corresponding to D_4 under the McKay correspondence), that the other spaces make up a semiuniversal deformation of this singularity, and that they have nice descriptions as symplectic quotients.

Remark 1. Extrinsically the Okamoto transformations of $\mathrm{P_{VI}}$ were explained in [3] in terms of the algebraic geometry of the moduli spaces, and intrinsically they were explained in [7,8] in terms of natural operations on connections, the Fourier–Laplace transformation and tensoring by connections on line bundles (which turned out [20] to be equivalent to the complex analytic version of Katz's middle convolution operation [30]) — the appendix to the present article will extend this viewpoint.

3. Kronheimer's ALE quiver varieties

We will recall the complex symplectic approach to Kronheimer's ALE spaces [32], which are noncompact complex surfaces admitting asymptotically locally Euclidean hyperKähler metrics and which arise as deformations of Kleinian singularities. Subsequently these varieties were generalised by Nakajima [34]. (The article [16] of Crawley-Boevey has been useful to us here.)

Let $\mathcal{Q}$ be a quiver (i.e. an oriented graph) whose underlying graph is a simply laced affine Dynkin diagram corresponding to a (finite) root system R of rank r. Thus R is of type A_r with $r = 1, 2, 3, \ldots$, of type D_r with $r = 4, 5, 6, \ldots$ or of type E_r with $r = 6, 7, 8$. Let I be the set of nodes and let $\mathcal{Q}$ denote the set of edges. The set I corresponds to the simple roots $\varepsilon_1, \ldots, \varepsilon_r$ of R plus ε_0 corresponding to the extending node of the diagram.

Following McKay we associate a finite-dimensional complex vector space W_i to each node. Namely McKay observed that $\mathcal{Q}$ may be identified with the McKay graph of a finite subgroup Γ of $\mathrm{SL}_2(\mathbb{C})$ defined as follows. The nodes of the McKay graph correspond to the irreducible representations V_i of Γ (for each node $i \in I$) with V_0 being the trivial one-dimensional representation. Then nodes i and j are connected by a_{ij} edges where a_{ij} is the multiplicity of V_j appearing in the tensor product of V_i with the two dimensional defining representation U of Γ:

$$U \otimes V_i \cong \bigoplus_j a_{ij} V_j.$$

(Since U is self-dual $a_{ij} = a_{ji}$.) Let $n_i := \dim(V_i)$ be the dimension of the representation V_i and let $\delta \in \mathbb{Z}^I$ denote the vector of these dimensions. Thus we may associate a vector space $W_i := \mathbb{C}^{n_i}$ to each node. (Intrinsically W_i should be thought of as the multiplicity space for V_i in the regular representation of Γ — cf. [32].) In Lie theoretic terms the dimensions n_i also arise as the coefficients expressing the longest root of R in terms of the simple roots. That is the longest root

$$\rho = \sum_1^r n_i \varepsilon_i$$

and $n_0 = 1$. These coefficients n_i are listed in [11] or [28, p. 54]. For example for the root system of type D_4 the dimensions are as in Figure 1.

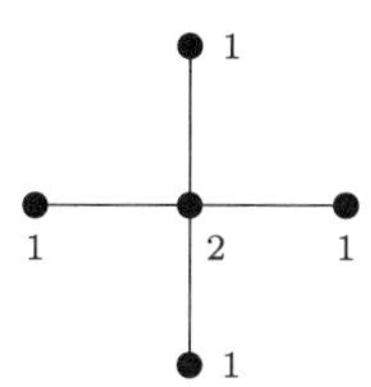

FIGURE 1. Extended D_4 Dynkin diagram.
McKay graph of the quaternion group $\{\pm 1, \pm \mathrm{i}, \pm \mathrm{j}, \pm \mathrm{k}\}$

Now to construct the quiver variety we consider for each edge of $\mathcal{Q}$ a homomorphism in each direction between the vector spaces at each end:

$$W = W(\mathcal{Q}) := \bigoplus_{e \in \mathcal{Q}} \operatorname{Hom}(W_{t(e)}, W_{h(e)}) \oplus \operatorname{Hom}(W_{h(e)}, W_{t(e)})$$

where, for a directed edge $e \in \mathcal{Q}$, $h(e), t(e) \in I$ are its head and tail respectively.

The vector space W inherits a linear complex symplectic structure if we identify the direct sum $\operatorname{Hom}(W_{t(e)}, W_{h(e)}) \oplus \operatorname{Hom}(W_{h(e)}, W_{t(e)})$ with the cotangent bundle $T^* \operatorname{Hom}(W_{t(e)}, W_{h(e)})$ in the obvious way.

Now consider the group

$$\widetilde{G} = \widetilde{G}(\mathcal{Q}) := \prod_{i=0}^{r} \operatorname{GL}(W_i)$$

which acts on the sum of the vector spaces W_i and thus has an induced action on W. (If $\phi \in \operatorname{Hom}(W_i, W_j)$ and $g = (g_0, \ldots, g_r) \in \widetilde{G}$ then $g \cdot \phi = g_j \circ \phi \circ g_i^{-1}$.)

The group $\widetilde{G}$ has a central subgroup $\mathbb{T}$ isomorphic to $\mathbb{C}^*$ which acts trivially on W (simply take each g_i to be the same nonzero scalar). Define $G = \widetilde{G}/\mathbb{T}$ to be the quotient group (which acts effectively on W). Thus G has a centre Z of dimension r (obtained by setting each g_i to be an arbitrary nonzero scalar).

A moment map $\mu = (\mu_0, \ldots, \mu_r) \colon W \to \prod_{i=0}^{r} \mathfrak{gl}(W_i)$ for the action of $\widetilde{G}$ on W is given by

$$\mu_i(\{\phi\}) := \sum_{\substack{e \in Q \\ h(e)=i}} \phi_e \circ \phi_{e^*} - \sum_{\substack{e \in Q \\ t(e)=i}} \phi_{e^*} \circ \phi_e$$

where $\phi_e \in \operatorname{Hom}(W_{t(e)}, W_{h(e)})$ and e^* is the edge e with reversed orientation.

Taking the sum of the traces of the components of μ clearly yields zero (since $\operatorname{Tr}(AB) = \operatorname{Tr}(BA)$ for any compatible rectangular matrices A, B) and the subspace on which this sum of traces is zero is naturally identified with the (dual of the) Lie algebra of G, and so μ is also a moment map for G.

Finally the corresponding quiver varieties are obtained by performing the complex symplectic quotient of W by G at any central value λ of the moment map μ:

$$N_{\mathcal{Q}}(\lambda) = N_{\mathcal{Q}}(\lambda, \delta) = W \underset{\lambda}{/\!\!/} G = \mu^{-1}(\lambda)/G$$

where δ is the dimension vector whose components were the dimensions n_i. (We may also view this as $W/\!\!/_{\lambda}\widetilde{G}$ since $\mathbb{T}$ acts trivially.)

Let $\mathfrak{h}$ be the set of such coadjoint invariant elements (i.e. central elements) λ of the dual of the Lie algebra of G. Since the trace pairing relates the adjoint action and the coadjoint action, $\mathfrak{h}$ may be identified with the Lie algebra of the central subgroup Z of G and so is a complex vector space of dimension r. Concretely it is given as a subspace of $\prod_{i=0}^{r} \mathfrak{gl}(W_i)$ by the $(r+1)$-tuples of scalar matrices whose sum of traces is zero. In other words if each diagonal entry of the ith matrix is λ_i then the constraint is that

$$\delta \cdot \lambda := \sum_{0}^{r} n_i \lambda_i = \sum_{0}^{r} \operatorname{Tr}_{W_i}(\lambda_i) = 0. \tag{6}$$

Carrying this out for the case of D_4 should now be familiar from the previous section, and note in particular that the above constraint matches up with that of (4).

4. Quivers and matrices

Of course one can carry out the construction of the previous section for any quiver and any dimension vector, although in general these spaces will probably not be relevant to Painlevé equations or isomonodromy in general. However there are many cases which are: the phase spaces of the Schlesinger equations for $\mathrm{GL}_n(\mathbb{C})$ arise as quiver varieties for *star-shaped* quivers.

This relationship is by now well-known and has been used in an essential way by Crawley-Boevey (see [16]) in his work on the Deligne–Simpson problem.

The $\mathrm{GL}_n(\mathbb{C})$ Schlesinger phase spaces are just the moduli spaces of logarithmic connections on trivial rank n holomorphic vector bundles over the Riemann sphere, obtained by fixing the adjoint orbits of the residues (the name "moduli spaces of Fuchsian systems" seems most suitable). Choosing a trivialisation of the bundle and a coordinate z on $\mathbb{P}^1$, such a connection will be of the form

$$\mathrm{d} - A\mathrm{d}z, \quad A = \sum_{i=1}^{m-1} \frac{A_i}{z - a_i} \tag{7}$$

which is equivalent to the Fuchsian system $\mathrm{d}/\mathrm{d}z - A$. If we set $a_m = \infty$ and take A_m to be the residue of $A\mathrm{d}z$ at ∞ then the residue theorem says that

$$\sum_1^m A_i = 0.$$

If we fix the residues to be in chosen (co)adjoint orbits $A_i \in \mathcal{O}_i$ then we notice that $\sum_1^m A_i$ is just the moment map for the diagonal conjugation action of $\mathrm{GL}_n(\mathbb{C})$ on the product of orbits, and that this action corresponds to the gauge equivalence of such connections, so the moduli space is just the symplectic quotient

$$\mathcal{O}_1 \times \cdots \times \mathcal{O}_m /\!\!/ \mathrm{GL}_n(\mathbb{C}) \tag{8}$$

at the zero value of the moment map (cf. [5, 26]). The rank two, four pole case is the Painlevé VI case described in a slightly different way in Section 2, which we have now recognised as D_4 quiver varieties. However the more general case (8) can also be obtained from a star-shaped quiver, as follows (see [16]).

Let w_i be the degree of the minimal polynomial p_i of $A_i \in \mathcal{O}_i$ and let ξ_{ij} be the roots of p_i ($i = 1, \dots, m$ and $j = 1, \dots, w_i$). Let $\mathcal{Q}$ be the star-shaped quiver with a central vertex labelled $*$ and m legs of lengths $w_i - 1$ for $i = 1, \dots, m$. Label the jth vertex of the ith leg by ij with $1 \le j < w_i$ (with j increasing as we go away from $*$). Orient all the edges to point towards the central vertex $*$. The corresponding dimension vector is obtained by assigning dimension n to the central vertex and the dimension

$$n_{ij} := \mathrm{rank}(A_i - \xi_{i1})(A_i - \xi_{i2}) \cdots (A_i - \xi_{ij})$$

to the vertex ij. This specifies the quiver and dimension vector corresponding to the given orbits.

Exercise. Consider the following leg of a star-shaped quiver, with the given vector spaces at each node.

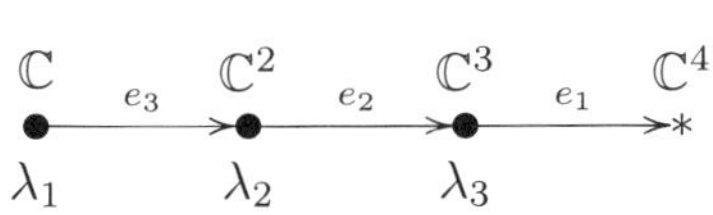

(i) Let $\lambda_i \in \mathrm{End}(\mathbb{C}^i)$ be a scalar matrix and write down the condition for each component μ_i of the moment map μ to take the value λ_i for $i = 1, 2, 3$.

(ii) If $\mu(\{\phi\}) = \lambda$, deduce the eigenvalues of the matrices

$$\phi_{e_i^*} \circ \phi_{e_i}, \quad \phi_{e_i} \circ \phi_{e_i^*}$$

for $i = 3, 2, 1$.

(iii) Thus observe that $\phi_{e_1} \circ \phi_{e_1^*} \in \mathrm{End}(\mathbb{C}^4)$ has eigenvalues

$$0, \ -\lambda_3, \ -\lambda_3 - \lambda_2, \ -\lambda_3 - \lambda_2 - \lambda_1$$

and so the λ_i's arise as eigenvalue *differences.*

Remark 2 (Eigenvalue conventions). A set of parameters λ for a star-shaped quiver satisfying $\lambda \cdot \delta = 0$ thus specifies (and orders) the eigenvalues of all of the residues of the corresponding Fuchsian system (7), modulo tensoring by a scalar system of the form $\mathrm{d} - \sum_1^{m-1} \Lambda_i \mathrm{d}z/(z - a_i)$ with $\Lambda_i \in \mathbb{C}$. In other words, up to adding a scalar Λ_i to A_i for $i = 1, \ldots, m-1$ (and varying the residue $-\sum_1^{m-1} A_i$ at infinity in the corresponding way). There are two standard ways of normalising the residues to remove this indeterminacy, and both are useful:

(1) "Trace zero": choose each Λ_i such that each residue has zero trace,

(2) "Determinant zero": choose Λ_i such that the first eigenvalue of each residue $A_1, \ldots, A_{m-1}$ is zero — it is this normalisation that arises in the first instance from the quiver viewpoint. (The residue at infinity is then $A_m - \nu$ where ν is minus the component of λ associated to the central node of $\mathcal{Q}$, $\sum_1^m A_i = \nu$, and the first eigenvalue of A_m is also zero.)

It is a triviality to go between these two normalisations, but one should be aware of them.

5. Star-shaped affine Dynkin diagrams

Now we will look at the area of intersection of the two previous sections, i.e. the case of a star-shaped affine Dynkin diagram. This leaves just the cases of D_4 and E_6, E_7, E_8 from the original ADE list. Inverting the construction of the previous section we can then view Kronheimer's E-type ALE spaces as moduli spaces of Fuchsian systems on $\mathbb{P}^1$, as we have already done for D_4.

The E-type affine Dynkin diagrams and dimension vectors, which are the same as the Platonic McKay graphs, are as in Figures 2 – 4.

In all these cases there are just three legs so they correspond to Fuchsian systems with just three poles on $\mathbb{P}^1$. Up to coordinate transformations we can

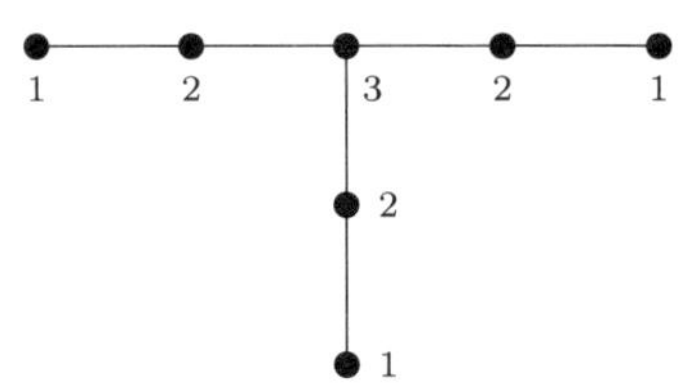

FIGURE 2. Extended E_6 Dynkin diagram.
McKay graph of the binary tetrahedral group

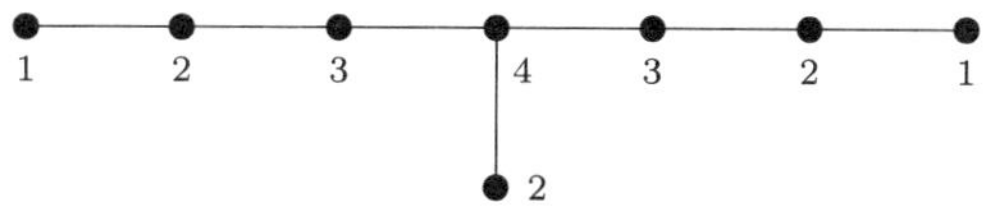

FIGURE 3. Extended E_7 Dynkin diagram.
McKay graph of the binary octahedral group

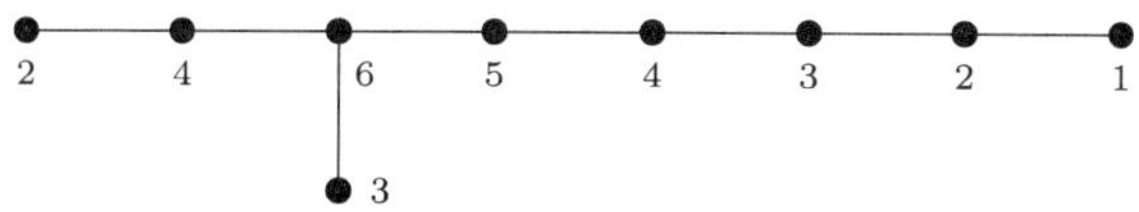

FIGURE 4. Extended E_8 Dynkin diagram.
McKay graph of the binary icosahedral group

TABLE 1.

	Size n	$\mathcal{O}_1$	$\mathcal{O}_2$	$\mathcal{O}_3$
E_6	3	$1,1,1$	$1,1,1$	$1,1,1$
E_7	4	$2,2$	$1,1,1,1$	$1,1,1,1$
E_8	6	$3,3$	$2,2,2$	$1,\dots,1$

always choose to put the poles at $0, 1, \infty$ and so the systems are of the form

$$\frac{\mathrm{d}}{\mathrm{d}z} - \left(\frac{A_1}{z} + \frac{A_2}{z-1}\right)$$

and the residue at infinity is $A_3 = -(A_1 + A_2)$. (Thus in general these systems have no nontrivial isomonodromic deformations, only discrete symmetries.) Now given any value of the parameters λ the prescription of the previous section applied to these graphs and dimension vectors determines the size and orbit type of the three matrices A_i that occur. When the parameters are generic the matrices are semisimple with eigenvalues distributed as in Table 1.

The last three columns list various partitions of the size n and indicate the eigenvalue multiplicities. Thus for example $2, 2, 2$ on the third row indicates orbits of 6×6 matrices with three distinct eigenvalues each repeated twice. Thus the corresponding quiver varieties are isomorphic to complex symplectic quotients

$$\mathcal{O}_1 \times \mathcal{O}_2 \times \mathcal{O}_3 /\!\!/ \mathrm{PGL}_n(\mathbb{C})$$

with the orbits $\mathcal{O}_i$ as indicated in the table.

In the E_6 case we recognise the generic three-poled 3×3 systems mentioned in [2, Remark 2.14] (it would be interesting to compute the Mellin transforms of the E_7 and E_8 systems in the table above to obtain "difference-difference" Lax pairs).

Exercise. Confirm directly in each case that:

(i) the dimension of the space of parameters (specifying the eigenvalues of the matrices) is equal to the rank of the finite root system,

(ii) that the above symplectic quotient does indeed have expected complex dimension 2, and

(iii) that as in the D_4 case above, if the parameters λ are not on any of the root hyperplanes then the triple A_1, A_2, A_3 (with $A_i \in \mathcal{O}_i$) is irreducible (i.e. cannot be

simultaneously conjugated into a block triangular form) and also there are no extra coincidences between the eigenvalues of each A_i.
(For example for E_8: (i) we need to choose 11 eigenvalues but there is one relation since the sum of the traces should be zero, and two more parameters are irrelevant since we can tensor with a scalar system with poles at $0, 1, \infty$. For (ii) the product of orbits has dimension $72 = 18 + 24 + 30$ and $\mathrm{PGL}_6(\mathbb{C})$ has dimension 35 so the expected dimension is $72 - 2 \times 35 = 2$. For (iii) write the three sets of eigenvalues as

$$(a, a, a, b, b, b), \quad (c, c, d, d, e, e), \quad (f, g, h, i, j, k).$$

If the A_i can be put in a block triangular form with an *irreducible* $r \times r$ block on the diagonal, then taking the sum of the traces of these blocks gives a linear relation amongst a subset of the eigenvalues: For $r = 1$ or 5 there are $2 \times 3 \times 6 = 36$ possibilities, for $r = 2$ or 4 there are $1 \times \binom{3}{2} \times \binom{6}{2} = 45$ possibilities, and for $r = 3$ there are $2 \times 1 \times \binom{6}{3} = 40$ possibilities,[1] giving altogether 202 linear relations (each appearing together with its minus so yielding 101 hyperplanes in the parameter space). If the parameters are off of these hyperplanes then clearly no such block triangular decomposition is possible. Further, for the eigenvalues to not have extra coincidences yields $1 + 3 + 15 = 19$ further hyperplanes, or 38 linear relations taking both possible signs of each relation. These $202 + 38 = 240$ linear relations are the E_8 roots.)

Remark 3. This list has also appeared in work of Kostov [31] (which we learnt of from Simpson [42]), but the link with root systems was not made.

Remark 4. Observe (as in [32]) that since the vector spaces attached to the nodes are of the same dimensions as the irreducible representations of the corresponding binary group, there is a simple formula for the complex dimension of the vector space W from which the quiver variety was constructed: its complex dimension is twice the order of the binary group (e.g., 240 for E_8).

Exercise (Non-star-shaped examples and irregular singularities). Relate the symplectic quotient descriptions of the quiver varieties of affine Dynkin types A_3, A_2 and A_1 to the symplectic quotient descriptions ([5] section 2) of the moduli spaces of rank two (irregular) linear systems on the Riemann sphere, with poles of orders $(2, 1, 1)$, $(3, 1)$ and (4) respectively. Hints:

(i) Use an extended orbit (denoted $\widetilde{\mathcal{O}} \cong T^*G \times \mathcal{O}_B$ in [5]) at each irregular singularity. Thus after performing the $G := \mathrm{GL}_2(\mathbb{C})$ symplectic quotient the spaces are respectively the symplectic quotients of

$$\mathcal{O}_B^{(2)} \times \mathcal{O}_1 \times \mathcal{O}_2, \quad \mathcal{O}_B^{(3)} \times \mathcal{O}_1, \quad \mathcal{O}_B^{(4)}$$

by the diagonal conjugation action of the torus $T = \left\{\begin{pmatrix} * & \\ & 1 \end{pmatrix}\right\} \subset G$, where the $\mathcal{O}_i$ are coadjoint orbits of G, and the $\mathcal{O}_B^{(k)}$ are coadjoint orbits of a nilpotent group (the group B_k of based k-jets of bundle automorphisms).

(ii) Show directly that $\mathcal{O}_B^{(k)} \cong T^*\mathbb{C}^{k-2}$ as Hamiltonian T-spaces, where $T \cong \mathbb{C}^*$ acts on $\mathbb{C}^{k-2}$ by scaling. (This is tricky for $k = 4$, but possible.)

(iii) Write out the corresponding ALE symplectic quotients explicitly in these cases, and symplectic quotient by a suitable codimension one subtorus such that the remaining $\mathbb{C}^*$ quotient is precisely that of (1) above (up to tensoring to make the $\mathcal{O}_i$ orbits of rank one matrices).

[1]Note for example that a 3×3 block with eigenvalues of the form $(a, a, b), (c, c, d), (f, g, h)$ is necessarily reducible so not counted in the $r = 3$ case.

6. Affine Weyl group action

The next task is to show that the Fuchsian systems appearing in the previous section do indeed have the corresponding affine Weyl group symmetries. Fortunately the finite Weyl group action is now quite well-known (and was known to Kronheimer) and the translations come from Schlesinger transformations.

Let $\mathcal{Q}$ be a star-shaped quiver (for example as considered in Section 4). (This is a more general context than strictly necessary here but these more general symmetries, for example changing the dimension vector, will give many interesting identifications between various spaces of Fuchsian systems.)

One can then define a Weyl group as follows. Let I be the set of nodes and let $r+1 = \#I$ be the number of nodes. Define the $(r+1)\times(r+1)$ Cartan matrix to be

$$C = 2\,\mathrm{Id}. - A$$

where the i,j entry of A is the number of edges connecting the ith and jth nodes. The *root lattice* $\mathbb{Z}^I = \bigoplus_0^r \mathbb{Z}\varepsilon_i$ inherits a bilinear form defined by

$$(\varepsilon_i, \varepsilon_j) = C_{ij}.$$

In turn we can define simple reflections s_i acting on the root lattice by the formula

$$s_i(\beta) := \beta - (\beta, \varepsilon_i)\varepsilon_i$$

for any $i \in I$. The Weyl group of the quiver is then the group generated by these simple reflections. We can also define dual reflections r_i acting on the vector space $\mathbb{C}^I$ by the formula

$$r_i(\lambda) = \lambda - \lambda_i \alpha_i$$

where $\lambda = \sum_0^r \lambda_i \varepsilon_i \in \mathbb{C}^I$ with $\lambda_i \in \mathbb{C}$ and where for each $i \in I$

$$\alpha_i := \sum_{j\in I} (\varepsilon_i, \varepsilon_j)\varepsilon_j \in \mathbb{C}^I.$$

By construction one has that $s_i(\beta)\cdot r_i(\lambda) = \beta\cdot\lambda$, where the dot denotes the pairing given by $\varepsilon_i \cdot \varepsilon_j = \delta_{ij}$. The main result we wish to quote is then

Theorem 1 ([17,34,35,39]). *Let $\mathcal{Q}$ be a fixed quiver as above. Then, if $\lambda_i \neq 0$, there is a natural isomorphism between the the quiver variety with dimension vector β and parameters $\lambda = \sum \lambda_i \varepsilon_i$ and that with dimension vector $s_i(\beta)$ and parameters $r_i(\lambda)$:*

$$N_{\mathcal{Q}}(\lambda, \beta) \cong N_{\mathcal{Q}}(r_i(\lambda), s_i(\beta)).$$

The desired reflection functors are constructed in [17, Section 5] (see also [34, 35, 39]) and that they induce isomorphisms of the moduli spaces is stated explicitly [14, Lemma 2.1] and proved in [13, Lemma 2.2]. (Apparently they were first written down independently by Nakajima [33] and Rump [39].)[2]

Now let us specialise to one of the four cases D_4, E_6, E_7, E_8 we are interested in. (This Weyl group action will then give all the desired symmetries except the translations.) The generating reflections will be described geometrically in the appendix; here we will check that the group acting is indeed the finite Weyl group of the corresponding type (i.e. D_4, E_6, E_7, or E_8).

[2]Of course for D_4 they were in some sense found earlier by Okamoto [36] (but even the most interesting Okamoto transformation was detected earlier at a quantum level by Regge [38], cf. [9]).

Thus $\mathcal{Q}$ is an extended Dynkin diagram and we label the nodes by $I = \{0, \ldots, r\}$ with 0 labelling the extending node (i.e. any of the extremal nodes with dimension 1). Let $\delta \in \mathbb{Z}^I$ be the corresponding dimension vector (i.e. the dimensions written by the nodes in Figures 1 – 4).

Then δ spans the kernel of the bilinear form (,) and so in particular all the reflections s_i fix δ. Recall from (6) that the space $\mathfrak{h}$ of coadjoint invariants is just $\mathfrak{h} = \{\lambda \mid \lambda \cdot \delta = 0\}$. Thus if $\lambda_i \neq 0$ the above construction yields an isomorphism $N_{\mathcal{Q}}(\lambda) \cong N_{\mathcal{Q}}(r_i(\lambda))$. We will check that this generates the finite Weyl group, so that we have an action of the Weyl group on the family of quiver varieties parameterised by the regular values of λ.

First note that each $\alpha_i := \sum_{j \in I}(\varepsilon_i, \varepsilon_j)\varepsilon_j$ is actually in $\mathfrak{h} \subset \mathbb{C}^I$ and that $\alpha_1, \ldots, \alpha_r$ form a basis of $\mathfrak{h}$. Then define a bilinear form on $\mathfrak{h}$ by declaring

$$(\!(\alpha_i, \alpha_j)\!) = C_{ij} \quad \text{for } i, j = 1, \ldots, r.$$

This is nondegenerate since C (with indices ≥ 1) is the Cartan matrix of a (non-affine) Dynkin diagram, and so $\alpha_1, \ldots, \alpha_r$ make up the simple roots of a copy of the corresponding finite root system. Now by definition the simple reflections of this root system are of the form

$$\lambda \mapsto \lambda - (\!(\lambda, \alpha_i)\!)\alpha_i.$$

But this is just the restriction of r_i to $\mathfrak{h}$ because:

Lemma 1. *For all* $\lambda \in \mathfrak{h}$ *we have* $(\!(\lambda, \alpha_i)\!) = \lambda \cdot \varepsilon_i =: \lambda_i$.

PROOF. On basis vectors: $(\!(\alpha_j, \alpha_i)\!) = C_{ji} = (\sum_k C_{jk}\varepsilon_k) \cdot \varepsilon_i = \alpha_j \cdot \varepsilon_i$. □

Hence the group of automorphisms of $\mathfrak{h}$ induced by the Weyl group generated by $r_1, \ldots, r_r$ is the desired finite Weyl group. (Note that r_0 gives nothing more since α_0 is minus the longest root; the fact that δ is in the kernel of (,) implies $\sum_0^r n_i\alpha_i = 0$.)

Next we will verify that the translations arising from Schlesinger transformations generate the translation subgroup of the corresponding *extended* affine Weyl group. Recall that for a simply laced root system R (as we have here) the translation subgroup of the affine Weyl group is the root lattice $Q(R) = \langle \alpha_1, \ldots, \alpha_r \rangle_{\mathbb{Z}} \subset \mathfrak{h}$ and the translation subgroup of the extended affine Weyl group is the finer lattice

$$P(R) = \{\lambda \in \mathfrak{h} \mid (\!(\lambda, \alpha_i)\!) \in \mathbb{Z} \text{ for all } i = 1, \ldots, r\} \supset Q(R).$$

Now by definition the Schlesinger transformations are the transformations of the Fuchsian systems induced by rational gauge transformations with poles at the poles of the system (see [27] for more details and some explicit formulae). Thus they vary the eigenvalues of the systems by integers and induce birational isomorphisms between the moduli spaces. Of course we wish to obtain systems with eigenvalue distributions of the same type, so the possible integer shifts are restricted. In the notation we are using the Schlesinger transformations yield the integer shifts given by the lattice

$$\{\lambda \mid \lambda_i \in \mathbb{Z}, \lambda \cdot \delta = 0\}.$$

Clearly, by Lemma 1, this is the desired lattice $P(R)$.

Thus altogether we obtain the full extended affine Weyl group of birational symmetries of the corresponding Fuchsian systems. (In the D_4 and E_6 cases there are further symmetries corresponding to the automorphisms of the corresponding nonextended Dynkin diagram, which may be obtained by suitable automorphisms of the underlying Riemann sphere on which the Fuchsian system lives.)

Remark 5. Another way of thinking of the Schlesinger transformations, at least for generic parameters (no nonzero integer eigenvalue differences), is as follows. Consider the moduli space $\mathcal{M}(\lambda)$ of logarithmic connections on degree zero vector bundles over $\mathbb{P}^1$, with poles and residue types as for the Fuchsian systems above. Thus the Fuchsian system moduli space $\mathcal{M}^*(\lambda) \cong \mathcal{O}_1 \times \mathcal{O}_2 \times \mathcal{O}_3 /\!\!/ \mathrm{PGL}_n(\mathbb{C})$ is a subset of $\mathcal{M}(\lambda)$ (and presumably arguments with determinant line bundles will show it is the complement of a divisor). Then the Riemann – Hilbert map furnishes an analytic isomorphism

$$\mathcal{M}(\lambda) \xrightarrow{\cong} M(\lambda) := \{(M_1, M_2, M_3) \in \mathcal{C}_1 \times \mathcal{C}_2 \times \mathcal{C}_3 \mid M_1 M_2 M_3 = 1\}/\,\mathrm{PGL}_n(\mathbb{C})$$

with the corresponding space of monodromy data. (Here $\mathcal{C}_j \subset \mathrm{GL}_n(\mathbb{C})$ is the conjugacy class $\exp(2\pi \mathrm{i} \mathcal{O}_j)$.) But it is immediate that if $\nu \in P(R)$ then $M(\lambda) = M(\lambda + \nu)$ and so using the Riemann – Hilbert map we obtain analytic isomorphisms

$$\mathcal{M}(\lambda) \cong \mathcal{M}(\lambda + \nu).$$

But we know this map is algebraic on the open subset $\mathcal{M}^*(\lambda)$ and so is a (biregular) algebraic isomorphism. This map extends the Schlesinger transformations since they clearly preserve the monodromy data. (These isomorphisms may also be described algebraically in terms of elementary transformations, which are described for example in [21].)

7. Blow-ups of the projective plane in 6, 7, 8 or 9 points

Finally, to tie things together, we will describe how the E_6, E_7 and E_8 ALE spaces of Kronheimer described above are related to the corresponding spaces of Sakai [40], on which the difference equations were originally defined. The key point is that one is essentially adding an affine line (and the biregular Weyl group action then extends to a biregular affine Weyl group action).

First recall, in the case of E_r, that Kronheimer's construction above yields a family of affine surfaces parameterised by $\mathbb{C}^r$, for $r = 6, 7, 8$, and that these surfaces will be nonsingular if the parameters do not lie on one of the reflection hyperplanes of the (finite) Weyl group action on $\mathbb{C}^r$. Kronheimer shows ([32, Section 4]) that the quotient of this family of varieties (denoted $Y \to Z \otimes \mathbb{C}$ in [32]) by the Weyl group is the semi-universal deformation of the corresponding Kleinian singularity (with $\mathbb{C}^*$ action). In particular the nonsingular fibres are isomorphic to the corresponding fibres of the simultaneous resolution of this deformation constructed by Brieskorn, Slodowy and Tjurina. Tjurina's approach to these fibres [43] was described by Pinkham ([37, pp. 190 – 197]): they are obtained by blowing up r points in the smooth locus of a cuspidal cubic in $\mathbb{P}^2$ (and then removing the strict transform of the cubic). Since this smooth locus is isomorphic to $\mathbb{C}$ the parameter space is again a copy of $\mathbb{C}^r$. The picture to have in mind is that if all r points are equal to the inflection point of the cubic (where the tangent line has third order contact) then there will be a tree of r (-2)-curves in the blow-up, intersecting according to the E_r Dynkin diagram, since the tangent line becomes a (-2)-curve on the third blow-up. (This tree then becomes the E_r singularity upon taking the anti-canonical model of the surface, and deforming the r points gives the semi-universal deformation.)

On the other hand the corresponding spaces of Sakai are obtained as follows:[3]

- E_8-case: blow-up 9 points on the smooth locus of a cuspidal cubic in $\mathbb{P}^2$,
- E_7-case: blow-up 9 points on the smooth locus of the degenerate cubic in $\mathbb{P}^2$ formed by a line *touching* a conic (3 points on the line and 6 on the conic),
- E_6-case: blow-up 9 points on the smooth locus of the degenerate cubic in $\mathbb{P}^2$ formed by three lines meeting in a single point (3 points on each line).

In each case one should finally remove the strict transform of the cubic, and there is a genericity condition (that there should be no other cubic through the same 9 points). Quotienting by the (identity component) of the subgroup of $\mathrm{Aut}(\mathbb{P}^2) = \mathrm{PGL}_3(\mathbb{C})$ preserving these cubics again identifies the spaces of parameters with $\mathbb{C}^r$ in each case (see [40] for explicit coordinates). In each case the resulting family of surfaces admits a biregular action of the corresponding (extended) affine Weyl group (at least over the complement of the reflection hyperplanes), described explicitly in terms of Cremona transformations — in other words the action just changes the way a given surface is expressed as a blow-up of $\mathbb{P}^2$.

Now in the E_8 case it should be clear how to relate the surfaces of Sakai and Tjurina: one is either blowing up 8 or 9 points in the smooth locus of a cuspidal cubic and then removing the strict transform of the cubic. Thus, in general, the difference is the affine line given by the exceptional curve of the ninth point minus its intersection point with the strict transform of the cubic. (In this E_8 case the genericity condition is equivalent to the 9 points *not* summing to zero in the group law of the cubic and up to automorphisms one may fix the value of the sum of the 9 points to be any arbitrary nonzero value, and label it 1. Thus up to automorphisms we see that the space of parameters is just the choice of the first 8 points. Similar statements hold in the other cases — see [40].)

In the E_7 and E_6 cases it is trickier to relate the surfaces of Sakai and Tjurina, since we are blowing up different numbers of points on different cubics. However in the end the difference still amounts to an affine line in the exceptional divisor of the blow-up of one extra point, as follows.

A useful fact we will recall first is that if a cuspidal cubic is blown-up at the cusp then one obtains two rational curves touching at a point. If this point is further blown-up then one obtains three rational curves intersecting at a single point (cf. [25, 3.9.1]).

Now for the E_7 case the key point is to see how 7 points on the smooth locus of a cuspidal cubic lead to 6 points on a conic in $\mathbb{P}^2$ and two points on a line that touches it. (Then to get to Sakai's picture we need only choose one further point on the line, and modulo automorphisms there is actually no choice involved here.) To do this one does the standard quadratic Cremona transform based at the cusp c and at two of the seven points (call them a, b say). In other words blow-up a, b, c and then contract the three (-1)-curves which are the strict transforms of the lines through ab, bc, ac. See Figure 5. The cubic becomes the desired conic; it still has 5 of the original points on it — the sixth point corresponds to the third intersection point of the line through ab with the cubic. Finally the line touching the conic is the image of the exceptional curve over the cusp c, and this obtains two points

---

[3]The cases E_8, E_7, E_6 are labelled by "Add. 1 – 3" or by $A_0^{(1)**}$, $A_1^{(1)*}$, $A_2^{(1)*}$ respectively in [40].

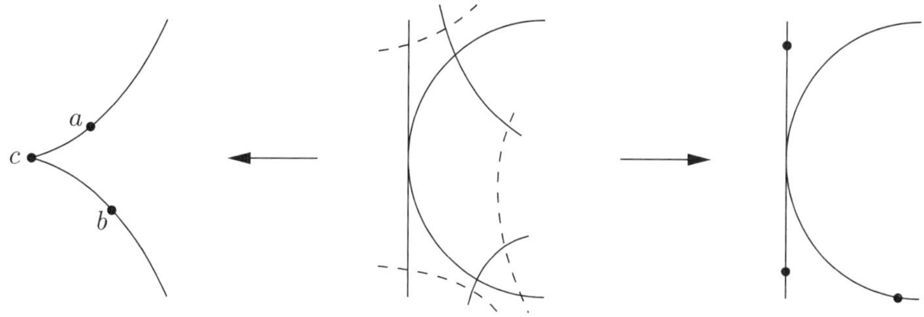

FIGURE 5.

as the images of the strict transforms of the lines through bc and ac — the points correspond to the directions these lines went in to the cusp.

(Notice that, said differently, we have explained how to identify two eight point blow-ups of $\mathbb{P}^2$, the points being either the cusp and seven points on a cuspidal cubic, or six points on a conic and two points on a line touching the conic.)

For the E_6 case the key point is to see how 6 points on the smooth locus of a cuspidal cubic lead to 8 points on three lines intersecting at a point, with 2 or 3 points on each line. To do this we blow-up the cusp twice (as mentioned above to get the desired three lines) and the first three of the six points. Then we contract the following five (-1)-curves:

(1) the strict transform of the line tangent to the cusp,
(2) the strict transform of the (unique) conic through the first three points and tangent to the cubic at the cusp,
(3) the strict transforms of the three lines from each of the first three points to the cusp.

Then one finds the desired 8 points on the resulting 3 lines as follows: On the first line we still have the last 3 of the 6 original points. On the second line are the images of the three lines in step (3), and on the third line there are the images of the two (-1)-curves in (1) and (2). Thus to get to Sakai's picture we need only choose one further point on the third line. (Again this can be rephrased as identifying two eight point blow-ups of $\mathbb{P}^2$, and also, modulo automorphisms, there really is no further choice involved.)

7.1. Weyl group actions. Finally we will describe the Weyl group actions that occur from the present point of view (it is in these terms — i.e. Cremona transforms — that Sakai originally defined the difference equations). Basically we will recall (from, e.g., [37]) the natural Weyl group action on r-tuples $p = (u_1, \dots, u_r)$ of points of the smooth locus $U \cong \mathbb{C}$ of a cuspidal cubic C in $\mathbb{P}^2$, for $r = 6, 7, 8, 9$. We will see how this relates to the standard Weyl group action defined in terms of the root systems and thus to the action of the previous section. (Recall that one has a canonical isomorphism $U \cong \mathrm{Pic}^1(C)$; $u \mapsto [u]$ and if 0 is the inflection point of C then $\mathrm{Pic}^1(C) \cong \mathrm{Pic}^0(C)$; $[u] \mapsto [u] - [0]$. In turn $\mathrm{Pic}^0(C) \cong \mathbb{C}$ as an abelian group, although this isomorphism is only determined up to the action of $\mathbb{C}^*$.)

Given $p = (u_1, \dots, u_r)$ let $X = X(p)$ denote the blow-up of $\mathbb{P}^2$ at the r points specified by p. (This makes sense even if some points coincide: one blows up the points in sequence, and at each step the remaining points determine unique points on the strict transform of the cubic, cf. [18].) Then $\mathrm{Pic}(X) \cong \bigoplus_0^r \mathbb{Z}\mathcal{E}_i$ has a uniquely determined basis $\{\mathcal{E}_i\}$, where $\mathcal{E}_0$ is the class of the inverse image of a line in $\mathbb{P}^2$, and $\mathcal{E}_i$ is the class of the *total* inverse image of $u_i \in U$. With respect to the intersection form on $\mathrm{Pic}(X)$ one has

$$\mathcal{E}_0^2 = 1, \qquad \mathcal{E}_i^2 = -1,\ \mathcal{E}_i \cdot \mathcal{E}_j = 0 \quad \text{for all } i > 0 \text{ and } j \neq i.$$

(Thus over U^r we have a canonical trivialisation of the relative Picard lattice.) The anticanonical class of X is $\delta := -\omega = 3\mathcal{E}_0 - \sum_1^r \mathcal{E}_i$ and is represented by the strict transform of the the cubic C. Let $Q = \delta^\perp$ be the orthogonal complement of δ in $\mathrm{Pic}(X)$; it is a rank r lattice isomorphic to the root lattice of type E_6, E_7, E_8, $\mathrm{E}_9 = \mathrm{E}_8^{(1)}$ for $r = 6,7,8,9$ respectively. A $\mathbb{Z}$-basis α of Q is given as follows (these are the simple roots of the corresponding root system):

$$\alpha = \{\mathcal{E}_0 - \mathcal{E}_1 - \mathcal{E}_2 - \mathcal{E}_3, \mathcal{E}_i - \mathcal{E}_{i+1} \text{for } i = 1, \dots, r-1\}$$

The corresponding Weyl group acts on $\mathrm{Pic}(X)$ preserving Q, and is generated by the reflections corresponding to these roots:

$$\mathcal{F} \in \mathrm{Pic}(X) \mapsto s_{\alpha_i}(\mathcal{F}) := \mathcal{F} + (\mathcal{F} \cdot \alpha_i)\alpha_i.$$

Below, a second basis

$$\beta = \{\mathcal{E}_i - \mathcal{E}_0/3, i = 1, \dots, r\}$$

of $Q \otimes \mathbb{C}$ will also be useful.

Thus, given the choice of p, we obtain a homomorphism

$$\chi_p \colon Q \to \mathrm{Pic}^0(C); \quad \mathcal{L} \mapsto \mathcal{L}|_C$$

which may be written explicitly in the α basis as

$$\chi(\mathcal{E}_0 - \mathcal{E}_1 - \mathcal{E}_2 - \mathcal{E}_3) = 3[0] - [u_1] - [u_2] - [u_3]; \quad \chi(\mathcal{E}_i - \mathcal{E}_{i+1}) = [u_i] - [u_{i+1}].$$

In other words we have a map

$$U^r \to \mathrm{Hom}(Q, \mathrm{Pic}^0(C)); \quad p \mapsto \chi_p.$$

The key point is that this is an isomorphism and it factorises through $\mathrm{Pic}^0(C)^r$. Specifically we identify U^r with $\mathrm{Pic}^0(C)^r$ using the above isomorphism $U \cong \mathrm{Pic}^0(C)$ and identify $\mathrm{Pic}^0(C)^r$ with $\mathrm{Hom}(Q, \mathrm{Pic}^0(C)) = \mathrm{Hom}(Q \otimes \mathbb{C}, \mathrm{Pic}^0(C))$ using the basis β. This just amounts to observing that (the linear extension to $Q \otimes \mathbb{C}$ of) χ_p satisfies

$$\chi_p(\mathcal{E}_i - \mathcal{E}_0/3) = [u_i] - [0],$$

thus motivating the definition of the basis β. In particular the Weyl group action on Q naturally induces an action on $\mathrm{Hom}(Q, \mathrm{Pic}^0(C))$ and thus gives an action on U^r via the above isomorphism. Note that for $r = 6,7,8$ the reflection hyperplanes in U^r correspond to configurations p such that $X(p)$ has extra (-2)-curves (which contract to singularities upon taking the anti-canonical model). In terms of the group law on $U \cong \mathbb{C}$ these occur if

(1) two points are equal: $u_i = u_j$
(2) three points are collinear: $u_i + u_j + u_k = 0$
(3) six points are on a conic: $\sum_{j=1}^6 u_{i_j} = 0$
(4) eight points are on a cubic with a double point: $2u_{i_1} + \sum_{j=2}^7 u_{i_j} = 0$.

(For $r = 9$ one can add any integer multiple of $\sum_1^9 u_i$ to the equations for these hyperplanes—noting that this corresponds to adding multiples of δ to the equations for the hyperplanes in $\mathrm{Hom}(Q,\mathbb{C})$.) Geometrically the generators of these Weyl group actions arise as follows. For roots $\mathcal{E}_i - \mathcal{E}_j$ one is simply swapping the labels of the points u_i and u_j thereby changing the order they are blown-up in. For the root $\mathcal{E}_0 - \mathcal{E}_1 - \mathcal{E}_2 - \mathcal{E}_3$ one is performing the quadratic Cremona transform based at the three points u_1, u_2, u_3. One may verify this gives the expected change of basis of $\mathrm{Pic}(X)$ and that the action on U^r is given (in terms of the group law on $U \cong \mathbb{C}$) by

$$u_i \mapsto u_i - 2\varepsilon \qquad i = 1,2,3,$$
$$u_i \mapsto u_i + \varepsilon \qquad i > 3$$

where $\varepsilon = (u_1 + u_2 + u_3)/3$. (This may be computed explicitly—after the Cremona transform the cubic C becomes another cuspidal cubic which may be mapped back to C via an automorphism of $\mathbb{P}^2$ and then scaled appropriately, cf. [37, pp. 196–197], or more recently [29, equation (139)]. The point is that these formulae then coincide with the reflection of U^r specified above using the β basis.)

Thus from this viewpoint it is the reflection corresponding to the single node on the shortest leg of the Dynkin diagram which is exceptional, whereas in the quiver framework it is the reflection corresponding to the central node which is unusual, as we will see in the appendix.

Remark 6. Note that a priori we have defined *two* partial compactifications of the semi-universal deformation, one by including connections on nontrivial holomorphic bundles and one by blowing up a ninth point. One may argue they are the same, at least off of the affine root hyperplanes, as follows. (This was suggested by D. Arinkin.) In both cases one has a biregular affine Weyl group action on a family of surfaces over the complement of the hyperplanes. One then checks in both cases that any point of the partial compactification divisor may be moved into the interior using this action. In terms of blowing up points this is elementary and in terms of connections this is essentially one of the statements of the Riemann–Hilbert problem actually established by Plemelj (see [1]), since off the hyperplanes at least one of the local monodromies will be semisimple.

Appendix A. Derivation of the reflections

The aim of this appendix is to write down a way to obtain geometrically (in terms of connections/Fuchsian systems) the reflections generating the finite Weyl group actions (the translation subgroup of the affine Weyl group coming from Schlesinger transformations). In brief the reflections corresponding to the nodes on the legs of the (star-shaped affine Dynkin) quiver amount to permuting the eigenvalues of the residues of the corresponding Fuchsian system (after one normalises each residue of the system to be traceless), so for example in the E_8 case one immediately obtains an action of $\mathrm{Sym}_2 \times \mathrm{Sym}_3 \times \mathrm{Sym}_6$. The trickier part is to obtain the reflection corresponding to the central node. In the D_4 (Painlevé VI) case this was obtained in [7] by passing to a system of higher rank (increased by one), permuting the eigenvalues of one of the residues of this "incremented" system and then mapping back down to the usual rank two framework (see also [8]). Here we will describe the natural extension of this to the E_6, E_7 and E_8 cases (one still passes to systems with incremented rank, permutes the eigenvalues there and maps back down).[4] To deal with all the cases at the same time we will work in a more general context as follows. (In particular this gives many more examples of unexpected isomorphisms between spaces of Fuchsian systems.)

Incremented almost affine quivers. Let $\mathcal{Q}$ be a star-shaped quiver with $m \geq 3$ legs and dimension vector δ. Define an integer N such that the dimension associated to the central node is $N-1$. We will say that $(\mathcal{Q}, \delta)$ is *almost affine* if it satisfies the following three conditions:

- The dimensions strictly decrease going down each leg,
- One of the legs is *full*, i.e. has dimensions $N-1, N-2\dots, 2, 1$. Label the legs such that this is the last (mth) leg.
- If n_i is the dimension at the node adjacent to the central node ($i = 1,\dots,m$, so $n_m = N-2$) then

$$\sum_1^m n_i = 2(N-1). \tag{9}$$

Since they are McKay graphs, any star-shaped affine Dynkin diagram (with the previously given dimension vector) satisfies (9) and one easily observes they are almost affine. Note also that the condition (9) implies that the reflection s_1 corresponding to the central node of $\mathcal{Q}$ fixes the dimension vector δ.

Any such almost affine quiver may be "incremented" to give a new quiver as follows: lengthen the mth leg by one node and increase by 1 all the dimensions on this leg (including the dimension of the central node, which will now have dimension N). Let $\mathcal{Q}^+$ denote the incremented quiver obtained from an almost affine quiver $\mathcal{Q}$ in this way, and let Δ be its dimension vector.

For example in the D_4 and E_7 cases the incremented quivers so obtained are:

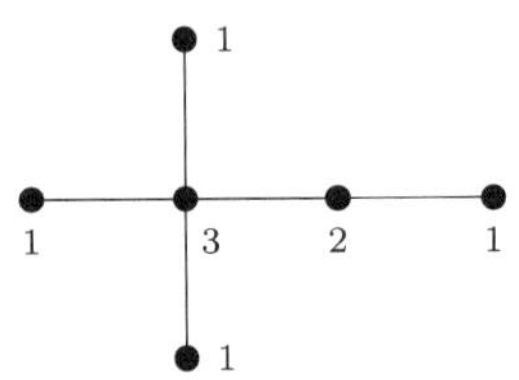

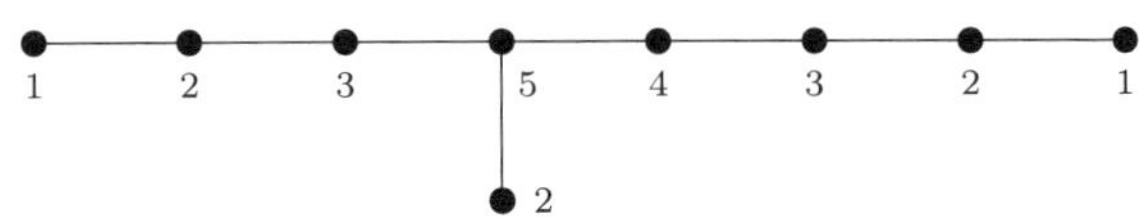

Note that an incremented quiver again has a full leg (with dimensions $N, N-1, \dots, 2, 1$). Also clearly the dimension of the node adjacent to the central node on the ith leg again has dimension n_i, for $i = 1, \dots, m-1$. The relation (9) immediately implies that the central dimension N partitions into these dimensions n_i, i.e. we have the relation:

$$\sum_1^{m-1} n_i = N \tag{10}$$

which will be important below.

---

[4]This is related to Katz's middle-convolution functor [30] — however the direct approach here is different to that of [19] and should be viewed as the complex analytic analogue of Katz's approach using the l-adic Fourier transform. (We only learnt of a relation to middle convolution via [20] after writing [7] — cf. [10, diagram 1] for more on the origins of our approach. In turn we hope this sheds more light on Katz's functor.) Note also that [15, Theorem 3.2] independently relates the central reflection to middle convolution.

Dual description. Now we will give another description of the quiver varieties associated to an incremented quiver $\mathcal{Q}^+$, which may be viewed as 'dual' to the Fuchsian description of section 4. This is a case of the duality studied by Harnad [24].

Choose a set of parameters λ for $\mathcal{Q}^+$ (a complex scalar for each node) satisfying the condition

$$\lambda \cdot \Delta = 0$$

where Δ is the given dimension vector of $\mathcal{Q}^+$. Then we may construct the quiver variety $N_{\mathcal{Q}^+}(\lambda)$ as described in Section 3. Firstly, as in Section 4, $N_{\mathcal{Q}^+}(\lambda)$ may be described as a space of Fuchsian systems; it is isomorphic to the symplectic quotient of a product of complex (co)adjoint orbits:

$$\begin{aligned} N_{\mathcal{Q}^+}(\lambda) &\cong \widehat{\mathcal{O}}_1 \times \cdots \times \widehat{\mathcal{O}}_m /\!\!/ \operatorname{GL}_N(\mathbb{C}) \\ &= \left\{ (B_1, \ldots, B_m) \;\middle|\; B_i \in \widehat{\mathcal{O}}_i, \sum_1^m B_i = 0 \right\} \Big/ \operatorname{GL}_N(\mathbb{C}). \end{aligned}$$

Here for $i < m$ we take $\widehat{\mathcal{O}}_i \subset \mathfrak{gl}_N(\mathbb{C})$ to be the orbit associated to the ith leg of $\mathcal{Q}^+$ (see Section 4). For $i = m$ if we set $\mathcal{O}'_m \subset \mathfrak{gl}_N(\mathbb{C})$ to be the orbit associated to the last leg of $\mathcal{Q}^+$, then we define $\widehat{\mathcal{O}}_m \subset \mathfrak{gl}_N(\mathbb{C})$ to be $\mathcal{O}'_m$ shifted by the scalar ν (= minus the component of λ associated to the central vertex of $\mathcal{Q}^+$):

$$\widehat{\mathcal{O}}_m = \{A + \nu \mid A \in \mathcal{O}'_m\}.$$

Then the relation $\lambda \cdot \Delta = 0$ translates into the statement $\sum_1^m \operatorname{Tr} B_i = 0$. Note for example that the quiver varieties for the incremented D_4 quiver thus parameterise the 3×3 Fuchsian systems with four poles used in [6, 7], having rank one residues at three of the poles. For simplicity we will impose the genericity assumption on λ that all the orbits $\widehat{\mathcal{O}}_i$ are semisimple (and so are closed). In particular $\widehat{\mathcal{O}}_m$ is then regular semisimple.

For the dual description first define orbits

$$\breve{\mathcal{O}}_i \subset \mathfrak{gl}_{n_i}(\mathbb{C})$$

for $i = 1, \ldots, m-1$ as follows: Remove the central vertex of $\mathcal{Q}^+$ and let $\mathcal{O}'_i$ be the orbit associated to what remains of the ith leg (the top end of which now has dimension n_i, so that $\mathcal{O}'_i \subset \mathfrak{gl}_{n_i}(\mathbb{C})$). These orbits will all have zero as an eigenvalue. Define $\breve{\mathcal{O}}_i$ by shifting by the scalar λ_i (where λ_i is the scalar associated to the vertex with dimension n_i):

$$\breve{\mathcal{O}}_i := \mathcal{O}'_i - \lambda_i \subset \mathfrak{gl}_{n_i}(\mathbb{C}).$$

(Thus for example if the full ith leg looked as in Exercise 4 of Section 4 then $n_i = 3$, elements of $\mathcal{O}'_i$ will have eigenvalues $0, -\lambda_2, -\lambda_2 - \lambda_1$ and elements of $\breve{\mathcal{O}}_i$ will have eigenvalues $-\lambda_3, -\lambda_3 - \lambda_2, \lambda_3 - \lambda_2 - \lambda_1$.) In the D_4 case these orbits are single points in $\mathfrak{gl}_1(\mathbb{C}) = \mathbb{C}$.

Now observe that since $\sum_1^{m-1} n_i = N$, if we define the group

$$H := \prod_1^{m-1} \operatorname{GL}_{n_i}(\mathbb{C})$$

then H embeds block diagonally into $\mathrm{GL}_N(\mathbb{C})$. Taking Lie algebras gives the block-diagonal embedding

$$\mathrm{Lie}(H) = \mathfrak{gl}_{n_1}(\mathbb{C}) \oplus \cdots \oplus \mathfrak{gl}_{n_{m-1}}(\mathbb{C}) \hookrightarrow \mathfrak{gl}_N(\mathbb{C}).$$

In particular this embeds the product of the orbits $\breve{\mathcal{O}}_i$ into $\mathfrak{gl}_N(\mathbb{C})$. Dually projecting onto the block-diagonal entries gives a map $\pi : \mathfrak{gl}_N(\mathbb{C}) \to \mathrm{Lie}(H)$. The restriction $\mu_H := \pi|_{\widehat{\mathcal{O}}}$ of π to an orbit $\widehat{\mathcal{O}} \subset \mathfrak{gl}_N(\mathbb{C})$ is a moment map for the conjugation action of $H \subset \mathrm{GL}_N(\mathbb{C})$ on $\widehat{\mathcal{O}}$ (where as usual for any k we identify $\mathfrak{gl}_k(\mathbb{C}) \cong \mathfrak{gl}_k^*(\mathbb{C})$ using the trace pairing $\mathrm{Tr}(AB)$). Here we will take

$$\widehat{\mathcal{O}} := -\widehat{\mathcal{O}}_m = \{B \mid -B \in \widehat{\mathcal{O}}_m\} \subset \mathfrak{gl}_N(\mathbb{C})$$

to be minus the regular semisimple coadjoint orbit $\widehat{\mathcal{O}}_m$ defined above.

Proposition 1. *The quiver variety $N_{\mathcal{Q}^+}(\lambda)$ is also isomorphic to the symplectic quotient of $\widehat{\mathcal{O}} \subset \mathfrak{gl}_N(\mathbb{C})$ by the action of $H = \prod_1^{m-1} \mathrm{GL}_{n_i}(\mathbb{C})$ at the coadjoint orbit $\breve{\mathcal{O}} := \prod_1^{m-1} \breve{\mathcal{O}}_i$ of H:*

$$N_{\mathcal{Q}^+}(\lambda) \cong \widehat{\mathcal{O}} \underset{\breve{\mathcal{O}}}{/\!/} H = \mu_H^{-1}(\breve{\mathcal{O}})/H.$$

Proof. This arises since $\widehat{\mathcal{O}}/\!/_{\breve{\mathcal{O}}} H$ and $\prod_1^m \widehat{\mathcal{O}}_i /\!/ \mathrm{GL}_N(\mathbb{C})$ are just two descriptions of the symplectic quotient of $T^* \mathfrak{gl}_N(\mathbb{C})$ by the group $H \times \mathrm{GL}_N(\mathbb{C})$ (at the coadjoint orbit $(-\breve{\mathcal{O}}) \times \widehat{\mathcal{O}})$—one may first do the H reduction and then the $\mathrm{GL}_N(\mathbb{C})$ reduction to obtain one description or do the reductions in the other order for the other description). Note that $\mu_H^{-1}(\breve{\mathcal{O}}) \subset \widehat{\mathcal{O}}$ will be empty unless the trace of an element of $\widehat{\mathcal{O}}$ equals the trace of any element of $\breve{\mathcal{O}}$—however the equality of these two traces is equivalent to the condition $\lambda \cdot \Delta = 0$.

This also fits directly into the original definition of the quiver variety, as f1qollows. Let $\mathcal{Q}' \subset \mathcal{Q}^+$ be the subquiver consisting of the central node (with dimension N) plus the $m-1$ adjacent nodes on the first $m-1$ legs (with dimensions $n_1, \ldots, n_{m-1}$ resp.), so that $\mathcal{Q}'$ has exactly m nodes. Notice that the group of automorphisms $\widetilde{G}(\mathcal{Q}')$ of the vector spaces at the nodes of $\mathcal{Q}'$ is just $H \times \mathrm{GL}_N(\mathbb{C})$. Moreover the big vector space $W(\mathcal{Q}')$ associated to $\mathcal{Q}'$ (taking a map in each direction for each edge):

$$W(\mathcal{Q}') = \bigoplus_1^{m-1} \left(\mathrm{Hom}(\mathbb{C}^{n_i}, \mathbb{C}^N) \oplus \mathrm{Hom}(\mathbb{C}^N, \mathbb{C}^{n_i})\right) \cong \mathfrak{gl}_N(\mathbb{C}) \times \mathfrak{gl}_N(\mathbb{C}) \cong T^* \mathfrak{gl}_N(\mathbb{C})$$

is isomorphic to $T^* \mathfrak{gl}_N(\mathbb{C})$, once we make an identification $\bigoplus_1^{m-1} \mathbb{C}^{n_i} \cong \mathbb{C}^N$ (as we did above embedding H in $\mathrm{GL}_N(\mathbb{C})$ block diagonally).

Now we may do the symplectic quotient $N_{\mathcal{Q}^+}(\lambda) = W(\mathcal{Q}^+)/\!/_\lambda \widetilde{G}(\mathcal{Q}^+)$ in stages, first quotienting by the subgroup K of $\widetilde{G}(\mathcal{Q}^+)$ corresponding to the nodes of $\mathcal{Q}^+ \setminus \mathcal{Q}'$, so that $\widetilde{G}(\mathcal{Q}^+) = K \times H \times \mathrm{GL}_N(\mathbb{C})$. This yields the isomorphism:

$$W(\mathcal{Q}^+)/\!/_\lambda K \cong \breve{\mathcal{O}}_1 \times \cdots \times \breve{\mathcal{O}}_{m-1} \times \widehat{\mathcal{O}}_m \times T^* \mathfrak{gl}_N(\mathbb{C}) = \breve{\mathcal{O}} \times \widehat{\mathcal{O}}_m \times T^* \mathfrak{gl}_N(\mathbb{C})$$

on which we still have the residual action of $H \times \mathrm{GL}_N(\mathbb{C})$. Suppose $P_i \in \mathrm{Hom}(\mathbb{C}^N, \mathbb{C}^{n_i})$ and $Q_i \in \mathrm{Hom}(\mathbb{C}^{n_i}, \mathbb{C}^N)$ for $i = 1, \dots, m-1$ and let us write

$$P := \begin{pmatrix} P_1 \\ \vdots \\ P_{m-1} \end{pmatrix}, \qquad Q := \begin{pmatrix} Q_1 & \dots & Q_{m-1} \end{pmatrix} \in \mathfrak{gl}_N(\mathbb{C}).$$

Then $(P, Q) \in T^* \mathfrak{gl}_N(\mathbb{C})$ and $\mathrm{GL}_N(\mathbb{C})$ acts on this via left or right multiplication on $\mathfrak{gl}_N(\mathbb{C})$. Specifically $g \in \mathrm{GL}_N(\mathbb{C})$ acts as $g(P,Q) = (Pg^{-1}, gQ)$ with moment map

$$\mu_L(P,Q) = QP = \sum_1^{m-1} Q_i P_i$$

and $h \in H \subset \mathrm{GL}_N(\mathbb{C})$ acts as $h(P,Q) = (hP, Qh^{-1})$ with moment map

$$\mu_R(P,Q) = -\pi(PQ) = (-P_1Q_1, -P_2Q_2, \dots) \in \mathfrak{gl}_{n_1}(\mathbb{C}) \times \mathfrak{gl}_{n_2}(\mathbb{C}) \times \cdots .$$

Thus we have immediately that

$$\begin{aligned} N_{\mathcal{Q}^+}(\lambda) &\cong \big(\breve{\mathcal{O}} \times \widehat{\mathcal{O}}_m \times T^* \mathfrak{gl}_N(\mathbb{C})\big) /\!\!/ \big(H \times \mathrm{GL}_N(\mathbb{C})\big) \\ &\cong \{(P,Q) \in T^* \mathfrak{gl}_N(\mathbb{C}) \mid P_iQ_i \in \breve{\mathcal{O}}_i, QP \in \widehat{\mathcal{O}}\} / \big(H \times \mathrm{GL}_N(\mathbb{C})\big) \\ &\cong \mu_R^{-1}(-\breve{\mathcal{O}}) \cap \mu_L^{-1}(\widehat{\mathcal{O}}) / \big(H \times \mathrm{GL}_N(\mathbb{C})\big) \\ &= T^* \mathfrak{gl}_N(\mathbb{C}) \underset{(-\breve{\mathcal{O}}) \times \widehat{\mathcal{O}}}{/\!\!/} \big(H \times \mathrm{GL}_N(\mathbb{C})\big). \end{aligned} \tag{11}$$

Now we claim that there is an isomorphism

$$T^* \mathfrak{gl}_N(\mathbb{C}) /\!\!/_{\widehat{\mathcal{O}}} \mathrm{GL}_N(\mathbb{C}) \cong \widehat{\mathcal{O}}.$$

(which will be an isomorphism of symplectic manifolds upon negating the symplectic form on the left-hand side). This claim will complete the proof of the proposition since, from (11), we then obtain $N_{\mathcal{Q}^+}(\lambda) \cong \widehat{\mathcal{O}} /\!\!/_{\breve{\mathcal{O}}} H$.

To establish the claim we need to identify the affine GIT quotient

$$\{(P,Q) \in T^* \mathfrak{gl}_N(\mathbb{C}) \bigm| QP \in \widehat{\mathcal{O}}\} / \mathrm{GL}_N(\mathbb{C})$$

with $\widehat{\mathcal{O}}$. The first step is to note that the invariant functions are the matrix entries of $B := PQ \in \mathfrak{gl}_N(\mathbb{C})$. Now if zero is not an eigenvalue of elements of $\widehat{\mathcal{O}}$, then $\mathrm{GL}_N(\mathbb{C})$ acts freely and PQ is conjugate to QP, and the result follows easily.

Next suppose zero is an eigenvalue of elements of $\widehat{\mathcal{O}}$, and we have (P,Q) with $QP \in \widehat{\mathcal{O}}$, so $\det(QP) = \det(PQ) = 0$. Label the nonzero eigenvalues of QP as $\nu_1, \dots, \nu_{N-1}$ and let $E_i \subset \mathbb{C}^N$ be the ν_i-eigenspace of QP for $i = 1, \dots, N-1$, and let $E_0 = \ker(QP)$ (thus $E_i \cong \mathbb{C}$ as $\widehat{\mathcal{O}}$ is regular semisimple). Let $F_i := P(E_i) \subset \mathbb{C}^N$ for $i > 0$ and observe $F_i \cong \mathbb{C}$, F_i is the ν_i-eigenspace of PQ, and that $Q(F_i) = E_i$. Let $F_0 := \ker(PQ)$. Counting dimensions implies $\dim(F_0) = 1$ and so $PQ \in \widehat{\mathcal{O}}$.

Thus the map $(P,Q) \mapsto B := PQ$ is a $\mathrm{GL}_N(\mathbb{C})$-invariant map to $\widehat{\mathcal{O}}$, which is clearly surjective. It is now straightforward to show that each fibre of this map contains exactly three $\mathrm{GL}_N(\mathbb{C})$ orbits: two non-closed orbits (with $\det(P) = 0 \neq \det(Q)$ and $\det(P) \neq 0 = \det(Q)$) and one closed orbit (with $\det(P) = \det(Q) = 0$) in the closure of both of the non-closed orbits. Thus the GIT quotient (identifying points if the closures of their orbits intersect) is $\widehat{\mathcal{O}}$. □

Remark 7. For use later we note that if $(P,Q) \in T^* \mathfrak{gl}_N(\mathbb{C})$ and $QP \in \widehat{\mathcal{O}}$ (and $\det(QP) = 0$) then we have shown that the closure of the $\mathrm{GL}_N(\mathbb{C})$ orbit of (P,Q) contains elements with $\det(P) = \det(Q) = 0$ — and so any point of the GIT quotient may be represented by such (P,Q). Now if $\det(P) = \det(Q) = 0$ we may deduce that

$$\ker(P) = \ker(QP) =: E_0, \quad \ker(Q) = \ker(PQ) =: F_0.$$

(E.g., for the first statement, a dimension count shows the image of P is $\bigoplus_1^{N-1} F_i$. Then if $v \in E_0$ we have $P(v) = \sum_1^{N-1} w_i$ with $w_i \in F_i$. Thus $0 = QPv = \sum_1^{N-1} v_i$ with $v_i = Qw_i \in E_i$, and $v_i = 0$ if and only if $w_i = 0$. But the E_is sum directly so each $v_i = 0$.) It then follows that, if we define $V = \bigoplus_1^{N-1} E_i \cong \mathbb{C}^{N-1}$, then $V \oplus E_0 = \mathbb{C}^N$ and each matrix $B_i := Q_i P_i$ preserves this direct sum decomposition (since $E_0 \subset \ker P_i$ and the image of Q_i is contained in the image $Q(\mathbb{C}^N) = V$ for all i).

Remark 8. As in [4, 24] we may interpret $\widehat{\mathcal{O}}/\!\!/_{\breve{\mathcal{O}}} H$ as a moduli space of certain irregular connections of the form

$$\mathrm{d} - \left(\frac{A_0}{z^2} + \frac{B}{z}\right)\mathrm{d}z$$

on the trivial rank N vector bundle over $\mathbb{P}^1(\mathbb{C})$, where $B = PQ$ is in $\widehat{\mathcal{O}}$ and $A_0 \in \mathfrak{gl}_n(\mathbb{C})$ is a diagonal element with eigenvalues of multiplicities $n_1, n_2, \ldots, n_{m-1}$ respectively (and so having centraliser H).

Scalar shift. Now we consider the effect of tensoring the above irregular connections by a logarithmic connection $\Lambda \mathrm{d}z/z$ on the trivial line bundle over $\mathbb{P}^1$. In other words we replace B by $B + \Lambda$ where $\Lambda \in \mathbb{C}$ is a scalar. The effect on the parameters λ is clear since we see how the eigenvalues of B change: only the components of λ associated to the nodes of $\mathcal{Q}'$ change, viz. λ_0 (associated to the central node with dimension N) increases by Λ and each λ_i (associated to the node with dimension n_i) decreases by Λ. Let us denote this new set of parameters by $\lambda(\Lambda)$, so that $\lambda = \lambda(0)$. Note that the relation $\sum n_i = N$ implies we have $\lambda(\Lambda) \cdot \Delta = 0$ for all Λ. The key point is that this map $B \mapsto B + \Lambda$ induces an isomorphism

$$N_{\mathcal{Q}^+}(\lambda) \cong N_{\mathcal{Q}^+}(\lambda(\Lambda))$$

for any $\Lambda \in \mathbb{C}$. This is immediate in this dual description but is less trivial in the original Fuchsian description — indeed for generic Λ this is (our approach to) Katz's middle convolution. (At least if zero is not an eigenvalue of elements of $\widehat{\mathcal{O}}$, one may describe this scalar shift on the Fuchsian side by using the $\mathrm{GL}_N(\mathbb{C})$ action to move to points where $Q = \mathrm{Id}.$ — then we see how to vary $P = B$ with Λ and in turn how $B_i = Q_i P_i \in \widehat{\mathcal{O}}_i(\Lambda)$ varies. In general one may move to a GIT-equivalent pair with $Q = 1$, and proceed in the same way.)

Projection to original quiver variety. Suppose $(\mathcal{Q}, \delta)$ is an almost affine quiver with incremented quiver $(\mathcal{Q}^+, \Delta)$ and that we have chosen some generic parameters λ for $\mathcal{Q}^+$. Let ν be minus the component of λ associated to the central vertex of $\mathcal{Q}^+$. Then we may shift λ by $\Lambda = \nu$ so that the central component $\lambda(\Lambda)$ associated to the central node of $\mathcal{Q}^+$ is zero. Thus $\lambda(\Lambda)$ constitutes a set of parameters for the original quiver $\mathcal{Q}$ (since $\mathcal{Q}$ may be obtained by deleting the central vertex of $\mathcal{Q}^+$ and then contracting the highest joint of the mth leg of $\mathcal{Q}^+$). Writing $\mathrm{pr}(\lambda)$ for the set of parameters of $\mathcal{Q}$ obtained from λ in this way, we have

$$\mathrm{pr}(\lambda)\cdot\delta = \lambda(\Lambda)\cdot\Delta = \lambda\cdot\Delta = 0.$$

Lemma 2. *The quiver variety of $\mathcal{Q}^+$ with parameters λ is isomorphic to the quiver variety of $\mathcal{Q}$ with parameters* $\mathrm{pr}(\lambda)$:

$$N_{\mathcal{Q}^+}(\lambda) \cong N_{\mathcal{Q}}\big(\mathrm{pr}(\lambda)\big).$$

SKETCH. By performing the scalar shift we may assume that the component of λ on the central node is zero. Then if $B = PQ \in \widehat{\mathcal{O}}$ represents a point of $N_{\mathcal{Q}^+}(\lambda)$ we know that B has exactly one zero eigenvalue. From Remark 7 we can move to a GIT-equivalent pair (P,Q) with $\det(P) = \det(Q) = 0$ and then the matrices $B_i = Q_iP_i$ simultaneously preserve a direct sum decomposition of the form $\mathbb{C}^N = V \oplus \mathbb{C}$. Taking $A_i = B_i|_V \in \mathcal{O}_i$ gives the desired set of Fuchsian residues representing a point of $N_{\mathcal{Q}}(\mathrm{pr}(\lambda))$. (In reverse we may take $B_i \in \widehat{\mathcal{O}}_i$ to be the block diagonal matrix $\mathrm{diag}(A_i, 0) \in \mathfrak{gl}_N(\mathbb{C})$.) □

Example. To help keep track of the various orbits we have defined let us write out the eigenvalues in the example of E_7. If we number the nodes of $\mathcal{Q}^+$ as follows

(12)

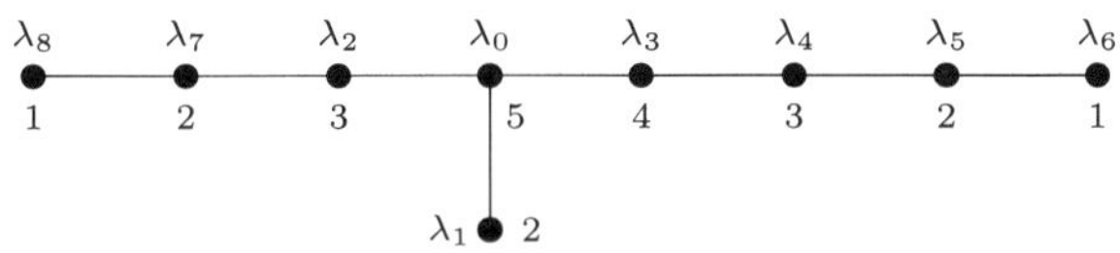

so that the first leg is the shortest leg, and the longest leg is last, then the eigenvalues of the various orbits are:

• Any element of $\widehat{\mathcal{O}}_1$ has eigenvalues $(0,0,0,-\lambda_1,-\lambda_1)$ and $\breve{\mathcal{O}}_1 = \{-\mathrm{diag}(\lambda_1,\lambda_1)\}$ which is a single point of $\mathfrak{gl}_2(\mathbb{C})$,

• Any element of $\widehat{\mathcal{O}}_2$ has eigenvalues $(0,0,-\lambda_2,-\lambda_2-\lambda_7,-\lambda_2-\lambda_7-\lambda_8)$ and any element of $\breve{\mathcal{O}}_2 \subset \mathfrak{gl}_3(\mathbb{C})$ has eigenvalues $(-\lambda_2,-\lambda_2-\lambda_7,-\lambda_2-\lambda_7-\lambda_8)$.

• Any element of $\widehat{\mathcal{O}}_3$ has eigenvalues $-\lambda_0$ and $-\lambda_0 - \sum_{i=3}^{j}\lambda_i$ for $j = 3,4,5,6$ and any element of $\widehat{\mathcal{O}}$ is minus an element of $\widehat{\mathcal{O}}_3$.

The parameters of the original affine Dynkin diagram obtained by projecting λ are then as follows:

(13)

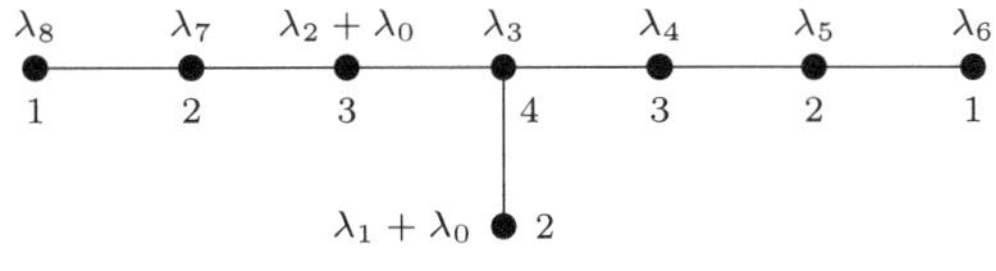

Thus we have identified $N_{\mathcal{Q}}(\mathrm{pr}(\lambda))$ with $N_{\mathcal{Q}^+}(\lambda(\Lambda))$ for any $\Lambda \in \mathbb{C}$, and so with spaces of Fuchsian systems of the form

$$\widehat{\mathcal{O}}_1 \times \cdots \times \widehat{\mathcal{O}}_m /\!\!/ \mathrm{GL}_N(\mathbb{C})$$

of incremented rank. In particular the size of the 'obvious' symmetry group (permuting the eigenvalues of the residues) is larger — namely we may permute all N of the eigenvalues of elements of $\widehat{\mathcal{O}}_m$ compared to the $N-1$ eigenvalues of $\mathcal{O}_m$. In particular suppose we consider the permutation swapping the first two eigenvalues of elements of $\widehat{\mathcal{O}}_m$. If the parameters on the mth leg of $\mathcal{Q}^+$ were labelled as follows:

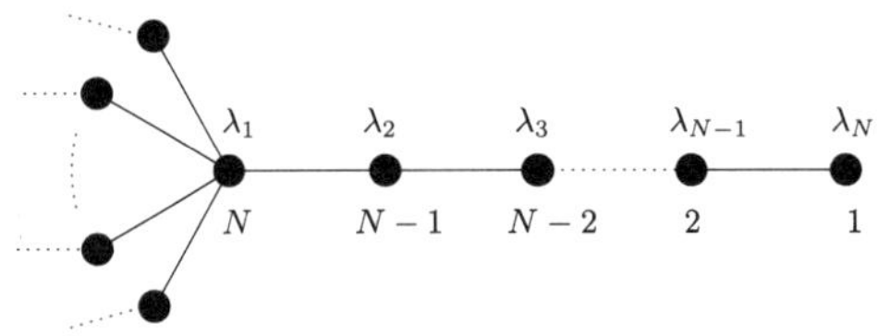

then the eigenvalues would be $-\sum_{j=1}^{i}\lambda_j$ for $i = 1, 2, \ldots, N$. Let $\mathrm{perm}(\lambda)$ denote the change in the parameters λ corresponding to this permutation swapping the first two eigenvalues. Explicitly we find that

$$\mathrm{perm}(\lambda): \qquad \lambda_1' = \lambda_1 + \lambda_2, \quad \lambda_2' = -\lambda_2, \quad \lambda_3' = \lambda_2 + \lambda_3$$

with all other components of λ not changing. The key fact is that this permutation is a lift of the desired 'central reflection' of $\mathcal{Q}$, as follows.

Lemma 3. *Let r_1 denote the reflection corresponding to the central node of the almost affine quiver $\mathcal{Q}$. Then, upon projecting to $\mathcal{Q}$, the eigenvalue permutation* perm *induces the reflection r_1. In other words*:

$$r_1(\mathrm{pr}(\lambda)) = \mathrm{pr}\big(\mathrm{perm}(\lambda)\big)$$

for any set of parameters λ for $\mathcal{Q}^+$ (with $\lambda \cdot \Delta = 0$).

PROOF. Let ν be minus the component of $\mathrm{pr}(\lambda)$ associated to the central vertex of $\mathcal{Q}$. Then r_1 acts by subtracting ν from all the components of $\mathrm{pr}(\lambda)$ on nodes of $\mathcal{Q}$ adjacent to the central node (and negating ν itself). (In terms of eigenvalues — in the 'determinant zero' normalisation — this amounts to subtracting ν from all of the not-necessarily-zero eigenvalues of each residue. E.g., for D_4 one has $\nu = \sum_1^4 \theta_i/2$ and r_1 maps each θ_i to $\theta_i - \nu$.)

The proof is now straightforward — we will illustrate it for the E_7 case, the general case being no more difficult.

From (13) we see that $r_1(\mathrm{pr}(\lambda))$ is

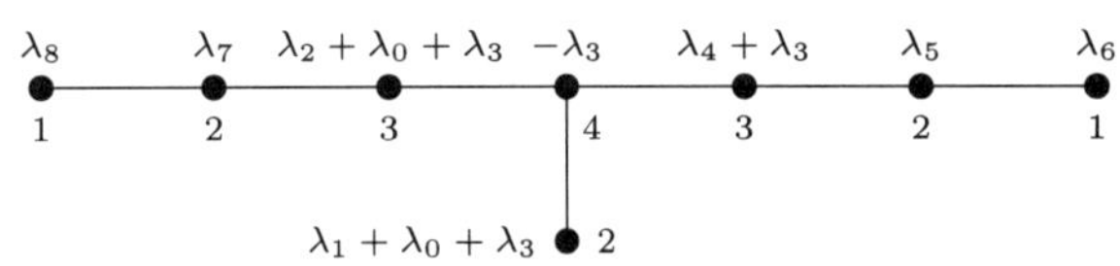

Moreover from (12), we see that $\mathrm{perm}(\lambda)$ is as follows:

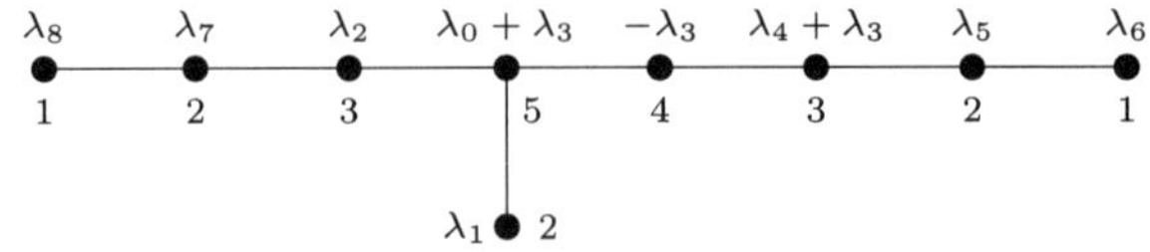

Now observe that these parameters project to those above, as desired. □

Of course relabelling the eigenvalues of elements of $\widehat{\mathcal{O}}_m$ does not change the spaces, so we have an isomorphism

$$N_{\mathcal{Q}^+}(\lambda) \cong N_{\mathcal{Q}^+}(\mathrm{perm}(\lambda))$$

(which is transparent when the quiver varieties are viewed as spaces of Fuchsian systems). Putting the above results together, we thus see how to obtain the desired isomorphism $N_{\mathcal{Q}}(\mathrm{pr}(\lambda)) \cong N_{\mathcal{Q}}\Big(r_1(\mathrm{pr}(\lambda))\Big)$ as the composition

$$N_{\mathcal{Q}}(\mathrm{pr}(\lambda)) \cong N_{\mathcal{Q}^+}(\lambda) \cong N_{\mathcal{Q}^+}(\mathrm{perm}(\lambda)) \cong N_{\mathcal{Q}}\Big(\mathrm{pr}(\mathrm{perm}(\lambda))\Big) = N_{\mathcal{Q}}\Big(r_1(\mathrm{pr}(\lambda))\Big)$$

as promised by passing to the spaces of incremented systems, permuting eigenvalues and projecting back down.

Acknowledgments. The author is grateful to E. Looijenga for useful conversations and to D. Arinkin for Remark 6.

References

1. D. V. Anosov and A. A. Bolibruch, *The Riemann – Hilbert problem*, Aspects Math., vol. E22, Friedr. Vieweg & Sohn, Braunschweig, 1994.
2. D. Arinkin and A. Borodin, *Moduli spaces of d-connections and difference Painlevé equations*, Duke Math. J. **134** (2006), no. 3, 515 – 556.
3. D. Arinkin and S. Lysenko, *Isomorphisms between moduli spaces of* SL(2)*-bundles with connections on* $\mathbb{P}^1 \setminus \{x_1, \ldots, x_4\}$, Math. Res. Lett. **4** (1997), no. 2-3, 181 – 190.
4. P. Boalch, *Stokes matrices, Poisson Lie groups and Frobenius manifolds*, Invent. Math. **146** (2001), no. 3, 479 – 506.
5. ———, *Symplectic manifolds and isomonodromic deformations*, Adv. Math. **163** (2001), no. 2, 137 – 205.
6. ———, *Painlevé equations and complex reflections*, Ann. Inst. Fourier (Grenoble) **53** (2003), no. 4, 1009 – 1022.
7. ———, *From Klein to Painlevé via Fourier, Laplace and Jimbo*, Proc. London Math. Soc. (3) **90** (2005), no. 1, 167 – 208, available at `arXiv:math/0308221`.
8. ———, *Six results on Painlevé* VI, Théories asymptotiques et équations de Painlevé (Angers, 2004) (É. Delabaere and M. Loday-Richaud, eds.), Sémin. Congr., vol. 14, Soc. Math. France, Paris, 2006, pp. 1 – 20.
9. ———, *Regge and Okamoto symmetries*, Comm. Math. Phys. **276** (2007), no. 1, 117 – 130.
10. ———, *Towards a nonlinear Schwarz's list*, available at `arXiv:0707.3375`.
11. N. Bourbaki, *Groupes et algèbres de Lie: Chapitres 4, 5 et 6*, Masson, Paris, 1981.
12. É. Brézin and V. A. Kazakov, *Exactly solvable field theories of closed strings*, Phys. Lett. B **236** (1990), no. 2, 144 – 150.
13. W. Crawley-Boevey, *Geometry of the moment map for representations of quivers*, Compositio Math. **126** (2001), no. 3, 257 – 293.
14. ———, *Decomposition of Marsden – Weinstein reductions for representations of quivers*, Compositio Math. **130** (2002), no. 2, 225 – 239.
15. ———, *Indecomposable parabolic bundles and the existence of matrices in prescribed conjugacy class closures with product equal to the identity*, Publ. Math. Inst. Hautes Études Sci. **100** (2004), 171 – 207.
16. ———, *Quiver algebras, weighted projective lines, and the Deligne – Simpson problem*, Proceedings of the International Congress of Mathematicians (Madrid, 2006), Volume II (M. Sanz-Solé, J. Soria, J. L. Varona, and J. Verdera, eds.), European Math. Soc., Zürich, 2007, pp. 117 – 129.
17. W. Crawley-Boevey and M. P. Holland, *Noncommutative deformations of Kleinian singularities*, Duke Math. J. **92** (1998), no. 3, 605 – 635.
18. M. Demazure, *Surfaces de del Pezzo*, Séminaire sur les singularités des surfaces (Palaiseau, 1976 – 1977) (M. Demazure, H. C. Pinkham, and B. Tessier, eds.), Lecture Notes in Math., vol. 777, Springer, Berlin, 1980, pp. 21 – 69.
19. M. Dettweiler and S. Reiter, *An algorithm of Katz and its application to the inverse Galois problem*, J. Symbolic Comput. **30** (2000), no. 6, 761 – 798.

20. ______, *Painlevé equations and the middle convolution*, Adv. Geom. **7** (2007), no. 3, 317–330. See also the preprint dated October 1st, 2004.
21. H. Esnault and E. Viehweg, *Semistable bundles on curves and irreducible representations of the fundamental group*, Algebraic Geometry: Hirzebruch 70 (Warsaw, 1998) (P. Pragacz, M. Szurek, and J. Wiśniewski, eds.), Contemp. Math., vol. 241, Amer. Math. Soc., Providence, RI, 1999, pp. 129–138.
22. B. Grammaticos and A. Ramani, *Discrete Painlevé equations: a review*, Discrete Integrable Systems (B. Grammaticos, Y. Kosmann-Schwarzbach, and T. Tamizhmani, eds.), Lecture Notes in Phys., vol. 644, Springer, Berlin, 2004, pp. 245–321.
23. D. J. Gross and A. A. Migdal, *A nonperturbative treatment of two-dimensional quantum gravity*, Nuclear Phys. B **340** (1990), no. 2-3, 333–365.
24. J. Harnad, *Dual isomonodromic deformations and moment maps to loop algebras*, Comm. Math. Phys. **166** (1994), no. 2, 337–365.
25. R. Hartshorne, *Algebraic geometry*, Grad. Texts in Math., vol. 52, Springer, New York, NY, 1977.
26. N. J. Hitchin, *Frobenius manifolds*, Gauge Theory and Symplectic Geometry (Montréal, 1995) (J. Hurtubise, F. Lalonde, and G. Sabidussi, eds.), NATO Adv. Sci. Inst. Ser. C Math. Phys. Sci., vol. 488, Kluwer Acad. Publ., Dordrecht, 1997, pp. 69–112.
27. M. Jimbo and T. Miwa, *Monodromy preserving deformations of linear differential equations with rational coefficients.* II, Phys. D **2** (1981), no. 3, 407–448.
28. V. G. Kac, *Infinite-dimensional Lie algebras*, 3rd ed., Cambridge Univ. Press, Cambridge, 1990.
29. K. Kajiwara, T. Masuda, M. Noumi, Y. Ohta, and Y. Yamada, *Point configurations, Cremona transformations and the elliptic difference Painlevé equation*, Théories asymptotiques et équations de Painlevé (Angers, 2004) (É. Delabaere and M. Loday-Richaud, eds.), Sémin. Congr., vol. 14, Soc. Math. France, Paris, 2006, pp. 169–198.
30. N. M. Katz, *Rigid local systems*, Ann. of Math. Stud., vol. 139, Princeton Univ. Press, Princeton, NJ, 1996.
31. V. P. Kostov, *The Deligne–Simpson problem for zero index of rigidity*, Perspectives of Complex Analysis, Differential Geometry and Mathematical Physics (St. Konstantin, 2000) (S. Dimiev and K. Sekigawa, eds.), World Sci. Publ., River Edge, NJ, 2001, pp. 1–35.
32. P. B. Kronheimer, *The construction of ALE spaces as hyper-Kähler quotients*, J. Differential Geom. **29** (1989), no. 3, 665–683.
33. H. Nakajima, *Instanons on ALE spaces and canonical bases*, Symposium on Representation Theory (Yamagata, 1992) (Japansese).
34. ______, *Instantons on ALE spaces, quiver varieties, and Kac–Moody algebras*, Duke Math. J. **76** (1994), no. 2, 365–416.
35. ______, *Reflection functors for quiver varieties and Weyl group actions*, Math. Ann. **327** (2003), no. 4, 671–721.
36. K. Okamoto, *Studies on the Painlevé equations.* I: *Sixth Painlevé equation* P_{VI}, Ann. Mat. Pura Appl. (4) **146** (1987), 337–381.
37. H. Pinkham, *Résolution simultanée de points doubles rationnels*, Séminaire sur les singularités des surfaces (Palaiseau, 1976–1977) (M. Demazure, H. C. Pinkham, and B. Tessier, eds.), Lecture Notes in Math., vol. 777, Springer, Berlin, 1980, pp. 179–204.
38. T. Regge, *Symmetry properties of Racah's coefficients*, Nuovo Cimento **11** (1959), 116–117.
39. W. Rump, *Doubling a path algebra, or: how to extend indecomposable modules to simple modules*, An. Ştiinţ. Univ. Ovidius Constanţa Ser. Mat. **4** (1996), no. 2, 174–185.
40. H. Sakai, *Rational surfaces associated with affine root systems and geometry of the Painlevé equations*, Comm. Math. Phys. **220** (2001), no. 1, 165–229.
41. ______, *Problem: Discrete Painlevé equations and their Lax forms*, Algebraic, Analytic and Geometric Aspects of Complex Differential Equations and Their Deformations. Painlevé Hierarchies (Kyoto, 2006) (Y. Takei, ed.), RIMS Kôkyûroku Bessatsu, vol. B2, Res. Inst. Math. Sci. (RIMS), Kyoto, 2007, pp. 195–208.
42. C. T. Simpson, *Katz's middle convolution algorithm*, available at `arXiv:math.AG/0610526`.

43. G. N. Tjurina, *Resolution of singularities of flat deformations of double rational points*, Funkcional. Anal. i Priložen. **4** (1970), no. 1, 77–83 (Russian); English transl., G. N. Tjurina, *Resolution of singularities of flat deformations of double rational points*, Funct. Anal. Appl. **4** (1970), 68–73.

École Normale Supérieure and CNRS, 45 rue d'Ulm, 75005 Paris, France
E-mail address: boalch@dma.ens.fr
URL: www.dma.ens.fr/~boalch

Centre de Recherches Mathématiques
CRM Proceedings and Lecture Notes
Volume **47**, 2009

Families of Ramanujan Graphs and Quaternion Algebras

Denis X. Charles, Eyal Z. Goren, and Kristin E. Lauter

1. Introduction

Expander graphs are graphs in which the neighbors of any given "not too large" set of vertices X form a large set relative to the size of X — rumors tend to spread very fast. Among those, the Ramanujan graphs are extremal in their expansion properties. To be precise, the eigenvalues of the adjacency matrix have an extremal property that guarantees good expansion properties. Expanders, and hence Ramanujan graphs, have many applications, practical and theoretical, to computer science, coding theory, cryptography and network construction, besides numerous purely mathematical applications. Some applications are briefly indicated in the last section of this paper; for a thorough overview see [14, 38] and the references therein.

Quaternion algebras make an appearance in many constructions of Ramanujan graphs. The constructions of Lubotzky – Phillips – Sarnak [24, 25] and Pizer [31] have used definite quaternion algebras over $\mathbb{Q}$, Pizer's construction allowing a more general setting, while making the arithmetic of quaternion algebras more dominant. The construction by Jordan – Livné [18, 23] makes use of quaternion algebras over totally real fields, but in essence is built out of the LPS (for Lubotzky, Phillips, Sarnak) graphs. In each of these cases the Ramanujan property follows from the Ramanujan conjecture for a suitable space of automorphic representations. This much is true also for a related construction by Li [21]. In hindsight, given recent research into Ramanujan complexes (see, e.g., [4, 26]), the reason for the appearance of quaternion algebras is that they supply one with discrete co-compact subgroups of $\mathrm{PGL}_2(F)$, where F is a finite extension of $\mathbb{Q}_p$. The combinatorial properties of the graphs, or, more generally, complexes, constructed from the Bruhat – Tits buildings associated $\mathrm{PGL}_n(F)$, are intimately related to automorphic forms on the appropriate group.

The construction presented in this paper generalizes some of Pizer's work from definite quaternion algebras over $\mathbb{Q}$ to totally definite quaternion algebras over totally real fields. It is, in essence, a special case of the construction by Jordan – Livné (JL), though our main examples are different from theirs as our emphasis

2000 *Mathematics Subject Classification.* Primary 05C25, 05C50, 14K02; Secondary 05C12, 11G10.

This is the final form of the paper.

is either on the case where the class number of the quaternion algebra is large, or on the case when the order is not a maximal order in the quaternion algebra. A particular feature of these graphs, which indeed was our initial motivation for their construction, is that for a chain of totally real fields $L_1 \subset L_2 \subset \cdots \subset L_n$ of strict class number one, and for distinct primes p and l, where p is unramified in all the fields L_i, one gets a chain (or "nested family") of Ramanujan graphs $G(L_1; p, l) \to G(L_2; p, l) \to \cdots \to G(L_n; p, l)$, where the arrows are morphisms of graphs in either the strict sense or in a modified sense that we define below. (We expect that the class number one assumption can be removed.) We can guarantee that the ratio of the size of the graph to the degree goes to infinity with n. We remark that this is a feature that can be obtained for LPS graphs (using the work of [18]) in great generality. See Section 5.3. At this point we are not able to decide if the maps are injective (perhaps under suitable additional hypotheses). We remark that having a nested family of Ramanujan graphs is appealing for certain applications where one desires to *augment* pre-existing graphs to construct larger graphs while retaining the Ramanujan property, and it raises many new questions we hope others will also find appealing, among them determining the situation for the family $G(L_1; p, l) \to G(L_2; p, l) \to \cdots \to G(L_n; p, l)$.

One of the main reasons for discussing such particular cases of the LPS or JL graphs is that, as in Pizer's work, the arithmetic of quaternion algebras is more prevalent. In addition, for such a totally real field L, the graph $G(L; p, l)$ is associated to a very interesting set of points on the Hilbert modular variety of L in characteristic p—the superspecial points. Thus, the connection between graphs and supersingular elliptic curves appearing in Pizer's work is generalized to a connection between graphs and superspecial abelian varieties with real multiplication. The Ramanujan property, appearing in an abstract representation theory language in the general construction, now takes the pleasant face of estimates for Fourier coefficients of theta series of quadratic forms valued in a totally real field L. To make these connections we make essential use of the thesis of Nicole [27]. We remark that this connection is appealing from the point of view of arithmetic geometry, but is not essential to the construction of the graphs. The whole construction can be done for any totally definite quaternion algebra B over a totally real field L, not necessarily of class number one, and very possibly for a larger family of orders than considered in this paper.

To our knowledge, families of "nested" Ramanujan graphs were not studied systematically before and many questions arise. For example, for any family of connected k-regular graphs (necessarily not nested) the second largest eigenvalue of the adjacency matrix is, by a theorem of Alon and Boppana, asymptotically at least $2\sqrt{k-1}-\epsilon$ for any $\epsilon > 0$. The bound $2\sqrt{k-1}$ is called the *Ramanujan bound.* However, since for a particular graph the Ramanujan bound can be broken, it raises the question whether one can construct a nested family of Ramanujan graphs all breaking the Ramanujan bound, where the degree is always small relative to the size of the graph. (Without this proviso the answer is easily "yes," see Section 5.) Other constructions of Ramanujan graphs also lend themselves to creating nested families of graphs. Examples include the Paley graphs and the Terras graphs. We discuss those examples, and explain the relation between our construction and [18, 25] in Section 5.

There are other nice features of the constructions discussed in this paper. For one, it is a general feature of the JL graphs that they allow the construction of essentially different Ramanujan graphs of the same degree on the same vertex set. Indeed, for example, for a prime l we can can get $l+1$-regular graphs from any totally real field of class number one in which l splits completely. The number of vertices of such a graph is the class number of a specific quaternion algebra over L and, if p splits completely in L, is of the order of magnitude $(p-1)^{[L:\mathbb{Q}]}\zeta_L(-1)$. By varying L and p we expect graphs of the same number of vertices to appear many times, while there is no reason to expect all such graphs to even have the same spectrum.

On the other hand, one can also fix the field L and the prime p, and varying l one obtains different graphs on the same vertex set. In that case, one can ask how likely it is that two vertices which are close to each other in one of the graphs are also close in the other graph. In Section 7 we introduce a notion of independence of graphs in this setting, and argue why in many cases the Pizer graphs arising from a fixed p and different l should be independent.

Another interesting feature of the graphs we construct is that one can study the number of closed walks of length n in the graphs we construct as sums of class numbers. Since the characteristic polynomial of the adjacency matrix A is determined by the sequence $\mathrm{tr}(A^n), n=1,2,3,\ldots$, either through Newton's formula or through the identity $\exp(\sum_{n=1}^{\infty}\mathrm{tr}(A^n)t^n/n)=\det(1-tA)^{-1}$, we are getting in that way information on the spectrum of the adjacency matrix.

2. Quaternion algebras and superspecial points

2.1. Elliptic curves and the quaternion algebra $B_{p,\infty}$. Let p be a prime. This section contains no original material; it recalls the well-known connection between supersingular elliptic curves over $\overline{\mathbb{F}}_p$ and orders in the quaternion algebra $B_{p,\infty}$ — the rational quaternion algebra ramified precisely at p and ∞. Good references for this section are [13, 42]. It prepares the ground for a more general theory to follow. Explicit descriptions of $B_{p,\infty}$ and a maximal order in it can be found in [31, Section 4].

Let $E=E_1,\ldots,E_h$ be representatives for the isomorphism classes of supersingular elliptic curves over $\overline{\mathbb{F}}_p$. We fix an identification $\mathrm{End}(E)\otimes\mathbb{Q}=B_{p,\infty}$ and let $\mathcal{O}_i=\mathrm{End}(E_i)$. Then, every order $\mathcal{O}_i$ is isomorphic to a maximal order of $B_{p,\infty}$ and each maximal order of $B_{p,\infty}$ is isomorphic to some $\mathcal{O}_i$. The class number of any maximal order of $B_{p,\infty}$ is h, where h is given by the following formula: $h=[p/12]+\epsilon(p)$, where $\epsilon(2),\epsilon(3)$ are equal to 1 and for $p\geq 5$, $\epsilon(p)=0,1,1,2$ if $p\equiv 1,5,7,11 \pmod{12}$, respectively. See [34, Chapter 5, Theorem 4.1]]. Moreover, fix representatives $I_1,\ldots,I_h$ for the right ideal classes of $\mathcal{O}=\mathcal{O}_1$ with $I_1=\mathcal{O}$. We can choose the ideals I_i in such a way so that $\mathrm{Hom}(E,E_i)\cong I_i$ as projective $\mathcal{O}$-modules. Furthermore, the two quadratic forms

$$\deg\colon \mathrm{Hom}(E,E_i)\to\mathbb{Z},\quad f\mapsto \deg(f)=f^t\circ f,$$

(the degree) and

$$\mathrm{Norm}_c\colon I_i\to\mathbb{Z},\quad f\mapsto\frac{\mathrm{Norm}(f)}{\mathrm{Norm}(I_i)},$$

(the calibrated norm) agree under that isomorphism. In fact, one may define E_i as $I_i\otimes_{\mathcal{O}}E$. We then have that $\mathcal{O}_i$ is the left order of I_i and $\mathrm{Hom}(E_i,E_j)=I_jI_i^{-1}$.

Let n be an integer. We define the *Brandt matrix* $\mathrm{B}(n)$, whose ijth entry $\mathrm{B}(n)_{ij}$ is the number of subgroup schemes H of order n of E_i with $E_i/H \cong E_j$. We note that $\mathrm{B}(n)$ has the following properties:

(1) The sum of every row of $\mathrm{B}(n)$ is equal to $\sigma'(n) := \sum_{d|n,(d,p)=1} d$.
(2) Let $w_i = \frac{1}{2}|\mathrm{Aut}(E_i)|$ then $w_j\,\mathrm{B}(n)_{ij} = w_i\,\mathrm{B}(n)_{ji}$.

Let l be a prime different from p. The matrix $\mathrm{B}(l)$ defines the adjacency matrix of a directed $l+1$ regular graph $G(\mathbb{Q}; p, l)$. Here and throughout the paper "graph" is taken in the loose sense: loops and multiple edges are allowed. If we want to emphasize that there are no loops or multiple edges we say "simple graph."

2.1.1. *Connectedness.* To prove that the graph is connected one can appeal to a theorem on definite quadratic forms in 4 variables. If such a form represents each integer locally at every completion of $\mathbb{Q}$ then it represents any large enough integer. Once the local conditions are verified, one can thus conclude that l^n is represented by the degree map on I_i for every i, provided $n \gg 0$, and so the graph is connected. Other proofs could be given using strong approximation (cf. § 5.3), or by decomposing the associated theta series, which is of weight 2 and level $\Gamma_0(p)$, into a nontrivial Eisenstein component, say $\sum b_n q^n$, and a cusp form and using a "Ramanujan type bound" to show that the coefficients $b_n, (n,p) = 1$, of the Eisenstein series component grow faster than those of the cusp form component. In particular, for n large enough, the coefficients, say c_n, of the associated theta series, which are the representation numbers for the norm forms for the ideal classes, will be nonzero at least of $(n,p) \neq 0, n \gg 0$. Elements of the ideal class of a given norm correspond to isogenies of that same degree and so one concludes that any two supersingular elliptic curves admit isogenies of degree l^n for any prime $l \neq p$ and any large enough n. The Ramanujan type bound referred to here says that if $f(q) = q + \sum_{n>1} a_n q^n$ is a normalized weight two eigenform of level $\Gamma_0(p)$, then $|a_l| \leq 2\sqrt{l}$. In fact, in the case at hand, the bound on the eigenvalues of Hecke operators, equivalently, on the Fourier coefficients of cusp forms, follows from the Eichler – Shimura isomorphism and Weil's own work on the Weil conjectures in the case of curves and abelian varieties.

Let $p \equiv 1 \pmod{12}$. Then each $w_i = 1$ and so the matrices $\mathrm{B}(n)$ are symmetric. In particular, we can pass to the $l+1$-regular undirected graph defined by $\mathrm{B}(l)$. In [31, Proposition 4.7] Pizer proves that this graph is Ramanujan. This *also* implies that the graph is connected. Pizer's work is more general than the case we consider, and in our case it can be simply explained: by classical work of Eichler, the spectrum of the Brandt matrix $\mathrm{B}(l)$ is equal to the spectrum of the Hecke operator T_l in its action on the weight two modular forms of level $\Gamma_0(p)$. The Ramanujan type bound for cusp forms thus gives that every eigenvalue λ of $\mathrm{B}(l)$, apart from $l+1$, satisfies $|\lambda| \leq 2\sqrt{l}$ and so that the graph is Ramanujan. Note, incidently, that this implies that the graphs are not bipartite, because a bipartite k-regular graph has $-k$ as an eigenvalue.

2.2. Abelian varieties with real multiplication and superspecial points. Let L be a totally real field of degree g over $\mathbb{Q}$ of strict class number 1. The moduli problem of classifying triples $(A, \iota\colon \mathcal{O}_L \to \mathrm{End}(A), \lambda)$, where A is a g-dimensional abelian variety, ι is a ring embedding (inducing a ring embedding $\mathcal{O}_L \to \mathrm{End}(A^t)$, where A^t is the dual abelian variety) and λ is an $\mathcal{O}_L$-equivariant principal polarization, has a coarse moduli scheme $\mathscr{M} \to \operatorname{Spec}(\mathbb{Z})$ of relative dimension g. One has $\mathscr{M}(\mathbb{C}) \cong \mathrm{SL}_2(\mathcal{O}_L)\backslash\mathfrak{H}^g$, where $\mathfrak{H}$ is the upper half plane $\{z \in \mathbb{C} : \mathrm{Im}(z) > 0\}$. We let $\mathscr{M}_p = \mathscr{M} \times_{\operatorname{Spec}(\mathbb{Z})} \operatorname{Spec}(\overline{\mathbb{F}}_p)$. The principal polarization λ on (A, ι) is determined uniquely by (A, ι) up to $\mathcal{O}_L$-automorphism. Indeed, the set of $\mathcal{O}_L$-linear symmetric homomorphisms $A \to A^t$ is a projective rank 1 $\mathcal{O}_L$-module and the polarizations are a positive cone in it. The strict class number being one, this module is isomorphic to $\mathcal{O}_L$ and the polarizations to $\mathcal{O}_L^+$. In particular, the principal polarizations are now given by $\mathcal{O}_L^{\times,+}$. Under this identification, the principally polarized abelian variety with real multiplication $(A, \iota, 1)$ is isomorphic to $(A, \iota, \epsilon^*1) = (A, \iota, \epsilon^2)$ for any $\epsilon \in \mathcal{O}_L^\times$. Strict class number one implies that any totally positive unit is a square of a unit thus proving our claim.

A *superspecial abelian variety* A of dimension g over an algebraically closed field of positive characteristic is equivalently:

(1) an abelian variety A isomorphic to a product of g supersingular elliptic curves;
(2) an abelian variety A isomorphic to E^g, where E is a supersingular elliptic curve;
(3) an abelian variety A such that the absolute Frobenius acts as zero on $H^1(A, \mathcal{O}_A)$.

The implications (2) $\Rightarrow$ (1) $\Rightarrow$ (3) are obvious but the full equivalence is far from obvious. The equivalence of (1) and (2) is a theorem of Deligne (see [33]) and the equivalence with (3) is a theorem of Oort [29]. If $g > 1$ then Deligne's theorem also says that $E^g \cong E'^g$ for any two supersingular elliptic curves over an algebraically closed field.

Definition 2.1. We call a superspecial principally polarized abelian variety with RM by L an L-superspecial variety.

2.2.1. Assume henceforth that L has strict class number one and fix a supersingular elliptic curve E over $\overline{\mathbb{F}}_p$ with endomorphism ring $\mathcal{O}$; we identify once and for all $B_{p,\infty}$ with $\mathcal{O} \otimes_{\mathbb{Z}} \mathbb{Q}$. A general reference for what follows is Nicole's thesis [27]. See also [28]. We assume that p is unramified in L.

A concrete example of an L-superspecial variety is given by $E \otimes_{\mathbb{Z}} \mathcal{O}_L$. Its dual is naturally $E^t \otimes D_{L/\mathbb{Q}}^{-1}$ but since $D_{L/\mathbb{Q}}^{-1}$ has a totally positive generator α, $E \otimes_{\mathbb{Z}} \mathcal{O}_L$ has a principal $\mathcal{O}_L$-linear polarization $\lambda \otimes \alpha$, where λ is the canonical principal polarization $\lambda : E \to E^t$. One can prove that $R := \operatorname{End}_{\mathcal{O}_L}(E \otimes \mathcal{O}_L)$ is equal to $\mathcal{O} \otimes_{\mathbb{Z}} \mathcal{O}_L$. This is an order of discriminant $p\mathcal{O}_L$ in the quaternion algebra $B_{p,L} := B_{p,\infty} \otimes_{\mathbb{Q}} L$. Let $p = \mathfrak{p}_1 \dots \mathfrak{p}_a$ be the decomposition of p into prime ideals in $\mathcal{O}_L$. Then $B_{p,L}$ is ramified precisely at the infinite places of L and at the primes $\mathfrak{p}_i$ such that $f(\mathfrak{p}_i/p) \equiv 1 \pmod 2$, where $f(\mathfrak{p}_i/p)$ is the residue degree $\dim_{\mathbb{F}_p}(\mathcal{O}_L/\mathfrak{p}_i)$.

Example 2.2. Let L be a real quadratic field. If p is inert then $B_{p,L}$ is ramified exactly at the infinite places and the order R is not maximal. On the other hand, if p is split then $B_{p,L}$ is ramified at the infinite places and at the two places above p and the order R is maximal.

The order R is not always maximal and so some care has to be taken with its ideal theory. Our main reference is Brzezinski [3]. The order R is an Eichler order of square-free level, though, and so is a hereditary order. An ideal of $B_{p,L}$ is by definition a finitely generated $\mathcal{O}_L$-module that contains a basis for $B_{p,L}$. An ideal I

is called a right R-ideal (or a right ideal of R) if $R = \{f \in B_{p,L} : If \subseteq I\}$. For such an ideal I, the properties projective, locally principal, invertible (in the sense that $II^{-1}, I^{-1}I$ are the trivial ideals, where $I^{-1} := \{f \in B_{p,L} : fI \subseteq I\}$) or satisfying $[R : I] = \mathrm{Norm}(I)^2$, are all equivalent. For a hereditary order one has that every ideal is projective. See *loc. cit.* for all this.

Let $R = I_1, \ldots, I_h$ be representatives for the right ideal classes of R. The isomorphism classes of L-superspecial varieties are in bijection with the abelian varieties $A_i := I_i \otimes_R (E \otimes_{\mathbb{Z}} \mathcal{O}_L)$, $A = A_1$. One has $\mathrm{Hom}_{\mathcal{O}_L}(A, A_i) = I_i$ and, more generally, $\mathrm{Hom}_{\mathcal{O}_L}(A_i, A_j) = I_j I_i^{-1}$. The left order R_i of I_i is thus naturally isomorphic to $\mathrm{End}(A_i)$. Every order that is everywhere locally conjugate to R is isomorphic to one of the orders R_i and each order R_i is everywhere locally conjugate to R. One can characterize the orders that are everywhere locally conjugate to R simply as Eichler orders of level $p\mathcal{O}_L$. See [27, 28]. Following *loc. cit.* we call such orders *superspecial orders*.

For completeness, we indicate the key point that goes into proving some of these assertions. If $A/\mathbb{F}_{p^r}$ is L-superspecial, then over some $\mathbb{F}_{p^{r'}}$ the abelian variety A is isomorphic to E^g, where E is a supersingular elliptic curve over $\mathbb{F}_p$. In particular, the Weil number of E is $\sqrt{-p}$. It follows that the characteristic polynomial of Frobenius on E (acting on any Tate module and also as an endomorphism) is just $x^2 + p$. Thus, the relative Frobenius over $\mathbb{F}_{p^{2r'}}$ is nothing else than multiplication by $p^{r'}$ on "anything in sight". It is not hard to conclude then that after a finite field extension $T_l(A) \cong \mathcal{O}_L^{2g}$ as $\mathcal{O}_L$-modules where the Galois action is given as above (and is independent of A). The proof that the same holds for the Dieudonné modules appears in [12]. To determine the local structure of homomorphisms, one uses now a variant of Tate's theorem: for abelian varieties A, B with RM by $\mathcal{O}_L$, defined over a finite field $\mathbb{F}_q$, $\mathrm{Hom}_{\mathcal{O}_L}(A, B) \otimes \mathbb{Z}_l \cong \mathrm{Hom}_{\mathcal{O}_L \otimes \mathbb{Z}_l}(T_l(A), T_l(B))^{\mathrm{Gal}(\overline{\mathbb{F}}_q/\mathbb{F}_q)}$ and the similar statement for Dieudonné modules (see [43]).

On the ideals I_i we have *calibrated norm* maps. Choose a totally positive generator α of $\mathrm{Norm}(I_i)$ and define

$$\mathrm{Norm}_c \colon I_i \to \mathcal{O}_L, \quad \mathrm{Norm}_c(f) = \mathrm{Norm}(f)/\alpha.$$

This is a positive definite quarternary $\mathcal{O}_L$-valued quadratic form, well defined up to a totally positive unit. On the other hand, choose principal $\mathcal{O}_L$-linear polarizations λ, λ_i on A, A_i and define

$$\deg_L \colon \mathrm{Hom}(A, A_i) \to \mathcal{O}_L, \quad \deg_L(f) = \lambda^{-1} f^t \lambda_i f.$$

This is also a positive definite quarternary $\mathcal{O}_L$-valued quadratic form, well defined up to a totally positive unit. As defined, $\deg(f)$ is clearly an element of R. It takes values in $\mathcal{O}_L$ because it is fixed by the Rosati involution. The projective R-modules I_i and $\mathrm{Hom}(A, A_i)$ are isomorphic. Moreover, the isomorphism can be chosen as to take Norm_c to $\deg_L$. Note that if $A_i = A$ there are canonical choices for both Norm_c and $\deg_L$ and they are equal. The above generalizes in the expected manner to $\mathrm{Hom}_{\mathcal{O}_L}(A_i, A_j)$ and $I_j I_i^{-1}$.

2.2.2. Let $H \subseteq A$ be a finite $\mathcal{O}_L$-invariant group scheme, where A is an abelian variety with RM, and assume that H is étale. Thus, over an algebraic closure we can write a composition series $0 \subsetneqq H_1 \subsetneqq H_2 \subsetneqq \cdots \subsetneqq H_n = H$ for H as an $\mathcal{O}_L$-module; the quotients H_i/H_{i-1} are $\mathcal{O}_L$-modules of the form $\mathcal{O}_L/\mathfrak{l}_i$, where $\mathfrak{l}_i$ is a prime ideal of $\mathcal{O}_L$ which is prime to p.

Definition 2.3. In the notation above, define the $\mathcal{O}_L$-degree of H, $\deg_L(H)$, to be $\mathfrak{l}_1\mathfrak{l}_2\cdots\mathfrak{l}_n$.

Lemma 2.4. *Let A be an L-superspecial variety, $R = \mathrm{End}(A)$ and $J \lhd R$ a left ideal whose norm is relatively prime to p. Let $A[J] = \bigcap_{f\in J}\mathrm{Ker}(f)$ then*

$$\deg_L(A[J]) = \mathrm{Norm}(J).$$

PROOF. Let $\mathfrak{l}$ be a prime ideal of $\mathcal{O}_L$ not dividing p. We have the primary decomposition $A[J] = \bigoplus_{(\mathfrak{l},p)=1} A[J]_{\mathfrak{l}}$ and $\deg_L(A[J]) = \prod_{(\mathfrak{l},p)=1}\deg_L(A[J]_{\mathfrak{l}})$. On the other hand, the group $A[J]_{\mathfrak{l}}$ is isomorphic over an algebraic closure to $T_{\mathfrak{l}}(A)/\sum_{f\in J} fT_{\mathfrak{l}}(A) = T_{\mathfrak{l}}(A)/\sum_{f\in J_{\mathfrak{l}}} fT_{\mathfrak{l}}(A)$, where, using that $T_l(A)$ is a free $\mathcal{O}_L\otimes_{\mathbb{Z}}\mathbb{Z}_l$-module of rank 2 [32], we have decomposed the Tate module $T_l(A)$ as $\bigoplus_{\mathfrak{l}|l} T_{\mathfrak{l}}(A) \cong \oplus_{\mathfrak{l}|l}{\mathcal{O}_L}_{\mathfrak{l}}^2$. Since all ideals are locally principal, we have $J_{\mathfrak{l}} = M_2(\mathcal{O}_{L\mathfrak{l}})j_{\mathfrak{l}}$ for a matrix $j_{\mathfrak{l}} \in M_2(\mathcal{O}_{L\mathfrak{l}})$, and $A[J]_{\mathfrak{l}}$ is isomorphic to $(\mathcal{O}_{L\mathfrak{l}})^2/j_{\mathfrak{l}}(\mathcal{O}_{L\mathfrak{l}})^2$. The product of the composition factors of this cokernel as an $\mathcal{O}_{L\mathfrak{l}}$-module is precisely $\det(j_{\mathfrak{l}})$, which is just the component at $\mathfrak{l}$ of $\mathrm{Norm}(J)$ (the local norm *is* the determinant). □

2.2.3. A very useful tool is the notion of kernel ideals studied in the context of general abelian varieties by Waterhouse in [42, Section 3.2]. Let A be a principally polarized superspecial abelian variety with RM over an algebraically closed field, $R = \mathrm{End}(A)$. For a subgroup scheme $H \subset A$ let $I(H) = \{f \in R : f(H) = 0\}$. One always has $J \subseteq I(A[J])$ and we call J a *kernel ideal* if $J = I(A[J])$. If A is a superspecial abelian variety with RM, p unramified in L, $R = \mathrm{End}(A)$ and H is étale then $I(H)$ is a left R-ideal; indeed it is easy to check that it is everywhere a locally principal ideal. It is proved in [27], following the proof of [42, Theorem 3.15], that every ideal of R is a kernel ideal. In particular, the map $J \mapsto A[J]$, taking ideals of norm prime-to-p contained in R to $\mathcal{O}_L$-subgroup schemes, is injective. More is true. This map is bijective when restricted to group schemes of order prime-to-p. To show that one needs only show that if $H_1 \subsetneqq H_2$ are two distinct $\mathcal{O}_L$-invariant group schemes of order prime-to-p then there is an endomorphism of A that vanishes on H_1 but not on H_2 (it then follows that $I(H_1) \neq I(H_2)$ and one concludes that we must have $A[I(H)] = H$ for every finite $\mathcal{O}_L$-subgroup scheme). We can thus consider only the case where both H_i are $\mathfrak{l}$-primary for some prime ideal $\mathfrak{l} \lhd \mathcal{O}_L$, $(\mathfrak{l},p) = 1$. We have then that $R\otimes_{\mathcal{O}_L}\mathcal{O}_{L\mathfrak{l}} \cong \mathrm{End}_{\mathcal{O}_{L\mathfrak{l}}}(T_{\mathfrak{l}}(A)) \cong M_2(\mathcal{O}_{L\mathfrak{l}})$. Using the correspondence between $\mathfrak{l}$-primary finite groups schemes $H \subset A$ and lattices $\Lambda \supset T_{\mathfrak{l}}(A) \cong \mathcal{O}_{L_{\mathfrak{l}}}^2$, $H \mapsto \Lambda(H) = \pi_H^*(T_{\mathfrak{l}}(A/H))$, we see that it is enough to prove that if $\Lambda_2 \supsetneqq \Lambda_1 \supsetneqq \mathcal{O}_{L_{\mathfrak{l}}}^2$ then there is a matrix $\gamma \in M_2(\mathcal{O}_{L_{\mathfrak{l}}})$ such that $\gamma(\Lambda_1) = \mathcal{O}_{L_{\mathfrak{l}}}^2$. This is of course immediate using, say, elementary divisors. One then approximates γ well enough $\mathfrak{l}$-adically, using the isomorphism $R\otimes_{\mathcal{O}_L}\mathcal{O}_{L_{\mathfrak{l}}} \cong \mathrm{End}_{\mathcal{O}_{L_{\mathfrak{l}}}} T_{\mathfrak{l}}(A) = M_2(\mathcal{O}_{L_{\mathfrak{l}}})$.

2.3. The situation at p. The subtle point in defining the analog at p is that we want the quotient A/H to have a principal $\mathcal{O}_L$-polarization and be superspecial. An abelian variety A/k with RM and principal polarization satisfies the Rapoport condition: the tangent space to A at the origin is a free $\mathcal{O}_L\otimes_{\mathbb{Z}} k$-module of rank 1. In our case, since the class number of L is one, every abelian variety satisfying the Rapoport condition has a principal $\mathcal{O}_L$-linear polarization.

If H is étale then A/H satisfies the Rapoport condition, because $A \to A/H$ induces an $\mathcal{O}_L$-equivariant isomorphism on tangent spaces at the origin. Thus A/H will also have a principal $\mathcal{O}_L$-polarization and be superspecial; in general, one can give examples of $\mathcal{O}_L$-invariant subgroup schemes $H \subset A[p]$ such that A/H does not satisfy the Rapoport condition and so will not have a principal $\mathcal{O}_L$-polarization. Therefore, the situation is more subtle when H is not étale.

Consider the following diagram

$$\begin{array}{ccc} A & \xrightarrow{f^*\mu} & A^t \\ \Big\downarrow f & & \Big\uparrow f^t \\ A/H & \xrightarrow{\mu} & (A/H)^t \end{array}$$

where $f\colon A \to A/H$ is the canonical projection. We are seeking a principal $\mathcal{O}_L$-polarization μ such that $f^*\mu := f^t \circ \mu \circ f$ has kernel in which H is maximal isotropic with respect to the Mumford pairing induced by $f^*\mu$. Let λ be a principal $\mathcal{O}_L$-polarization of A then any $\mathcal{O}_L$-polarization, in particular $f^*\mu$, is of the form $a\lambda$, where $a \in \mathcal{O}_L$ is a totally positive element. The kernel is just $A[a]$, which fits into an exact sequence of $\mathcal{O}_L$ group schemes $1 \to H \to A[a] \to H^t \to 1$. There is an additional condition: we want A/H to be superspecial.

Since our wish is to maintain the correspondence between ideals and subgroup schemes, we first examine "the situation at p" for superspecial orders. Let $\mathfrak{p} \mid p$ be a prime of $\mathcal{O}_L$. There are two cases:

(1) *The quaternion algebra $B_{p,L}$ is ramified at $\mathfrak{p}$.* In this case the completion at $\mathfrak{p}$ of $B_{p,L}$ is the unique quaternion algebra over $L_\mathfrak{p}$. If R is a superspecial order then R is a maximal order at $\mathfrak{p}$ and its completion at $\mathfrak{p}$ is uniquely determined. One can give a model for $R_\mathfrak{p}$. Let Q be an unramified extension of degree two of $L_\mathfrak{p}$, V its valuation ring and σ the nontrivial automorphism of Q/L. Let π be a uniformizer of $\mathcal{O}_{L_\mathfrak{p}}$. Then $B_{p,L} \otimes_L L_\mathfrak{p} \cong \left\{\left(\begin{smallmatrix} a & b \\ -\pi b^\sigma & a^\sigma \end{smallmatrix}\right) : a, b \in Q\right\}$ and $R_\mathfrak{p} = \left\{\left(\begin{smallmatrix} a & b \\ -\pi b^\sigma & a^\sigma \end{smallmatrix}\right) : a, b \in V\right\}$ with the trace and norm given by the trace and determinant of matrices. It has a unique maximal ideal, given by the matrices $\left\{\left(\begin{smallmatrix} \pi a & b \\ -\pi b^\sigma & \pi a^\sigma \end{smallmatrix}\right) : a, b \in V\right\} = R_\mathfrak{p}\left(\begin{smallmatrix} \pi & 1 \\ -\pi & \pi \end{smallmatrix}\right)$, which has norm equal to $\mathfrak{p}\mathcal{O}_{L_\mathfrak{p}}$. In fact, all ideals are two-sided and every ideal is a power of the maximal ideal [40, [Chapter II, §1]]. We thus conclude that there is a unique ideal $J \lhd R$ such that $\mathrm{Norm}(J) = \mathfrak{p}$. Clearly $A[J] \subset A[p]$; we claim that $A[J] = (\mathrm{Ker}(\mathrm{Fr}\colon A \to A^{(p)}))_\mathfrak{p}$. Indeed, the uniqueness of the principally polarized superspecial crystal with RM [12] allows us to assume that the action of $B_{p,L} \otimes_\mathbb{Q} \mathbb{Q}_p$ on $H^1_{\mathrm{crys}}(A/W(\overline{\mathbb{F}}_p))$ is induced from the action of $B_{p,\infty}$ on a supersingular elliptic curve E. Thus, in essence we are working with the abelian variety $E \otimes \mathcal{O}_L$ and our assertion follows from the case of elliptic curves, where the uniqueness of a subgroup scheme of order p makes the claim straightforward.

In this case it is clear that the only subgroups H should have as component at $\mathfrak{p}$ are of the form, $\mathrm{Ker}(\mathrm{Fr}^n\colon A \to A^{(p^n)})_\mathfrak{p}$ for some n.

We can take $\pi \in \mathcal{O}_L$ such that $(\pi) = \mathfrak{p}$. Then we take $a = \pi^n$. It is a direct verification that $H \subset A[a]$ is totally isotropic with respect to the pairing defined by the polarization $a\lambda$. Moreover the $\mathfrak{p}$-primary part of the tangent space $T_{A/H,0}$ of A/H at the origin is naturally isomorphic to the $\mathfrak{p}$-primary part of $T_{A^{(p^n)},0}$ as an $\mathcal{O}_L$-module and so is a locally free $\mathcal{O}_{L_\mathfrak{p}} \otimes \overline{\mathbb{F}}_p$-module of rank 1, while the $\mathfrak{q}$-primary part of $T_{A/H,0}$ for every other prime $\mathfrak{q} \mid p$ of $\mathcal{O}_L$ is naturally isomorphic to the $\mathfrak{q}$-primary part of $T_{A,0}$. Those isomorphisms commute with the action of Fr and Ver and so A/H is also superspecial.

(2) *The quaternion algebra $B_{p,L}$ is split at $\mathfrak{p}$.* In this case the completion at $\mathfrak{p}$ of $B_{p,L}$ is isomorphic to $M_2(L_\mathfrak{p})$. If R is a superspecial order then R is an Eichler

order of level $p\mathcal{O}_L$ and under a suitable isomorphism its completion $R_\mathfrak{p}$ at $\mathfrak{p}$ is $\left\{\left(\begin{smallmatrix} a & b \\ \pi c & d\end{smallmatrix}\right) : a,b,c,d \in O_{L_\mathfrak{p}}\right\}$, where π is a uniformizer of $O_{L_\mathfrak{p}}$. The Jacobson radical $J(R_\mathfrak{p})$ is of finite index in $R_\mathfrak{p}$ and one thus concludes that it contains the matrices $\left(\begin{smallmatrix} \pi & 0 \\ 0 & 0\end{smallmatrix}\right), \left(\begin{smallmatrix} 0 & 0 \\ 0 & \pi\end{smallmatrix}\right)$ and the nilpotent elements $\left(\begin{smallmatrix} 0 & 1 \\ 0 & 0\end{smallmatrix}\right), \left(\begin{smallmatrix} 0 & 0 \\ \pi & 0\end{smallmatrix}\right)$ and so

$$J(R_\mathfrak{p}) = \left\{\begin{pmatrix} \pi a & b \\ \pi c & \pi d\end{pmatrix} : a,b,c,d \in \mathcal{O}_{L_\mathfrak{p}}\right\}.$$

Since $R_\mathfrak{p}$ is an Eichler order it has Eichler symbol 1 at $\mathfrak{p}$ which means that $R_\mathfrak{p}/J(R_\mathfrak{p})$ is isomorphic to $\mathcal{O}_L/\mathfrak{p} \oplus \mathcal{O}_L/\mathfrak{p}$ and indeed this isomorphism is visibly induced from $\left(\begin{smallmatrix} a & b \\ \pi c & d\end{smallmatrix}\right) \mapsto (a,d) \mod \pi$. Thus, $R_\mathfrak{p}$ has two maximal left lattices that are

$$J_1 = \left\{\begin{pmatrix} a & b \\ \pi c & \pi d\end{pmatrix} : a,b,c,d \in \mathcal{O}_{L_\mathfrak{p}}\right\}, \quad J_2 = \left\{\begin{pmatrix} \pi a & b \\ \pi c & d\end{pmatrix} : a,b,c,d \in \mathcal{O}_{L_\mathfrak{p}}\right\}.$$

Those lattices are *not* $R_\mathfrak{p}$-ideals; J_1 is an ideal for the order $\left\{\left(\begin{smallmatrix} a & \pi^{-1}b \\ \pi c & d\end{smallmatrix}\right) : a,b,c,d \in \mathcal{O}_{L_\mathfrak{p}}\right\}$ and J_2 is an ideal for $M_2(\mathcal{O}_{L_\mathfrak{p}})$, and indeed one can verify that they are not locally principal for the order $R_\mathfrak{p}$. However, $J(R_\mathfrak{p})$ also has norm $\mathfrak{p}$ and is locally principal, equal to $R_\mathfrak{p}\left(\begin{smallmatrix} \pi & 1 \\ \pi & \pi\end{smallmatrix}\right)$. We find again that there is a unique R-ideal J of norm $\mathfrak{p}$, whose kernel, as before, is equal to $\mathrm{Ker}(\mathrm{Fr})_\mathfrak{p}$. It should be noted that there are other principal ideals of $R_\mathfrak{p}$ of norm $\mathfrak{p}$. Indeed, any matrix M with $(\det(M)) = \mathfrak{p}$ provides such an ideal $R_\mathfrak{p}M$ and this constitutes all such examples. The ideal $J(R)$ can be characterized in two different ways: (i) It is the collection of matrices M such that $M^2 \equiv 0 \pmod{\pi}$; (ii) If $J(\mathfrak{O}_p)$ is the maximal ideal of a maximal order $\mathcal{O}$ of $B_{p,\infty}$ then, choosing an isomorphism $\mathcal{O}_p \otimes_{\mathbb{Z}_p} \mathcal{O}_{L_\mathfrak{p}} \cong R_\mathfrak{p}$, we have $J(R_\mathfrak{p}) = J(\mathcal{O}) \otimes_\mathbb{Z} \mathcal{O}_{L_\mathfrak{p}}$. To see that, note that $J(\mathcal{O}) \otimes_\mathbb{Z} \mathcal{O}_{L_\mathfrak{p}}$ is an ideal of $R_\mathfrak{p}$ containing $\pi R_\mathfrak{p} = pR_\mathfrak{p}$ and the quotient $R_\mathfrak{p}/J(\mathcal{O}) \otimes_\mathbb{Z} \mathcal{O}_{L_\mathfrak{p}}$ is isomorphic to $\mathcal{O}/J(\mathcal{O}) \otimes \mathcal{O}_{L\mathfrak{p}} \cong \mathbb{F}_{p^2} \otimes \mathcal{O}_{L\mathfrak{p}} \cong (\mathcal{O}_L/\mathfrak{p})^2$. Here we have used that p is unramified in $\mathcal{O}_L$ and that $B_{p,L}$ splits at $\mathfrak{p}$ if and only if $2 \mid f(\mathfrak{p}/p)$.

Let E be a supersingular elliptic curve with $\mathrm{End}(E) = \mathcal{O}$. The isomorphism $\mathcal{O}_p \otimes_{\mathbb{Z}_p} \mathcal{O}_{L_\mathfrak{p}} \cong R_\mathfrak{p}$ can be chosen so that the action of $R_\mathfrak{p}$ on $H^1_{\mathrm{crys}}(A/W(\overline{\mathbb{F}}_p))$ is the action of $\mathcal{O}_p \otimes_{\mathbb{Z}_p} \mathcal{O}_{L_\mathfrak{p}}$ on $H^1_{\mathrm{crys}}(E/W(\overline{\mathbb{F}}_p)) \otimes_{\mathbb{Z}_p} \mathcal{O}_{L_\mathfrak{p}}$, by the uniqueness of the superspecial crystal with RM. This shows that $J(R_\mathfrak{p})$ kills the $\mathfrak{p}$-part of the kernel of Frobenius. In the same way as in the first case one concludes that A/H is superspecial and satisfies the Rapoport condition.

We summarize our discussion.

Definition 2.5. Call an $\mathcal{O}_L$-invariant subgroup scheme *admissible* if for every $\mathfrak{p} \mid p$ the $\mathfrak{p}$-primary component of H is equal to the $\mathfrak{p}$-primary component of $\mathrm{Ker}(\mathrm{Fr}^{(n)} \colon A \to A^{(p^n)})$ for some $n = n(\mathfrak{p}) \geq 0$. Call a left ideal I *admissible* if for every prime $\mathfrak{p} \mid p$ the local component $I_\mathfrak{p}$ is a power of the Jacobson ideal.

Writing $p = \mathfrak{p}_1 \cdots \mathfrak{p}_a$, the p-primary part of admissible subgroups is in bijection with vectors $(b_1, \ldots, b_a)$ with $b_i \in \mathbb{Z}_{\geq 0}$. Given an admissible subgroup H, we write $H = H' \oplus H''$, where H' is the prime-to-p part of H and H'' is classified by a vector $(b_1, \ldots, b_a)$. We then let $\deg_L(H) = \deg_L(H') \cdot \prod_{i=1}^a \mathfrak{p}_i^{b_i}$.

Proposition 2.6. *Let A be a superspecial abelian variety with RM by L and having a principal $\mathcal{O}_L$-polarization. The admissible subgroups H correspond bijectively to admissible locally principal ideals $I \lhd R$, by $J \mapsto A[J]$, $H \mapsto I(H)$ and $\deg_L(H) = \mathrm{Norm}\big(I(H)\big)$.*

Example 2.7. Let L be a real quadratic field. Let A be a superspecial abelian surface with RM by L. Let p be inert in L, then there is a unique admissible $\mathcal{O}_L$-invariant group scheme in $A[p]$; it is equal to the kernel of $\mathrm{Fr}\colon A \to A^{(p)}$.

If, on the other hand, $p = \mathfrak{p}_1\mathfrak{p}_2$ is split in L, then there are precisely two admissible $\mathcal{O}_L$-invariant subgroup schemes of order p in A. These are the $\mathfrak{p}_i$-primary subgroups of the kernel of Frobenius.

3. Hecke operators, Brandt matrices and superspecial graphs

We keep our assumption that L is a totally real field of degree g of strict class number 1. The prime p is unramified in L and decomposes as $p = \mathfrak{p}_1 \cdots \mathfrak{p}_a$.

3.1. Brandt matrices. Let R be a superspecial order in $B_{p,L} = B_{p,\infty} \otimes_{\mathbb{Q}} L$ and let h be its class number. Let A be a superspecial abelian variety with an identification $\mathrm{End}(A) \cong R$. Index the right ideal classes of R as before by $R = I_1, \dots, I_h$. We denote equivalence in the class group by $\sim$. Let R_i be the left order of I_i and let $A_i = I_i \otimes_R A$ be the corresponding superspecial abelian variety.

Definition 3.1. Let $\mathfrak{n}$ be an ideal of $\mathcal{O}_L$. We define the Brandt matrix $\mathrm{B}(\mathfrak{n})$ to be the $h \times h$ matrix whose ijth entry is the number of admissible integral left ideals $J \lhd R_i$ such that $\mathrm{Norm}(J) = \mathfrak{n}$ and $J \sim I_i I_j^{-1}$.

Proposition 3.2. *The Brandt matrices* $\mathrm{B}(\mathfrak{n})$ *have the following properties*:

(1) *The entry* $\mathrm{B}(\mathfrak{n})_{ij}$ *is equal to the number of admissible* $\mathcal{O}_L$*-invariant subgroup schemes* H *of* A_i *with* $\deg_L(H) = \mathfrak{n}$ *and* $A_i/H \cong A_j$.
(2) *We have* $\sum_j \mathrm{B}(\mathfrak{n})_{ij} = \sigma'(\mathfrak{n})$, *where* $\sigma'(\mathfrak{n}) = \sum_{\mathfrak{d}|\mathfrak{n},(\mathfrak{d},p)=1} \mathrm{Norm}_{L/\mathbb{Q}}\,\mathfrak{d}$
(3) *Let* $w_i = |R_i^\times/\mathcal{O}_L^\times|$. *Then* w_i *is finite and* $w_j\,\mathrm{B}(\mathfrak{n})_{ij} = w_i\,\mathrm{B}(\mathfrak{n})_{ji}$.
(4) *The matrices* $\mathrm{B}(\mathfrak{n})$ *commute. If* $(\mathfrak{n},\mathfrak{m}) = 1$ *then* $\mathrm{B}(\mathfrak{m})\,\mathrm{B}(\mathfrak{n}) = \mathrm{B}(\mathfrak{mn})$.
(5) *For a prime ideal* $\mathfrak{p}$ *above* p *we have* $\mathrm{B}(\mathfrak{p})^n = \mathrm{B}(\mathfrak{p}^n)$. *The matrix* $\mathrm{B}(\mathfrak{p})$ *is a permutation matrix of order* 2.
(6) *For a prime ideal* $\mathfrak{l}$ *not above* p *we have*

$$\mathrm{B}(\mathfrak{l})\,\mathrm{B}(\mathfrak{l}^n) = \mathrm{B}(\mathfrak{l}^{n+1}) + \mathrm{Norm}\,\mathfrak{l} \cdot \mathrm{B}(\mathfrak{l}^{n-2}).$$

PROOF. By symmetry it is enough to consider the case of $i = 1$, and so $R = R_1$. Every such right ideal $J \lhd R$ defines a subgroup scheme $A[J]$, giving a bijection between admissible ideals of norm $\mathfrak{n}$ and admissible subgroup schemes H such that $\deg_L(H) = \mathfrak{n}$. The abelian variety $A/A[J]$ is isomorphic to the abelian variety $\mathrm{Hom}_R(J, A) \cong J^{-1} \otimes A$, which is isomorphic to $I_j \otimes A = A_j$ if J is in the class of I_j^{-1} [42, Corollary A.4]. Since we have a bijection between admissible ideals and admissible group schemes, we conclude the first statement.

The number of admissible $\mathcal{O}_L$-invariant group schemes of degree $\mathfrak{n}$ is a multiplicative function in $\mathfrak{n}$ and so is σ'. We may therefore assume that $\mathfrak{n} = \mathfrak{l}^k$ is a power of a prime ideal. The case of $\mathfrak{l} \mid p$ is trivial and we assume thus that $\mathfrak{l} \nmid p$. We now argue by induction on k. Since the number of $\mathcal{O}_L$-group schemes of $\deg_L = \mathfrak{l}$ is the number of lines in the $\mathbb{F}_\mathfrak{l} := \mathcal{O}_L/\mathfrak{l}$-vector space $A[\mathfrak{l}] \cong \mathbb{F}_\mathfrak{l}^2$, which is equal to $\sharp\mathbb{P}^1(\mathbb{F}_\mathfrak{l}) = \mathrm{Norm}(\mathfrak{l}) + 1$, we see that the cases $k = 0, 1$ hold. We have $\sigma'(\mathfrak{l}^k) - \sigma'(\mathfrak{l}^{k-2}) = \mathrm{Norm}\,\mathfrak{l}^k + \mathrm{Norm}\,\mathfrak{l}^{k-1} = l^{kf(\mathfrak{l})} + l^{(k-1)f(\mathfrak{l})}$. On the other hand, $\sum_j(\mathrm{B}(\mathfrak{l}^k)_{ij} - \mathrm{B}(\mathfrak{l}^{k-2})_{ij})$ is exactly the number of $\mathcal{O}_L$-subgroup schemes H of A_i such that $\deg_L(H) = \mathfrak{l}^k$ and $H \not\supseteq A_i[\mathfrak{l}]$. Passing to $T_\mathfrak{l}(A)/(\mathfrak{l}^k)$, we see that this is the number of cyclic $\mathcal{O}_L$-modules of $(\mathcal{O}_L/\mathfrak{l}^k)^2$ isomorphic to $\mathcal{O}_L/\mathfrak{l}^k$. This number is just the number of elements (a,b) of $(\mathcal{O}_L/\mathfrak{l}^k)^2$ such that at least one of a,b is not divisible by $\mathfrak{l}$, taken modulo $(\mathcal{O}_L/\mathfrak{l}^k)^\times$, a group of order $(l^{f(\mathfrak{l})}-1)l^{(k-1)f(\mathfrak{l})}$. On the other hand the number of such generators (a,b) is clearly $l^{2kf(\mathfrak{l})}-l^{2(k-1)f(\mathfrak{l})}$. We conclude that there are $\left(l^{2kf(\mathfrak{l})}-l^{2(k-1)f(\mathfrak{l})}\right)/\left((l^{f(\mathfrak{l})}-1)l^{(k-1)f(\mathfrak{l})}\right)=l^{kf(\mathfrak{l})}+l^{(k-1)f(\mathfrak{l})}$ such $\mathcal{O}_L$-modules and we are done.

To show finiteness of $R^\times/\mathcal{O}_L^\times$, let R^1 be the elements of norm 1 in R and consider the injective group homomorphism $R^\times/\mathcal{O}_L^\times \to R^1$ given by $z\mapsto z/\bar{z}$. The positive definiteness of the norm map implies that R^1 is finite. To get the symmetry property consider $\operatorname{Hom}(A_i,A_j)$ and $\operatorname{Hom}(A_j,A_i)$ and the $\deg_L$ map on both, viewed as taking values in $\mathcal{O}_L/\mathcal{O}_L^\times$. If $\phi\in\operatorname{Hom}(A_i,A_j)$ is of degree prime to p and $\deg(\phi)=\mathfrak{n}$ then $\phi^t\in\operatorname{Hom}(A_j^t,A_i^t)=\operatorname{Hom}(A_j,A_i)$ has the same degree. The map from $\operatorname{Hom}(A_i,A_j)$ to group schemes $\phi\mapsto\operatorname{Ker}(\phi)$ has fibers that are principal homogeneous spaces under $R_j^\times$ and the result follows.

If $(\mathfrak{m},\mathfrak{n})=1$ then the identity $\mathrm{B}(\mathfrak{m})\,\mathrm{B}(\mathfrak{n})=\mathrm{B}(\mathfrak{mn})$ is just the decomposition of $\mathcal{O}_L$-group schemes of $\deg_L$ equal $\mathfrak{mn}$ into their $\mathfrak{m}$-primary and $\mathfrak{n}$-primary components. This also implies that such matrices commute. The formula $\mathrm{B}(\mathfrak{p}^n)=\mathrm{B}(\mathfrak{p})^n$ is immediate from the definition. We have $\mathrm{B}(\mathfrak{p}^2)_{ii}=1$, because the subgroup $A[p]_\mathfrak{p}=\operatorname{Ker}(\pi)$, where $(\pi)=\mathfrak{p}$ and $A/\operatorname{Ker}(\pi)\cong A$. It follows that $\mathrm{B}(\mathfrak{p})^2=\mathrm{Id}$. The formula $\mathrm{B}(\mathfrak{l})\,\mathrm{B}(\mathfrak{l}^n)=\mathrm{B}(\mathfrak{l}^{n+1})+\operatorname{Norm}\mathfrak{l}\cdot\mathrm{B}(\mathfrak{l}^{n-2})$ is the usual dévissage argument, similar to the counting arguments we did above. Note that now the commutativity of all the matrices follows. □

3.1.1. *Symmetry.* Our next goal is to find conditions that guarantee the symmetry of the Brandt matrices. This will later allow us to pass from a directed graph to an undirected graph.

Proposition 3.3. *Let $L_1,L_2,\dots,L_n$ be totally real fields. There is a positive density of rational primes p such that for every superspecial order R of B_{p,L_i} we have $R^\times=\mathcal{O}_L^\times$.*

PROOF. Let L be a totally real field. Let R be a maximal order of $B_{p,L}$ and suppose that $R^\times \supsetneqq \mathcal{O}_L^\times$. Let $\alpha\in R^\times-\mathcal{O}_L^\times$. Then α defines a CM field $L(\alpha)=L[x]/(x^2-(\alpha+\bar\alpha)x+\alpha\bar\alpha)$ with an embedding into $B_{p,L}$. Note that α is an algebraic integer.

Let us now consider CM fields $M\supset L$ and the group of units $\mathcal{O}_M^\times\supset\mathcal{O}_L^\times$. There is an exact sequence

$$1\to\mathcal{O}_L^\times\to\mathcal{O}_M^\times\to W_M,$$

where W_M is the roots of unity in M. The map $\mathcal{O}_M^\times\to W_M$ is given by $\alpha\mapsto\alpha/\bar\alpha$. According to [41, Theorem 4.12], we have

$$[\mathcal{O}_M^\times{:}W_M\mathcal{O}_L^\times]=\begin{cases}2 & \text{if } \mathcal{O}_M^\times\twoheadrightarrow W_M,\\ 1 & \text{otherwise.}\end{cases}$$

We need the following lemma:

Lemma 3.4. *There are finitely many such CM extensions M of L with $W_M=\{\pm1\}$ and $[\mathcal{O}_M^\times:\mathcal{O}_L^\times]=2$. These extensions can be effectively enumerated.*

PROOF OF LEMMA 3.4. If M is such an extension then there is a unit α in M such that $\alpha/\bar\alpha=-1$ and so $M=L(\alpha)$ and α has minimal polynomial over L given by $x^2+\alpha\bar{\alpha}$. The discriminant of M is thus bounded by $\mathrm{Norm}(\mathrm{disc}(\alpha))\,\mathrm{disc}(L/\mathbb{Q})^2 = 4^{[L:\mathbb{Q}]}\,\mathrm{disc}(L/\mathbb{Q})^2$. By Hermite – Minkowski (which is effective) there are only finitely many such fields M. □

We return to the proof of the proposition. Let us now consider all CM fields M containing L such that $\mathcal{O}_M^\times \supsetneqq \mathcal{O}_L^\times$. If $W_M \neq \{\pm 1\}$ then $M = L\cdot\mathbb{Q}(W_M)$ and $\varphi(|W_M|)$ divides $2[L:\mathbb{Q}]$. This shows that there are only finitely many such fields M (that can be effectively enumerated). If $W_M = \{\pm 1\}$ then by Lemma 3.4 there are only finitely many fields M such that $[\mathcal{O}_M^\times : \mathcal{O}_L^\times] = 2$.

Now, there is a positive density of prime ideals p such that p splits completely in the finitely many CM fields M such that $[M{:}L_i] = 2$ for some $1 \leq i \leq n$ and $\mathcal{O}_M^\times \supsetneqq \mathcal{O}_{L_i}^\times$. If A is an abelian variety with CM by an order of such a field M then A is ordinary. Indeed, by passing to an isogenous abelian variety we may assume that $\mathcal{O}_M \subseteq \mathrm{End}(A)$ and that A is obtained as a reduction from characteristic 0, hence that $H^1_{\mathrm{crys}}(A/W(\overline{\mathbb{F}}_p))$ is a free $\mathcal{O}_M \otimes_{\mathbb{Z}} W(\overline{\mathbb{F}}_p)$-module of rank 1. Let σ be the Frobenius homomorphism on $\overline{\mathbb{F}}_p$ and consider the corresponding decomposition of $H^1_{\mathrm{dR}}(A/\overline{\mathbb{F}}_p)$ as $\bigoplus H^1_{\mathrm{dR}}(A/\overline{\mathbb{F}}_p)^\chi$, where χ runs over all ring homomorphisms $\mathcal{O}_M \to \overline{\mathbb{F}}_p$ and $\mathcal{O}_M$ acts on $H^1_{\mathrm{dR}}(A/\overline{\mathbb{F}}_p)^\chi$ by χ. The absolute Frobenius Fr induces a σ-linear map on $H^1_{\mathrm{dR}}(A/\overline{\mathbb{F}}_p)$ taking $H^1_{\mathrm{dR}}(A/\overline{\mathbb{F}}_p)^\chi$ to $H^1_{\mathrm{dR}}(A/\overline{\mathbb{F}}_p)^{\sigma\circ\chi}$. The information on $H^1_{\mathrm{crys}}(A/W(\overline{\mathbb{F}}_p))$ implies that each $H^1_{\mathrm{dR}}(A/\overline{\mathbb{F}}_p)^\chi$ is one dimensional. Since p is split completely we get in our case that $\sigma\circ\chi = \chi$ for every χ. One concludes that on a subset Φ of g homomorphisms χ (the "CM-type") the map Fr is the zero map and that it is an isomorphism on each χ-typical component for $\chi \notin \Phi$. Moreover, the same considerations applied to the Verschiebung map show that we have a corresponding decomposition of group schemes $A[p] = \bigoplus_\chi A[p]^\chi$, where each $A[p]^\chi$ is a group scheme of rank p. It follows that the group schemes $A[p]^\chi$ for $\chi \notin \Phi$ are étale and so that A is ordinary.

We conclude that for such primes, for every superspecial order R in B_{p,L_i} we have $R^\times = \mathcal{O}_L^\times$. □

Remark 3.5. If in Proposition 3.3 the list of fields consists only of the field $\mathbb{Q}$, then the fields arising in its proof are the cyclotomic field $\mathbb{Q}(i)$ and $\mathbb{Q}(\omega)$, and the primes guaranteed in the Proposition are just the primes p congruent to 1 modulo 12, which is the condition for p to split in both $\mathbb{Q}(i)$ and $\mathbb{Q}(\omega)$.

3.2. Superspecial graphs. Let $\mathfrak{l}$ be a prime ideal of L not dividing p. Let h be the class number of a superspecial order R in $B_{p,L}$. The Brandt matrix $\mathrm{B}(\mathfrak{l})$ defines a directed graph on h vertices indexed by ideal classes $I_1,\dots,I_h$ of R. We denote this graph by $G(L;p,\mathfrak{l})$ and call it a *superspecial graph*. By our previous results it can be viewed as the graph of isogenies between L-superspecial abelian varieties in characteristic p (modulo a suitable equivalence relation) and that explains our choice of terminology.

Proposition 3.6. *The graph $G(L;p,\mathfrak{l})$ has the following properties*:

(1) *It is a directed graph which is regular of degree $l^{f(\mathfrak{l}/l)}+1$.*

(2) *For fixed g, the number of vertices h is approximately of the order*

$$2^{1-g}p^g\cdot|\zeta_L(-1)|.$$

More precisely, let

$$H = 2^{1-g} \cdot |\zeta_L(-1)| \cdot \prod_{\substack{\mathfrak{p}|p \\ f(\mathfrak{p})\ odd}} (\mathrm{Norm}(\mathfrak{p}) - 1) \prod_{\substack{\mathfrak{p}|p \\ f(\mathfrak{p})\ even}} (\mathrm{Norm}(\mathfrak{p}) + 1).$$

Then

$$H \leq h \leq C(g)H,$$

where

$$C(g) = \begin{cases} 2^{g+3}g^2 & g \geq 3 \\ 240 & g = 2 \\ 48 & g = 1. \end{cases}$$

(In fact, for $g = 1$ *we have* $h = [p/12] + \epsilon_p$ *where* ϵ_p*, where* $\epsilon_2 = \epsilon_3 = 1$ *and for* $p \geq 5$ *we have* $\epsilon_p = 0, 1, 1, 2$ *if* $p \equiv 1, 5, 7, 11 \pmod{12}$*).)*

(3) *The number of edges from* I_i *to* I_j *is the number of integral ideals* J *of* $\mathcal{O}_i$ *such that* $\mathrm{Norm}(J) = l$ *and* $J \sim I_i I_j^{-1}$.

(4) *There is a positive density of primes* p *such that the Brandt matrices* $\mathrm{B}(\mathfrak{n})$ *(relative to* p *and* L*) are symmetric for every* $\mathfrak{n}$.

PROOF. This follows immediately from Propositions 3.2, 3.3, except for the estimate for the class number, which we proceed to explain. The order R is an Eichler order of discriminant p, which we write as $D \cdot N$, where D is the discriminant of $B_{p,L}$ (so $D = \prod_{\{i: f(\mathfrak{p}_i/p) \text{ is odd}\}} \mathfrak{p}_i$). Using [40, Chapter V, Corollary 2.3], we find the following mass formula:

$$\sum_{i=1}^{h} \frac{1}{[\mathcal{O}_i^* : \mathcal{O}_L^*]} = 2^{1-g} \cdot |\zeta_L(-1)| \cdot \prod_{\mathfrak{p}|D} (\mathrm{Norm}(\mathfrak{p}) - 1) \prod_{\mathfrak{p}|N} (\mathrm{Norm}(\mathfrak{p}) + 1).$$

The right-hand side is of the order $2^{1-g}p^g \cdot |\zeta_L(-1)|$. To analyze the left hand side we let $\mu(R)$ denote the torsion subgroup of $R^\times$. If $u \in \mu(R)$ then, viewed as an element of the field $L(u)$, u is a root of unity. One has an exact sequence

$$1 \to \mu(R) \to R^\times \xrightarrow{\mathrm{Norm}} \mathcal{O}_L^\times,$$

which induces an inclusion $R^\times/\mathcal{O}_L^\times \mu(R) \subseteq \mathcal{O}_L^\times/\mathcal{O}_L^{\times,2}$. Thus, $|R^\times/\mathcal{O}_L^\times| \leq 2^{g-1}|\mu(R)|$. By the discussion in Section 3.1, any element of $\mu(R)$ is a root of unity of order r with $\phi(r) \mid 2g$. Since $\phi(r) \geq \sqrt{r}/2$, this implies that r is less than $8g^2$ (it is easy to improve on that for any given g). Now, since the quaternion algebra is ramified at infinity, $\mu(R)/\pm 1 \subseteq \mathbb{H}_\mathbb{R}/\pm 1 \cong \mathrm{SO}_3(\mathbb{R})$. If the image is dihedral of order $2n$ or cyclic of order n then $\mu(R)$ will contain an element of order $2n$ and so in this case $|\mu(R)| \leq 4n \leq 16g^2$. Else, the image is isomorphic to A_4, S_4 or A_5 and so of order at most 60 and $|\mu(R)| \leq 120$. One can now deduce the upper bound for $g \geq 3$.

The case $g = 1$ is classical and the formula is in [34, Section V.4.1]. However, in the spirit of the argument above, we get the bound 48 for the size of $\mu(R)$, though the correct bound is 24, using that every root of unity must have order $1, 2, 3, 4$ or 6 in this case, thus ruling out the A_5 case. For $g = 2$, roots of unity can have order r with $\phi(r) \mid 4$ implying $r = 1, 2, 3, 4, 5, 6, 8, 10$. Our methods give only the bound $|\mu(R)| \leq 120$. □

3.3. Hecke operators for $B_{p,L}$. The Ramanujan conjecture is often phrased and proven in the language of automorphic representations. To use this literature we connect the Hecke operator at $\mathfrak{l}$ defined in the language of Brandt matrices as $\mathrm{B}(\mathfrak{l})$ with a Hecke operator at $\mathfrak{l}$ defined in adelic language.

Let $\mathfrak{l}$ be a prime of L. We define another $h \times h$ matrix $C(\mathfrak{l})$ by the following data. Fix a superspecial order R. Let $\mathscr{B}$ be the algebraic group over $\mathcal{O}_L$ such that for every $\mathcal{O}_L$-algebra S we have $\mathscr{B}(S) = (R \otimes_{\mathcal{O}_L} S)^\times$. Thus, for example, $\mathscr{B}(\mathcal{O}_L) = R^\times$ and $\mathscr{B}(L) = B_{p,L}^\times$. The right ideal classes for R are in natural correspondence with $\mathscr{B}(L)\backslash\mathscr{B}(\mathbb{A}_{L,f})/\mathscr{B}(\widehat{\mathcal{O}_L})$, where $\mathbb{A}_{L,f}$ is the ring of finite adèles of L and $\widehat{\mathcal{O}_L}$ is its maximal open-compact subring, the profinite completion of $\mathcal{O}_L$.

We may therefore view the complex valued functions on the superspecial points as functions on the double cosets space $\mathscr{B}(L)\backslash\mathscr{B}(\mathbb{A}_{L,f})/\mathscr{B}(\widehat{\mathcal{O}_L})$. Let $\mathfrak{l} = (\pi_\mathfrak{l})$, $a_\mathfrak{l}$ the adèle of $\mathscr{B}(\mathbb{A}_{L,f})$ whose components are 1 at every place different from $\mathfrak{l}$ and $\left(\begin{smallmatrix}\pi_\mathfrak{l} & \\ & 1\end{smallmatrix}\right)$ at $\mathfrak{l}$. Consider the double coset $Ua_\mathfrak{l}U = \coprod x_i U$, where $U = \mathscr{B}(\widehat{\mathcal{O}_L})$. The Hecke operator $T_\mathfrak{l}$ is now defined as the averaging operator

$$f \mapsto T_\mathfrak{l}(f), \quad T_\mathfrak{l}(f)(x) = \sum_i f(xx_i).$$

Lemma 3.7. *The operator $T_\mathfrak{l}$ with respect to the basis consisting of δ functions is equal to the Brandt matrix* $\mathrm{B}(\mathfrak{l})$.

PROOF. We choose representatives $x_1, \ldots, x_h$ for $\mathscr{B}(L)\backslash\mathscr{B}(\mathbb{A}_{L,f})/\mathscr{B}(\widehat{\mathcal{O}_L})$ corresponding to the right ideal classes $I_1, \ldots, I_h$. The component at a place $\mathfrak{q}$ of I_i is $(x_i)_\mathfrak{q} R_\mathfrak{q}$. Identifying a function f with the column vector ${}^t(f(x_1), \ldots, f(x_h))$, the operator defined by $Ua_\mathfrak{l}U$ is given by the matrix $C(\mathfrak{l})$ whose ijth entry is the number of x_n such that the ideal associated to $x_i x_n$ is in the same ideal class as x_j.

The elementary divisors theorem implies that the decomposition $Ua_\mathfrak{l}U = \coprod x_i U$ can be done so that the representatives x_i are the image of the $l^{f(\mathfrak{l})} + 1$ matrices in $\mathrm{GL}_2(L_\mathfrak{l})$ given by $\left(\begin{smallmatrix}1 & \\ & \pi_\mathfrak{l}\end{smallmatrix}\right)$ and $\left(\begin{smallmatrix}\pi_\mathfrak{l} & i \\ & 1\end{smallmatrix}\right)$, where i runs over a set of representatives for $\mathcal{O}_L/\mathfrak{l}$. Note that this gives us the set of all left ideals $J \lhd R$ of norm $\mathfrak{l}$. Let us denote the ideal corresponding to J_i by x_i. Since all superspecial orders are locally conjugate, this also produces such a set for any superspecial order. We may therefore concern ourselves just with the order R, i.e., just with the point x corresponding to the trivial ideal class R. The effect of passing from x to xx_i is passing to the ideal class of J_i. The number of times we get the ideal class I_j is the number of ideals J_i such that $J_i \sim I_j$, which is exactly the $1i$ entry in the Brandt matrix $\mathrm{B}(\mathfrak{l})$. □

4. Properties of superspecial graphs

In this section we investigate the properties of the graphs $G(L; p, \mathfrak{l})$ we have constructed. In particular, we prove that they are Ramanujan graphs and that they are "nested" in a suitable sense. In proving the Ramanujan property we use what has become a common technique: we deduce it from the Ramanujan conjecture for a suitable class of automorphic representations.

4.1. Connectivity.

Theorem 4.1. *The graph $G(L; p, \mathfrak{l})$ is connected.*[1]

[1]This also follows from the Ramanujan property in a straight-forward manner; one just looks at the expansion of the set of vertices of a connected component of minimal size.

PROOF. We fix a maximal order and describe the vertices of the graph as the right ideal classes $I_1, \dots, I_h$ for this order. The problem translates into showing that the quadratic form Norm_c on $I_j I_i^{-1}$ represents λ^n for some $n \gg 0$, where λ is a totally positive generator of $\mathfrak{l}$. To see why, we find it convenient to think geometrically. Using Propositions 2.6, 3.2, we conclude that to show that for every i, j there is a path from I_i to I_j in the graph it is enough to show that the entry $\mathrm{B}(\mathfrak{l}^n)_{ij}$ is not zero. That is, it is enough to show that for every i, j there is an isogeny between the abelian varieties corresponding to I_i, I_j, whose kernel H satisfies $\deg_L(H) = \mathfrak{l}^n$. Such an isogeny corresponds to an element of $I_j I_i^{-1}$ with Norm_c a totally positive generator of $\mathfrak{l}^n$.

One possible proof is to consider the associated theta series, writing it as a sum of Eisenstein series and cusp forms to deduce that eventually all its coefficients are positive integers. Another proof can be obtained using the strong approximation theorem. See Section 5.3. We choose to appeal to a theorem of quadratic modular forms in 4 variables over $\mathcal{O}_L$. The theorem states that if such a quadratic form locally represents every element of $\mathcal{O}_{L\mathfrak{q}}$ for every prime ideal $\mathfrak{q}$ of $\mathcal{O}_L$ then it represents every totally positive element of $\mathcal{O}_L$ that is large enough.

We now verify the local conditions (cf. [11]) — that there are no local obstructions. The ideal $I_i I_j^{-1}$ is locally principal everywhere and in fact an isomorphism with the trivial local ideal can be chosen such that the function Norm_c becomes the norm form on $R_\mathfrak{q}$. All superspecial orders being locally conjugate, we may moreover assume that $R = \mathcal{O} \otimes \mathcal{O}_L$, where $\mathcal{O}$ is a maximal order in $B_{p,\infty}$. If $\mathfrak{q}$ doesn't divide p then $R_\mathfrak{q} \cong M_2(\mathcal{O}_{L\mathfrak{q}})$ with the norm being the determinant and we are done. If $\mathfrak{q}$ divides p then either we are dealing with an Eichler order of conductor $\mathfrak{q}$ in $M_2(\mathcal{O}_{L\mathfrak{q}})$, or with the maximal order of the unique division quaternion algebra over $\mathcal{O}_{L\mathfrak{q}}$. In the first case, the order is conjugate to $\left\{\left(\begin{smallmatrix} a & b \\ \pi c & d\end{smallmatrix}\right) : a, b, c, d \in \mathcal{O}_{L\mathfrak{q}}\right\}$, where π is a uniformizer of $\mathcal{O}_{L_\mathfrak{q}}$. The norm form is just the determinant and clearly there are no local obstructions. In the second case, we can represent $\mathcal{O}_\mathfrak{q}$ as the ring $\left\{\left(\begin{smallmatrix} a & b \\ -qb^\sigma & a^\sigma\end{smallmatrix}\right) : a, b \in W(\mathbb{F}_{q^2})\right\}$, where q is the rational prime below $\mathfrak{q}$ and σ is the Frobenius automorphism. Note that being in this case implies that $f(\mathfrak{q}/q)$ is odd and so $W(\mathbb{F}_{q^2}) \otimes \mathcal{O}_{L\mathfrak{q}} \cong W(\mathbb{F}_{q^{2f}})$, where $f = f(\mathfrak{q}/q)$, and $\mathrm{Gal}(\mathbb{F}_{q^2}/\mathbb{F}_q) = \mathrm{Gal}(\mathbb{F}_{q^{2f}}/\mathbb{F}_{q^f}) = \{1, \sigma\}$. We conclude that we need to deal with the order $\left\{\left(\begin{smallmatrix} a & b \\ -qb^\sigma & a^\sigma\end{smallmatrix}\right) : a, b \in W(\mathbb{F}_{q^{2f}})\right\}$. Again, the norm form is just the determinant and we need to show that one can write $\lambda \in W(\mathbb{F}_{q^f})^\times$ as $aa^\sigma - qbb^\sigma$. In fact, one can take $b = 0$ by local class field theory. We see that there are no local obstructions. □

4.2. The Ramanujan property. We assume now that the suitable conditions as in Proposition 3.6(4) hold so that in particular the Brandt matrix $\mathrm{B}(\mathfrak{l})$ is symmetric for every $\mathfrak{l}$ and the units of every superspecial order are just $\mathcal{O}_L^\times$. The superspecial graphs $G(L; p, \mathfrak{l})$ (associated to a totally real field L of strict class number one, a rational prime p unramified in L and a prime $\mathfrak{l}$ of L) are then well-defined.

Theorem 4.2. *The graphs $G(L; p, \mathfrak{l})$ are (connected) Ramanujan graphs of degree $\sharp\mathbb{P}^1(\mathcal{O}_L/\mathfrak{l})$. Let l be a rational prime and let $\mathfrak{l}_1, \dots, \mathfrak{l}_a$ be prime ideals dividing l in $\mathcal{O}_L$. Let $G(L; p, \mathfrak{l}_1, \dots, \mathfrak{l}_a)$ be the graph whose adjacency matrix is $\mathrm{B}(\mathfrak{l}_1) + \mathrm{B}(\mathfrak{l}_2) + \dots + \mathrm{B}(\mathfrak{l}_a)$ then $G(L; p, l)$ is a (connected) graph of degree $d = \sum_{i=1}^a (l^{f(\mathfrak{l}_i)} + 1)$ whose eigenvalues are d and the rest are bounded in absolute value by $2\sum_{i=1}^a \sqrt{l^{f(\mathfrak{l}_i)}}$.*

PROOF. There are two ways to argue. The first approach uses the Jacquet–Langlands correspondence to connect the Brandt matrices with Hecke operators on a Hilbert modular group. The second approach, which is historically a precursor of the first, uses theta series to do the same. Eventually, the Ramanujan property follows from the Ramanujan property for a suitable space of Hilbert modular forms and is due, in the case we need, to Livné [23].

FIRST PROOF. Let R be a maximal order of $B_{p,L}$. Consider the space S_R of complex functions on the double coset space $\mathscr{B}(L)\backslash\mathscr{B}(\mathbb{A}_{L,f})/\mathscr{B}(\widehat{\mathcal{O}_L})$ in the notation of Section 3.3, endowed with the action of the prime-to-p Brandt matrices. The Brandt matrices act either via the identification of the double coset space with the L-superspecial abelian varieties in characteristic p, or, equivalently, through the identification of the Brandt matrix $\mathrm{B}(\mathfrak{l})$ with the adelic Hecke operator $T_{\mathfrak{l}}$ defined by the double coset $Ua_{\mathfrak{l}}U$ (*loc. cit.*).

The argument now is as in [16–18]; see also [7, 8, 35]. Consider the subspace of S_R consisting of functions f such that, for $x \in B(L\otimes_{\mathbb{Q}}\mathbb{A}_f)$, $f(x)$ depends only on $\mathrm{Norm}(x) \in \mathbb{A}_{L,f}^{\times}/(\mathcal{O}_L\otimes\widehat{\mathbb{Z}_f})^{\times}$. Since $\mathrm{Norm}\colon B_{p,L}^{\times} \to L^{\times,+}$ is surjective and $\mathbb{A}_{L,f}^{\times} = L^{\times,+}(\mathcal{O}_L\otimes\widehat{\mathbb{Z}_f})^{\times}$ (due to strict class number 1) we conclude that such a function f is constant. On the other hand, consider the space S_L of Hilbert modular newforms of level $\Gamma_0(p)$ on $\mathrm{PSL}_2(L)$ of weight 2. The Jacquet–Langlands correspondence (see [10, 15]) gives an isomorphism $S_R/\mathbb{C} \to S_L$, which is Hecke-equivariant. In particular, the eigenvalues of $\mathrm{B}(\mathfrak{l})$, besides the eigenvector $(1,\dots,1)$ are precisely those of the Hecke operator $T_{\mathfrak{l}}$. By the Ramanujan conjecture for such Hilbert modular forms (see [23]), we conclude that the graph defined by $\mathrm{B}(\mathfrak{l})$ is Ramanujan.

SECOND PROOF. The second proof makes use of Eichler's result on theta series and Hecke operators [9]. L.c. assumes that the order is maximal, however the results carry through with the obvious modifications (the level of the theta series changes of course, however as long as one avoids the Hecke operators at primes dividing p everything goes through). The Brandt matrices appearing in l.c. §7 (25)-(26) agree with the Brandt matrices as defined in this paper. The theta functions of l.c. §9 (for the constant function 1 as a spherical polynomial) are now holomorphic weight 2 Hilbert modular forms of level $\Gamma_0(p)$ (we note that since the strict class number is 1 we also have that the groups $\mathrm{SL}_2(\mathcal{O}_L)$ and $\mathrm{SL}(\mathcal{O}_L\oplus\mathfrak{d}^{-1}) = \left\{\left(\begin{smallmatrix}\alpha&\beta\\ \gamma&\delta\end{smallmatrix}\right) : \alpha,\delta\in\mathcal{O}_L, \beta\in\mathfrak{d}^{-1}, \gamma\in\mathfrak{d}\right\}$ are conjugate in $\mathrm{SL}_2(L)$). Theorem 7 of l.c. implies that the spectrum of $\mathrm{B}(\mathfrak{l})$ in its action on the theta series thus constructed is that of $V_2(\mathfrak{l}^{-1})T_2(\mathfrak{l})$ in the notation of that reference. Here $T_2(\mathfrak{l})$ is the Hecke operator $T_2(\mathfrak{l},\mathfrak{J}/\mathfrak{J}\mathfrak{l}^{-1})$ defined there in (14) and $V_2(\mathfrak{l}^{-1})$ is $V_2(\mathfrak{l},\mathfrak{J}/\mathfrak{J}\mathfrak{l}^{-2})$ of (17), which by l.c., Proposition 3, is trivial in the case of strict class number one. It remains to verify that the Hecke operators used in l.c. have the correct normalization (see, e.g., [39, Section VI.1]) leading to an Euler product associated to a normalized eigenform, and that is a simple matter of comparing definitions. □

4.3. "Cleaner" Ramanujan graphs. Following Pizer, for a given field L, we shall explain here how to get simple undirected graphs $G(L;p,l)$ by posing conditions on the prime p. In Proposition 3.6 we saw that if p splits in finitely many quadratic CM extensions of L then the Brandt matrices are symmetric. We shall find below conditions that guarantee that there are no loops or multiple edges;

those conditions are the requirement that p splits in finitely many number fields M_i that depend on l. It will thus follow that there are infinitely many primes $\mathfrak{l} \lhd \mathcal{O}_L$ such that the graphs defined by the Brandt matrix $\mathrm{B}(\mathfrak{l})$ relative to $B_{p,L}$ are simple Ramanujan graphs.

Instead of developing a criterion in the greatest generality, we take a particular example where $l = 2$ is inert in L and L is quadratic . The reader will readily see that the argument generalizes for any prime $\mathfrak{l} \nmid p$, where l may have any splitting behavior and L need not be quadratic.

We consider the Brandt matrix $\mathrm{B}(2)$. We assume p already satisfies congruences so that $\mathrm{B}(2)$ is symmetric and we thus get a $5 = 2^2 + 1$ regular graph. To say that there are no self loops is to say that there are no isogenies $f\colon A \to A$ of degree 4, for any L-superspecial abelian surface A in characteristic p. If f is such an isogeny then $f \notin \mathcal{O}_L$ and so f induces an embedding of a CM order $\mathcal{O}_L[x]/(x^2 - (f + \bar{f}) + 2\epsilon) \hookrightarrow \mathrm{End}(A)$. Note that $\epsilon \in \mathcal{O}_L^{\times,+}$ and that we may replace f by uf for $u \in \mathcal{O}_L^{\times}$. We may thus assume that we have the orders $\mathcal{O}_L[x]/(x^2 - (f + \bar{f}) + 2) \hookrightarrow \mathrm{End}(A)$, whose field of quotients are CM fields, quadratic over L. This implies that $(f + \bar{f})^2 - 8$ is totally negative and so the element $a(f) = (f + \bar{f})^2$ is a totally positive element of $\mathcal{O}_L$ which is bounded by 8 under any embedding into $\mathbb{R}$. It follows that, given L, there are only finitely many such orders arising, regardless of p. Let K_i be their fields of quotients.

Multiple edges, say $f_1, f_2\colon A \to A'$ of degree 4, imply an endomorphism $f_2^t f_1 \in \mathrm{End}(A)$ of degree $16 = 2^4$, which is verified to be not multiplication by 2. The same argument as above gives a finite list of CM orders that may arise out of $f_2^t f_1$ and we let N_i be their field of quotients.

Fixing L, we have a positive density of primes splitting in all the finitely many fields $\{M_i\}, \{K_i\}, \{N_i\}$. For such a prime p we get a simple undirected 5-regular graph $G(L; p, 2)$.

4.4. Sequences of Ramanujan graphs. An interesting feature of our construction is the existence of natural maps between the graphs we have constructed. For every inclusion of totally real fields $L \subset M$, a rational prime p and prime ideals $\mathfrak{m}_1\mathfrak{m}_2 \cdots \mathfrak{m}_a = \mathfrak{l}$, where $\mathfrak{l}$ is a prime of L and $\mathfrak{m}_i$ of M, there is a natural "map"

$$G(L; p, \mathfrak{l}) \to G(M; p, \mathfrak{m}_1, \mathfrak{m}_2, \ldots, \mathfrak{m}_a).$$

The map is canonical on vertices. If A is an abelian variety with RM by L then $A \otimes_{\mathcal{O}_L} \mathcal{O}_M$ is an abelian variety with RM by M. In the language of quaternion algebras, once we have fixed a superspecial order R of $B_{p,L}$ and a right ideal I of R, we get a superspecial order $R' = R \otimes_{\mathcal{O}_L} \mathcal{O}_M$ and an ideal $I' = I \otimes_{\mathcal{O}_L} \mathcal{O}_M$ of R'. This process is compatible with calculating left orders. If $\phi : A \to A'$ is an $\mathcal{O}_L$-isogeny with $(\deg_L(\phi)) = \mathfrak{l}$ then $\phi \otimes 1 : A \otimes_{\mathcal{O}_L} \mathcal{O}_M \to A \otimes_{\mathcal{O}_L} \mathcal{O}_M$ is an $\mathcal{O}_M$-isogeny with $(\deg_M(\phi)) = \mathfrak{l}\mathcal{O}_M$. We can decompose this isogeny into a sequence of $\mathcal{O}_M$-isogenies $\psi_1 \circ \cdots \circ \psi_a$ so that $(\deg_M \psi_i) = \mathfrak{m}_i$. Thus, the single edge from A to A' is replaced by a path of length a from $A \otimes_{\mathcal{O}_L} \mathcal{O}_M$ to $A' \otimes_{\mathcal{O}_L} \mathcal{O}_M$. In that sense we have a map of graphs $G(L; p, \mathfrak{l}) \to G(M; p, \mathfrak{m}_1, \mathfrak{m}_2, \ldots, \mathfrak{m}_a)$. Note that if $a = 1$, that is, if $\mathfrak{l}$ is inert in M, then this is a map of graphs in the usual sense.

The question of whether these maps of graphs are *injective* on vertices is more subtle. Consider such an extension of totally real fields $L \subset M$ and two L-superspecial abelian varieties A, A'. One approach could be to consider $\mathrm{Hom}_{O_M}(A \otimes_{O_L} O_M, A' \otimes_{O_L} O_M)$ with the O_M-valued degree map. This module is isomorphic to $\mathrm{Hom}_{O_L}(A, A') \otimes_{O_L} O_M$ with the degree map being the O_M-linear extension of the O_L degree map on $\mathrm{Hom}_{O_L}(A, A')$. If $A \otimes_{O_L} O_M \cong A' \otimes_{O_L} O_M$, then the lattices $\mathrm{End}_{O_L}(A)$ and $\mathrm{Hom}_{O_L}(A, A')$ become isomorphic after extension to O_M. Such issues are considered in [20] in detail for the case $L = \mathbb{Q}$. Based on the discussion there (see also the remarks in the end of the book), it doesn't seem far-fetched to ask whether such an isomorphism implies that $\mathrm{End}_{O_L}(A)$ is isomorphic to $\mathrm{Hom}_{O_L}(A, A')$. If so, this implies that 1 is represented by the degree map on $\mathrm{Hom}_{O_L}(A, A')$ and so that A and A' are isomorphic as L-superspecial abelian varieties. This, in turn, implies that the map of superspecial graphs $G(L; p, l) \to G(M; p, l)$ is injective on vertices. This question, though interesting and relevant to our topic, is a subject for independent research, as the discussion in [20] (which only deals with $L = \mathbb{Q}$) indicates.

5. Families of nested Ramanujan graphs

Our discussion in this section does not claim to be exhaustive. Our purpose is to discuss certain nested families of Ramanujan graphs, in particular we consider our superspecial graphs from another point of view, and pose questions that we find intriguing.

5.1. Paley graphs. Let $\mathbb{F}_q$ be a finite field of order $q \equiv 1 \pmod 4$. The vertices of the Paley graph $P(q)$ are the elements of $\mathbb{F}_q$, and $x \neq y$ are connected if $x-y$ is a square in $\mathbb{F}_q$. This is a $(q-1)/2$ regular graph on q vertices. Recall that a graph is strongly regular with parameters (k, a, b) if it is a k-regular incomplete graph, any two adjacent vertices have $a \geq 0$ common neighbors and any two nonadjacent vertices have $b \geq 1$ common neighbors. One can show that a Paley graph is a strongly regular graph with q vertices and parameters $((q-1)/2, (q-5)/4, (q-1)/4)$. Cf. [24, Section 8.3]. The eigenvalues of the adjacency matrix are therefore $(q-1)/2$ and the roots of $x^2 + x - (q-1)/4$, namely $(-1 \pm \sqrt{q})/2$. Note that these graphs beat the bound $2\sqrt{(q-1)/2 - 1}$; for $q \gg 0$ their ratio is about $2\sqrt{2}$.

We have an inclusion $P(q) \hookrightarrow P(q^n)$ for any n and so, for concreteness, we can take the chain of Ramanujan graphs $P(p) \to P(p^3) \to P(p^9) \to \cdots$. We make three remarks:

(1) The degree is very large compared to the size of the graph, a fact which renders this graph inappropriate for most applications.

(2) The possible advantage of the sequence $P(p) \to P(p^3) \to P(p^9) \to \cdots$ over, say, the sequence $P(p) \to P(p^2) \to P(p^4) \to \cdots$ is that any two vertices of $P(p^{2^i})$ become adjacent in $P(p^{2^{i+1}})$, while in $P(p) \to P(p^3) \to P(p^9) \to \cdots$ nonadjacent vertices remain nonadjacent. Since the diameter of all these graphs is 2 we conclude that the arrows in $P(p) \to P(p^3) \to P(p^9) \to \cdots$ are isometries.

(3) An even simpler construction can be made with the complete graphs K_n. There are (noncanonical) inclusions $K_1 \hookrightarrow K_2 \hookrightarrow K_3 \hookrightarrow \cdots$. Trivially, those inclusions are isometries. The graph K_n has eigenvalues $n-1$ and -1 with multiplicities 1 and $n-1$ respectively and so is Ramanujan for $n \geq 2$. The same comments as to the interest in those examples apply.

5.2. Terras graphs. We discuss the graphs defined by Terras. See [36, 37] and the references therein.

Consider a finite field $\mathbb{F}_q$ with q elements, q odd. Let $\delta \in \mathbb{F}_q$ be a nonsquare. Then $\mathbb{F}_{q^2} = \mathbb{F}_q \oplus \mathbb{F}_q\sqrt{\delta}$ as an $\mathbb{F}_q$-vector space. Terras defines the finite upper half plane $\mathfrak{H}_q$ as

$$\mathfrak{H}_q = \{x + y\sqrt{\delta} : x, y \in \mathbb{F}_q, y \neq 0\} = \mathbb{F}_{q^2} - \mathbb{F}_q.$$

The group $\mathrm{PGL}_2(\mathbb{F}_q)$ acts transitively on $\mathfrak{H}_q$ by $z \mapsto (az+b)/(cz+d)$ for $\left(\begin{smallmatrix} a & b \\ c & d \end{smallmatrix}\right) \in \mathrm{GL}_2(\mathbb{F}_q)$. Note that the action does not depend on δ. The stabilizer of $\sqrt{\delta}$ is $K_q = \{\left(\begin{smallmatrix} a & b\delta \\ b & a \end{smallmatrix}\right) : a, b \in \mathbb{F}_q, a^2 - b^2\delta \neq 0\}$. Thus $\mathfrak{H}_q = \mathrm{GL}_2(\mathbb{F}_q)/K_q$. The elements of $\mathfrak{H}_q$ are the vertices $V(T(q))$ of the Terras graph $T(q)$.

Define a function,

$$d\colon V(T(q)) \to \mathbb{F}_q, \quad d(z,w) = \frac{N(z-w)}{\mathrm{Im}(z)\,\mathrm{Im}(w)},$$

where $\mathrm{Im}(x + y\sqrt{\delta}) = y$, $\overline{x + y\sqrt{\delta}} = x - y\sqrt{\delta}$ and $N(z) = z\bar{z}$. Choose an element $a \in \mathbb{F}_q$, $a \neq 0, 4\delta$ (in characteristic different from 3, a canonical choice could be $a = \delta$). Define z, w to be adjacent if $d(z,w) = a$. One can show that this gives a $q+1$-regular graph $T(q) = T(q, \delta, a)$ on $q^2 - q$ vertices. The group $\mathrm{GL}_2(\mathbb{F}_q)$ acts as isometries on this graph. Such graphs were proven to be Ramanujan by Katz.

The neighbors of $\sqrt{\delta}$ are the set $S = \{x + y\sqrt{\delta} : x^2 = ay + \delta(y-1)^2\}$. One can show that $T(q)$ is the Cayley graph of the group $\left\{\left(\begin{smallmatrix} y & x \\ 0 & 1 \end{smallmatrix}\right) : x, y \in \mathbb{F}_q, y \neq 0\right\}$ with respect to the set of generators S. In [2, Theorems 1 and 2] the diameter of the graph $T(q, \delta, a)$ is determined: *If $a \notin \{0, 2\delta, 4\delta\}$ then the diameter is 3 if $\delta - a$ is a square in $\mathbb{F}_q$ and 4 otherwise. If $a = 2\delta$ then the diameter is 3 unless $q = 3, 5$ in which case the diameter is 2.*

Consider the question of which vertices in $T(q)$ are at distance 2 from a given vertex. Since $\mathrm{GL}_2(\mathbb{F}_q)$ acts transitively on the set of vertices, we may consider the question just for the vertex $\sqrt{\delta}$. Given a point $x_0 + y_0\delta$ we are interested in the set of points $x + y\sqrt{\delta}$ that solve the two equations:

$$(x - x_0)^2 - (y - y_0)^2\delta = ayy_0, \quad x^2 - (y-1)^2\delta = ay.$$

The number of such points is the number of distinct paths of length 2 from $x_0 + y_0\delta$ to $\sqrt{\delta}$.

Viewing x_0, y_0 as fixed and x, y as variables and homogenizing we get two quadratic curves in $\mathbb{P}^2$:

$$(x - x_0z)^2 - (y - y_0z)^2\delta = ay_0yz, \quad x^2 - (y-z)^2\delta = ayz.$$

By Bezout's theorem they intersect at 4 points (counted with multiplicity) on which $\mathrm{Gal}(\mathbb{F}_q^{\mathrm{alg}}/\mathbb{F}_q)$ act. At infinity, the curves intersect at the two points $\{(x : y : 0) : x^2 - \delta y^2 = 0\}$ which form a Galois orbit for $\mathrm{Gal}(\mathbb{F}_q^{\mathrm{alg}}/\mathbb{F}_q)$. It follows that there are the following possibilities for the intersection points that lie in $\mathbb{A}^2$ (i.e., solutions to the original system of equations) and no others:

(1) The intersection points in $\mathbb{A}^2$ are not defined over $\mathbb{F}_q$. In that case, there are two of them and they form a single $\mathrm{Gal}(\mathbb{F}_q^{\mathrm{alg}}/\mathbb{F}_q)$-orbit and are defined over $\mathbb{F}_{q^2}$.

(2) The intersection points in $\mathbb{A}^2$ are defined over $\mathbb{F}_q$. In this case, there are two different paths of length two from $x_0 + y_0\sqrt{\delta}$ to $\sqrt{\delta}$ if the intersection points are distinct, and one path if they are the same.

To fix ideas, assume now that $p > 3$. We may then choose $a = \delta \in \mathbb{F}_p$ a nonsquare and get a directed system of Terras graphs

$$T(p) \to T(p^3) \to T(p^9) \to \cdots.$$

The maps are injective on vertices and are isometries locally in the sense that if two vertices in $T(p^{3^n})$ are adjacent in $T(p^{3^{n+1}})$ then they are already adjacent in $T(p^{3^n})$. Since the diameter of all the graphs is 3 the maps are globally isometries. Indeed, if two points of $T(q)$, $q = p^{3^n}$, are in distance 2 they stay in distance 2 since the distance function d on $T(q^3)$ extends the one on $T(q)$. If they are at distance 3 in $T(q)$ they cannot be in distance 1 in $T(q^3)$ (for the same reason) and also cannot be in distance 2 in $T(q^3)$, because that would mean that the solutions $x + y\sqrt{\delta}$ appearing in the first conclusion above, are only defined over $\mathbb{F}_{q^2}$ (because the points are not in distance 2 in $T(q)$) yet belong to $\mathbb{F}(q^3)$, which is absurd.

Question. What is the limit distribution of the eigenvalues of $T(p^n)$ for fixed p and $n \to \infty$?

To the best of our knowledge this question is open. See [37] for some discussion of this and [19] for the failure of the "naive conjecture."

5.3. LPS graphs. The graphs defined by Lubotzky – Phillips – Sarnak [24,25] and Jordan – Livné [18], that we call *LPS graphs*, can be described conceptually as follows.

Let l be a prime and consider the Bruhat – Tits tree $\mathscr{T}_K$ of $\mathrm{PGL}_2(K)$, where K is a finite extension of $\mathbb{Q}_l$. Let $\mathcal{O}_K$ be its valuation ring, π_K a uniformizer and κ the residue field. The tree has vertices corresponding to lattices in K^2 modulo homothety. To describe a lattice up-to-homothety is to give a basis, modulo change of basis and then modulo re-scaling, and so the vertices correspond to $K^\times \backslash \mathrm{GL}_2(K)/\mathrm{GL}_2(\mathcal{O}_K) = \mathrm{PGL}_2(K)/\mathrm{PGL}_2(\mathcal{O}_K)$. We say that two classes of lattices are adjacent if we can find representatives L_0, L_1 such that $L_1 \subset L_0$ and $[L_0/L_1] \cong \kappa$ as $\mathcal{O}_K$ modules. If κ has l^f elements, since the lattices adjacent to L_0 are parameterized by $\mathbb{P}^1(\kappa)$, the tree is $l^f + 1$-regular. Also, to give an edge is to give a lattice L_0 together with a cyclic $\mathcal{O}_K$-module of order l^f in L_0/lL_0 and thus the edges are parameterized by $\mathrm{PGL}_2(K)/\overline{I(K)}$, where $I(K) = \left\{\left(\begin{smallmatrix} a & b \\ c & d \end{smallmatrix}\right) : a, b, c, d \in \mathcal{O}_K, c \in (\pi_K)\right\}$ is the standard Iwahori subgroup and $\overline{I(K)}$ its projection to $\mathrm{PGL}_2(K)$.

Trees are the best expanders one can hope for. They are infinite, though. One therefore looks for a subgroup $\Gamma \subset \mathrm{Aut}(\mathscr{T}_K)$ such that the quotient $\Gamma\backslash\mathscr{T}_K$ is a finite graph (and conversely, the universal covering space of any $l^f + 1$-regular tree is isomorphic to $\mathscr{T}_K$); to get a finite graph, Γ needs to be a discrete co-compact subgroup and there is an art to finding such groups. One method is to use quaternion algebras.

Let L be a number field, l a rational prime and $\mathfrak{l}|l$ an unramified prime factor of l in $\mathcal{O}_L$. For simplicity of exposition we assume L is totally real. Let B be a definite quaternion algebra over L, split at $\mathfrak{l}$, $K = L_\mathfrak{l}$. Let $\mathcal{O}$ be an order (over $\mathcal{O}_L$) of B and let $\Gamma = (\mathcal{O}[l^{-1}])^\times$. Let $\mathcal{O}^\dagger$ be a maximal order of B containing $\mathcal{O}$. Through an identification $B(L_\mathfrak{l}) \cong \mathrm{GL}_2(K)$, such that $\mathcal{O}^\dagger \otimes \mathcal{O}_{L_\mathfrak{l}} = \mathrm{GL}_2(\mathcal{O}_K)$, Γ is a discrete co-compact subgroup of $\mathrm{GL}_2(O_K)$ and one obtains a finite graph (possibly with multiple edges and self-loops) $\Gamma\backslash\mathscr{T}_K$. See [24, Section 7.3].

To illustrate the construction of *nested families* of LPS graphs, we make the simplifying assumption that $\mathcal{O}$ is a maximal order in $B_{p,\infty}$ — the rational quaternion algebra B ramified at p and ∞; the construction works for any definite quaternion algebra over $\mathbb{Q}$. For every field $L \supseteq \mathbb{Q}$, let $B_{p,L} = B_{p,\infty} \otimes_\mathbb{Q} L$. Choose a sequence of totally real fields $\mathbb{Q} \subset L_1 \subset L_2 \subset \cdots$ and a sequence of compatible primes $\cdots \mathfrak{l}_3 \mid \mathfrak{l}_2 \mid \mathfrak{l}_1 \mid l$ and assume that p is split and l is unramified in each of the fields L_i. Then the discriminant of B_{p,L_i} is $p\mathcal{O}_{L_i}$ and hence $\mathcal{O} \otimes_{\mathbb{Z}} \mathcal{O}_{L_i}$ is a maximal order of B_{p,L_i}. We let $\Gamma_i = (\mathcal{O} \otimes_{\mathbb{Z}} \mathcal{O}_{L_i}[l^{-1}])^\times$. Let $K_i = (L_i)_{\mathfrak{l}_i}$ then $\mathbb{Q}_l \subset K_1 \subset K_2 \subset \cdots$ is a sequence of unramified l-adic fields. There are canonical injections of trees $\mathscr{T}_{\mathbb{Q}} \hookrightarrow \mathscr{T}_{K_1} \hookrightarrow \mathscr{T}_{K_2} \hookrightarrow \cdots$, corresponding to the natural map of lattices $\Lambda \subset K_i^2 \mapsto \Lambda \otimes_{\mathcal{O}_{K_i}} \mathcal{O}_{K_{i+1}} \subset K_{i+1}^2$. In terms of double cosets, this is the map taking $K_i^\times \gamma \operatorname{GL}_2(\mathcal{O}_{K_i})$ to $K_{i+1}^\times \gamma \operatorname{GL}_2(\mathcal{O}_{K_{i+1}})$. If $K_{i+1}^\times \gamma_1 \operatorname{GL}_2(\mathcal{O}_{K_{i+1}}) = K_{i+1}^\times \gamma_2 \operatorname{GL}_2(\mathcal{O}_{K_{i+1}})$ then $\gamma_2^{-1}\gamma_1 = l\delta$ for some $l \in K_{i+1}^\times, \delta \in \operatorname{GL}_2(\mathcal{O}_{K_{i+1}})$. If $\gamma_1, \gamma_2 \in \operatorname{GL}_2(K_i)$ then $l\delta \in K_{i+1}^\times \operatorname{GL}_2(\mathcal{O}_{K_{i+1}}) \cap \operatorname{GL}_2(K_i)$. However, $K_{i+1}^\times \operatorname{GL}_2(\mathcal{O}_{K_{i+1}}) \cap \operatorname{GL}_2(K_i) = K_i^\times \operatorname{GL}_2(\mathcal{O}_{K_i})$. First, the inclusion $\supseteq$ is obvious. To see the other inclusion, write an element $l\delta$ as above in the form $\pi^n l'\delta$, where π is a uniformizer of K_i and $l' \in \mathcal{O}_{K_{i+1}}^\times$. It is enough to show that $l'\delta \in \operatorname{GL}_2(\mathcal{O}_{K_i})$, but $l'\delta \in \operatorname{GL}_2(\mathcal{O}_{K_{i+1}}) \cap \operatorname{GL}_2(K_i) = \operatorname{GL}_2(\mathcal{O}_{K_i})$. We therefore conclude that $K_i^\times \gamma_1 \operatorname{GL}_2(\mathcal{O}_{K_i}) = K_i^\times \gamma_2 \operatorname{GL}_2(\mathcal{O}_{K_i})$. Note that the map of trees that we get are not just injection on vertices. It is an isometry (namely, it is truly a map of graphs), because the extensions are unramified.

To get a map of graphs $\Gamma\backslash\mathscr{T}_{\mathbb{Q}} \hookrightarrow \Gamma_1\backslash\mathscr{T}_{K_1} \hookrightarrow \Gamma_2\backslash\mathscr{T}_{K_2} \hookrightarrow \cdots$, we first need that $\Gamma_{i+1} \cap \operatorname{GL}_2(K_i) \supseteq \Gamma_i$ and that is clear. Next note that if two vertices of $\mathscr{T}_{K_i}$ (i.e., two lattices in K_i^2, up to homothety) are equivalent under an element γ of Γ_{i+1} then we may suppose $\gamma \in \operatorname{GL}_2(K_i)$. To get actual injections $\Gamma_i\backslash\mathscr{T}_{K_i} \hookrightarrow \Gamma_{i+1}\backslash\mathscr{T}_{K_{i+1}}$, we at least need $(\mathcal{O} \otimes_{\mathbb{Z}} \mathcal{O}_{L_{i+1}}[l^{-1}])^\times \cap B_{L_i} = (\mathcal{O} \otimes_{\mathbb{Z}} \mathcal{O}_{L_i}[l^{-1}])^\times$, which is not hard to verify. This is not sufficient for injectivity, however, as pointed out in Section 4.4 above. Whatever the case may be, this construction gives a direct family of graphs $\Gamma\backslash\mathscr{T}_{\mathbb{Q}} \to \Gamma_1\backslash\mathscr{T}_{K_1} \to \Gamma_2\backslash\mathscr{T}_{K_2} \to \cdots$ that are all Ramanujan; The Ramanujan property follows from the deep results [23, Theorem 2.4; 24, *loc. cit.* and Corollary 5.5.3], using the fact that the fields are totally real. Finally, we remark that the assumption that p splits in all the fields L_i is not essential. One can simply assume that p is not ramified; the same arguments apply, the only difference being that the orders $\mathcal{O} \otimes \mathcal{O}_{L_i}$ are not necessarily maximal.

5.3.1. *The connection between the work of Lubotzky – Phillips – Sarnak and Pizer.* In this section we explain the relation between the work of Lubotzky – Phillips – Sarnak and the work of Pizer. This is well understood by the experts, though only cryptic remarks appear in the literature. We thus find it worthwhile to explain that, and in so doing to explain the relation between the constructions appearing in this paper and the work of Jordan – Livné.

Let L be a totally real field of strict class number 1, B a totally definite quaternion algebra over L and $\mathcal{O}$ a hereditary $\mathcal{O}_L$-order of B contained in a maximal order $\mathcal{O}^\dagger$. The order $\mathcal{O}$ gives us a ring scheme over $\mathcal{O}_L$ whose value for every $\mathcal{O}_L$-algebra S is $\mathcal{O} \otimes S$, and a group scheme $\mathscr{B}$ over $\mathcal{O}_L$ which is the units in the ring scheme. Let $\mathfrak{l}|l$ be a prime of L such that $(l, \operatorname{disc}_L) = 1$ and which splits B.

Since $\mathcal{O}$ is hereditary, every $\mathcal{O}$-ideal is locally principal and therefore the class group of $\mathcal{O}$ is given by $\mathscr{B}(L)\backslash\mathscr{B}(\mathbb{A}_{L,f})/\mathscr{B}(\mathcal{O}_L \otimes \widehat{\mathbb{Z}})$, where $\mathbb{A}_{L,f}$ are the finite adèles of L. The inclusion $L_\mathfrak{l} \to \mathbb{A}_{L,f}$ induces inclusions $\mathscr{B}(L_\mathfrak{l}) \hookrightarrow \mathscr{B}(\mathbb{A}_{L,f})$ and $\mathscr{B}(\mathcal{O}_L[l^{-1}])\backslash\mathscr{B}(L_\mathfrak{l})/\mathscr{B}(\mathcal{O}_{L_\mathfrak{l}}) \hookrightarrow \mathscr{B}(L)\backslash\mathscr{B}(\mathbb{A}_{L,f})/\mathscr{B}(\mathcal{O}_L \otimes \widehat{\mathbb{Z}})$. The last map is in fact a surjection. We need to show that $\mathscr{B}(\mathbb{A}_{L,f}) = \mathscr{B}(L)\mathscr{B}(L_\mathfrak{l})\mathscr{B}(\mathcal{O}_L \otimes \widehat{\mathbb{Z}})$. This follows from strong approximation. (This only requires that B is ramified at least in one place at infinity.) Recalling that B is split at $\mathfrak{l}$ we conclude a bijection

$$(1) \qquad \mathcal{O}[l^{-1}]^\times \backslash \mathrm{GL}_2(L_\mathfrak{l})/\mathrm{GL}_2(\mathcal{O}_{L_\mathfrak{l}}) \cong \mathscr{B}(L)\backslash\mathscr{B}(\mathbb{A}_{L,f})/\mathscr{B}(\mathcal{O}_L \otimes \widehat{\mathbb{Z}}) = \mathrm{Cl}(\mathcal{O}).$$

If $B = B_{p,\infty}$ is the rational quaternion algebra ramified at p and ∞ alone, $\mathcal{O}$ a maximal order, then $\mathrm{Cl}(\mathcal{O})$ is in bijection with supersingular elliptic curves. Taking a nonmaximal order gives the cases considered by Pizer (only that his orders are not necessarily hereditary). For L a totally real field of class number one and a prime p unramified in L, and the order $\mathcal{O}\otimes\mathcal{O}_L$ in $B_{p,L} := B_{p,\infty}\otimes_\mathbb{Q} L$, we have $\mathrm{Cl}(\mathcal{O}\otimes\mathcal{O}_L)$ is in bijection with the superspecial abelian varieties with real multiplication by $\mathcal{O}_L$. The bijection (1) is not only connecting the constructions of [18, 25] with ours, but also shows the connectivity of the graph of $\mathfrak{l}$-isogenies.

5.4. A question. The above examples motivate the following problem. Construct a long or infinite series of Ramanujan graphs $G_0 \to G_1 \to G_2 \to \cdots$, where G_i has n_i vertices and degree d_i, and the maps, if not injective, should at least be "nondegenerate" in some well-quantified sense. We are particularly interested in a construction meeting one or more of the following extra requirements:

- That d_i be bounded from above by $\log(n_i)^r$ for some positive r.
- That $n_i = o(i^{1+\epsilon})$ for some $\epsilon < 1$.
- That the morphisms $G_i \to G_{i+1}$ be isometries.
- That each G_i be a Cayley graph.
- That each G_i beat the Ramanujan bound, e.g., in the sense that we have $\lambda(G_i) < 2\sqrt{d_i - 1} - \epsilon$ for some $\epsilon > 0$ independent of i (or an even stronger requirement). We remark that for every d-regular simple graph G we have $\lambda(G) \geq \sqrt{d}\cdot\sqrt{(n-d)/(n-1)}$, which puts a bound on by how much one can beat the Ramanujan bound.

6. Applications and implementation

There are many interesting and well-known applications of Ramanujan graphs. Due to the fact that random walks on expanders (and hence on Ramanujan graphs) mix very rapidly, expanders are used as pseudo-random number generators, for approximating the average value of a function, and in the design of low-density parity check codes in coding theory. See [14] for a comprehensive exposition of the many applications in different branches of mathematics and computer science. Here we point out some new applications that are enabled by our construction, and raise some related questions.

6.1. Reliable networked storage. One (possibly new) application of graphs with good expansion properties is to use the graph to build a network of users who share storage of files and content on each others' machines, for example using network coding (see for example [6]). In this scenario, a network of participants is built by modeling the participants as nodes of a graph, and forming a participant's neighbor set (with whom it shares storage) as the set of neighbors of that node in the graph. Given an existing network of participants, one may wish to add a new collection of participants to the network while preserving the existing connections. For this purpose, a nested family of Ramanujan graphs could be used. Any user's files are then distributed to its neighbors via network coding, and subsequently stored. If a node fails, the data can be reconstructed from some subset of the participants through the error correcting capacity of network coding.

6.2. Cryptographic hash functions. In [5], the idea was proposed that Ramanujan graphs can also be used to construct cryptographic hash functions. Hash functions used in cryptographic protocols need to be efficiently computable and collision resistant, at a minimum. A hash function can be constructed from a graph by specifying a starting vertex, using the input to the hash function as directions for walking around the graph (without backtracking), and then returning the final vertex of the walk as the output of the hash function. If the graph has good expansion properties, the output of the hash function will appear random, since walks on expander graphs quickly approximate the uniform distribution. Given a graph with suitable labels for its edges and vertices, finding collisions in such a hash function is equivalent to finding cycles in the graph. Thus this construction can be applied to construct collision-resistant hash functions from any expander graph in which finding cycles is a hard problem. For more details see [5]. We give there two families of Ramanujan graphs, constructed by Pizer and Lubotzky-Phillips-Sarnak, and report the efficiency and collision resistance properties further.

When constructing a hash function from the Ramanujan graph of supersingular elliptic curves over $\mathbb{F}_{p^2}$ with l-isogenies, l a prime different from p, finding collisions is at least as hard as computing isogenies between supersingular elliptic curves. This is believed to be a hard problem, and the best algorithm currently known solves the problem in $O(\sqrt{p}\log^2 p)$ time ([5]). Thus the prime p can be taken to be a 256-bit prime, to get 128 bits of security from the resulting hash function. To compute the hash function from Pizer's graph when $l = 2$ requires roughly $2\log(p)$ field multiplications per bit of input to the hash function. This is roughly the same efficiency as a provable hash based on the Elliptic Curve Discrete Logarithm Problem.

The Pizer graph is the $g = 1$ case of our general construction of superspecial graphs. One feature of such graphs is that the same vertex set (fixing p) can give rise to many different graphs by varying l, the degree of the isogenies. One can ask, given two different edge sets on the same vertex set, whether there is any correlation of distances between vertices in the two graphs. This question is relevant to attacks on the hash function. Indeed, starting at two vertices which are at distance one from each other in one graph, and taking a walk from each vertex using the same directions in the two different graphs, it should not happen with high probability that the two ending vertices are close to each other in the other graph. This application raises an interesting question about the independence of graphs which will be considered in Section 7.

7. Metrics and independence of graphs

Fix L and p, where L is a totally real field of strict class number 1, and p is an unramified prime in L. Taking various prime ideals $\mathfrak{l}$ (not above p) we get a collection of Ramanujan graphs $G(L; p, \mathfrak{l})$. Let V be the vertex set of the graphs $G(L; p, \mathfrak{l})$. As $\mathfrak{l}$ varies, one can view these graphs as defining a sequence of metrics $d_{\mathfrak{l}} \colon V \times V \to \mathbb{N}$, where $d_{\mathfrak{l}}(u, v)$ is the length of the shortest path between the vertices u and v in $G(L; p, \mathfrak{l})$. It is natural to wonder what, if any, relations exist between these various metrics. It turns out that even for the Pizer graphs this question is already difficult to investigate. In this section we treat these metrics as defining random variables on $V \times V$ and study some properties of these random variables.

Let V be a set with N elements. Given two connected graphs $G_1 = (V, E_1)$ and $G_2 = (V, E_2)$ that are k_1 and k_2 regular, respectively, our goal for this section is to study some properties of the random variables d_1 and d_2 that define the distance metric on these graphs. In the cases of interest (to us) G_1 and G_2 will be Ramanujan graphs and hence we may specialize our results to this situation. We begin with a discussion on the average distance between two vertices in a graph.

7.1. Average distance. Let G be a connected k-regular graph on N vertices. Define a random variable $d\colon V \times V \to \mathbb{N}$ as follows: Pick two vertices u, v uniformly (and independently) from V and set $d(u, v)$ to be the length of the shortest path in G between u and v. We study the expectation of this random variable in this section.

Proposition 7.1. *Suppose $G = (V, E)$ is a k-regular graph ($k > 2$) that is a Ramanujan graph and let $|V| = N$. Then $d(u, v) \leq 2\lceil \log_{1+c} N/2 \rceil + 1$ for any pair of vertices u and v, where $c = (1 - 2\sqrt{(k-1)/k^2})/2$.*

Proof. Let $U \subseteq V$, $\overline{U}$ be the complement of U in V, and let $\Gamma(U) := \{v : v \notin U \text{ and } (u, v) \in E \text{ for some } u \in U\}$. Since G is a Ramanujan graph $E(U, \overline{U}) = \{(u, v) \in E : u \in U, v \notin U\}$ satisfies

$$|E(U, \overline{U})| \geq \frac{k - 2\sqrt{k-1}}{2} \min\{|U|, N - |U|\}$$

(see [1]). Since the graph is k-regular, this means that

$$|\Gamma(U)| \geq c|U| \quad \text{if } |U| \leq \frac{N}{2}, \tag{2}$$

where $c = (1 - 2\sqrt{(k-1)/k^2})/2$ in the following discussion. Let u and v be any two vertices in the graph. We start with the sets $U_0 = \{u\}$ and $V_0 = \{v\}$ and define $U_i := U_{i-1} \cup \Gamma(U_{i-1})$ and $V_i := V_{i-1} \cup \Gamma(V_{i-1})$ for $i > 0$. Let $\tilde{\imath}$ and $\tilde{\jmath}$ be a pair (i, j) such that $|U_i| + |V_j| > N$ and that the sum $i + j$ is minimal, then $\tilde{\imath} + \tilde{\jmath}$ is the distance between the vertices u and v. From (2), we find that $|U_i| \geq (1+c)^i$ as long as $|U_i| \leq N/2$ and similarly for $|V_i|$. In any case, $|U_i| > N/2$ if $i \geq \lceil \log_{1+c} N/2 \rceil + 1$. Thus,

$$d(u, v) \leq 2\lceil \log_{1+c} N/2 \rceil + 1. \qquad \square$$

The following corollary is immediate:

Corollary 7.2. *Let G_n be a family of connected k-regular Ramanujan graphs ($k > 2$). Let $\mathrm{E}[d_n]$ be the expectation of the random variable d_n that gives the distance between vertices in G_n. Then $\mathrm{E}[d_n] \leq 2\lceil \log_{1+c} |V(G_n)|/2 \rceil + 1$ where $c = (1 - 2\sqrt{(k-1)/k^2})/2$.*

The next result shows that the above upper bound is quite close to the truth.

Proposition 7.3. *Let G_n be a family of connected k-regular ($k > 2$) graphs such that $|V(G_n)|$ is unbounded and let d_n be the random variable giving the distance between vertices in G_n. Then*

$$\liminf_{n \to \infty} \frac{\mathrm{E}[d_n]}{\log_{k-1} |V(G_n)|} \geq 1.$$

PROOF. For a k-regular graph, one can find an upper bound on the probability $\Pr_{u,v}[d(u,v) \le i]$ as follows. Define, for any vertex u, the set $\Gamma^{\le i}(u) := \{v \mid d(u,v) \le i\}$. One sees that

$$\Pr_{u,v}[d(u,v) \le i] \le \max_{w} \frac{|\Gamma^{\le i}(w)|}{N},$$

where N is the number of vertices in the graph. For a k-regular graph one has

$$|\Gamma^{\le i}(w)| \le 1 + \sum_{1 \le j \le i} k(k-1)^{j-1}.$$

Consequently, one has

$$\Pr_{u,v}[d(u,v) \le i] \le \frac{k((k-1)^i - 1) + (k-2)}{(k-2)N}.$$

Let $a < 1$, then

$$\begin{aligned} \Pr_{u,v}[d(u,v) \le a \log_{k-1} N] &\le \frac{k((k-1)^{a \log_{k-1} N} - 1) + (k-2)}{(k-2)N} \\ &\le \frac{k(N^a - 1) + (k-2)}{(k-2)N} \\ &\le \frac{kN^a}{(k-2)N} + \frac{1}{N} \\ &\le \frac{k}{k-2}(N^{a-1} + N^{-1}). \end{aligned}$$

In particular, $\lim_{N \to \infty} \Pr_{u,v}[d(u,v) \le a \log_{k-1} N] = 0$.

Suppose $\mathrm{E}[d] \le b \log_{k-1} N$ where $b < a$. Then by Markov's inequality

$$\Pr_{u,v}[d(u,v) > a \log_{k-1} N] \le \frac{b \log_{k-1} N}{a \log_{k-1} N} = \frac{b}{a}$$

and b/a is bounded away from 1. On the other hand, we showed earlier that if N is large enough $\Pr_{u,v}[d(u,v) > a \log_{k-1} N] > b/a$, a contradiction.

Thus, for large enough graphs, G_n, $\mathrm{E}[d] > b \log_{k-1} |V(G_n)|$ for any $b < 1$. □

7.2. Independence of graphs. Let $G_1 = (V, E_1)$ and $G_2 = (V, E_2)$ be two graphs on the same vertex set that are k_1 and k_2 regular respectively. Define d_i, for $i = 1, 2$, to be the random variable giving the distance in the graph G_i. One natural definition of independence of graphs relates to the independence of these random variables, more precisely:

Definition 7.4. We call the graphs G_1 and G_2 *independent* if for each $i \le \mathrm{Diam}(G_1)$ and $j \le \mathrm{Diam}(G_2)$ we have that

$$\Pr_{u,v}[d_1(u,v) = i \text{ and } d_2(u,v) = j] = \Pr_{u,v}[d_1(u,v) = i] \cdot \Pr_{u,v}[d_2(u,v) = j].$$

We note that it is perhaps unreasonable to expect independence of graphs under this strong notion of independence. The definition, however, allows one to quantify "approximate independence" in terms of the size of the difference between the two quantities. In the following we investigate the Pizer graphs to see how closely the above condition holds.

7.2.1. *Independence of Pizer graphs.* Let l_1 and l_2 be two different primes and consider the Pizer graphs $G_1 = G(p, l_1)$ and $G_2 = G(p, l_2)$. For simplicity, we will assume $p \equiv 1 \mod 12$ so that edges on the graphs correspond to isogenies, up to a sign. We first analyze the case where $i = 1$ and j is allowed to be any positive integer less than the diameter of G_2. To say that $d_1(u, v) = 1$ and $d_2(u, v) = j$ means that there are two isogenies $\phi : E_u \to E_v$ and $\psi : E_u \to E_v$ such that $\deg(\phi) = l_1$ and $\deg(\psi) = l_2^j$. Taking $\hat{\psi} \circ \phi$ we get an endomorphism in $\mathrm{End}(E_u)$ of degree $l_1 l_2^j$ while taking $\psi \circ \hat{\phi}$ we get an endomorphism of degree $l_1 l_2^j$ in $\mathrm{End}(E_v)$ (here $\hat{\phi}$ and $\hat{\psi}$ refer to the dual isogeny of ϕ and ψ respectively); both endomorphisms are well-defined up to a sign. In this way we get embeddings of a quadratic imaginary order $\mathbb{Z}[x]/(x^2 - ax + l_1 l_2^j)$ into $B_{p,\infty}$. Now

$$\begin{aligned}\sharp\{(u,v) \mid d_1(u,v) = 1 \text{ and } d_2(u,v) = j\} &= \frac{1}{2}\sum_E \sharp\{\phi \in \mathrm{End}(E) \mid \deg(\phi) = l_1 l_2^j\} \\ &= \sum_{\substack{0 \le a \le 2\sqrt{l_1 l_2^j} \\ p \text{ inert or ramified in } \mathbb{Q}(\sqrt{a^2 - 4 l_1 l_2^j})}} H(a^2 - 4 l_1 l_2^j)\end{aligned}$$

where $H(m)$ is the Hurwitz class number of the order of discriminant m in $\mathbb{Q}(\sqrt{m})$. Using the estimate $H(m) \le |m|^{\frac{1}{2}+\epsilon}$ (for every $\epsilon > 0$) we get

$$\begin{aligned}\sharp\{(u,v) \mid d_1(u,v) = 1 \text{ and } d_2(u,v) = j\} &\le \sum_{0 \le a \le 2\sqrt{l_1 l_2^j}} (4 l_1 l_2^j - a^2)^{\frac{1}{2}+\epsilon} \\ &\le \frac{\pi}{4}(l_1 l_2^j)^{1+\epsilon},\end{aligned}$$

where we have used approximation of the sum by an integral in the second step. Finally, we get the bound

$$\mathrm{Pr}_{u,v}[d_1(u,v) = 1 \text{ and } d_2(u,v) = j] \le \frac{(l_1 l_2^j)^{1+\epsilon}}{N^2}. \tag{3}$$

One can obtain a lower bound for this probability for many primes p. Indeed, if p remains inert or ramified in at least a constant proportion of the fields $\mathbb{Q}\big(\sqrt{a^2 - 4 l_1 l_2^j}\big)$, we can use the Brauer – Siegel ineffective lower bound for the class number to get

$$\mathrm{Pr}_{u,v}[d_1(u,v) = 1 \text{ and } d_2(u,v) = j] \gg \frac{(l_1 l_2^j)^{1-\epsilon}}{N^2}. \tag{4}$$

The Chebotarev density theorem implies that the lower bound holds for at least a constant proportion of the primes p for fixed l_1, l_2 and j. On the other hand there are also a constant proportion of primes for which the lower bound does not hold.

Meanwhile, $\mathrm{Pr}_{u,v}[d_1(u,v) = 1] = (l_1 + 1)/N$. Since G_2 is $l_2 + 1$ regular we have that $\mathrm{Pr}_{u,v}[d_2(u,v) = j] \le (l_2 + 1) l_2^{j-1}/N$. On the other hand, the expansion property shows us that for any u, $\sharp\{v \mid d(u,v) = j\} = \Gamma(\Gamma^{\le j-1}(\{u\})) \ge c(1+c)^{j-1}$ for small j (adopting the notation introduced in Section 7.1). Thus we get the bounds

$$\begin{aligned}\frac{(l_1+1)c(1+c)^{j-1}}{N^2} &\le \mathrm{Pr}_{u,v}[d_1(u,v) = 1] \cdot \mathrm{Pr}_{u,v}[d_2(u,v) = 1] \\ &\le \frac{(l_1+1)(l_2+1) l_2^{j-1}}{N^2}.\end{aligned} \tag{5}$$

Equations (3) – (5) together imply that the graphs are close to being independent (for many primes p). We expect that a similar result holds for $\Pr_{u,v}[d_1(u,v) = i$ and $d_2(u,v) = j]$ for $i > 1$.

References

1. N. Alon and V. D. Milman, λ_1 *isoperimetric inequalities for graphs, and superconcentrators*, J. Combin. Theory Ser. B **38** (1985), no. 1, 73 – 88.
2. J. Angel and R. Evans, *Diameters of finite upper half plane graphs*, J. Graph Theory **23** (1996), no. 2, 129 – 137.
3. J. Brzeziński, *On orders in quaternion algebras*, Comm. Algebra **11** (1983), no. 5, 501 – 522.
4. D. I. Cartwright, P. Solé, and A. Żuk, *Ramanujan geometries of type* $\tilde{A}_n$, Discrete Math. **269** (2003), no. 1-3, 35 – 43.
5. D. X. Charles, E. Z. Goren, and K. E. Lauter, *Cryptographic hash functions from expander graphs*, J. Cryptology, to appear. DOI: 10.1007/s00145-007-9002-x.
6. D. X. Charles, K. Jain, and K. E. Lauter, *Signatures for network coding*, CISS 2006, 40th Conference on Information Sciences and Systems, available at `http://www288.pair.com/ciss/ciss/numbered/340.pdf`.
7. C. Consani and J. Scholten, *Arithmetic on a quintic threefold*, Internat. J. Math. **12** (2001), no. 8, 943 – 972.
8. L. Dembélé, *Quaternionic Manin symbols, Brandt matrices, and Hilbert modular forms*, Math. Comp. **76** (2007), no. 258, 1039 – 1057.
9. M. Eichler, *On theta functions of real algebraic number fields*, Acta Arith. **33** (1977), no. 3, 269 – 292.
10. S. S. Gelbart, *Automorphic forms on adèle groups*, Ann. of Math. Stud., vol. 83, Princeton Univ. Press, Princeton, NJ, 1975.
11. E. Z. Goren and K. E. Lauter, *Evil primes and superspecial moduli*, Int. Math. Res. Not. **2006**, Art. ID 53864.
12. E. Z. Goren and F. Oort, *Stratifications of Hilbert modular varieties*, J. Algebraic Geom. **9** (2000), 111 – 154.
13. B. H. Gross, *Heights and the special values of L-series*, Number Theory (Montréal, 1985) (H. Kisilevsky and J. Labute, eds.), CMS Conf. Proc., vol. 7, Amer. Math. Soc., Providence, RI, 1987, pp. 115 – 187.
14. S. Hoory, N. Linial, and A. Wigderson, *Expander graphs and their applications*, Bull. Amer. Math. Soc. (N.S.) **43** (2006), no. 4, 439 – 561.
15. H. Jacquet and R. P. Langlands, *Automorphic forms on* GL(2), Lecture Notes in Math., vol. 114, Springer, Berlin, 1970.
16. B. W. Jordan and R. Livné, *Integral Hodge theory and congruences between modular forms*, Duke Math. J. **80** (1995), no. 2, 419 – 484.
17. ———, *Ramanujan local systems on graphs*, Topology **36** (1997), no. 5, 1007 – 1024.
18. ———, *The Ramanujan property for regular cubical complexes*, Duke Math. J. **105** (2000), no. 1, 85 – 103.
19. A. Katamoto, *On 3rd and 4th moments of finite upper half plane graphs*, Finite Fields Appl. **13** (2007), no. 2, 249 – 258.
20. Y. Kitaoka, *Arithmetic of quadratic forms*, Cambridge Tracts in Math., vol. 106, Cambridge Univ. Press, Cambridge, 1993.
21. W.-C. W. Li, *Character sums and abelian Ramanujan graphs*, J. Number Theory **41** (1992), no. 2, 199 – 217.
22. ———, *Ramanujan hypergraphs*, Geom. Funct. Anal. **14** (2004), no. 2, 380 – 399.
23. R. Livné, *Communication networks and Hilbert modular forms*, Applications of Algebraic Geometry to Coding Theory, Physics and Computation (Eilat, 2001) (C. Ciliberto, F. Hirzebruch, R. Miranda, and M. Teicher, eds.), NATO Sci. Ser. II Math. Phys. Chem., vol. 36, Kluwer Acad. Publ., Dordrecht, 2001, pp. 255 – 270.
24. A. Lubotzky, *Discrete groups, expanding graphs and invariant measures*, with an appendix by J. D. Rogawski, Progr. Math., vol. 125, Birkhäuser, Basel, 1994.
25. A. Lubotzky, R. S. Phillips, and P. Sarnak, *Ramanujan graphs*, Combinatorica **8** (1988), no. 3, 261 – 277.

26. A. Lubotzky, B. Samuels, and U. Vishne, *Ramanujan complexes of type* $\tilde{A}_d$, Israel J. Math. **149** (2005), 267–299.
27. M.-H. Nicole, *Superspecial abelian varieties, Theta series and the Jacquet–Langlands correspondence*, Ph.D. thesis, McGill University, 2005.
28. ———, *Superspecial abelian varieties and the Eichler basis problem for Hilbert modular forms*, J. Number Theory, to appear.
29. F. Oort, *Which abelian surfaces are products of elliptic curves?*, Math. Ann. **214** (1975), 35–47.
30. A. K. Pizer, *An algorithm for computing modular forms on* $\Gamma_0(N)$, J. Algebra **64** (1980), no. 2, 340–390.
31. ———, *Ramanujan graphs*, Computational Perspectives on Number Theory (Chicago, IL, 1995) (D. A. Buell and J. T. Teitelbaum, eds.), AMS/IP Stud. Adv. Math., vol. 7, Amer. Math. Soc., Providence, RI, 1998, pp. 159–178.
32. M. Rapoport, *Compactifications de l'espace de modules de Hilbert-Blumenthal*, Compositio Math. **36** (1978), no. 3, 255–335.
33. T. Shioda, *Supersingular K3 surfaces*, Algebraic Geometry (Copenhagen, 1978) (K. Lønsted, ed.), Lecture Notes in Math., vol. 732, Springer, Berlin, 1979, pp. 564–591.
34. J. H. Silverman, *The arithmetic of elliptic curves*, Grad. Texts in Math., vol. 106, Springer, New York, 1986.
35. J. Socrates and D. Whitehouse, *Unramified Hilbert modular forms, with examples relating to elliptic curves*, Pacific J. Math. **219** (2005), no. 2, 333–364.
36. A. Terras, *Are finite upper half plane graphs Ramanujan?*, Expanding Graphs (Princeton, NJ, 1992) (J. Friedman, ed.), DIMACS Ser. Discrete Math. Theoret. Comput. Sci., vol. 10, Amer. Math. Soc., Providence, RI, 1993, pp. 125–142.
37. ———, *Survey of spectra of Laplacians on finite symmetric spaces*, Experiment. Math. **5** (1996), no. 1, 15–32.
38. A. Valette, *Graphes de Ramanujan et applications*, Séminaire Bourbaki, Vol. 1996/97, Astérisque, vol. 245, Exp. No. 829, 4, Soc. Math. France, Paris, 1997, pp. 247–276.
39. G. van der Geer, *Hilbert modular surfaces*, Ergeb. Math. Grenzgeb. (3), vol. 16, Springer, Berlin, 1988.
40. M.-F. Vignéras, *Arithmétique des algèbres de quaternions*, Lecture Notes in Math., vol. 800, Springer, Berlin, 1980.
41. L. C. Washington, *Introduction to cyclotomic fields*, 2nd ed., Grad. Texts in Math., vol. 83, Springer, New York, 1997.
42. W. C. Waterhouse, *Abelian varieties over finite fields*, Ann. Sci. École Norm. Sup. (4) **2** (1969), 521–560.
43. W. C. Waterhouse and J. S. Milne, *Abelian varieties over finite fields*, 1969 Number Theory Institute (Stony Brook, 1969) (D. J. Lewius, ed.), Proc. Sympos. Pure Math., vol. 20, Amer. Math. Soc., Providence, RI, 1971, pp. 53–64.

Microsoft Live Labs, One Microsoft Way, Redmond, WA 98052, USA
E-mail address: `cdx@microsoft.com`

Department of Mathematics and Statistics, McGill University, 805 Sherbrooke St. W, Montreal, QC H3A 2K6, Canada
E-mail address: `goren@math.mcgill.ca`

Microsoft Research, One Microsoft Way, Redmond, WA 98052, USA
E-mail address: `klauter@microsoft.com`

Centre de Recherches Mathématiques
CRM Proceedings and Lecture Notes
Volume **47**, 2009

Normal Forms, K3 Surface Moduli, and Modular Parametrizations

Adrian Clingher, Charles F. Doran, Jacob Lewis, and Ursula Whitcher

For John McKay

1. Introduction

The geometric objects of study in this paper are K3 surfaces which admit a polarization by the unique even unimodular lattice of signature $(1, 17)$. This lattice can be realized as the orthogonal direct sum

$$\mathrm{M} = \mathrm{H} \oplus \mathrm{E}_8 \oplus \mathrm{E}_8,$$

where H is the standard rank-two hyperbolic lattice and E_8 is the unique even, negative-definite and unimodular lattice of rank eight. A lattice-polarization by the lattice M imposes severe constraints on the geometry of a K3 surface X. In particular, the Picard rank of X has to be 18, 19, or 20 (see [25]).

A standard Hodge-theoretic observation about this special class of K3 surfaces is that the polarized Hodge structure of an M-polarized K3 surface is identical with the polarized Hodge structure of an abelian surface $A = \mathrm{E}_1 \times \mathrm{E}_2$ realized as a cartesian product of two elliptic curves. Since both types of surfaces involved admit appropriate versions of the Torelli theorem, Hodge theory implies a well-defined correspondence, giving rise to a canonical analytic isomorphism between the moduli spaces on the two sides. By employing a modern point of view from the frontier of algebraic geometry with string theory, one can regard this correspondence as a Hodge-theoretic *duality map*, a correspondence that relates two seemingly different types of surfaces sharing similar Hodge-theoretic information.

In [4], the first two authors showed that that the resemblance of the two Hodge structures involved in the duality correspondence is not fortuitous, but rather is merely a consequence of a quite interesting geometric relationship. We show in

2000 *Mathematics Subject Classification.* 14J15, 14J28, 11F11, 11G05.

A. C. acknowledges support from a University of Missouri Research Board Grant. He would also like to thank the Department of Mathematics of the University of Washington for its hospitality and financial support during the 2007 summer school on K3 Surfaces and String Duality.

C. F. D. was supported in part by the University of Washington Royalty Research Fund.

J. L. was supported in part by a VIGRE Graduate Fellowship.

U. W. was supported in part by a McKibben and Merner Fellowship.

This is the final form of the paper.

[4, Theorem 1.1] that: the surface X possesses a canonical involution defining a Shioda – Inose structure; the minimal resolution of X quotiented by this involution is a new K3 surface Y endowed with a canonical Kummer structure; this structure realizes Y as the Kummer surface of an abelian surface A which is canonically represented as a Cartesian product of two elliptic curves (i.e., $A = \mathrm{E}_1 \times \mathrm{E}_2$); and the construction induces a canonical Hodge isomorphism between the M-polarized Hodge structure of X and the natural H-polarized Hodge structure of the abelian surface A.

One important feature of the special class of M-polarized K3 surfaces is that they turn out to be completely classified by two modular invariants σ and π in $\mathbb{C}$, much in the same way as elliptic curves over the field of complex numbers are classified by the j-invariant. However, the two modular invariants σ and π are not geometric in origin. They are defined Hodge-theoretically, and the result leading to the classification is a consequence of the appropriate version of the Global Torelli Theorem for lattice polarized K3 surfaces. In the context of the duality map, the two invariants can be seen as the standard symmetric functions on the j-invariants of the dual elliptic curves, σ being their sum and π their product. This interpretation suggests that the modular invariants of an M-polarized K3 surface can be computed by determining the two elliptic curves.

Explicit M-polarized K3 surfaces can be constructed by various geometrical procedures. One such method, introduced in 1977 by Inose [19], constructs a two-parameter family $X(a,b)$ of M-polarized K3 surfaces by taking minimal resolutions of the projective quartics in $\mathbb{P}^3$ associated with the special equations:

$$y^2zw - 4x^3z + 3axzw^2 - \tfrac{1}{2}(z^2w^2 + w^4) + bzw^3 = 0 \tag{1}$$

for a, b in $\mathbb{C}$. In Section 3 below, this construction is generalized so as to compactify the affine (a,b)-parametric family to a weighted projective space with covering coordinates $[a,b,d]$. The coarse moduli space for M-polarized K3 surfaces is then given by the locus where $d \neq 0$ in this weighted projective space. One can thus regard the Inose quartic as a *normal form for* M*-polarized* K3 *surfaces*. In the second part of [4], we use the geometric correspondence to explicitly describe the j-invariants of the two elliptic curves E_1 and E_2 associated to the Inose surface $X(a,b)$. The result is reproduced below (Theorem 3.3) in the context of the $X(a,b,d)$ family. As a consequence, the coarse moduli space for M-polarized K3 surfaces is explicitly identified with the classical Hilbert modular surface $\mathcal{H}_1$ for the elliptic modular group.

In this paper, we use explicit computations of Picard – Fuchs differential equations to explore (and exploit) the "universal" property of the moduli space of M-polarized K3 surfaces with respect to enhancement to *non-generic* Picard lattices. The conditions on the hypersurfaces for enhancement to Picard lattices of rank 19 are of particular interest, as these conditions determine one-parameter subfamilies of the coarse moduli space. As we work throughout in parallel with the case of the Weierstrass normal form for elliptic curves, this case is reviewed in Section 2. The approach is a natural extension of the second author's work on Picard – Fuchs uniformization for families of elliptic curves and lattice-polarized K3 surfaces [11, 12].

The Inose family normal form is realized as projective hypersurfaces with just ADE type singularities. This provides a crucial advantage to using this normal form, as now the Griffiths – Dwork technique (see Section 2.2.1 for a general discussion) can be applied to compute the differential equations satisfied by periods of the holomorphic 2-forms directly. We do this first in Section 3 with independent variables taken to be the free coefficients in the normal form hypersurface equations; this is the analog of taking partial derivatives of the Weierstrass elliptic periods with respect to the parameters g_2 and g_3. It proves convenient to work in the affine chart $[1, b, d]$ for these computations. Upon substitution of the modular invariants from [4], this partial differential system completely decouples, reducing to a pair of ODEs. Each of these is the Picard – Fuchs ODE of a corresponding "factor" family of elliptic curves, reflecting the geometric fact that our normal form K3 surfaces are Shioda – Inose surfaces of products of pairs of elliptic curves.

Next we apply the Griffiths – Dwork method with independent variable t; this is the analog of treating $g_2(t)$ and $g_3(t)$ as rational functions of a single complex variable t. In the case of Weierstrass normal form elliptic curves, this modification simply allows one to compute the Picard – Fuchs ODE for arbitrary one parameter families of elliptic curves, as the moduli space of elliptic curves is itself one-dimensional. In our case, this modification allows one to "probe" the two-dimensional moduli space of $\mathrm{H} \oplus \mathrm{E}_8 \oplus \mathrm{E}_8$ lattice polarized K3 surfaces with rational curves, computing the Picard – Fuchs ODE for the restricted period functions.[1] The modular invariants of [4] can once again be substituted in, yielding a fourth order ODE whose coefficients are differential rational functions of $j_1(t)$ and $j_2(t)$.

Here is where the universal property for the moduli space with respect to lattice enhancement makes its impact felt. Enhancement of polarization from M to

$$\mathrm{M}_n = \mathrm{H} \oplus \mathrm{E}_8 \oplus \mathrm{E}_8 \oplus \langle -2n \rangle$$

is equivalent to the condition that this fourth order ODE reduces to a third order ODE (a condition on ODEs studied by Gino Fano over a century ago). Under the hypothesis that $j_1(t)$ and $j_2(t)$ are nonconstant, this reduction occurs precisely when these two functions satisfy a particular nonlinear ODE. This nonlinear ODE is the *master equation for modular parametrizations* in that any parametrization of the modular curve $X_0(n)$ by rational functions $j_1(t)$ and $j_2(t)$ solves this ODE. What's more, for any solutions $j_1(t)$ and $j_2(t)$ not themselves rational functions of t, but such that their sum and product are still rational functions, the pair parametrizes the modular curve $X_0(n) + n$.

The master equation can even be rewritten in a form which makes manifest its relationship with pairs of n-isogenous elliptic curves. First, all of the terms involving $j_1(t)$ can be moved to one side of the equation and all those involving $j_2(t)$ to the other. Of course, up to exchanging $j_1(t)$ with $j_2(t)$, the expressions on each side are then the same. Moreover, they are recognizable as the coefficients of projective normalized Picard – Fuchs ODEs for a pair of elliptic curves. As such, they are each the sum of one term involving the rational function characterizing the uniformizing differential equation for the elliptic modular parameter j, and another term consisting of the Schwarzian derivative of j with respect to t (when set equal to zero, this expression is also known as the "Schwarzian differential equation" for j). Once the master equation for modular parametrizations has been expressed in this form, it is natural to ask whether an analogous equation, based on Hauptmoduls other

[1] Computation of Picard – Fuchs equations for particular one-parameter families of K3 surfaces of Picard rank 19 dates back at least to [26]. The approach taken in the present paper, however, results in a universal expression for the Picard – Fuchs equations for *all* one-parameter families of K3 surfaces admitting (at least) an M-polarization. The families are explicitly realized as subfamilies of the single two-parameter family of (1).

than the elliptic modular parameter j, will similarly characterize parametrizations of genus zero quotients of the upper half plane by the moonshine groups. We address this question in Section 3.3.1, and illustrate there explicit parametrizations of some modular equations studied by Cohn – McKay.

Currently under investigation are an extension of the method to handle modular curves of genus greater than zero (by modifying the Griffiths – Dwork technique) and a generalization to parametrizing Humbert surfaces and Shimura curves based now on universal properties of the moduli space of $\mathrm{H} \oplus \mathrm{E}_8 \oplus \mathrm{E}_7$ lattice polarized K3 surfaces studied by Clingher and Doran in [5, 6].

2. Elliptic curves

2.1. Normal form and moduli space. As is well-known, the classification of elliptic curves over the field of complex numbers is based on the following classical features. First, every elliptic curve E can be realized through an explicit *normal form* given, for instance, by the projective version of the Weierstrass cubic:

$$y^2 z - 4x^3 + g_2 x z^2 + g_3 z^3 = 0$$

for some pair $(g_2, g_3) \in \mathbb{C}^2$ with $g_2^3 - 27g_3^2 \neq 0$. The Weierstrass form is not unique. Nevertheless, the weighted projective point $[g_2, g_3] \in \mathbb{WP}(2,3)$ classifies E up to isomorphism. This leads to a nice (coarse) *moduli space* for complex elliptic curves, taken as the one-dimensional open variety:

$$\mathcal{M} = \{[g_2, g_3] \in \mathbb{WP}(2,3) \mid g_2^3 - 27g_3^2 \neq 0\}.$$

In this context, the *j-invariant*:

$$j(\mathrm{E}) = \frac{g_2^3}{g_2^3 - 27g_3^2}$$

realizes an identification between $\mathcal{M}$ and $\mathbb{C}$.

From a Hodge theoretic point of view, an elliptic curve can be seen as an element (period) in the complex upper half-plane $\mathbb{H}$. The classifying space for such periods is the *modular quotient*

$$\mathcal{F}_1 = \Gamma_1 \backslash \mathbb{H}$$

where $\Gamma_1 = \mathrm{PSL}(2, \mathbb{Z})$ with the standard action on $\mathbb{H}$.

The link between the two points of view is realized by the period map:

$$\mathrm{per} \colon \mathcal{M} \to \mathcal{F}_1 \tag{2}$$

which makes an isomorphism of analytic spaces. The inverse of (2) can be described in terms of Γ_1-modular forms as:

$$\mathrm{per}^{-1} = [60 \cdot \mathrm{E}_4, 140 \cdot \mathrm{E}_6]$$

where $\mathrm{E}_4, \mathrm{E}_6 \colon \mathbb{H} \to \mathbb{C}$ are the classical Eisenstein series of weights four and six, respectively. Using the inverse period map, one may view the j-invariant as a map $j \colon \mathbb{H} \to \mathbb{C}$.

2.2. Picard – Fuchs differential equation. When considering the periods of varieties in families, it is often easier to compute the regular singular differential equation satisfied by all the period functions on the base of the family — the *Picard – Fuchs differential equation* — than to explicitly describe the period functions themselves. We first review in Section 2.2.1 the method, due to Griffiths and Dwork, for computing the Picard – Fuchs differential equation for families of hypersurfaces in projective space via residues. In Section 2.2.2 this technique is applied to families of elliptic curves in $\mathbb{P}^2$ in Weierstrass normal form.

2.2.1. *The Griffiths – Dwork technique.* Let X be a hypersurface in $\mathbb{P}^n$ given by a homogeneous polynomial Q in coordinates $[x_0, \dots, x_n]$, and let $\iota\colon X \to \mathbb{P}^n$ be the inclusion map. Let $\mathcal{H}(X)$ be the de Rham cohomology of rational n-forms on $\mathbb{P}^n - X$. We may write any representative of $\mathcal{H}(X)$ as $P\Omega_0/Q^k$, where $\Omega_0 = \sum_{i=0}^{n}(-1)^i x^i \mathrm{d}x^0 \wedge \cdots \widehat{\mathrm{d}x^i} \cdots \wedge \mathrm{d}x^n$ is the usual holomorphic form on $\mathbb{P}^n$ and P is a homogeneous polynomial of degree $\deg P = k \deg Q - (n+1)$.

Let the Jacobian ideal $J(Q)$ be the ideal generated by the partial derivatives $\partial Q/\partial x_i$. If we have an element of $\mathcal{H}(X)$ of the form $(K/Q^{k+1})\Omega_0$ where $K = \sum_i A_i \partial Q/\partial x_i$ is a member of the Jacobian ideal, then we may reduce the order of the pole:

$$\frac{\Omega_0}{Q^{k+1}} \sum_i A_i \frac{\partial Q}{\partial x_i} = \frac{1}{k}\frac{\Omega_0}{Q^k} \sum_i \frac{\partial A_i}{\partial x_i} + \text{exact terms.} \tag{3}$$

Let γ be a cycle in X, and let $T(\gamma)$ be a small tubular neighborhood of γ in $\mathbb{P}^n - X$. Then we may define the *residue map* $\mathrm{Res}\colon \mathcal{H}(X) \to H^{n-1}(X, \mathbb{C})$ by

$$\frac{1}{2\pi\mathrm{i}} \int_{T(\gamma)} \frac{P\Omega_0}{Q^k} = \int_\gamma \mathrm{Res}\left(\frac{P\Omega_0}{Q^k}\right). \tag{4}$$

Let H be the hyperplane class in $H^{n-1}(\mathbb{P}^n, \mathbb{C})$. We refer to the perpendicular complement of $\iota^*(H)$ in $H^{n-1}(X, \mathbb{C})$ as the *primitive cohomology* of X, and denote it by $PH(X)$. The residue map is an isomorphism onto the primitive cohomology [15].

Now, consider a family of hypersurfaces $X_{t_1 \cdots t_j}$ given by polynomials $Q_{t_1 \cdots t_j}$, where $t_1, \dots, t_j$ are independent parameters. We may define a corresponding family of cycles $\gamma(t_1, \cdots, t_j)$. For $(t_1, \dots, t_j)$ in a sufficiently small neighborhood of a fixed parameter value $(t_1', \dots, t_j')$, $T(\gamma(t_1, \dots, t_j))$ is homologous to $T(\gamma(t_1', \dots, t_j'))$ in $H_n(\mathbb{P}^n - X, \mathbb{C})$. Thus, we may differentiate as follows:

$$\begin{aligned} \frac{\partial}{\partial t_i} \int_{T(\gamma(t_1,\dots,t_j))} \frac{P\Omega_0}{Q(t)^k} &= \frac{\partial}{\partial t_i} \int_{T(\gamma(t_1',\dots,t_j'))} \frac{P\Omega_0}{Q(t)^k} \\ &= -k \int_{T(\gamma(t_1',\dots,t_j'))} \frac{P\Omega_0}{Q(t)^{k+1}} \frac{\partial Q}{\partial t_i}. \end{aligned} \tag{5}$$

If $r = \dim_{\mathbb{C}}(H_{n-1}(X)) = \dim_{\mathbb{C}}(H^{n-1}(X, \mathbb{C}))$, only $r - 1$ derivatives can be linearly independent. Therefore the periods must satisfy a linear differential equation with coefficients in $\mathbb{Q}(t_1, \dots, t_j)$ of order at most r — this is a Picard – Fuchs differential equation. One may compute the Picard – Fuchs equation by systematically taking derivatives of $\int_{T(\gamma(t_1,\dots,t_j))} P\Omega_0/Q(t)^k$ with respect to the various parameters and using (3) to rewrite the results in terms of a standard basis for $H^{n-1}(X, \mathbb{C})$. This method is known as the *Griffiths – Dwork technique.* (See [9] or [13] for a more detailed discussion.)

2.2.2. *Griffiths – Dwork for the Weierstrass form.* Consider the hypersurface

$$Q = y^2z - 4x^3 + g_2xz^2 + g_3z^3,$$

the Weierstrass form for a family of elliptic curves. We illustrate here the Griffiths–Dwork technique, first treating g_2 and g_3 as independent parameters. Equation (5) tells us that we may differentiate under the integral sign:

$$\frac{\partial}{\partial g_2}\int\frac{\Omega_0}{Q} = -\int\frac{xz^2\Omega_0}{Q^2}, \quad \frac{\partial}{\partial g_3}\int\frac{\Omega_0}{Q} = -\int\frac{z^3\Omega_0}{Q^2}. \tag{6}$$

A Gröbner basis computation shows that xz^2 and z^3 are equivalent modulo the Jacobian ideal $J(Q)$. Using (3) to reduce the pole order, we find that

$$\frac{\partial}{\partial g_2}\int\frac{\Omega_0}{Q} - \frac{3g_3}{2g_2}\frac{\partial}{\partial g_3}\int\frac{\Omega_0}{Q} = \frac{-1}{4g_2}\int\frac{\Omega_0}{Q}. \tag{7}$$

Now, suppose g_2 and g_3 are both functions of a single parameter t. We compute:

$$\begin{aligned}
\frac{\mathrm{d}}{\mathrm{d}t}\int\frac{\Omega_0}{Q} &= \left(\frac{\partial}{\partial g_2}\int\frac{\Omega_0}{Q}\right)\frac{\partial g_2}{\partial t} + \left(\frac{\partial}{\partial g_3}\int\frac{\Omega_0}{Q}\right)\frac{\partial g_3}{\partial t} \\
&= -g_2'(t)\int\frac{xz^2\Omega_0}{Q^2} - g_3'(t)\int\frac{z^3\Omega_0}{Q^2},
\end{aligned} \tag{8}$$

$$\begin{aligned}
\frac{\mathrm{d}^2}{\mathrm{d}t^2}\int\frac{\Omega_0}{Q} &= 2g_2'(t)\int\frac{xz^2(g_2'(t)xz^2 + g_3'(t)z^3)}{Q^3}\Omega_0 - g_2''(t)\int\frac{xz^2}{Q^2}\Omega \\
&\quad + 2g_3'(t)\int\frac{z^3(g_2'(t)xz^2 + g_3'(t)z^3)}{Q^3}\Omega_0 - g_3''(t)\int\frac{z^3}{Q^2}\Omega \\
&= 2\big(g_2'(t)\big)^2\int\frac{(xz^2)^2}{Q^3}\Omega_0 + 4g_2'(t)g_3'(t)\int\frac{(xz^2)(z^3)}{Q^3}\Omega_0 \\
&\quad + 2\big(g_3'(t)\big)^2\int\frac{(z^3)^2}{Q^3}\Omega_0 - g_2''(t)\int\frac{xz^2}{Q^2}\Omega_0 - g_3''(t)\int\frac{z^3}{Q^2}\Omega_0.
\end{aligned} \tag{9}$$

We may use (3) together with a Gröbner basis computation to rewrite $\mathrm{d}^2(\int\Omega_0/Q)/\mathrm{d}t^2$ as a sum of integrals of expressions with Q^2 in the denominator:

$$\begin{aligned}
\frac{\mathrm{d}^2}{\mathrm{d}t^2}\int\frac{\Omega_0}{Q} &= 2\big(g_2'(t)\big)^2\int\frac{\alpha_1xz^2 + \beta_1z^3}{Q^2}\Omega_0 + 4g_2'(t)g_3'(t)\int\frac{\alpha_2xz^2 + \beta_2z^3}{Q^2}\Omega_0 \\
&\quad + 2\big(g_3'(t)\big)^2\int\frac{\alpha_3xz^2 + \beta_3z^3}{Q^2}\Omega_0 - g_2''(t)\int\frac{xz^2}{Q^2}\Omega_0 - g_3''(t)\int\frac{z^3}{Q^2}\Omega_0.
\end{aligned} \tag{10}$$

Here the α_j and β_j are rational functions in g_2 and g_3. Note that we have expressed $\mathrm{d}^2(\int\Omega_0/Q)/\mathrm{d}t^2$ entirely in terms of $\int(xz^2/Q^2)\Omega_0 = -\partial(\int\Omega_0/Q)/\partial g_2$ and $\int(z^3/Q^2)\Omega_0 = -\partial(\int\Omega_0/Q)/\partial g_3$. Since $\mathrm{d}(\int\Omega_0/Q)/\mathrm{d}t$ is also written in terms of $\int(xz^2/Q^2)\Omega_0$ and $\int(z^3/Q^2)\Omega_0$, we might hope to relate $\mathrm{d}(\int\Omega_0/Q)\mathrm{d}t$ and $\mathrm{d}^2(\int\Omega_0/Q)\mathrm{d}t^2$. If such a relationship is to exist for an arbitrary choice of $g_2(t)$ and $g_3(t)$, $\int(xz^2/Q^2)\Omega_0$ and $\int(z^3/Q^2)\Omega_0$ cannot be independent. In fact, they are not: $xz^2 \cong -3g_3/(2g_2)z^3 \bmod J(Q)$, so applying (3) we find that

$$\int\frac{xz^2}{Q^2}\Omega_0 = \frac{-3g_3}{2g_2}\int\frac{z^3}{Q^2}\Omega_0 + \frac{1}{4g_2}\int\frac{\Omega_0}{Q}. \tag{11}$$

Combining (8), (10), and (11), and setting $\Delta = g_2^3 - 27g_3^2$, we obtain the Picard – Fuchs differential equation for a one-parameter family of elliptic curves in

Weierstrass form:

$$A_2\frac{\mathrm{d}^2}{\mathrm{d}t^2}\int\frac{\Omega_0}{Q}+A_1\frac{\mathrm{d}}{\mathrm{d}t}\int\frac{\Omega_0}{Q}+A_0\int\frac{\Omega_0}{Q}=0 \tag{12}$$

where

$$\begin{aligned}A_2&=16\Delta(3g_2'g_3-2g_2g_3')\\A_1&=16\big(9g_2^2g_3(g_2')^2-(7g_2^3+135g_3^2)g_2'g_3'+108g_2g_3(g_3')^2\\&\qquad\qquad+\Delta(-3g_3g_2''+2g_2g_3'')\big)\\A_0&=21g_2g_3(g_2')^3-18g_2^2(g_2')^2g_3'+8g_3'(15g_2(g_3')^2-\Delta g_2'')\\&\qquad\qquad-4g_2'(27g_3(g_3')^2-2\Delta g_3'').\end{aligned} \tag{13}$$

If we make the substitution $j=g_2^3/\Delta$, then (12) reduces to the standard Picard – Fuchs equation for a one-parameter family of elliptic curves in Weierstrass form, described for example in [27]:

$$\frac{\mathrm{d}^2}{\mathrm{d}t^2}\int\frac{\mathrm{d}x}{y}+B_1\frac{\mathrm{d}}{\mathrm{d}t}\int\frac{\mathrm{d}x}{y}+B_0\int\frac{\mathrm{d}x}{y}=0 \tag{14}$$

where

$$\begin{aligned}B_1&=\frac{g_3'}{g_3}-\frac{g_2'}{g_2}+\frac{j'}{j}-\frac{j''}{j'}\\B_0&=\frac{(j')^2}{144j(j-1)}+\frac{\Delta'}{12\Delta}\left(B_1+\frac{\Delta''}{\Delta'}-\frac{13\Delta'}{12\Delta}\right).\end{aligned} \tag{15}$$

2.3. Relationship to toric geometry. Another model for the Weierstrass family of elliptic curves comes from toric geometry. This one-parameter family can be realized as the family of *anticanonical hypersurfaces in the toric variety* $X=\mathbb{WP}(1,2,3)$. This family K has defining equation

$$f_{(\lambda_0,\lambda_1,\lambda_2,\lambda_3)}(x_0,x_1,x_2)=\lambda_0x_0x_1x_2+\lambda_1x_1^3+\lambda_2x_0^6+\lambda_3x_2^2=0$$

in the global homogeneous coordinate ring $\mathbb{C}[x_0,x_1,x_2]$ of X (where x_0, x_1, x_2 have weights 1, 2, 3 respectively).

The four parameters $(\lambda_0,\ldots,\lambda_3)$ are redundant. In [9], the authors define a "simplified polynomial moduli space" $\mathcal{M}_{\mathrm{simp}}$ for K with the property that $\mathcal{M}_{\mathrm{simp}}$ is a finite-to-one cover (generically) of the actual moduli space $\mathcal{M}$ for K (which in the present case is just $\mathbb{WP}(2,3)\simeq\mathbb{P}^1$). Moreover, following [9], we have that $\mathcal{M}_{\mathrm{simp}}$ is a one-dimensional toric variety, and $t=\lambda_1^2\lambda_2\lambda_3^3/\lambda_0^6$ is a coordinate on the torus $\mathbb{C}^*\subset\mathcal{M}_{\mathrm{simp}}$. We can use the fact that the defining equation is defined only up to a nonzero constant to set $\lambda_0=1$. Then we can use the natural action of $T\simeq(\mathbb{C}^*)^2$ to set $\lambda_1=4$, $\lambda_3=-1$. Then our simplified polynomial modulus becomes $t=-16\lambda_2$.

Let ϕ be the rational map $X\to\mathbb{P}^2$ defined by

$$(x_0,x_1,x_2)\mapsto\left(\frac{x_1}{x_0^2}+\frac{1}{48},\frac{x_2}{x_0^3}-\frac{x_1}{2x_0^2},1\right)=(x,y,z).$$

The image of K under ϕ has defining equation

$$y^2z=4x^3-\frac{1}{192}xz^2+\left(\lambda_2+\frac{1}{13824}\right)z^3.$$

Using the j-invariant of an elliptic curve as an affine coordinate on $\mathcal{M} \simeq \mathbb{WP}(2,3)$, we have a map $\mathcal{M}_{\text{simp}} \to \mathcal{M}$ given in affine coordinates by

$$t \mapsto \frac{1}{1728t(t-432)} = j.$$

The Picard – Fuchs differential equation for K is derived by factoring a Gel′fand – Kapranov – Zelevinsky hypergeometric equation in [22, Section 5.1]. The ODE produced there is

$$\begin{aligned} (16) \qquad 0 &= \big(\theta^2 - 12t(6\theta+5)(6\theta+1)\big) f(t) \\ &= t\big(t(432t-1)f''(t) + (864t-1)f'(t) + 60f(t)\big) \end{aligned}$$

where $\theta = td/dt$. Plugging in $g_2(t) = 1/192$, $g_3(t) = (864t-1)/13824$ into (12) gives

$$(17) \qquad 0 = \frac{1}{393216}\big(t(432t-1)f''(t) + (864t-1)f'(t) + 60f(t)\big) \, .$$

So the methods of GKZ and Griffiths – Dwork yield the same equation in this case (up to an inconsequential overall factor).

3. K3 surfaces polarized by $\mathbf{H} \oplus \mathbf{E_8} \oplus \mathbf{E_8}$

3.1. Normal form and moduli space. In [4], the authors presented the full classification of K3 surfaces (X, i) polarized by the unimodular even lattice of rank eighteen $\mathrm{M} = \mathrm{H} \oplus \mathrm{E}_8 \oplus \mathrm{E}_8$. The contents of this section are adapted from [4], where proofs of the results may be found. The crucial ingredient in this classification is the existence of a *normal form.* That is:

Theorem 3.1 ([4]). *Let (X, i) be an M-polarized K3 surface. Then, there exists a triple $(a,b,d) \in \mathbb{C}^3$, with $d \neq 0$ such that (X, i) is isomorphic to the minimal resolution of the quartic surface:*

$$(18) \qquad \mathrm{Q}(a,b,d)\colon y^2zw - 4x^3z + 3axzw^2 + bzw^3 - \tfrac{1}{2}(dz^2w^2 + w^4) = 0.$$

The equation in Theorem 3.1 can be regarded as the analog, for M-polarized K3 surfaces, of the Weierstrass form in the classical theory of elliptic curves. It is a compactification of Inose's (a,b) family from (1).

Two distinct quartics as in (18) may correspond to isomorphic polarized K3 surfaces.

Theorem 3.2 ([4]). *Two quartics $\mathrm{Q}(a_1,b_1,d_1)$ and $\mathrm{Q}(a_2,b_2,d_2)$ determine isomorphic M-polarized K3 surfaces as their minimal resolutions if and only if:*

$$(a_2,b_2,d_2) = (\lambda^2 a_1, \lambda^3 b_1, \lambda^6 d_1)$$

for some parameter $\lambda \in \mathbb{C}^*$.

One obtains therefore a *coarse moduli space* for M-polarized K3 surfaces in the form of the open variety:

$$(19) \qquad \mathcal{M}_{\mathrm{M}} = \{[a,b,d] \in \mathbb{WP}(2,3,6) \mid d \neq 0\}$$

as well as a pair of *fundamental $\mathcal{W}$-invariants* $(\mathcal{W}_1, \mathcal{W}_2)$:

$$\mathcal{W}_1 = \frac{a^3}{d}, \quad \mathcal{W}_2 = \frac{b^2}{d}.$$

We turn to Hodge theory. Denote by T the rank-four even indefinite lattice $\mathrm{H} \oplus \mathrm{H}$. Then, following the general framework of [10], one notes that the Hodge structure of an M-polarized K3 surface (X, i) can be seen as a point in the classifying space:

$$\Gamma \backslash \Omega \tag{20}$$

where

$$\Omega = \{\omega \in \mathbb{P}(\mathrm{T} \otimes \mathbb{C}) \mid (\omega, \omega) = 0, (\omega, \overline{\omega}) > 0\}$$

and Γ is the group of integral isometries of the lattice T, with its standard action on Ω. Upon further inspection, the quotient (20) turns out to be identical to the *classical Hilbert modular surface*:

$$\mathcal{H}_1 = \big(\mathrm{PSL}(2, \mathbb{Z}) \times \mathrm{PSL}(2, \mathbb{Z})\big) \rtimes \mathbb{Z}/2\mathbb{Z} \backslash \mathbb{H} \times \mathbb{H} \tag{21}$$

where $\mathbb{H}$ denotes a copy of the complex upper-half plane. This reinterpretation reflects, at a Hodge theoretic level, the geometric one-to-one correspondence existing between M-polarized K3 surfaces and abelian surfaces that split as a product of two elliptic curves.

The Global Torelli Theorem (see [10]) asserts then that the period map:

$$\mathrm{per}\colon \mathcal{M}_{\mathrm{M}} \to \mathcal{H}_1$$

is an analytic isomorphism. This isomorphism can be made completely explicit. Let $j\colon \mathbb{H} \to \mathbb{C}$ be the classical elliptic modular function. There are two important modular functions on the Hilbert surface $\mathcal{H}_1$, given, on the cover $\mathbb{H} \times \mathbb{H}$, by the two symmetric functions:

$$\sigma(\tau_1, \tau_2) = j(\tau_1) + j(\tau_2), \quad \pi(\tau_1, \tau_2) = j(\tau_1) \cdot j(\tau_2).$$

They can be seen naturally as coordinates on the surface $\mathcal{H}_1$ realizing an identification:

$$(\sigma, \pi)\colon \mathcal{H}_1 \xrightarrow{\simeq} \mathbb{C}^2.$$

Theorem 3.3 ([4]). *The inverse period map* $\mathrm{per}^{-1}\colon \mathcal{H}_1 \to \mathcal{M}_{\mathrm{M}}$ *is given by:*

$$\mathrm{per}^{-1} = [\pi^{1/3}, (\pi - \sigma + 1)^{1/2}, 1].$$

In other words, the $\mathcal{W}$*-invariants of an* M*-polarized* K3 *surface* (X, i) *are linked to the periods of* X *by the formulas:*

$$\mathcal{W}_1 = \pi, \quad \mathcal{W}_2 = \pi - \sigma + 1.$$

3.2. Loci supporting lattice enhancements. Since the rank 18 lattice M is unimodular, the only possible enhancements to M′-polarized lattices with $\mathrm{M} \subset \mathrm{M}'$ and M′ of rank 19 are those for which $\mathrm{M}' = \mathrm{M}_n = \mathrm{M} \oplus \langle -2n \rangle = \mathrm{H} \oplus \mathrm{E}_8 \oplus \mathrm{E}_8 \oplus \langle -2n \rangle$ for some $n \in \mathbb{N}$. The M_n polarized K3 surfaces for $n > 1$ correspond, via the Shioda-Inose construction, to products of elliptic curves with an n-isogeny between them. The extra algebraic cycle on the K3 corresponds to the graph of the n-isogeny on the product of the two elliptic curves. Determining the subloci of $\mathbb{WP}(2, 3, 6)$ on which these enhancements occur thus reduces to the problem of finding relations between the j-invariants of pairs of elliptic curves. This, of course, is a classical problem with a rich history.

Consider the map $\phi_n\colon \mathbb{H} \to \mathbb{P}^2$ given by $\tau \mapsto [j(\tau), j(n\tau), 1] = [x, y, z]$, and denote by $X_0(n)$ the closure of the image of ϕ_n. $\phi_n(\tau) = \phi_n(\tau')$ if and only if

$$\tau' = \frac{a\tau + b}{c\tau + d}, \quad n\tau' = \frac{\alpha n\tau + \beta}{\gamma n\tau + \delta}$$

for some

$$\begin{pmatrix} a & b \\ c & d \end{pmatrix}, \quad \begin{pmatrix} \alpha & \beta \\ \gamma & \delta \end{pmatrix} \in \mathrm{SL}_2(\mathbb{Z})$$

Examining these equations, we see that

$$\begin{pmatrix} na & nb \\ c & d \end{pmatrix} = \pm \begin{pmatrix} n\alpha & \beta \\ n\gamma & \delta \end{pmatrix}$$

which is possible if and only if $\tau' = (a\tau + b)/(c\tau + d)$, for

$$\begin{pmatrix} a & b \\ c & d \end{pmatrix} \in \Gamma_0(n).$$

Thus $X_0(n)$ is a compactification of $\Gamma_0(n)\backslash\mathbb{H}$, and parametrizes (ordered) pairs of n-isogenous elliptic curves.

The genus of $X_0(n)$ for many small n can be found in [7]. The equation $\Phi_n(x, y) = 0$ for $X_0(n)$ on the affine patch $z = 1$ is the classical modular equation for $\Gamma_0(n)$.[2] Because it is symmetric in x and y, it can be written as a polynomial in terms of the elementary symmetric functions $\pi = xy$, $\sigma = x + y$. Let $\Phi_n^{+n}(\pi, \sigma)$ be the corresponding polynomial such that $\Phi_n^{+n}(xy, x + y) = \Phi_n(x, y)$. Note that

$$\big(j(\tau)j(n\tau), j(\tau) + j(n\tau)\big) = \big(j(\tau')j(n\tau'), j(\tau') + j(n\tau')\big)$$

if and only if $\big(j(\tau), j(n\tau)\big) = \big(j(\tau'), j(n\tau')\big)$ or $\big(j(n\tau'), j(\tau')\big)$, which will occur if and only if $\tau' = (a\tau + b)/(c\tau + d)$ for some

$$\begin{pmatrix} a & b \\ c & d \end{pmatrix} \in \Gamma_0(n) + n.$$

We may thus view $\Phi_n^{+n}(\pi, \sigma) = 0$ as the modular relation for $\Gamma_0(n) + n$. Then, using the relationships $\pi = a^3$, $\sigma = a^3 - b^2 + 1$ from [4], we can rewrite $\Phi_n(x, y)$ as a polynomial $\Psi_n^{+n}(a^3, b^2)$, and view $\Psi_n^{+n}(a^3, b^2) = 0$ as the equation for a curve $Y_0(n) + n \subset \mathbb{WP}(2, 3, 6)$ on the affine patch $d = 1$ of the weighted projective plane with homogeneous coordinates $[a, b, d]$ of weights 2, 3, and 6 respectively. Note $X_0(n) + n \simeq Y_0(n) + n$. $Y_0(n) + n$ is exactly the moduli space of M_n-polarized K3 surfaces as a submoduli space of the moduli space of M-polarized K3 surfaces.

Thus the theory of modular curves and modular equations can be used to analyze the moduli of M_n-polarized K3 surfaces. For example, the genus of $Y_0(n)+n$ can be found in the literature, and in the case where $Y_0(n) + n$ has genus zero, it can be parametrized by the Hauptmodul for $\Gamma_0(n) + n$.

[2]The j-invariant we use is normalized so that $j(i) = 1$, not 1728 as in many works on modular equations in the literature.

3.2.1. *Examples: $n = 2$, 3, and 6.* Since $\Gamma_0(6)$ is contained in $\Gamma_0(3)$ and $\Gamma_0(2)$, it is natural to work over $X_0(6)$. Choosing as affine coordinate on $X_0(6) \simeq \mathbb{P}^1$ the Hauptmodul t for $\Gamma_0(6)$, the family of elliptic curves over $X_0(6)$ was shown in [2] to form a rational elliptic surface S with four singular fibers, of Kodaira types I_1, I_2, I_3, I_6. Beauville gave a model for this surface:

$$S = \{([x, y, z], t) \in \mathbb{P}^2 \times \mathbb{P}^1 : (x + y)(y + z)(z + x) + txyz = 0\}$$

with the singular fibers of types I_1, I_2, I_3, I_6 over the points

$$t = -8, 1, 0, \infty$$

respectively. Another model for this surface is given in [16]:

$$S' = \{([x, y, z], [\alpha, \beta]) \in \mathbb{P}^2 \times \mathbb{P}^1 : y^2z - 2(\alpha + \beta)xyz + 2\alpha\beta^2yz^2 + x^3 = 0\}$$

An explicit isomorphism between S and S' is

$$\begin{array}{ccc} S & \xrightarrow{f} & S' \\ \downarrow & & \downarrow \\ \mathbb{P}^1 & \xrightarrow{g} & \mathbb{P}^1 \end{array}$$

where f is defined by

$$([x, y, z], t) \mapsto \big(\big[tyz, t\big(z^2 + (2 + t)yz + y^2 + x(z + y)\big), -z(z + x)\big], [t/2, 1]\big)$$

and g is defined by

$$t \mapsto [t/2, 1]$$

To describe the 2, 3, and 6-isogenies explicitly, it is more convenient to use S'. We will denote by $S'_{[\alpha,\ \beta]}$ the fiber of S' over the point $[\alpha, \beta] \in \mathbb{P}^1$. From [16], we have that $S'_{[\alpha,\beta]}$ is 2-isogenous to $S'_{[-2\beta,\ \alpha]}$, 3-isogenous to $S'_{[-\alpha-4\beta,-2\alpha+\beta]}$, and 6-isogenous to $S'_{[4\alpha-2\beta,-\alpha-4\beta]}$. The isogenies are not given in [16], but [3] gives an explicit 2-isogeny, and [20] gives an explicit 3-isogeny for families of elliptic curves. Comparing these families with S' and noting that the 6-isogeny $S'_{[\alpha,\ \beta]} \to S'_{[4\alpha-2\beta,-\alpha-4\beta]}$ is the composition of the 2-isogeny $S'_{[\alpha:\beta]} \to S'_{[-2\beta,\alpha]}$ with the 3-isogeny $S'_{[-2\beta,\alpha]} \to S'_{[4\alpha-2\beta,-\alpha-4\beta]}$, we can derive explicitly what the isogenies $\phi_n \colon S' \to S'$ are for $n = 2, 3, 6$. Denote $\phi_n(([x, y, z], [\alpha, \beta])$ by $([x'_n, y'_n, z'_n], [\alpha'_n, \beta'_n])$. Then $x'_n, y'_n, z'_n, \alpha'_n, \beta'_n$ are given below for $n = 2, 3$. (In the interest of space, we omit the corresponding data for $n = 6$, but we note that it can easily be deduced, since ϕ_6 is the composition of ϕ_2 and ϕ_3.)

$$\begin{aligned}
x'_2 &= x(\beta^2z - x)(2\alpha\beta z - x)\\
y'_2 &= (\beta - 2\alpha)x^3 + 2\alpha\beta^3yz^2 - 2\beta^2xyz + x^2y\\
z'_2 &= (\beta^2z - x)^2z\\
\alpha'_2 &= -2\beta\\
\beta'_2 &= \alpha
\end{aligned}$$

$$\begin{aligned}
x'_3 &= x\big((4\beta^2z - 3x)x^2 + 4\alpha\beta xz(-3\beta^2z + 2x) + 4\alpha^2z(3\beta^4z^2 - 3\beta^2xz + x^2)\big)\\
y'_3 &= -8\alpha^3z(3\beta^6z^3 - 3\beta^2x^2z + x^3) + x^3(-8\beta^3z + 12\beta x - 3y)\\
&\quad - 12\alpha^2\beta z(2\beta^2z - x)(-2x^2 + \beta yz) + 6\alpha x(4\beta^4xz^2 - 9\beta^2x^2z + 2x^3 + 2\beta^3yz^2)
\end{aligned}$$

$$
\begin{aligned}
&\quad - 6\zeta_3\big(-4\alpha\beta^3xz^2 + x^3 + \alpha^2 z(8\beta^4 z - 4\beta^2 x)\big)\big(\alpha(\beta^2 z - x) - \beta x + y\big)\\
z_3' &= x^3 z\\
\alpha_3' &= -\alpha - 4\beta\\
\beta_3' &= -2\alpha + \beta
\end{aligned}
$$

where ζ_3 is a primitive third root of unity.

In order to understand $Y_0(n)+n$ for $n = 2, 3, 6$ as a sublocus of $\mathbb{WP}[2,3,6]$, there are two ways one could proceed. One is to find the defining equation $\Psi_n^{+n}(a,b,d) = 0$ for $Y_0(n)+n$, and another — since in these cases $Y_0(n)+n$ is genus zero — is to find a parametrization $[a(t), b(t), d(t)]$ for the curve. Of course, these two points of view are closely related. Given a parametrization one can eliminate the parameter to obtain a defining equation, and conversely given the defining equation for a genus zero curve one can employ standard algorithms (for example in [1]) to obtain a parametrization.

Defining equations for $X_0(n)$ for small n are readily available (for example, in MAGMA), and from these we can obtain defining equations $\Psi_n^{+n}(a,b,d) = 0$ for $Y_0(n) + n$. We give $\Psi_n^{+n}(a,b,d)$ for $n = 2,3$ here. $\Psi_6^{+6}(a,b,d)$ is a degree 72 polynomial with large coefficients, so we will omit it.

$$
\begin{aligned}
\Psi_2^{+2}(a,b,d) &= 64a^9 - 192b^2a^6 - 21360da^6 + 192b^4a^3 + 1792857d^2a^3\\
&\quad - 83424b^2da^3 - 64b^6 - 1771561d^3 - 175692b^2d^2 - 5808b^4d\\
\Psi_3^{+3}(a,b,d) &= 729a^{12} - 2916b^2a^9 - 129551076da^9 + 4374b^4a^6\\
&\quad + 5754777529878d^2a^6 - 1298340252b^2da^6 - 2916b^6a^3\\
&\quad + 77703185570076d^3a^3 - 3974452231068b^2d^2a^3\\
&\quad - 733336092b^4da^3 + 729b^8 + 262254607552729d^4\\
&\quad - 262365230658916b^2d^3 + 110638660374b^4d^2 - 15554916b^6d
\end{aligned}
$$

Given a parametrization $(x(t_n), y(t_n))$ of $X_0(n)$ on the affine patch $z \neq 0$, using coordinates (b^2, d) on the affine patch $a \neq 0$ of $\mathbb{WP}(2,3,6)$, we can construct a parametrization of $Y_0(n) + n$ by setting

$$
\begin{aligned}
b^2(t_n) &= \frac{(x(t_n) - 1)(y(t_n) - 1)}{x(t_n)y(t_n)},\\
d(t_n) &= \frac{1}{x(t_n)y(t_n)}.
\end{aligned}
$$

[24] gives such a parametrization $(x(t_n), y(t_n))$ by the Hauptmodul t_n for $\Gamma_0(n)$ for $n \leq 25$ such that $X_0(n)$ is a genus 0 curve. These parametrizations for $n = 2,3,6$ give

$$
\begin{aligned}
b^2(t_2) &= \frac{(t_2 - 512)^2(t_2 - 8)^2(t_2 + 64)^2}{(t_2 + 16)^3(t_2 + 256)^3}\\
d(t_2) &= \frac{2^{12}3^6t^3}{(t + 16)^3(t_2 + 256)^3}\\
b^2(t_3) &= \frac{(t_3^2 - 486t_3 - 19683)^2(t_3^2 + 18t_3 - 27)^2}{(t_3 + 27)^2(t_3^2 + 246t_3 + 729)^3}
\end{aligned}
$$

$$d(t_3) = \frac{2^{12}3^6 t_3^4}{(t_3+3)^3(t_3+27)^2(t_3+243)^3}$$

$$\begin{aligned} b^2(t_6) &= (t_6^2+12t_6+24)^2(t_6^2+36t_6+216)^2 \\ &\quad \times (-t_6^4+504t_6^3+13824t_6^2+124416t_6+373248)^2 \\ &\quad \times \frac{(t_6^4+24t_6^3+192t_6^2+504t_6-72)^2}{\left((t_6+6)(t_6+12)(t_6^3+18t_6^2+84t_6+24)(t_6^3+252t_6^2+2^4 3^5 t_6+2^6 3^5)\right)^3} \\ d(t_6) &= \frac{2985984 t_6^7(t_6+8)^5(t_6+9)^5}{\left((t_6+6)(t_6+12)(t_6^3+18t_6^2+84t_6+24)(t_6^3+252t_6^2+2^4 3^5 t_6+2^6 3^5)\right)^3} \end{aligned}$$

Another interesting parametrization of $Y_0(n)+n$ can be derived from the "two-valued" modular equations derived, following Fricke, in [7].

3.3. Picard – Fuchs differential equations. Let us set $a = 1$ in (18) and consider the resulting polynomial $Q = y^2zw - 4x^3z + 3xzw^2 + bzw^3 - \frac{1}{2}(dz^2w^2 + w^4)$. (We have simply reduced to the affine patch $a \neq 0$ of the parameter space $\mathbb{WP}(2,3,6)$.) Applying the Griffiths – Dwork technique to $\int \Omega_0/Q$ yields a pair of second-order Picard – Fuchs equations:

$$\text{(22)} \quad \frac{\partial^2}{\partial b^2}\int\frac{\Omega_0}{Q} - 4\Big(d\frac{\partial^2}{\partial d^2}\int\frac{\Omega_0}{Q} + \frac{\partial}{\partial d}\int\frac{\Omega_0}{Q}\Big) = 0$$

$$\text{(23)} \quad (-1+b^2+d)\frac{\partial^2}{\partial b^2}\int\frac{\Omega_0}{Q} + 2b\frac{\partial}{\partial b}\int\frac{\Omega_0}{Q} + 4bd\frac{\partial^2}{\partial b\partial d}\int\frac{\Omega_0}{Q} + 2d\frac{\partial}{\partial d}\int\frac{\Omega_0}{Q} + \frac{5}{36}\int\frac{\Omega_0}{Q} = 0.$$

We can use the relationship between b, d and the j-invariants of elliptic curves from Theorem 3.3 to write $b^2 = (j_1-1)(j_2-1)/(j_1j_2)$ and $d = 1/(j_1j_2)$. Here j_1 and j_2 are the j-invariants of the two elliptic curves E_1 and E_2 whose product corresponds to $X(1,b,d)$. Let E_i have affine Weierstrass model

$$y^2 = 4x^3 - g_2^{(i)}x - g_3^{(i)}$$

for $i = 1, 2$. Then we can rewrite (22) and (23) in terms of j_1 and j_2. The resulting system decouples (no mixed partials appear). The system reduces (after taking appropriate linear combinations of the resulting equations) to

$$\begin{aligned} 0 &= 72j_1\big((2j_1-1)F^{(1,0)}(j_1,j_2) + 2(j_1-1)j_1F^{(2,0)}(j_1,j_2)\big) - 5F(j_1,j_2) \\ 0 &= 72j_2\big((2j_2-1)F^{(0,1)}(j_1,j_2) + 2(j_2-1)j_2F^{(0,2)}(j_1,j_2)\big) - 5F(j_1,j_2) \end{aligned}$$

where $F^{(i,j)}(j_1,j_2) = \partial^{i+j}F/\partial j_1^i \partial j_2^j$.

To solve this system, one need merely solve each ODE separately, then take products of the solutions. Each of these ODEs separately is a Picard – Fuchs differential equation satisfied by the periods of the form $\omega^{(i)} = \big(g_2^{(i)}\big)^{1/4}\mathrm{d}x/y$. Thus periods satisfying the Picard – Fuchs system arising via Griffiths – Dwork are simply products of periods of $\omega^{(1)}$ and $\omega^{(2)}$ (cf. [23, Theorem 1.1]).

Now, consider a one-parameter family $\mathcal{F}$ of M-polarized K3 surfaces obtained by treating b and d as functions of a single parameter t. We may use the Griffiths – Dwork technique to analyze this family, just as we computed the Picard – Fuchs equation for a one-parameter family of elliptic curves in Section 2.2.2. The result is generically a fourth-order ODE, which we do not reproduce in full here. The Picard–Fuchs equation for $\mathcal{F}$ will reduce to a third-order ODE precisely when $\mathcal{F}$ is an M_n-polarized family.

Let $j_1(t)$, $j_2(t)$ be two functions of a complex variable t such that $j_1(t) + j_2(t)$ and $j_1(t)j_2(t)$ are rational functions of t. In this case $b^2(t) = (j_1(t)-1)\times(j_2(t)-1)/(j_1(t)j_2(t))$ and $d(t) = 1/(j_1(t)j_2(t))$ are also rational functions of t, and we may write the Picard–Fuchs equation for $\mathcal{F}$ in terms of $j_1(t)$ and $j_2(t)$. The coefficient $r_4(t)$ of $\mathrm{d}^4(\int \Omega_0/Q)/\mathrm{d}t^4$ in the Picard–Fuchs ODE then becomes

$$144\big((j_1(t)-1)(j_2(t)-1)\big)^3\big(j_1(t)j_2(t)\big)^4\big(j_1(t)-j_2(t)\big)^7\big(j_1'(t)j_2'(t)\big)^2 \times \Big(\Box\big(j_2(t)\big) - \Box\big(j_1(t)\big)\Big)$$

where

$$\Box\big(j(t)\big) = j'(t)^2\frac{36j(t)^2 - 41j(t) + 32}{144(j(t)-1)^2 j(t)^2} + \frac{1}{2}\{j(t), t\}$$

and

$$\{j(t), t\} = \frac{2j'(t)j'''(t) - 3j''(t)^2}{2j'(t)^2}$$

is the Schwarzian derivative.

If $j_1(t)$ and $j_2(t)$ are both nonconstant, then $r_4(t)$ will vanish if and only if either $j_1(t) = j_2(t)$ (in which case the family of K3 surfaces is M_1-polarized) or $\Box\big(j_1(t)\big) = \Box\big(j_2(t)\big)$. This observation motivates the following theorem.

Theorem 3.4. *Let $j_1(t), j_2(t)$ be nonconstant functions of a complex variable t such that $j_1(t)+j_2(t)$ and $j_1(t)j_1(t)$ are rational functions of t. Then*

$$\Box\big(j_1(t)\big) = \Box\big(j_2(t)\big) \tag{24}$$

if and only if $(j_1(t), j_2(t))$ is a parametrization of $X_0(n)$ for some $n \geq 1$.

Proof. The Picard–Fuchs ODE for $\mathcal{F}$, suitably normalized, is the tensor product of the Picard–Fuchs ODE's of the two pencils of elliptic curves over $\mathbb{P}^1_t$ with functional invariants $j_1(t), j_2(t)$ respectively. If these second-order ODE's $L_1 = 0, L_2 = 0$ are in projective normal form

$$L_1 = \frac{\mathrm{d}^2 f}{\mathrm{d}t^2} + p_2(t)f, \quad L_2 = \frac{\mathrm{d}^2 g}{\mathrm{d}t^2} + q_2(t)g$$

then $p_2(t) = \Box\big(j_1(t)\big)$ and $q_2(t) = \Box\big(j_2(t)\big)$. Their tensor product is

$$\begin{aligned}0 = H^{(4)}(t) &+ \frac{q_2'(t)-p_2'(t)}{p_2(t)-q_2(t)}H'''(t) + 2(p_2(t)+q_2(t))H''(t)\\ &+ \frac{p_2(t)\big(p_2'(t)+5q_2'(t)\big) - q_2(t)\big(5p_2'(t)+q_2'(t)\big)}{p_2(t)-q_2(t)}H'(t)\\ &+ \left(\big(p_2(t)-q_2(t)\big)^2 + p_2''(t) + q_2''(t) + \frac{q_2'(t)^2 - p_2'(t)^2}{p_2(t)-q_2(t)}\right)H(t).\end{aligned} \tag{25}$$

According to [14], this fourth-order equation factorizes as a third-order equation times a first-order equation if and only if $p_2(t) = q_2(t)$, i.e., if and only if $\Box\big(j_1(t)\big) = \Box\big(j_2(t)\big)$. On the other hand, the Picard–Fuchs equation of $\mathcal{F}$ has third order if and only if $\mathcal{F}$ is M_n-polarized, and this occurs if and only if the two pencils of elliptic curves are fiberwise n-isogenous–or in other words, if and only if $\big(j_1(t), j_2(t)\big) \in X_0(n)$ for all t. □

For this reason, we call (24) *the master equation for modular parametrization* of modular equations for the elliptic modular function $j(\tau)$.

Corollary 3.5. *If two pencils of elliptic curves over $\mathbb{P}^1_t$ admit a fiberwise n-isogeny for n such that $X_0(n)+n$ is genus 0, then the projective normal forms of their Picard–Fuchs differential operators are identical.*

3.3.1. *Modular relations and differential identities for Hauptmoduls.* Theorem 3.4, which relates the existence of a parametrized modular relation between $j(\tau)$ and $j(n\tau)$ to a differential identity involving a parametrization, can be generalized to other Hauptmoduls. If h is a Hauptmodul for a genus 0 modular group Γ, then h will satisfy a Schwarzian differential equation

$$h'(\tau)^2 Q_\Gamma(h(\tau)) + \tfrac{1}{2}\{h(\tau),\tau)\} = 0 \tag{26}$$

where $Q_\Gamma(h)$ is a rational function we shall call the "Q-value for h." A list of "Q-values" for (suitably normalized) Hauptmoduls for all genus 0 modular groups are given in [21].

As was noted in [17, 18], if the Hauptmodul h for Γ can be expressed as a function $h_1(t)$ of a Hauptmodul t for a genus 0 group Γ', then

$$h_1'(t)^2 Q_\Gamma(h_1(t)) + \tfrac{1}{2}\{h_1(t),t\} = Q_{\Gamma'}(t) \tag{27}$$

This identity follows quite easily from formally writing $h_1(t) = h(\tau(t))$, and applying the chain rule to (26). If we can also write $h(n\tau)$ as a function $h_2(t)$, then we have an analogous result to (27) with $h_1(t)$ replaced by $h_2(t)$, and hence we have

Theorem 3.6. *If h and t are Hauptmoduls for genus zero modular groups Γ and Γ' respectively, and if $h(\tau), h(n\tau)$ can both be expressed as rational functions of t then*

$$h_1'(t)^2 Q_\Gamma(h_1(t)) + \tfrac{1}{2}\{h_1(t),t\} = h_2'(t)^2 Q_\Gamma(h_2(t)) + \tfrac{1}{2}\{h_2(t),t\} = Q_{\Gamma'}(t). \tag{28}$$

Such a situation will occur when $h(\tau), h(n\tau)$ satisfy a modular equation of genus zero. Theorem 3.4 covers the special case $\Gamma = \mathrm{PSL}(2,\mathbb{Z})$, $\Gamma' = \Gamma_0(n)$.

For example, [8] discusses the Hauptmoduls satisfying modular equations of levels two and three, and gives formulae for the modular equations. Parametrizing the curves defined by these equations, computing the left-hand side of (27), and comparing with list of "Q-values" in [21], we can verify Theorem 3.6 in these cases (and identify the group Γ' if it is not already known). We illustrate this for $\Gamma = \Gamma_0(3)+3$ below.

The Hauptmodul h for $\Gamma = \Gamma_0(3)+3$ has Q-value $(h^2-48h+7560)/(4(h^2-24h-2772)^2)$ and satisfies the level two modular equation $\Phi_2\big(h(\tau),h(2\tau)\big) = 0$ where

$$\Phi_2(h_1,h_2) = h_1^3 - h_2^2h_1^2 + 17343h_2h_1 + h_2^3 + 741474(h_1+h_2) + 1566(h_1^2+h_2^2) + 28166076.$$

The curve defined by $\Phi_2(h_1,h_2) = 0$ can be parametrized by setting

$$h_1(t) = \frac{-512t^3+804t^2-12t+1}{(1-6t)^2t}, \quad h_2(t) = \frac{244t^3-30t^2-6t+1}{t^2-6t^3}$$

Evaluating the left-hand side of (27), we see that

$$Q_{\Gamma'}(t) = \frac{2848t^4-800t^3+108t^2+4t+1}{4t^2(120t^3-68t^2+2t+1)^2}$$

and by comparing with [21] we see that t is a Hauptmodul for $\Gamma' = \Gamma_0(6)+3$.

3.4. Relationship to toric geometry. Another model for the two parameter family of M-polarized K3 surfaces comes from toric geometry. The 2-parameter family can be realized as the family of anticanonical hypersurfaces in the mirror (polar) toric variety X to $\mathbb{WP}(1,1,4,6)$. The family of anticanonical K3's K has defining equation

$$f_{(\lambda_0,\lambda_1,\ldots,\lambda_5)}(x_0,x_1,x_2,x_3) = \lambda_0 x_0x_1x_2x_3 + \lambda_1 x_0^{12} + \lambda_2 x_1^{12} + \lambda_3 x_2^3 + \lambda_4 x_3^2 + \lambda_5 x_0^6 x_1^6 = 0$$

in the global homogeneous coordinate ring $\mathbb{C}[x_0,x_1,x_2,x_3]$ of X (where x_0,x_1,x_2,x_3 have weights $1,1,4,6$ respectively).

As in Section 2.3, these six parameters $(\lambda_0,\lambda_1,\ldots,\lambda_5)$ are redundant. Let $\mathcal{M}_{\text{simp}}$ be the simplified polynomial moduli space for K. $\mathcal{M}_{\text{simp}}$ is a two-dimensional toric variety and is a finite-to-one cover (generically) of the actual moduli space $\mathcal{M}$ (which in the present case is $\mathbb{WP}(2,3,6)$). We will use $(z_1,z_2) = (\lambda_3^2\lambda_4^3\lambda_5/\lambda_0^6, \lambda_1\lambda_2/\lambda_5^2)$ as affine coordinates on the torus $(\mathbb{C}^*)^2 \subset \mathcal{M}_{\text{simp}}$. We can use the fact that the defining equation is only defined up to an overall nonzero constant to set $\lambda_4 = 1$. Then we can use the natural action of $T \simeq (\mathbb{C}^*)^3$ on X to set $\lambda_1 = \lambda_2 = -1/2$, and $\lambda_3 = -4$. Then our simplified polynomial moduli are $(z_1,z_2) = (16\lambda_5/\lambda_0^6, 64/\lambda_5$.

Let ϕ be the rational map $X \to \mathbb{P}^3$ defined by

$$(x_0,x_1,x_2,x_3) \mapsto \left(\frac{x_2}{x_0^2x_1^2} + \frac{\lambda_0^2}{48}, \frac{x_3}{x_0^3x_1^3} + \frac{\lambda_0 x_2}{2x_0^2x_1^2}, \frac{x_1^6}{x_0^6}, 1\right) = (x,y,z,w).$$

The image of K under ϕ has defining equation

$$0 = y^2zw - 4x^3z + \frac{\lambda_0^4}{192}xzw - \frac{1}{2}(z^2w^2 + w^4) + \left(\frac{-\lambda_0^6}{13824} + \lambda_5\right)zw^3.$$

Comparing this with the Inose normal form for M-polarized K3 surfaces, we get a map $\mathcal{M}_{\text{simp}} \to \mathbb{WP}(2,3,6)$ given in affine coordinates by

$$(z_1,z_2) \mapsto \left(\frac{1}{144^3z_1^2z_2}, \frac{864z_1 - 1}{144^3z_1^2z_2}\right) = (a^3, b^2).$$

The Picard–Fuchs equations for this family has been computed from GKZ methods (e.g., by Lian–Yau in [23]) as

$$0 = L_1 f(z_1,z_2) = \big(\theta_1(\theta_1 - 2\theta_2) - 12z_1(6\theta_1+5)(6\theta_1+1)\big)f(z_1,z_2) \tag{29}$$

$$0 = L_2 f(z_1,z_2) = \big(\theta_2^2 - z_2(2\theta_2 - \theta_1 + 1)(2\theta_2 - \theta_1)\big)f(z_1,z_2). \tag{30}$$

See also [29] for some related discussion of the Lian–Yau example.

Changing coordinates into the (b^2,d) affine patch on $\mathbb{WP}(2,3,6)$ gives

$$(z_1,z_2) \mapsto 864z_1 - 1, 144^3z_1^2z_2) = (b^2,d)$$

and plugging this into (22), (23) gives

$$0 = -\frac{1}{z_2}L_2F(z_1,z_2) \tag{31}$$

$$0 = \left(-\frac{1}{z_1}L_1 + 1728L_2\right)F(z_1,z_2) \tag{32}$$

Therefore, as in Section 2.3, the solution spaces of the Picard–Fuchs equations computed via GKZ and Griffiths–Dwork are identical.

References

1. S. S. Abhyankar and C. L. Bajaj, *Automatic parametrization of rational curves and surfaces. III: Algebraic plane curves*, Comput. Aided Geom. Design **5** (1988), no. 4, 309–321.
2. A. Beauville, *Les familles stables de courbes elliptiques sur* $\mathbb{P}^1$ *admettant quatre fibres singulières*, C. R. Acad. Sci. Paris Sér. I Math. **294** (1982), no. 19, 657–660.
3. J. W. S. Cassels, *Lectures on elliptic curves*, London Math. Soc. Stud. Texts, vol. 24, Cambridge Univ. Press, Cambridge, 1991.
4. A. Clingher and C. F. Doran, *Modular invariants for lattice polarized* K3 *surfaces*, Michigan Math. J. **55** (2007), no. 2, 355–393.
5. ———, *Lattice polarized* K3 *surfaces and Siegel modular forms*, preprint.
6. ———, *On a geometric isogeny of* K3 *surfaces*, preprint.
7. H. Cohn, *Fricke's two-valued modular equations*, Math. Comp. **51** (1988), no. 184, 787–807.
8. H. Cohn and J. McKay, *Spontaneous generation of modular invariants*, Math. Comp. **65** (1996), no. 215, 1295–1309.
9. D. Cox and S. Katz, *Mirror symmetry and algebraic geometry*, Math. Surveys Monogr., vol. 68, Amer. Math. Soc., Providence, RI, 1999.
10. I. V. Dolgachev, *Mirror symmetry for lattice polarized* K3 *surfaces*, J. Math. Sci. **81** (1996), no. 3, 2599–2630.
11. C. F. Doran, *Picard–Fuchs uniformization: modularity of the mirror map and mirror-moonshine*, The Arithmetic and Geometry of Algebraic Cycles (Banff, 1998) (B. B. Gordon, J. D. Lewis, S. Müller-Stach, S. Saito, and N. Yui, eds.), CRM Proc, Lecture Notes, vol. 24, Amer. Math. Soc., Providence, RI, 2000, pp. 257–281.
12. ———, *Picard–Fuchs uniformization and modularity of the mirror map*, Comm. Math. Phys. **212** (2000), no. 3, 625–647.
13. C. F. Doran, B. Greene, and S. Judes, *Families of quintic Calabi–Yau 3-folds with discrete symmetries*, available at `arXiv:hep-th/0701206v1`.
14. G. Fano, *Über lineare homogene Differentialgleichungen mit algebraischen Relationen zwischen den Fundamentallösungen*, Math. Ann. **53** (1900), no. 4, 493–590.
15. P. A. Griffiths, *On the periods of certain rational integrals.* I, Ann. of Math. (2) **90** (1969), 460–495.
16. T. Hadano, *Elliptic curves with a torsion point*, Nagoya Math. J. **66** (1977), 99–108.
17. J. Harnad, *Picard–Fuchs equations, Hauptmoduls and integrable systems*, Integrability: The Seiberg–Witten and Witham Equations (Edinburgh, 1998) (H. W. Braden and I. M. Krichever, eds.), Gordon and Breach, Amsterdam, 2000, pp. 137–152.
18. J. Harnad and J. McKay, *Modular solutions to generalized Halphen equations*, Proc. Roy. Soc. London Ser. A **456** (2000), no. 1994, 261–294.
19. H. Inose, *Defining equations of singular* K3 *surfaces and a notion of isogeny*, Proceedings of the International Symposium on Algebraic Geometry (Kyoto, 1977) (M. Nagata, ed.), Kinokuniya Book Store, Tokyo, 1978, pp. 495–502.
20. M. Kuwata, *Quadratic twists of an elliptic curve and maps from a hyperelliptic curve*, Math. J. Okayama Univ. **47** (2005), no. 1, 85–97.
21. Lian B. and J. Wiczer, *Genus zero modular functions*, available at `arXiv:math/0611291v1`.
22. B. Lian and S.-T. Yau, *Arithmetic properties of mirror map and quantum coupling*, Comm. Math. Phys. **176** (1996), no. 1, 163–191.
23. ———, *Mirror maps, modular relations, and hypergeometric series.* II, Nuclear Phys. B Proc. Suppl. **46** (1996), 248–262.
24. R. S. Maier, *On rationally parametrized modular equations*, available at `arXiv:math/0611041v3`.
25. D. R. Morrison, *On* K3 *surfaces with large Picard number*, Invent. Math. **75** (1984), no. 1, 105–121.
26. C. Peters, *Monodromy and Picard–Fuchs equations for families of* K3*-surfaces and elliptic curves*, Ann. Sci. École Norm. Sup. (4) **19** (1986), no. 4, 583–607.
27. U. Schmickler-Hirzebruch, Schriftenreihe Math. Inst. Univ. Munster 2. Ser., vol. 33, Univ. Münster, Münster, 1985.
28. J. Top and N. Yui, *Explicit equations of some elliptic modular surfaces*, Rocky Mountain J. Math. **37** (2007), no. 2, 663–687.

29. Y. Yang and N. Yui, *Differential equations satisfied by modular forms and* K3 *surfaces*, Illinois J. Math. **51** (2007), no. 2, 667–696.

Department of Mathematics and Computer Science, University of Missouri – St. Louis, St. Louis, MO 63121, USA
E-mail address: clingher@arch.umsl.edu

Department of Mathematics, University of Washington, Seattle, WA 98195, USA
E-mail address, C. F. Doran: doran@math.washington.edu
E-mail address, J. Lewis: jacobml@math.washington.edu
E-mail address, U. Whitcher: ursula@math.washington.edu

Centre de Recherches Mathématiques
CRM Proceedings and Lecture Notes
Volume **47**, 2009

Spontaneous Generation of Hilbert Modular Functions

Harvey Cohn

ABSTRACT. The method of McKay and the author for finding the modular function $j(z)$ uses only Newton polygon structure at infinity for several modular equations without knowledge of the coefficients. An outline is given here with minimal computational detail so that it may be viewed as a *method* and not as a special trick. It is also appropriate to view what is likely to be the simplest extension of this method, to the Hilbert modular functions. While the current limited computing techniques would make such an extension a stupefying challenge today, some comments on such a procedure can still be made. They will show how some of the computer algebraic techniques also rise to a higher order.

Designation of "spontaneous"

It is now almost classical that the modular function $j(z)$ may be deduced from the modular equation connecting $j(z)$ with $j(2z)$. A joint paper with John McKay (see [3]) showed that the exact coefficients of the (polynomial) modular equation are unnecessary (only the Newton polygon data are needed), if enough equations between $j(z)$ and $j(pz)$, (p prime), are used. If there are not enough, then invariants from replicability theory also occur. This construction is called "spontaneous generation" because it springs up independently of the elaborate history of Fuchsian groups.

Summary of basic method

The computation involved in this result is as follows: We start with a series for $j(z)$, based only on $j(z) = j(z+1)$, with coefficients to be determined,

$$j(z) = 1/q + c_0 + c_1 q + c_2 q^2 + \cdots, \quad q = \exp(2\pi i z), \tag{1}$$

(and we take $c_0 = 0$ for convenience). The other condition characterizing $j(z)$ is $j(z) = j(-1/z)$ and it leads to the conclusion that the general modular equation will be given by a polynomial $g_p(u, v)$ symmetric in

$$u = j(z), \qquad v = j(pz), \quad (p \text{ prime}). \tag{2}$$

2000 *Mathematics Subject Classification.* Primary 11F03; Secondary 11R11, 11-04.
This is the final form of the paper.

Symmetry follows for $g_p(u,v)$, since under $z \to -1/z$, $u(z)$ $(= j(z))$ is unchanged while $v \to j(-p/z)$ $(= j(z/p))$, so u and v become interchanged. The degree of $g_p(u,v)$ in each variable is $p+1$ because of the set of conjugates $\{j(pz), j((z+m)/p), 0 \le m < p-1\}$.

Therefore (different) branches at $\Im z \to \infty$ (or $q \to 0$) satisfy

$$(3) \qquad u \approx v^p, \quad v \approx u^p.$$

With such Newtonian polygon data, u and v satisfy a polynomial equation of the type

$$(4) \quad g_p(u,v) = u^{p+1} + v^{p+1} - u^p v^p + \sum_{i,j=0}^{p} a_{i,j} u^i v^j = 0 \quad (a_{i,j} = a_{j,i},\ a_{p,p} = 0).$$

Thus in addition to the unknown c_t, there are $p(p+3)/2$ additional unknown $a_{i,j}$. By substituting $j(z)$ in (1) into polynomial equation (4), we obtain a formal series vanishing identically (see [5]),

$$(5) \qquad g_p(u,v) = q^{-Q} \sum_0^{\infty} M_r q^r \quad (Q = p^2 + p - 1).$$

By eliminating the $a_{i,j}$ we obtain recursions expressing the unknown c_t in terms of a subset S_p of size $\le Q$. *What we do then is to use enough different values of* p, *to generate enough interrelations to determine the* c_t. This requires electronic computation at a level which was only recently achieved!

The computation

We start with $p = 2$, $(Q = 5)$ and $p = 3$, $(Q = 11)$ where these subsets are

$$(6) \qquad S_2 = (c_1, c_2, c_3, c_5), \quad S_3 = (c_1, c_2, c_3, c_4, c_5, c_7, c_8, c_{11}).$$

Note the gaps, e.g., in S_2, $c_4 = (-c_1 + 2c_3 + c_1^2)/2$. Obviously, a plethora of identities will be created by expressing c_t in terms of S_2 or S_3. We first express S_3 in terms of S_2 leading to four unknowns, and then we expand $g_3(u,v)$ into the formal series (5). All M_r must vanish, but they are algebraically dependent in a mysterious way, so that $M_9, M_{10}, M_{14}, M_{17}$ alone determine S_2. They are unfactorable, but they lead to factorable resultants. Ultimately, we obtain 29 sets S_2 (see [3]) which generate 29 series

$$(7) \qquad u = 1/q + c_1 q + c_2 q^2 + c_3 q^3 + \cdots.$$

These include the one for $j(z)$ where

$$(8) \quad (S_2 =)\ (c_1, c_2, c_3, c_5) = (196884, 21493760, 864299970, 333202640600).$$

There are obvious trivialities, $u = 1/q$ and $u = 1/q + q$, so that there are 26 nontrivial extraneous series to eliminate to specify $j(z)$.

Extraneous solutions

These 26 functions are very relevant as "replicable functions" or in a more classical context as extended modular functions due (at various times) to R. Fricke, E. Bessel-Hagen, A. O. L. Atkin, and J. Lehner. An extended modular group is a subgroup of the following group parametrized (see [2]) by N and T (where $N \geq 1$, $T \mid N$, $\gcd(N, N/T) = 1$):

(9) $G_{N,T} = \{(Az+B)/(Cz+D)\}$, where $AD - BC = T$, $N \mid B$, $T \mid \gcd(A, D)$.

Such groups act on the upper-half plane, and when the fundamental domain is of genus zero, then there is a one-valued function ("Hauptmodul" or "principal modulus") of type (7) which satisfies a modular equation of type (4), *except when* $p \mid N$. (Of course q must be taken as $\exp(2\pi \mathrm{i} z/N)$). It is enough to just list 26 such cases (other than the modular group) but individual identifications can be made using a list of replicable functions (see [6]).

Thus from the genus zero list, the Newtonian structures (4) needed to narrow down $j(z)$ by elimination may be limited to the primes

$$p \in \{2, 3, 5, 7, 11, 13, 17, 19, 23, 29, 31, 41, 47, 59, 71\}. \tag{10}$$

They are called variously the "Fricke" primes because of the genus zero property and "Monster" primes because they divide the order of the Monster group.

Hilbert modular functions

A reasonable challenge (of which there are not many to choose from) might be an extension to a field of Hilbert modular functions. Such a field exists for every real quadratic ring, but we use one illustration, for $\mathbb{Z}[\sqrt{2}]$ (see [4]). It is rational (when symmetrized), and more important, it has an illustrative modular equation (system) which can be calculated (see [1]) because the ring has an element of small enough norm (two).

The group $\mathrm{PSL}(2, \mathbb{Z}[\sqrt{2}])$ acts on the upper half-planes $\Im z > 0$, $\Im z' > 0$. It shall be symmetrized so that z and z' are interchangeable. Actions of this group at infinity are generated by

$$(z, z') \to (z+1, z'+1),\ (z+\sqrt{2}, z'-\sqrt{2}),\ \big((3+2\sqrt{2})z, (3-2\sqrt{2})z'\big), \tag{11}$$

(note the unit $3 \pm 2\sqrt{2}$). The group has the additional generator $(z, z') \to (-1/z, -1/z')$ and the symmetry $(z, z') \to (z', z)$.

The invariant functions are now ratios of modular forms which naturally have *double* series expansions at infinity. They may still be treated like a simple series because of a special compactification: Every series is indexed by $\alpha \in \mathbb{Z}[\sqrt{2}]$ where $\alpha = a + b\sqrt{2}$, $\alpha' = a - b\sqrt{2}$, and $\alpha > 0 > \alpha'$. The series is typically

$$\sum c_\alpha \exp(\pi(\alpha z - \alpha' z')/\sqrt{2}) = \sum c_\alpha q^b r^a, \tag{12a}$$

where

$$q = \exp\big(\pi \mathrm{i}(z + z')\big),\ r = \exp(\pi \mathrm{i}(z - z')/\sqrt{2}), \quad (\text{so } |q| < |r|^{\sqrt{2}}). \tag{12b}$$

By the unit condition in (11), α is restricted to nonassociates (ideals) making $|a| \leq b$. While $q \to 0$ does not make $r \to 0$, the terms of (12a) clearly vanish. They can be viewed as a power series primarily in q, multiplied by polyomials in r and $1/r$ (really in $r + 1/r$).

Such series describe the two "j-invariants" in two complex variables for the group $\mathrm{PSL}(2, \mathbb{Z}[\sqrt{2}])$ acting symmetrically. The modular equation now becomes replaced by a modular system of two equations for two invariants. The author used invariants $j(z, z')$, $k(z, z')$ whose derivation (from Eienstein series see [1,4]) is too lengthy to include here, but we decribe the modular system for the multiplier $2 \pm \sqrt{2}$ of norm 2 to obtain insight into the algebraic challenges.

Modular multiplier system for norm 2

Only one system (see [1]) is known explicitly. It relates (j, k) to (J, K) where

$$(13)\quad J(z, z') = j((2+\sqrt{2})z, (2-\sqrt{2})z'), \quad K(z, z') = k((2+\sqrt{2})z, (2-\sqrt{2})z').$$

The multipliers $2 \pm \sqrt{2}$ have norm 2 and ratio $3 \pm 2\sqrt{2}$ so J and K are also symmetric in (z, z'). Therefore the equations for (J, K) are cubic over (j, k) (see (15c) below), and symmetric in the sense that (j, k) and (J, K) are interchangeable. Thus we have the system

$$(14a)\qquad G(J; j, k) = 0,\ H(K; j, k) = 0; \quad G(j; J, K) = 0,\ H(k; J, K) = 0,$$

where

$$\begin{aligned}(14b)\quad G(J; j, k) = {} & J^3 j + J^2 j(432 + 156k - jk) \\ & + J(4j^2k^2 + 207j^2k + 1152jk^2 + 19008jk + 62208j + 82944k^2) \\ & + j(jk + 144j - 1728)^2,\end{aligned}$$

$$\begin{aligned}(14c)\quad H(K; j, k) = {} & K^3k^2(k + 4) - K^2k^2(j - 4k - 48) + Kjk(-5k + 108) \\ & + j(jk + 144k - 1728).\end{aligned}$$

The algebraic implications of this system are remarkable unto themselves. The system is an intersection of cubic cylinders. The symmetry of (j, k) and (J, K) is shown by the fact that $G(j; J, K)$ is a factor of the k-resultant of $G(J; j, k), H(K; j, k)$ (with an enormous nondescript cofactor!), and similarly for $H(k; J, K)$ from the j-resultant. This is not a computationally trivial verification of symmetry!

The modular invariants can conversely be deduced (with some restrictions) from the system (14) (see [2]), by using series (12). Here the process unfortunately uses the coefficients of (14) rather than just "Newtonian shapes."

Modular systems for prime norm

Modular systems like (14abc) are as yet unknown for other multipliers. "Newtonian shapes" are also unknown, but we consider what we know in advance of the general case of prime norm

$$(15a)\qquad p = f^2 - 2g^2 = N(\omega) = N(\omega') \quad (\omega = f + g\sqrt{2},\ \omega' = f - g\sqrt{2}).$$

Since ω/ω' is not a unit, the symmetry of $j(z, z')$ in z and z' does not carry over , e.g.,

$$(15b)\qquad (J =)\ j(\omega z, \omega' z') \neq (J' =)\ j(\omega z', \omega' z)\ \big(= j(\omega' z, \omega z')\big).$$

As in the case of the modular equation (4) for $j(z)$ and $j(pz)$, J (or J') is one of $p + 1$ conjugates. For instance, J is part of the set

$$(15c)\qquad \left\{ j(\omega z, \omega z'),\ j\left(\frac{z+m}{\omega}, \frac{z'+m}{\omega'}\right),\ m = 0, \dots, p-1 \right\}.$$

The same holds for $K = k(\omega z, \omega' z')$. Therefore the system like (14) for J (and K) becomes one of degree $2(p+1)$, and, as before, the symmetry of j, k with J, K must hold through enormous resultants.

Concluding remarks

It would be challenging to extend McKay's method of Newtonian shapes (equations of unknown coefficients) to determine the modular invariants $j(z, z')$, $k(z, z')$, by finding modular equations of the preceding type with a sufficient set of primes $\omega \mid p$ It would be interesting to characterize a minimal set of such primes (like the "Fricke – Monster" set (10)). In addition, of course, the "extraneous" solutions would surely have a life of their own.

References

1. H. Cohn, *An explicit modular equation in two variables and Hilbert's twelfth problem*, Math. Comp. **38** (1982), no. 157, 227 – 236.
2. ———, *The primary role of modular equations*, Number Theory (New York, 1991 – 1995) (D. V. Chudnovsky, G. V. Chudnovsky, and M. B. Nathanson, eds.), Springer, New York, 1996, pp. 19 – 41.
3. H. Cohn and J. McKay, *Spontaneous generation of modular invariants*, Math. Comp. **65** (1996), no. 215, 1295 – 1309.
4. K.-B. Gundlach, *Die Bestimmung der Funktionen zu einigen Hilbertschen Modulgruppen*, J. Reine Angew. Math. **220** (1965), 109 – 153.
5. K. Mahler, *On a class of non-linear functional equations connected with modular functions*, J. Austral. Math. Soc. Ser. A **22** (1976), no. 1, 65 – 118.
6. J. McKay and H. Strauss, *The q-series decompositions of monstrous moonshine and the decomposition of the head characters*, Comm. Algebra **18** (1990), no. 1, 253 – 278.

Center for Communications Research, Institute for Defense Analyses, 4320 Westerra Court, San Diego, CA 92121, USA

E-mail address: `hcohn@ccrwest.org`

Centre de Recherches Mathématiques
CRM Proceedings and Lecture Notes
Volume **47**, 2009

On a Class of Congruence Subgroups

C. J. Cummins and Noushin Sabetghadam Haghighi

ABSTRACT. In this note we give the signature and level of a parametrized class of congruence subgroups. The conjugacy classes of groups of genus zero and one are found and tabulated.

A subgroup H of $\Gamma = \mathrm{SL}(2,\mathbb{Z})$ is said to be a congruence subgroup if it contains one of the principal congruence subgroups $\Gamma(N)$. The smallest N such that $\Gamma(N)$ is contained in H is called the level of H.

Any congruence subgroup of Γ acts on the extended complex upper half-plane $\mathfrak{H}^* = \mathfrak{H} \cup \mathbb{Q} \cup \{\infty\}$ by fractional linear transformations and the number of orbits of H acting on $\mathbb{Q}^* = \mathbb{Q} \cup \{\infty\}$ is called the cusp number of H. Denote by $\overline{H}$ the image of H in $\overline{\Gamma} = \mathrm{PSL}(2,\mathbb{Z})$ and by $\overline{H}_\alpha$ the stabilizer of $\alpha \in \mathbb{Q}^*$ in $\overline{H}$. Then the cusp width of α is defined to be the index of $\overline{H}_\alpha$ in $\overline{\Gamma}_\alpha$.

Larcher in [3] defined the following congruence subgroups of level m:

$$\Gamma_\tau(m; m/d, \varepsilon, \chi) = \left\{ A \in \Gamma \,\middle|\, A = \pm\begin{pmatrix} 1 + k_1(m/\varepsilon\chi) & k_2 d \\ k_3(m/\chi) & 1 + k_4(m/\varepsilon\chi) \end{pmatrix}, k_3 \equiv \tau k_1 \pmod{\chi} \right\},$$

where $d \mid m$, $m/d = h^2 n$, with n square-free, $\varepsilon \mid h$ and $\chi \mid \gcd(d\varepsilon, m/d\varepsilon^2)$. We will call these groups Larcher groups. He used these groups to show that for any congruence subgroup H the set of cusp widths $C(H)$ is closed under taking greatest common divisors and least common multiples. In fact his result is somewhat stronger; he shows that every congruence subgroup H contains at least one Larcher subgroup L with the property that if m is an element of H which stabilizes some α in $\mathbb{Q}^*$, then m is also an element of L so that L is a "large" subgroup of H.

2000 *Mathematics Subject Classification.* 11F06, 20H05.

This is the final form of the paper.

With the exception of Sebbar [6], who used Larcher's results to classify the torsion-free genus-zero congruence subgroups, Larcher's groups have been little studied in the literature. In this note we report on work concerning a larger set of congruence subgroups:

$$H(p,q,r;\chi,\tau) = \left\{ \begin{pmatrix} 1+ap & bq \\ cr & 1+dp \end{pmatrix} \in \Gamma \;\middle|\; a,b,c,d \in \mathbb{Z}, c \equiv \tau a \pmod{\chi} \right\},$$

where $p \mid qr$ and $\chi \mid \gcd(p, qr/p)$. Together with the groups $\pm H(p,q,r;\chi,\tau)$ these include Larcher's groups. By $\pm H$ we mean the group generated by H and -1_2 where 1_2 is the identity of Γ.

Let H be one of the groups $H(p,q,r;\chi,\tau)$ or $\pm H(p,q,r;\chi,\tau)$. The signature of H is the tuple $(\mu,\nu_2,\nu_3,\nu_\infty,\nu'_\infty)$ where μ is the index of $\overline{H}$ in $\overline{\Gamma}$, ν_2 and ν_3 are the numbers of inequivalent elliptic fixed points of order 2 and 3 respectively, ν_∞ is the number of inequivalent regular cusps and ν'_∞ is the number of inequivalent irregular cusps of H. In [2] we computed these signatures and this result is stated below. In this note we also find the levels of $H(p,q,r;\chi,\tau)$ and $\pm H(p,q,r;\chi,\tau)$ and tabulate the conjugacy classes of the groups $\pm H(p,q,r;\chi,\tau)$ of genus zero and one. Equivalently these are the conjugacy classes of the images of these groups in $\mathrm{PSL}(2,\mathbb{Z})$. The latter result requires a bound found by Zograf [9] for the index of a congruence subgroup in terms of its genus. As noted above, the torsion-free, genus-zero groups were found by Sebbar.

Let $H(p,N;\chi) = H(p,N,1;\chi,1)$, then the following Lemma shows that we can reduce to computing the signature of $H(p,N;\chi)$:

Lemma 1 ([2]). *Suppose p,q,r,χ and τ are positive integers such that $p \mid qr$ and $\chi \mid \gcd(p,qr/p)$. Let $g = \gcd(\chi,\tau)$. Then the groups $H(p,q,r;\chi,\tau)$ and $H(p,gqr;\chi/g)$ have the same signature.*

To give the signatures of $H(p,N;\chi)$ we let $k = \mathrm{lcm}[\gcd(p^2,N)\chi, N]$ and define $c(p,N;\chi)$ by:

$$c(p,N;\chi) = \frac{\chi N \phi(p)}{\phi(N)} \sum_{d \mid k/\chi} \frac{\phi(d)\phi(d')}{\mathrm{lcm}[d,d',pk/N]},$$

where $d\,d' = k/\chi$ and ϕ is Euler's function. Also define $\nu_2(N)$ and $\nu_3(N)$ to be the number of inequivalent elliptic fixed points of order 2 and 3 respectively of $\Gamma_0(N)$. See, for example, [7, Prop. 1.43] for explicit formulas. Then the required signatures are given by the following Theorem:

Theorem 1 ([2]). *Suppose p,N and χ are positive integers such that $p \mid N$ and $\chi \mid \gcd(p,N/p)$. Let $c = c(p,N;\chi)$ and $\psi(N) = N\prod_{\substack{p \mid N \\ p \text{ prime}}}(1+1/p)$. The signature $(\mu,\nu_2,\nu_3,\nu_\infty,\nu'_\infty)$ of $H(p,N;\chi)$ is:*

$$\mu = \begin{cases} \chi\phi(p)\psi(N), & \text{if } p=2 \text{ and } \chi=1, \text{ or } p=1, \\ \frac{1}{2}\chi\phi(p)\psi(N), & \text{otherwise.} \end{cases}$$

$$\nu_2 = \begin{cases} \nu_2(N) & \text{if } p=1, \text{ or } p=2 \text{ and } 2 \parallel N, \\ 0 & \text{otherwise.} \end{cases}$$

$$\nu_3 = \begin{cases} \nu_3(N) & \textit{if } p = 1, \textit{ or } p = 3 \textit{ and } 3 \parallel N, \\ 0 & \textit{otherwise.} \end{cases}$$

$$(\nu_\infty, \nu'_\infty) = \begin{cases} (c, 0) & \textit{if } p = 2 \textit{ and } \chi = 1, \textit{ or } p = 1, \\ (\frac{2}{5}c, \frac{1}{5}c) & \textit{if } p = 2,\ \chi = 2,\ 2 \parallel N/p, \\ (\frac{1}{4}c, \frac{1}{2}c) & \textit{if } p = 2,\ \chi = 2,\ 2^k \parallel N/p,\ k \textit{ odd},\ k > 1, \\ (\frac{1}{3}c, \frac{1}{3}c) & \textit{if } p = 2,\ \chi = 2,\ 2^k \parallel N/p,\ k \textit{ even}, \\ (\frac{2}{5}c, \frac{1}{5}c) & \textit{if } p = 4,\ 2 \nmid N/p \ (\textit{so } \chi = 1), \\ (\frac{1}{2}c, 0) & \textit{otherwise.} \end{cases}$$

The signature of $\pm H$ *is* $(\mu, \nu_2, \nu_3, \nu_\infty + \nu'_\infty, 0)$.

Theorem 2. *The level of* $H(p, q, r; \chi, \tau)$ *is given by* $\operatorname{lcm}(\chi' p, q, \chi r)$ *where* $\chi' = \chi / \gcd(\chi, \tau)$.

Proof. Let $H = H(p, q, r; \chi, \tau)$. If $\Gamma(N)$ is a subgroup of H, then $\left(\begin{smallmatrix} 1 & N \\ 0 & 1 \end{smallmatrix}\right)$ is in H and so $q \mid N$. Similarly the fact that $\left(\begin{smallmatrix} 1 & 0 \\ N & 1 \end{smallmatrix}\right)$ is in H implies $r \mid N$ and that N/r is divisible by χ. Finally that $\left(\begin{smallmatrix} 1+N & -N \\ N & 1-N \end{smallmatrix}\right)$ is in H implies $p \mid N$ and also that $(N/r) \equiv \tau(N/p) \pmod{\chi}$. As χ divides N/r we deduce that $\chi' p \mid N$ where $\chi' = \chi / \gcd(\chi, \tau)$. Thus we have that $L = \operatorname{lcm}(\chi' p, q, \chi r)$ divides N. In particular L divides the level of H. But it is straightforward to verify that $\Gamma(L)$ is contained in H and so the level of H is L as required. □

Theorem 3. *Let* L *be the level of* $H(p, q, r; \chi, \tau)$. *Then* L *is also the level of* $\pm H(p, q, r; \chi, \tau)$ *unless the following conditions are satisfied, in which case the level is* $M = L/2$:

(1) *M is divisible by* 2.
(2) *q divides* M.
(3) *r divides* M.
(4) *Either*
 (a) $p = 2$, $\chi = 2$ *and* $1 \equiv \tau \pmod 2$ *or*
 (b) $p = 4$, $M \equiv 2 \pmod 4$.

Proof. Let $H = H(p, q, r; \chi, \tau)$. We first observe that by Wohlfahrt's Theorem [8] (see also [5, p. 37] or [4, Section 5]) if H is a congruence subgroup its level L is the lowest common multiple of the set of integers $\{k_m \mid m \in \Gamma\}$ where k_m is the smallest positive integer k such that $m^{-1}\left(\begin{smallmatrix} 1 & k \\ 0 & 1 \end{smallmatrix}\right)m$ is in H. Since the square of every parabolic element of $\pm H$ is in H this implies that the level of $\pm H$ is either L or $L/2$. In particular for the level to change L must be divisible by 2.

The group H contains -1 if and only if either $p = 1$ or $p = 2$ and $\chi = 1$ or $p = 2$, $\chi = 2$ and τ is even. So $\pm H = H$ in these cases and so the levels are equal. Thus we only have to consider the cases where L is even, p is not 1 and if $p = 2$ then $\chi = 2$ and τ is odd.

We first show that the conditions given in the theorem are necessary. So suppose $H = H(p, q, r; \chi, \tau)$ has level L and $\pm H(p, q, r; \chi, \tau)$ has level $M = L/2$. Then there must be some $P = m^{-1}\left(\begin{smallmatrix} 1 & k \\ 0 & 1 \end{smallmatrix}\right)m$ such that P is not in H, but $-P$ is in H. Otherwise by Wohlfahrt's Theorem H and $\pm H$ would have the same level. Now P has the form $\left(\begin{smallmatrix} 1-ack & a^2k \\ -c^2k & 1+ack \end{smallmatrix}\right)$ for some integers a and c. So if $-P$ is in H we must have $p \mid -2 + ack$ and $p \mid -2 - ack$. So p must divide 4. Since -1 is not in H this implies that either $p = 2$, $\chi = 2$ and τ is odd or $p = 4$.

If $\pm H$ has level M then either $\left(\begin{smallmatrix} 1 & M \\ 0 & 1 \end{smallmatrix}\right)$ is in H or $-\left(\begin{smallmatrix} 1 & M \\ 0 & 1 \end{smallmatrix}\right)$ is in H. In either case we must have $q \mid M$.

Similarly either $\left(\begin{smallmatrix} 1 & 0 \\ M & 1 \end{smallmatrix}\right)$ is in H or $-\left(\begin{smallmatrix} 1 & 0 \\ M & 1 \end{smallmatrix}\right)$ is in H. From which we must have $r \mid M$.

If M is odd then we conclude that qr and hence p is odd. But this a contradiction since either $p = 2$ or $p = 4$. Thus M is even.

This accounts for all the conditions of the theorem except for the second part of 4(b). If $p = 4$ then neither $-\left(\begin{smallmatrix} 1 & M \\ 0 & 1 \end{smallmatrix}\right)$ nor $-\left(\begin{smallmatrix} 1 & 0 \\ M & 1 \end{smallmatrix}\right)$ is in H. Thus $\left(\begin{smallmatrix} 1 & M \\ 0 & 1 \end{smallmatrix}\right)$ and $\left(\begin{smallmatrix} 1 & 0 \\ M & 1 \end{smallmatrix}\right)$ are in H. As M is even, $\Gamma(M)$ is generated over $\Gamma(L)$ by $\left(\begin{smallmatrix} 1 & M \\ 0 & 1 \end{smallmatrix}\right)$, $\left(\begin{smallmatrix} 1 & 0 \\ M & 1 \end{smallmatrix}\right)$ and an element J congruent to $\left(\begin{smallmatrix} 1+M & 0 \\ 0 & 1+M \end{smallmatrix}\right)$ modulo L. As $\pm H$ has level M either J or $-J$ is in H, but if J is in H, then H would have level M also. Thus we must have $-J$ in H. This implies that $-1 - M \equiv 1 \pmod{p}$ and so $M \equiv 2 \pmod{4}$ as required.

This shows the conditions of the theorem are necessary. Next we show they are sufficient.

Suppose conditions 1, 2, 3 and 4(a) hold. We need to show that $\pm H$ contains $\Gamma(M)$. Condition 2 implies that $\left(\begin{smallmatrix} 1 & M \\ 0 & 1 \end{smallmatrix}\right)$ is in H. Claim: M/r is odd and hence $-\left(\begin{smallmatrix} 1 & 0 \\ M & 1 \end{smallmatrix}\right)$ is in H. We have $L = \operatorname{lcm}(\chi' p, q, \chi r) = \operatorname{lcm}(4, q, 2r)$. Now the 2-part of L cannot be equal to the 2-part of q since q divides $L/2$. Also the 2-part of $2r$ cannot be less than 4 since then qr would have 2-part at most 2 which would imply that $\chi = 1$ since χ divides $\gcd(qr/p, p)$. Thus the 2-part of L must be equal to the 2-part of $2r$ and so M/r is odd. If $M/2$ is even then J is in H and if $M/2$ is odd then $-J$ is in H. This shows that $\pm H$ contains $\Gamma(M)$ as required.

Finally suppose conditions 1, 2, 3 and 4(b) hold. First observe that $\left(\begin{smallmatrix} 1 & M \\ 0 & 1 \end{smallmatrix}\right)$ is in H. Condition 4(b) implies 2 exactly divides M. Since both q and r divide M we must have that the 2-parts of both q and r are at most 2. As $\chi \mid \gcd(qr/p, p)$ and $p = 4$ we conclude that $\chi = 1$. Then $\left(\begin{smallmatrix} 1 & 0 \\ M & 1 \end{smallmatrix}\right)$ is in H since $r \mid M$. Also $p \mid 2 + M$ and so $-J$ is in H. So again $\Gamma(M)$ is contained in $\pm H$ as required. □

In order to find the groups $\pm H(p, q, r; \chi, \tau)$ of low genus we will make use of the following result of Zograf [9]:

Theorem 4. *For any congruence subgroup K of G we have:*

$$g(K) + 1 > \frac{3}{64} \operatorname{EC}(G) \ \operatorname{Index}(G{:}K).$$

In this theorem $\operatorname{EC}(G)$ is the Euler characteristic of G. In particular $\operatorname{EC}(\Gamma) = \frac{1}{6}$ and so for any subgroup H of Γ of genus g we have $\operatorname{Index}(H) < 128(g+1)$. The index of $H(p, q, r; \chi, \tau)$ is $\chi\phi(p)\psi(qr)$ which is bounded below by qr and so we have $qr < 128(g+1)$ for these groups.

The index of $\pm H(p, q, r; \chi, \tau)$ is also $\chi\phi(p)\psi(qr)$ if $p = 1$ or $p = 2$ and $\chi = 1$ or $p = 2$, $\chi = 2$ and τ is even. In these cases we get the same bound for qr. Otherwise the index is $\frac{1}{2}\chi\phi(p)\psi(qr)$. If $p \geq 3$ then $\phi(p) \geq 2$ and so again we get the bound $qr < 128(g+1)$. This leave the case $p = 2$, $\chi = 2$ and τ is odd. But since $\chi = 2$ we again get the same bound for qr. As both p and χ are bounded above by qr and we can take $1 \leq \tau \leq \chi$ this gives a finite set of possible values to search.

In Table 1 we give representatives of the Γ conjugacy class of the groups $\pm H(p, q, r, \chi, \tau)$ of genus zero and one. As a check this table should agree with Sebbar's list of genus-zero, torsion-free congruence subgroups. We find agreement, although in some cases we have selected different representatives of the conjugacy classes. The representatives were selected by first minimizing χ, p and then q. Note

TABLE 1. $\mathrm{SL}(2,\mathbb{Z})$ conjugacy classes of groups $\pm H(p,q,r;\chi,\tau)$ of genus zero and one

p	q	r	χ	τ	μ	ν_2	ν_3	ν_∞	label
1	1	2	1	1	3	1	0	2	$2\mathrm{B}_1^0$
1	2	2	1	1	6	0	0	3	$2\mathrm{C}_1^0$
1	1	3	1	1	4	0	1	2	$3\mathrm{B}_1^0$
1	3	3	1	1	12	0	0	4	$3\mathrm{D}_1^0$
1	1	4	1	1	6	0	0	3	$4\mathrm{B}_1^0$
1	2	4	1	1	12	0	0	4	$4\mathrm{E}_1^0$
1	4	4	1	1	24	0	0	6	$4\mathrm{G}_1^0$
1	1	5	1	1	6	2	0	2	$5\mathrm{B}_1^0$
5	1	5	1	1	12	0	0	4	$5\mathrm{D}_1^0$
1	5	5	1	1	30	2	0	6	$5\mathrm{G}_1^0$
5	5	5	1	1	60	0	0	12	$5\mathrm{H}_1^0$
1	1	6	1	1	12	0	0	4	$6\mathrm{F}_1^0$
1	2	6	1	1	24	0	0	6	$6\mathrm{I}_1^0$
1	3	6	1	1	36	0	0	8	$6\mathrm{K}_1^0$
1	1	7	1	1	8	0	2	2	$7\mathrm{C}_1^0$
7	1	7	1	1	24	0	0	6	$7\mathrm{G}_1^0$
1	1	8	1	1	12	0	0	4	$8\mathrm{D}_1^0$
1	2	8	1	1	24	0	0	6	$8\mathrm{H}_1^0$
8	2	4	1	1	24	0	0	6	$8\mathrm{J}_1^0$
8	1	8	1	1	24	0	0	6	$8\mathrm{K}_1^0$
8	4	4	1	1	48	0	0	10	$8\mathrm{P}_1^0$
8	2	8	1	1	48	0	0	10	$8\mathrm{Q}_1^0$
1	1	9	1	1	12	0	0	4	$9\mathrm{B}_1^0$
9	3	3	1	1	36	0	0	8	$9\mathrm{H}_1^0$
9	1	9	1	1	36	0	0	8	$9\mathrm{I}_1^0$
1	1	10	1	1	18	2	0	4	$10\mathrm{C}_1^0$
5	1	10	1	1	36	0	0	8	$10\mathrm{F}_1^0$
1	1	12	1	1	24	0	0	6	$12\mathrm{E}_1^0$
12	2	6	1	1	48	0	0	10	$12\mathrm{I}_1^0$
12	1	12	1	1	48	0	0	10	$12\mathrm{J}_1^0$
1	1	13	1	1	14	2	2	2	$13\mathrm{A}_1^0$
1	1	16	1	1	24	0	0	6	$16\mathrm{D}_1^0$
4	1	8	2	1	24	0	0	6	$16\mathrm{E}_1^0$
4	2	8	2	1	48	0	0	10	$16\mathrm{H}_1^0$
8	1	16	1	1	48	0	0	10	$16\mathrm{I}_1^0$
1	1	18	1	1	36	0	0	8	$18\mathrm{F}_1^0$
6	1	12	2	1	48	0	0	10	$24\mathrm{B}_1^0$
1	1	25	1	1	30	2	0	6	$25\mathrm{A}_1^0$
5	1	25	1	1	60	0	0	12	$25\mathrm{B}_1^0$
3	1	9	3	1	36	0	0	8	$27\mathrm{A}_1^0$
4	1	16	2	1	48	0	0	10	$32\mathrm{A}_1^0$
1	6	6	1	1	72	0	0	12	$6\mathrm{F}_1^1$
1	7	7	1	1	56	0	2	8	$7\mathrm{B}_1^1$
1	4	8	1	1	48	0	0	8	$8\mathrm{F}_1^1$
8	4	8	1	1	96	0	0	16	$8\mathrm{K}_1^1$
1	3	9	1	1	36	0	0	6	$9\mathrm{D}_1^1$
9	3	9	1	1	108	0	0	18	$9\mathrm{J}_1^1$
1	2	10	1	1	36	0	0	6	$10\mathrm{G}_1^1$
5	2	10	1	1	72	0	0	12	$10\mathrm{K}_1^1$
1	1	11	1	1	12	0	0	2	$11\mathrm{A}_1^1$
11	1	11	1	1	60	0	0	10	$11\mathrm{D}_1^1$
1	2	12	1	1	48	0	0	8	$12\mathrm{Q}_1^1$
1	3	12	1	1	72	0	0	12	$12\mathrm{T}_1^1$
12	2	12	1	1	96	0	0	16	$12\mathrm{W}_1^1$
1	1	14	1	1	24	0	0	4	$14\mathrm{E}_1^1$
7	1	14	1	1	72	0	0	12	$14\mathrm{L}_1^1$
1	1	15	1	1	24	0	0	4	$15\mathrm{C}_1^1$
5	1	15	1	1	48	0	0	8	$15\mathrm{G}_1^1$
15	1	15	1	1	96	0	0	16	$15\mathrm{I}_1^1$
1	2	16	1	1	48	0	0	8	$16\mathrm{E}_1^1$
8	2	16	1	1	96	0	0	16	$16\mathrm{M}_1^1$
1	1	17	1	1	18	2	0	2	$17\mathrm{A}_1^1$
1	2	18	1	1	72	0	0	12	$18\mathrm{J}_1^1$
1	1	19	1	1	20	0	2	2	$19\mathrm{A}_1^1$
1	1	20	1	1	36	0	0	6	$20\mathrm{E}_1^1$
5	1	20	1	1	72	0	0	12	$20\mathrm{J}_1^1$
1	1	21	1	1	32	0	2	4	$21\mathrm{C}_1^1$
1	1	24	1	1	48	0	0	8	$24\mathrm{I}_1^1$
12	1	24	1	1	96	0	0	16	$24\mathrm{L}_1^1$
1	1	27	1	1	36	0	0	6	$27\mathrm{A}_1^1$
9	1	27	1	1	108	0	0	18	$27\mathrm{C}_1^1$
1	1	32	1	1	48	0	0	8	$32\mathrm{A}_1^1$
8	1	32	1	1	96	0	0	16	$32\mathrm{E}_1^1$
1	1	36	1	1	72	0	0	12	$36\mathrm{C}_1^1$
1	1	49	1	1	56	0	2	8	$49\mathrm{A}_1^1$

that for genus one it is no longer the case that all torsion-free groups are (up to conjugacy) Larcher groups, nor is it true that they are all contained (up to conjugacy) in Γ as can be seen by examining the tables in [1].

Each entry in Table 1 consists of the five parameters p, q, r, χ, τ which specify a group $\pm H(p,q,r;\chi,\tau)$. The parameters ν, ν_2, ν_3 and ν_∞ are the signature of the group as given by Theorem 1 (there are no irregular cusps since these groups contain -1_2). The last entry is the label given to this conjugacy class of groups in the tables of [1]. This label is of the form $\text{level(label)}_f^{\text{genus}}$ where level is the level of the group, genus is its genus, f is a square-free integer (always 1 in this case) such that the group is contained in $\Gamma_0(f)^+$, and label is a letter labeling the group amongst all groups of the same level and genus. See [1] for more details.

References

1. C. J. Cummins, *Congruence subgroups of groups commensurable with* PSL(2, $\mathbb{Z}$) *of genus* 0 *and* 1, Experiment. Math. **13** (2004), no. 3, 361–382.
2. C. J. Cummins and N. Sabetghadam Haghighi, *On the signature of a class of congruence subgroups*, submitted.
3. H. Larcher, *The cusp amplitudes of the congruence subgroups of the classical modular group.* II, Illinois Math. J. **28** (1984), no. 2, 312–338.
4. R. A. Rankin, *The modular group and its subgroups*, The Ramanujan Institute, Madras, 1969.
5. ———, *Modular forms and functions*, Cambridge Univ. Press, Cambridge, 1977.
6. A. Sebbar, *Classification of torsion-free genus zero congruence groups*, Proc. Amer. Math. Soc. **129** (2001), no. 9, 2517–2527.
7. G. Shimura, *Introduction to the arithmetic theory of automorphic functions*, Publ. Math. Soc. Japan, vol. 11, Iwanami Shoten, Publishers, Tokyo, 1971. Kanô Memorial Lectures, No. 1; reprint, Princeton, Univ. Press, Princeton, NJ, 1994.
8. K. Wohlfahrt, *An extension of F. Klein's level concept*, Illinois Math. J. **8** (1964), 529–535.
9. P. Zograf, *A spectral proof of Rademacher's conjecture for congruence subgroups of the modular group*, J. Reine Angew. Math. **414** (1991), 113–116.

Department of Mathematics and Statistics, Concordia University, Montréal, QC H3G 1M8, Canada

E-mail address, C. J. Cummins: cummins@mathstat.concordia.ca

E-mail address, N. Sabetghadam Haghighi: n_sabet@mathstat.concordia.ca

Centre de Recherches Mathématiques
CRM Proceedings and Lecture Notes
Volume **47**, 2009

McKay's Correspondence for Cocompact Discrete Subgroups of SU(1, 1)

Igor V. Dolgachev

To John McKay

ABSTRACT. The classical McKay correspondence establishes an explicit link from the representation theory of a finite subgroup Π of $\mathrm{SU}(2)$ and the geometry of the minimal resolution of the affine surface $V = \mathbb{C}^2/\Pi$. In this paper we discuss a possible generalization of the McKay correspondence to the case when Π is replaced with a discrete cocompact subgroup of the universal cover of $\mathrm{SU}(1,1)$ such that its image Γ in $\mathrm{PSU}(1,1)$ is a fuchsian group of signature $(0, e_1, \dots, e_n)$. We establish a correspondence between a certain class of finite-dimensional unitary representations of Π and vector bundles on an open algebraic surface with trivial canonical class canonically associated to Γ.

1. Introduction

It has been known since the work of P. Du Val in the thirties that Coxeter–Dynkin diagrams of type ADE are in bijective correspondence with the conjugacy classes of finite subgroups Π of $\mathrm{SU}(2)$ in such a way that the intersection graph of a minimal resolution of $\mathbb{C}^2/\Pi$ is the diagram corresponding to the group Π. In the early eighties John McKay added more to this mysterious connection by introducing a certain graph attached to any finite group. When the group is equal to a binary polyhedral group Π the graph coincides with the affine extension of the Dynkin diagram attached to Π [28, 29]. The vertices of the graph correspond to irreducible representations of Π where the extended vertex corresponds to the trivial representation. The first geometric explanations of the McKay correspondence were given independently by G. Gonzalez-Sprinberg and J.-L. Verdier [18] and H. Knörrer [22]. Other, more algebraic, interpretations were given later by B. Kostant [23], T. Springer [40], R. Steinberg [41], M. Artin and J.-L. Verdier [1], H. Esnault and H. Knörrer [17], Y. Ito and I. Nakamura [20], and W. Crawley-Boevey [8].

Modern development reveals a more general context of the correspondence. The current slogan is that the McKay correspondence establishes an isomorphism of the Grothendieck group $K_G(X)$ of G-equivariant coherent sheaves on an algebraic

2000 *Mathematics Subject Classification.* Primary 14E15; Secondary 14H60, 20H10, 16G10.

The author was supported in part by NSF grant 0245203.

This is the final form of the paper.

variety X on which a finite group G acts and the Grothendieck group $K(Y)$ of coherent sheaves on a crepant resolution of the quotient X/G (when it exists), or more generally, an equivalence of the corresponding derived categories. For example, when $X = \mathbb{C}^n$, $n = 2, 3$ a crepant resolution exists if and only if G is a subgroup of $\mathrm{SU}(n)$ and an equivalence of the categories was established by M. Kapranov and E. Vasserot ($n = 2$) and T. Bridgeland, A. King and M. Reid ($n = 3$). We refer for all this and much more to an excellent survey of M. Reid [36].

In this paper we propose a generalization of the McKay correspondence to a certain class of discrete infinite groups Π. It is based on the following observations. For any finite subgroup Π of $\mathrm{SU}(2)$, the affine surface $\mathbb{C}^2/\Pi$ is isomorphic to the affine spectrum of the algebra of Γ-invariant sections of the tangent bundle on a simply connected Riemann surface $\mathbb{P}^1$, where Γ is the image of Π in $\mathrm{PSU}(2)$. Replacing $\mathbb{P}^1$ with the unit disk $\mathbb{D}$ and the tangent bundle with the cotangent bundle, we can consider the algebra of automorphic forms $A(\Gamma)$ with respect to any cocompact discrete subgroup Γ of $\mathrm{PSU}(1,1)$ of some signature $(g; e_1, \ldots, e_n)$. It is a finitely generated graded algebra and its affine spectrum is an affine surface V_Γ with an isolated singular point $0 \in V_\Gamma$ corresponding to the unique maximal homogeneous ideal. The fundamental group of $V_\Gamma^* = V_\Gamma \setminus \{0\}$ is isomorphic to a central extension Π of Γ equal to the pre-image of Γ in the universal cover of $\mathrm{PSU}(1,1)$. This suggests to consider linear representations of Π as the first side of the McKay correspondence. In fact, to keep the analogy closer, we have to restrict ourselves to finite-dimensional unitary linear representations of Π. Each such representation is isomorphic to the pull-back of a unitary representation of the fundamental group of a Deligne–Mumford stacky hyperbolic curve (see [6]). Note that a linear irreducible representation of a binary polyhedral group is isomorphic to the pullback of a unitary representation of a spherical DM-curve. Any unitary representation of Π is isomorphic to the direct sum of irreducible ones. We restrict ourselves to admissible representations. i.e., representations isomorphic to the direct sum of irreducible unitary representations whose image intersects $\mathrm{SL}(n,\mathbb{C})$ in a Zariski dense subgroup. In particular, we want to avoid representations that are factored through a subgroup of finite index of Π.

To get the second side of the McKay correspondence we replace a minimal resolution of V_Γ (never crepant) with an open subset $\widetilde{V}_\Gamma$ of a unique minimal smooth normal crossing compactification of V_Γ preserving the $\mathbb{C}^*$-action defined by the grading of $A(\Gamma)$. Its complement is the exceptional curve $\mathbf{E}_0$ of the minimal resolution of the singular point 0 of V_Γ. It has been known to the author for 30 years (as well as to some other experts) that, in the case when $\mathbb{D}/\Gamma \cong \mathbb{P}^1$, the curve $\mathbf{E}_\infty = \widetilde{V}_\Gamma \setminus V_\Gamma^*$ is equal to the union of smooth rational curves with self-intersection -2 and its dual graph is a star-shaped tree with n arms of length $e_1, \ldots, e_n$ (counting the center). In particular, the canonical class of $\widetilde{V}_\Gamma$ is zero. So the surface $\widetilde{V}_\Gamma$ seems to be a correct substitute for the minimal resolution of singularities of the surface V_Γ in the case of finite Γ.

Let ρ be an admissible irreducible representation of Π and m be the order of the image of the center of Π (the *level* of ρ). We choose a surface subgroup Γ' of Γ with $C = \mathbb{D}/\Gamma'$ of genus g and $m \mid 2g - 2$. The restriction of ρ to the pre-image Π' of Γ' in Π defines an irreducible local unitary system on the curve C. By a theorem of Narasimhan and Seshadri [33], it corresponds to a stable vector bundle E on C. We show that E admits a linearization with respect to some central extension $m.G$

of the group $G = \Gamma/\Gamma'$. We take the pull-back $\pi^*(E)$ of E to the ruled surface $\pi\colon X \to C$ defined by a m-root of the tangent bundle and then restrict $\pi^*(E)$ to the complement X^0 of the exceptional section S. The quotient $X^0/m.G$ has only cyclic quotient singularities of types A_{e_i}. The usual McKay correspondence assigns to $\pi^*(E)|X^0$ a unique vector bundle on $\widetilde{V}_\Gamma$. This is our generalized McKay correspondence.

In the case when Γ is of signature $(0; 2, 3, 7)$ one can choose C to be the Klein quartic curve of genus 3 and $G = \mathrm{PSL}(2, \mathbb{F}_7)$. We showed in [11] that there are two stable $\mathrm{SL}(2, \mathbb{F}_7)$-linearized bundles of rank 2 and also two of rank 3. Thus there are two admissible representations of Π of dimension 2 of level 2 and two of dimension 3 of level 1. It is clear that we can construct admissible representations of any dimensions, for example in the Klein case, by considering symmetric powers of the two-dimensional representations. We conjecture that there exists a certain set of "fundamental admissible representations" bijectively corresponding to the irreducible components of the curve $\mathbf{E}_\infty$ making an analogy with fundamental weights of a simple infinite-dimensional hyperbolic Lie algebra with the Dynkin diagram equal to the dual graph of the curve $\mathbf{E}_\infty$. For example, in the Klein case the diagram is E_{10} so we are looking for 10 fundamental admissible irreducible representations of Π.

I am grateful to Daniel Allcock for many useful conversations on the topic of this paper. I thank Brian Jurgelewicz for his help in making the manuscript more reader-friendly.

2. Canonical affine surfaces with a good $\mathbb{C}^*$-action

A familiar construction of ADE-singularities is as follows. Let

$$h = |z_1|^2 + |z_2|^2\colon \mathbb{C}^2 \to \mathbb{R}$$

be a positive definite Hermitian from, and

$$\mathcal{U} = h^{-1}(\mathbb{R}_{>0}) = \mathbb{C}^2 \setminus \{0\}.$$

The projection $\mathcal{U} \to \mathbb{P}^1$, $(z_1, z_2) \mapsto z_2/z_1$ is surjective. The fibres are isomorphic to $\mathbb{C}^*$. Let

$$\Pi \subset \mathrm{SU}(2)$$

be a finite subgroup acting naturally on $\mathcal{U}$. Its image Γ in $\mathrm{PSU}(2)$ is a polyhedral group, a finite group of automorphisms of $\mathbb{P}^1$. The kernel $\Pi \to \Gamma$ is the center $\{\pm 1\}$ of $\mathrm{SU}(2)$ (unless Γ is a cyclic group of odd order, in which case the kernel is trivial). Consider the orbit space $\mathcal{U}/\Pi$. Then there exists a normal affine surface V_Γ with a singular point $0 \in V_\Gamma$ such that

$$V_\Gamma^* = V_\Gamma \setminus \{0\} \cong \mathcal{U}/\Pi. \tag{2.1}$$

The singular point $(V_\Gamma, 0)$ represents the analytical class of a double rational point, or an ADE-singularity, or a Du Val singularity.

The surface V_Γ is naturally isomorphic to affine spectrum of the graded algebra

$$A(\Gamma) = \bigoplus_{k=0}^{\infty} H^0(\mathbb{P}^1, T_{\mathbb{P}^1}^{\otimes k})^\Gamma, \tag{2.2}$$

where $T_{\mathbb{P}^1}$ is the tangent bundle on $\mathbb{P}^1$. The surface $T^{\#}_{\mathbb{P}^1} = T_{\mathbb{P}^1} \setminus \{\text{zero section}\}$ has universal cover isomorphic to $\mathcal{U}$ with Galois group equal to the center of Π. The group Π is isomorphic to the fundamental group of the surface V_Γ^*.

Now we change the signature of h. Let

$$h = |z_1|^2 - |z_2|^2 \colon \mathbb{C}^2 \to \mathbb{R}$$
$$\mathcal{U} = h^{-1}(\mathbb{R}_{>0}).$$

The projection $\mathcal{U} \to \mathbb{P}^1$, $(z_1, z_2) \mapsto z_2/z_1$ has the image equal to the unit disk $\mathbb{D} = \{z \in \mathbb{C} : |z| < 1\}$. It can be naturally identified with the complement of the zero section of the cotangent bundle $T_{\mathbb{D}}^*$. An analog of a finite subgroup of $\mathrm{PSU}(2)$ in this case is a *cocompact discrete subgroup* $\Gamma \subset \mathrm{PSU}(1,1)$. Let Π be its pre-image in the universal cover $\widetilde{\mathrm{SU}}(1,1)$ of $\mathrm{SU}(1,1)$ and Π' be the image of Π in $\mathrm{SU}(1,1)$. The group Π' is a central double extension of Γ. The group Π' acts discretely on $\mathcal{U}$ and we can take the quotient $\mathcal{U}/\Pi'$.

Theorem 2.1. *There exists a unique normal affine surface V_Γ with a good $\mathbb{C}^*$-action such that*

$$V_\Gamma^* = V_\Gamma \setminus \{0\} \cong \mathcal{U}/\Pi' \tag{2.3}$$

(2.1)*, where* 0 *is the unique closed orbit of the $\mathbb{C}^*$ action.*

We replace $\mathbb{P}^1$ with $\mathbb{D}$, another simply-connected complex manifold of dimension 1. The algebra (2.2) is now replaced with the graded algebra of automorphic forms with respect to Γ

$$A(\Gamma) = \bigoplus_{k=0}^{\infty} H^0(\mathbb{D}, T_{\mathbb{D}}^{*\otimes k})^\Gamma. \tag{2.4}$$

The surface V_Γ is defined to be the affine spectrum of $A(\Gamma)$. The group Γ acts on $(T_{\mathbb{D}}^*)^{\#} = T_{\mathbb{D}}^* \setminus \{\text{zero section}\}$ freely and the quotient is isomorphic to V_Γ^*. The universal cover of V_Γ^* is equal to the universal cover of $\mathcal{U}$ and the fundamental group of V_Γ^* coincides with the central extension Π of Γ, the pre-image of Γ in the universal cover of $\mathrm{PSU}(1,1)$.

We call the surface V_Γ the *canonical affine surface* defined by the group Γ. The grading of its ring of regular functions defines a good $\mathbb{C}^*$-action, i.e. the action of $\mathbb{C}^*$ with a unique closed orbit.

Remark 2.2. It is known that any normal affine surface V with a good $\mathbb{C}^*$-action is obtained in a similar way. One considers a holomorphic line bundle L on a simply connected Riemann surface P admitting an ample linearization with respect to a cocompact subgroup Γ of $\mathrm{Aut}(P)$ acting freely outside the zero section of L. Then one proves that the graded algebra

$$A(\Gamma; L) = \bigoplus_{k=0}^{\infty} H^0(P, L^{\otimes k})^\Gamma$$

is finitely generated and hence defines a normal affine surface with a good $\mathbb{C}^*$-action. Any such surface is obtained from a unique pair (Γ, L) [11, 35]. In the case when $P = \mathbb{P}^1$ and $L = T_P$, the algebra $A(P, L)$ is always generated by 3 elements and hence the surface V_Γ is isomorphic to a hypersurface in $\mathbb{C}^3$. In the case when $P = \mathbb{D}$ this could happen only if some power of L is Γ-isomorphic to T_P^* [13]. There are 31 possible signatures of Γ when it really happens (see [37, 42]). Fourteen of them have signature $(0; e_1, e_2, e_3)$ and the singular point of V_Γ in this case is analytically isomorphic to one of the 14 exceptional unimodal singularities of V. Arnold [10] (see nice accounts of these results in [4, 43]).

Recall that a cocompact discrete subgroup of $\mathrm{PSU}(1,1)$ is described by its signature $(q; e_1, \ldots, e_n)$, where q is the genus of the Riemann surface $\mathbb{D}/\Gamma$ and e_i are the orders of the non-trivial stabilizer subgroups of Γ. The cover $\mathbb{D} \to \mathbb{D}/\Gamma$ is a Galois cover ramified over n points $p_1, \ldots, p_n$ in $\mathbb{D}/\Gamma$ with ramification indices $e_1, \ldots, e_n$. One of the differences between the finite and infinite case is that $\mathcal{U}$ is simply-connected in the first case and is not simply-connected in the second case. In fact, the group $\mathrm{SU}(2)$ is simply-connected but $\mathrm{SU}(1,1)$ is not. The universal cover $\widetilde{\mathrm{SU}}(1,1)$ can be described as follows (where I am following a nice exposition from [4]). First identify a matrix

$$\begin{pmatrix} z & w \\ \bar{w} & \bar{z} \end{pmatrix} \in \mathrm{SU}(1,1)$$

with an element of the set $X = \{(z,w) \in \mathbb{C}^2 : |z| - |w|^2 = 1\}$. Then

$$\widetilde{X} = \{(w, \phi, r) \in \mathbb{C} \times \mathbb{R} \times \mathbb{R}_+ : |w|^2 = r^2 - 1\}$$

with the map $(w, \phi, r) \mapsto (re^{i\phi}, w)$ is the universal covering of X. Equivalently, we may replace the unit disk $\mathbb{D}$ with the upper half-plane $\mathbb{H}$, the group $\mathrm{SU}(1,1)$ with $\mathrm{SL}(2,\mathbb{R})$ and realize the universal cover $\widetilde{\mathrm{SL}}(2,\mathbb{R})$ as the subgroup

$$\left\{(\alpha, g) \in \mathcal{O}(\mathbb{H})^* \times \mathrm{SL}(2,\mathbb{R}) : \mathrm{e}^{2\pi\mathrm{i}\alpha} = \frac{\mathrm{d}g}{\mathrm{d}z}\right\}$$

of the semi-direct product $\mathcal{O}(\mathbb{H})^* \rtimes \mathrm{SL}(2,\mathbb{R})$, where $\mathrm{SL}(2,\mathbb{R})$ acts naturally on $\mathbb{H}$ and hence on the group of invertible holomorphic functions $\mathcal{O}(\mathbb{H})^*$.

Let $\widetilde{\mathrm{SU}}(1,1)$ be the universal cover. It is a 3-dimensional simply-connected Lie group given by the universal central extension

$$1 \to \mathbb{Z} \to \widetilde{\mathrm{SU}}(1,1) \to \mathrm{PSU}(1,1) \to 1.$$

The pre-image of Γ in $\widetilde{\mathrm{SU}}(1,1)$ is the group Π which fits in the central extension

$$1 \to \mathbb{Z} \to \Pi \to \Gamma \to 1. \tag{2.5}$$

Recall that

$$\pi_1(V_\Gamma^*) \cong \Pi.$$

It is known that the link space of the singularity $0 \in V_\Gamma$ is a Seifert 3-manifold with invariants $(q; (e_1, 1), \ldots, (e_n, 1))$. The group Γ has a standard set of generators $g_1, \ldots, g_n, a_1, \ldots, a_q, b_1, \ldots, b_q$ with defining relations

$$g_1^{e_1} = \cdots = g_r^{e_n} = g_1 \cdots g_n[a_1, b_1] \cdots [a_q, b_q] = 1.$$

The group Π has generators $c, \tilde{g}_1, \ldots, \tilde{g}_n, \tilde{a}_i, \tilde{b}_i$, $i = 1, \ldots, q$, where c is the generator of the center $Z(\Pi) = \mathrm{Ker}(\Pi \to \Gamma)$ with the remaining defining relations

$$\tilde{g}_1^{e_1} = \cdots = \tilde{g}_n^{e_n} = c, \ \tilde{g}_1 \cdots \tilde{g}_n[\tilde{a}_1, \tilde{b}_1] \cdots [\tilde{a}_q, \tilde{b}_q] = c^{2-2q-n}. \tag{2.6}$$

Under the canonical map $\Pi \to \Gamma$, the generators $\tilde{g}_i, \tilde{a}_j, \tilde{b}_j$ are mapped to the generators g_i, a_j, b_j (see [25, 38]).

From now on we will assume that

$$\mathbb{D}/\Gamma \cong \mathbb{P}^1$$

i.e., the signature of Γ is equal to $(0; e_1, \ldots, e_n)$, where

$$\mu_\Gamma := n - 2 - \sum_{i=1}^{n} \frac{1}{e_i} > 0.$$

Let $\Gamma' \subset \Gamma$ be a torsion-free normal subgroup of finite index (it always exists). The quotient $C = \mathbb{D}/\Gamma'$ is a compact Riemann surface of some genus $g > 1$ and Γ' is isomorphic to the fundamental group of C. We will call Γ' a *surface subgroup of* Γ *of genus* g. The group $G = \Gamma/\Gamma'$ acts faithfully on C by holomorphic automorphisms. Let

$$V(C) = \operatorname{Spec} \bigoplus_{m=0}^{\infty} H^0(C, \mathcal{O}_C(mK_C)).$$

If C is not hyperelliptic, this is the affine cone over C embedded canonically in $\mathbb{P}^{g-1}$. The group G acts naturally on $V(C)$ and we obtain

$$V_\Gamma \cong V(C)/G. \tag{2.7}$$

This isomorphism allows one to find a minimal smooth compactification of the surface V_Γ. To avoid technicalities, we assume that C is not hyperelliptic. First we replace $V(C)$ with the projective cone over $C \hookrightarrow \mathbb{P}^{g-1}$ and then take its natural minimal resolution. The resulting projective surface $\overline{V(C)}$ is isomorphic to the projectivization of the tangent line bundle T_C. It has the exceptional section S_0 with self-intersection $2 - 2g$ and the section at infinity S_∞. The group G acts on $\overline{V(C)}$ with finitely many fixed points: n orbits of fixed points $x_1^0, \ldots, x_n^0$ lying on S_0 and n orbits of fixed points $x_1^\infty, \ldots, x_n^\infty$ lying on S_∞. Each pair (x_i^0, x_i^∞) lies on the same fibre of the projection $\overline{V(C)} \to C$. Choose a generator g_i of G_{x_i}. Then g_i has eigenvalues η_{e_i}, η_{e_i} (resp. $\eta_{e_i}, \eta_{e_i}^{-1}$) at x_i^0 (resp. x_i^∞), where η_k denote a primitive kth root of the unity.

The quotient surface $\overline{V(C)}/G$ has n singular cyclic quotient singularities on each curve $\bar{S}_0 = S_0/G$ and $\bar{S}_\infty = S_\infty/G$. Let

$$\pi\colon \overline{V_\Gamma} \to \overline{V(C)}/G$$

be a minimal normal crossing resolution of singularities. Let $\mathbf{E}_0$ and $\mathbf{E}_\infty$ be the pre-images of the curves S_∞/G and S_0/G. Then

$$\overline{V_\Gamma} \setminus \mathbf{E}_\infty \to (\overline{V(C)} \setminus S_\infty)/G = V_\Gamma$$

is a minimal normal crossing resolution of the affine surface V_Γ with the exceptional curve equal to $\mathbf{E}_0$. We have

$$\overline{V}_\Gamma = V_\Gamma^* \cup \mathbf{E}_\infty \cup \mathbf{E}_0.$$

The surface $\overline{V_\Gamma}$ does not depend on the choice of Γ'. It a unique minimal normal crossing smooth $\mathbb{C}^*$-equivariant compactification of $V(\Gamma)^*$.

Each of the curves $\mathbf{E}_0, \mathbf{E}_\infty$ is the union of smooth rational curves intersecting each other transversally at most one point with star-shaped incidence graph with n arms and a central vertex (we draw only the case $n = 3$). The curve $\mathbf{E}_0$ looks as in Figure 1.

The numbers indicate the self-intersection of curves.

In $\mathbf{E}_\infty$ each arm consists of e_i smooth rational curves, and all curves have self-intersection equal to -2.

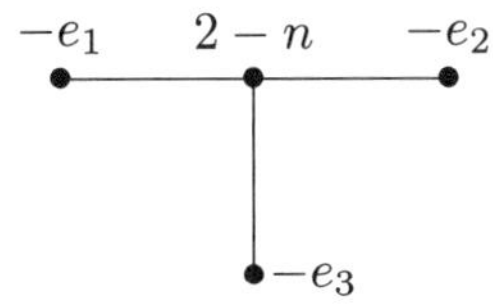

FIGURE 1. $\mathbf{E}_0$

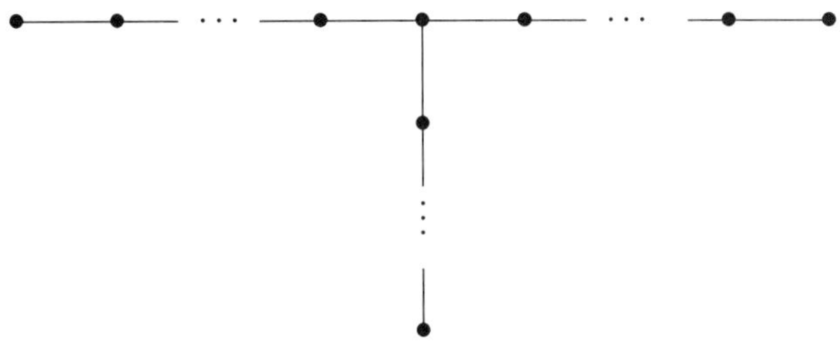

FIGURE 2. $\mathbf{E}_\infty$

Note that in the case when Γ is a polyhedral group, the exceptional curve of the minimal resolution looks like $\mathbf{E}_\infty$ and the graph is a Dynkin diagram. The curve at the infinity looks like $\mathbf{E}_0$.

The projection $\overline{V(C)} \to C$ induces a structure of a non-minimal ruled surface on $\overline{V}_\Gamma$:

$$f\colon \overline{V}_\Gamma = \overline{V(C)}/G \to C/G = \mathbb{P}^1.$$

Let $p_1, \ldots, p_n \in \mathbb{P}^1$ be the exceptional orbits of Γ on $\mathbb{D}$ with stabilizer subgroups of orders $e_1, \ldots, e_n$. The fibre over p_i in $\overline{V}_\Gamma$ is a chain of curves with incidence graph

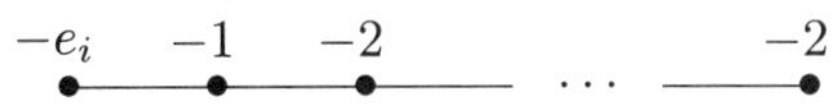

Here the number of (-2)-curves is equal to $e_i - 1$ (unless $e_i = 2$ in which case it is equal to 2). All other fibres are isomorphic to $\mathbb{P}^1$. The central components of the curves $\mathbf{E}_0$ and $\mathbf{E}_\infty$ are the disjoint special sections s_0 and s_∞ of the ruling.

Let F_i be the (-1)-curve in the fibre over p_i. Blow down the curves F_i and then blow down the components of $\mathbf{E}_\infty$ contained in the fibre. We arrive at a minimal ruled surface $\mathbf{F}_{n-2}$. Its exceptional section is the image of s_0. In the case $n = 3$, we can blow down the exceptional section to get a birational morphism

$$\sigma\colon \overline{V}_\Gamma \to \mathbb{P}^2.$$

The image of the section s_0 is a point p_0. The images of the fibres $f^{-1}(p_i)$ are members ℓ_1, ℓ_2, ℓ_3 of the pencil of lines through p_0. The image of the section s_∞ is a line ℓ not containing p_0. The surface $\overline{V}_\Gamma$ is obtained by a sequence of the blow-ups of the points $\ell_i \cap \ell$ and their infinitely near points.

3. The dual McKay correspondence

Recall that the "easy part" of the McKay correspondence for binary polyhedral groups is a natural bijection between the set $\mathcal{C}(\Pi)$ of conjugacy classes of non-trivial elements of the binary polyhedral group $\tilde{\Gamma}$ and the irreducible components of the curve $\mathbf{E}_0$. There are many ways to establish this correspondence. For example, we can use the description of the generators of Γ in terms of natural generators of the fundamental group of the boundary $\mathcal{T}$ of a tubular neighborhood of $\mathbf{E}_0$ on the surface $\overline{V}_\Gamma$ given by D. Mumford [31]. The boundary $\mathcal{T}$ is mapped to $\mathbf{E}_0$ and its restriction over each component E of $\mathbf{E}_0 \setminus$ singular points is a circle bundle and the image of $\pi_1(\text{fibre})$ in $\pi_1(\mathcal{T}) \cong \pi_1(V^*)$ is a certain element s_E of $\pi_1(\mathcal{T}) \cong \pi_1(V_\Gamma^*)$. Mumford proves that the element s_E corresponding to the central component can be taken for the generator c of the center of Π. The element s_E corresponding to an end-component E can be taken as generators γ_i. The element e_E corresponding to the component next to an end-component E_i is equal to $s_{E_i}^2$, and so on. Thus the bijective correspondence follows from the following.

Lemma 3.1 (J.-L. Brylinski [5]). *Let $\Pi \subset \mathrm{SL}(2,\mathbb{C})$ be a noncyclic binary polyhedral group generated by elements $\gamma_1, \dots, \gamma_r, c$ with the standard defining relations (2.6). Any non-trivial non-central element is conjugate to a unique power γ_i^t for some i and $t < e_i$.*

PROOF. Considering the action of Π on $\mathbb{P}^1$, it is easy to see that any element of finite order is conjugate to a power γ_i^t as in the assertion of the lemma. To prove the uniqueness, one considers the standard action of $\widetilde{\Gamma}$ on $\mathbb{C}^2$ and compares the characters of each power of γ_i. □

To extend the previous lemma to the hyperbolic case we first modify the definitions by considering everything modulo the center.

Lemma 3.2. *Let $\Pi \subset \widetilde{\mathrm{SU}}(1,1)$ be a subgroup generated by elements $\gamma_1, \dots, \gamma_r, c$ with standard relations (2.6). Any non-trivial element of finite order modulo the center is conjugate to a unique element $\gamma_i^v c^u$ for some $1 \le i \le r$ and $0 < v < e_i$ and $u \ge 0$.*

PROOF. Again considering the action of Π on the upper-half plane it is easy to see that any element in $\Gamma = \Pi/\langle c\rangle$ is conjugate to a power of the coset of a unique γ_i. It remains to see that the elements $\gamma_i^v c^u$ and $\gamma_i^{v'} c^{u'}$ are not conjugate if $(v,u) \ne (v',u')$. This is obvious. First we may assume that $u = 1$. Raising the elements to the e_ith power we obtain that c^v is conjugate to $c^{v'+e_iu'}$. Since $v < e_i$ this implies that $u' = 1$, $v = v'$. □

It follows from the last lemma that we can identify the irreducible components of the curve with conjugacy classes of elements γ_i^v, $0 < v < e_i$ and c such that each element of finite order modulo the center is conjugate to one of those modulo the center.

4. G-linearized vector bundles

Here we recall some known facts about G-linearized vector bundles. Let X be a scheme and $E \to X$ an algebraic vector bundle of rank r on X. Let G' be a finite group acting on X not necessarily faithfully. A G'-linearization of E is a lift of the action of G' on X to an action on the total space of E. Let G be the image of G in $\mathrm{Aut}(X)$. Obviously, E is G-invariant, i.e., for any $g \in G$ there is an isomorphism $\phi_g \colon g^*(E) \to E$. A G-linearization on E is a choice of compatible sets of isomorphisms ϕ_g. It may not exist, but it always exists if we replace G with some central extension G'. A well-known argument introduces the group $\mathcal{G}_E$ whose elements are the pairs (g,ϕ_g), $g \in G$, $\phi_g \colon g^*(E) \xrightarrow{\sim} E$. The group law is

$$(g,\phi_g)\cdot(g',\phi_{g'}) = \big(gg', \phi_{g'} \circ g'^*(\phi_g)\big).$$

The projection $(g, \phi_g) \mapsto g$ defines a surjective homomorphism $\mathcal{G}_E \to G$. Its kernel is the group of automorphisms of E identical on the zero section. For example, when E is simple, the kernel is equal to $\mathbb{C}^*$. A G'-linearization of E defines a homomorphism of G-extensions $G' \to \mathcal{G}_E$. If G' is a faithful action on E, then the homomorphism $G' \to \mathcal{G}_E$ is injective. Conversely, any finite subgroup of $\mathcal{G}_E$ that covers G defines a faithful linearization of E.

Assume that X is over $\mathbb{C}$ with $H^0(X, \mathcal{O}_X) = \mathbb{C}$, and E is a simple vector bundle. Then we have a central extension

$$1 \to \mathbb{C}^* \to \mathcal{G}_E \to G \to 1.$$

Its class ξ is an element of the finite group $M(G) := H^2(G, \mathbb{C}^*)$, the Schur multiplier of G. Let m be the order of ξ. The Kummer exact sequence

$$1 \to \mu_k \to \mathbb{C}^* \to \mathbb{C}^* \to 1$$

gives an exact sequence

$$0 \to \operatorname{Char}(G)/m \operatorname{Char}(G) \to H^2(G, \mu_m) \to H^2(G, \mathbb{C}^*)[m] \to 0,$$

where $\operatorname{Char}(G) = \operatorname{Hom}(G, \mathbb{C}^*)$ and $A[m]$ denotes the m-torsion part of an abelian group A. This shows that ξ is the image of an element of $H^2(G, \mu_m)$ that defines a non-trivial central extension

$$1 \to \mu_m \to m.G \to G \to 1 \tag{4.1}$$

such that $\mathcal{G}_E = \mathbb{C}^* \times_{\mu_m} m.G$. In particular, $m.G$ defines a linearization of E and E cannot be linearized by any smaller group. Note that the extension (4.1) is not uniquely defined.

Let E be a G-linearized rank r vector bundle over a nonsingular complex projective curve X. For any point x in X its stabilizer subgroup G_x acts naturally on the fibre E_x of E. The corresponding linear representation $G_x \to \operatorname{GL}(E_x)$ is called the *isotropy action*. We identify X with the zero section of the total space $\mathbb{V}(E)$ of E. The action of G_x on the tangent space $T_x(\mathbb{V}(E))$ decomposes into the direct sum of two invariant subspaces $T_x(X)$ and $T_x(E_x) = E_x$. Let $e(x) = |\bar{G}_x|$, where $\bar{G}_x$ is the image of G_x in $\operatorname{Aut}(X)$. Let $\xi_{e(x)}$ be a primitive $e(x)$-th root of unity. We may assume that the eigenvalue of a generator g_x of $\bar{G}_x$ on $T_x(X)$ is equal to $\xi_{e(x)}$ and the eigenvalues of g_x on $T_x(E_x)$ are equal to $\xi_{e(x)}^{q_i(x))}$ for some $0 \le q_1(x) \le \cdots \le q_r(x) < e$.

Let $Gx_1, \ldots, Gx_n$ be the orbits of G lying over the branch points $p_1, \ldots, p_n$ of the projection $\pi \colon X \to Y = X/G$. Let $e_j = e(x_j)$, $q_i^{(j)} = q_i(x_j)$, $1 \le i \le r$, $1 \le j \le n$. We call the vectors

$$\mathbf{s}_j = (q_1^{(j)}/e_j, \ldots, q_r^{(j)}/e_j), \quad j = 1, \ldots, n.$$

the *Seifert invariants* of E. Of course, an expert would rephrase this in terms of the degree of the corresponding parabolic bundle on $\mathbb{P}^1$ but we skip it.

Let $\pi_*(E)^G$ be the sheaf on Y whose sections on an open subset U are G-invariant sections of E over $\pi^{-1}(U)$. Since Y is a nonsingular curve, the sheaf $\pi_*(E)$ is locally free, hence $\pi_*(E)^G$ is locally free. Thus $\pi_*(E)^G$ is the sheaf of sections of a vector bundle on Y. We denote it by E^G. Note that its total space is not in general isomorphic to $\mathbb{V}(E)/G$. However it is isomorphic to this quotient over the subset of Y over which the quotient map is unramified. In particular the rank of E^G is equal to the rank of E.

Let U be the universal cover of X and $\pi_1(X, x)$ be the fundamental group of X and $\pi_1(X; G)$ be given by the extension

$$1 \to \pi_1(X, x) \to \pi_1(X; G) \to G \to 1. \tag{4.2}$$

It is a subgroup of $G \times \mathrm{Aut}(U)$ which consists of pairs $(g, \tilde{g})$, where $g \in G$ and $\tilde{g}$ is a lift of g to an automorphism of U.

Remark 4.1. To follow the modern trend one can translate everything from above into the language of stacks or orbifolds. A G-linearized vector bundle on X is a vector bundle on the smooth 1-dimensional DM-stack $[X/G]$. It is a reduced stack if G acts faithfully. The quotient variety X/G is the coarse moduli space of the stack $[X/G]$. The group $\pi_1(X; G)$ defined in (4.2) is the fundamental group of the stack $[X/G]$ (see [2]). Any smooth projective DM-curve is defined, up to isomorphism, by a smooth projective curve Y of genus q, a (maybe empty) collection of points $y_1, \ldots, y_n$ on Y, a collection of integers $e_1, \ldots, e_n > 1$ and the generic inertia group H. There are three types of such DM-curves:hyperbolic, euclidean and spherical. We will be interested only in hyperbolic DM-curves. A hyperbolic DM-curve is characterized by the condition

$$\chi(Y) := 2 - 2q - n + \sum_{i=1}^{n} e_i^{-1} < 0.$$

In the category of analytic stacks a hyperbolic DM-curve is isomorphic to the quotient stack $[C/H.G]$, where C is a compact Riemann surface of genus $g > 1$ and $H.G$ is a finite extension of a group G of automorphisms of C such that the projection $C \to C/G = Y$ is ramified over $p_1, \ldots, p_n$ with ramification indices $e_1, \ldots, e_n$. There is a quotient stack $[\mathbb{D}/H.\Gamma]$, where $H.\Gamma$ is an extension of a cocompact subgroup Γ of $\mathrm{PSU}(1,1)$ of signature $(q; e_1, \ldots, e_n)$. It is the universal cover of the DM-stack $[C/H.G]$, where $C = \mathbb{D}/\Gamma'$ for some surface subgroup of Γ of genus g. In this paper we will be dealing with the case $q = 0$ and $H.G$ a central extension by a finite cyclic group H.

Let $\rho\colon \pi_1(X; G) \to \mathrm{GL}(r)$ be a linear representation. The restriction of ρ to the subgroup $\pi_1(X)$ defines a vector bundle E_ρ on X equal to the quotient of the trivial bundle $U \times \mathbb{C}^r$ by the action $\gamma\colon (u, v) \mapsto \big(\gamma(u), \rho(\gamma)(v)\big)$. The factor group $G = \pi_1(X; G)/\pi_1(X)$ acts naturally on E_ρ and gives it the structure of a G-linearized vector bundle on X. The following result of A. Weil (see [19]) gives a necessary and a sufficient condition for a G-linearized bundle to be isomorphic to a G-linearized bundle of the form E_ρ.

Theorem 4.2 (A. Weil). *Let E be a G-linearized vector bundle on X and $E = \bigoplus_i E_i$ be its decomposition as a direct sum of indecomposable G-bundles. Then $E \cong E_\rho$ for some linear representation $\rho\colon \pi_1(X; G) \to \mathrm{GL}(r)$ if and only if for any E_i the sum of the coordinates of its Seifert vectors is equal to the minus the degree of the vector bundle E_i^G.*

It follows from this theorem that E_ρ is indecomposable if and only if ρ is an irreducible representation.

Definition 4.3. Let G be a group acting on X not necessarily faithfully. A G-linearized vector bundle E is called G-stable (resp. G-semi-stable) if for any G-linearized subbundle F of E of smaller rank one has $\mu(F) < \mu(E)$ (resp. $\mu(F) \leq$

$\mu(E)$), where as usual the slope μ is defined as the ratio $\deg(E)/\operatorname{rank}(E)$. If the equality takes place we say that E is G-semistable.

A G-stable G-linearized vector bundle is always semi-stable as a vector bundle, but not necessarily stable. Conversely, a semi-stable vector bundle but not stable vector bundle can be stable as a G-linearized bundle.

A theorem of Narasimhan and Seshadri [33] asserts that any irreducible unitary representation ρ defines a stable vector bundle E_ρ of degree 0. Conversely, any such bundle arises from an irreducible unitary representation of Γ'. A reducible unitary representation defines a semi-stable bundle such that each indecomposable summand is a stable bundle of degree 0. This correspondence makes an equivalence of the corresponding categories.

It is easy to extend the theorem of Weil to stable vector bundles (see [11]).

Theorem 4.4. *Let $\rho\colon \pi_1(X;G) \to \mathrm{U}(r)$ be an irreducible unitary linear representation of $\pi_1(X;G)$. Then the G-linearized vector bundle E_ρ is G-stable. It is stable if the restriction of ρ to $\pi_1(X)$ is irreducible. Conversely, a G-linearized G-stable bundle satisfying the condition on its Seifert vectors from the Weil theorem is isomorphic to a bundle E_ρ for some unitary irreducible representation of $\pi_1(X;G)$. It is stable if the restriction of the representation to $\pi_1(X)$ is irreducible.*

In the next section we will interpret the condition from the Weil Theorem as the vanishing of the first Chern class of a G-linearized bundle.

5. The second cohomology group

Most of the results in this section are known in the theory of Seifert 3-manifolds (see [25]). We recall them in a way convenient for an algebraic geometer.

Let E be a line bundle on a nonsingular projective curve X with the sheaf of sections $\mathcal{O}_X(D)$ for some divisor D on X. A G-linearization on E is a choice of a divisor D' in the linear equivalence class of D such that $g^*(D') = D'$. Two such divisors define isomorphic G-linearized line bundles if and only if they differ by the divisor of a G-invariant rational function on X. The isomorphism classes of G-linearized line bundles form the group $\operatorname{Pic}(X;G)$ (the Picard group of the DM-stack $[X/G]$) which fits in the exact sequence

$$0 \to \operatorname{Char}(G) \to \operatorname{Pic}(X;G) \to \operatorname{Pic}(X)^G \to H^2(G,\mathbb{C}^*) \to 1 \tag{5.1}$$

The image of the map $\operatorname{Char}(G) = \operatorname{Hom}(G,\mathbb{C}^*) \to \operatorname{Pic}(X;G)$ consists of isomorphic classes of trivial line bundles with the action defined by a character of G. The last non-trivial map defines an obstruction class for the existence of linearization on a G-invariant line bundle.

Also we have an exact sequence

$$0 \to \operatorname{Pic}(Y) \to \operatorname{Pic}(X;G) \to \bigoplus_{j=1}^{n} \operatorname{Char}(G_{x_i}) \to 0, \tag{5.2}$$

where the first nontrivial map is defined by taking the pre-image of a line bundle under the projection $\pi\colon X \to Y$. The next map is defined by the isotropy action.

We identify the degree homomorphism $\deg\colon \operatorname{Pic}(Y) \to \mathbb{Z}$ with the first Chern class of a line bundle taking for the generator of $H^2(Y,\mathbb{Z})$ the fundamental class of Y. One can define the first Chern class of a G-linearized bundle that gives a homomorphism $c_1^G\colon \operatorname{Pic}(X;G) \to H^2(\pi_1(X;G),\mathbb{Z})$. Assume G acts faithfully on X.

In this case $\Gamma = \pi_1(X;G)$ and coincides with the orbifold fundamental group of Y. One has an exact sequence similar to (5.2)

$$0 \to H^2(Y,\mathbb{Z}) \to H^2(\Gamma,\mathbb{Z}) \to \bigoplus_{j=1}^{n} H^2(G_{x_i},\mathbb{Z}) \to 0, \tag{5.3}$$

where the second map is defined by the restriction homomorphism for cohomology groups. The Chern class defines the commutative diagram

$$\begin{array}{ccccccccc}
 & & 0 & & 0 & & & & \\
 & & \downarrow & & \downarrow & & & & \\
0 & \longrightarrow & \mathrm{Pic}^0(Y) & \longrightarrow & \mathrm{Pic}^0(X;G) & \longrightarrow & 0 & & \\
 & & \downarrow & & \downarrow & & \downarrow & & \\
0 & \longrightarrow & \mathrm{Pic}(Y) & \longrightarrow & \mathrm{Pic}(X;G) & \longrightarrow & \bigoplus_{j=1}^n \mathrm{Char}(G_{x_i}) & \longrightarrow & 0 \\
 & & \downarrow{\scriptstyle c_1} & & \downarrow{\scriptstyle c_1^G} & & \downarrow{\scriptstyle \cong} & & \\
0 & \longrightarrow & H^2(Y,\mathbb{Z}) & \longrightarrow & H^2(\Gamma,\mathbb{Z}) & \longrightarrow & \bigoplus_{j=1}^n H^2(G_{x_i},\mathbb{Z}) & \longrightarrow & 0 \\
 & & \downarrow & & \downarrow & & \downarrow & & \\
 & & 0 & & 0 & & 0 & &
\end{array}$$

Here the isomorphism $\mathrm{Char}(G_{x_i}) \to H^2(G_{x_i},\mathbb{Z})$ is defined by using the Euler exact sequence

$$0 \to \mathbb{Z} \to \mathbb{C} \xrightarrow{\exp} \mathbb{C}^* \to 0.$$

In the case relevant to this paper, $Y = \mathbb{P}^1$ and hence $\mathrm{Pic}^0(Y) = 0$. Thus the Chern classes maps are isomorphisms. The group $H^2(\Gamma,\mathbb{Z})$ is easy to compute. We have

$$H^2(\Gamma,\mathbb{Z}) \cong \mathbb{Z} \oplus \mathrm{Tors}(H^2(\Gamma,\mathbb{Z})).$$

To compute the torsion group we use again the Euler sequence. It gives

$$\mathrm{Tors}\big(H^2(\Gamma,\mathbb{Z})\big) \cong \mathrm{Tors}\big(H^1(\Gamma,\mathbb{C}^*)\big) = \mathrm{Tors}\big(\mathrm{Hom}(\Gamma/[\Gamma,\Gamma],\mathbb{C}^*)\big).$$

Assume that $X = \mathbb{D}/\Gamma'$, where Γ' is a surface subgroup of Γ and $G = \Gamma/\Gamma'$. In this case G acts faithfully on C and $\pi_1(X;G) = \Gamma$. Using the presentation of Γ we find that

$$\Gamma/[\Gamma,\Gamma] = \mathrm{Coker}(\mathbb{Z}^{n+1} \xrightarrow{f} \mathbb{Z}^n),$$

where the map f is given by the matrix

$$\begin{pmatrix} 1 & e_1 & 0 & 0 & \dots & 0 \\ 1 & 0 & e_2 & 0 & \dots & 0 \\ \vdots & \vdots & \vdots & \vdots & & \vdots \\ 1 & 0 & 0 & 0 & \dots & e_n \end{pmatrix}.$$

Using the elementary theory of abelian groups we obtain

$$\Gamma/[\Gamma,\Gamma] \cong \mathbb{Z}/a_1\mathbb{Z} \oplus \dots \mathbb{Z}/a_n\mathbb{Z},$$

where $a_i = c_i/c_{i-1}$, $c_0 = 1$, and

$$c_k = \gcd\big((e_{i_1}\cdots e_{i_k})_{1\le i_1<\dots<i_k\le n}\big), \quad k = 1,\dots,n-1.$$

In particular, this group is finite and coincides with its torsion group. We have

$$\big|\mathrm{Tors}\big(H^2(\Gamma,\mathbb{Z})\big)\big| = |\Gamma/[\Gamma,\Gamma]| = c_{n-1}. \tag{5.4}$$

It is trivial if the numbers $e_1,\dots,e_n$ are coprime.

Let us identify $H^2(\mathbb{D}/\Gamma,\mathbb{Z})$ with $\mathbb{Z}$ such that 1 is the fundamental class of $\mathbb{D}/\Gamma$. Let γ_0 be a generator of a free part of $H^2(\Gamma,\mathbb{Z})$ such that

$$\alpha(1) = l\gamma_0, \quad l > 0.$$

Then

$$\mathbb{Z}/l\mathbb{Z} \oplus \mathrm{Tors}(\Gamma/[\Gamma,\Gamma]) \cong \bigoplus_{k=1}^{n} \mathbb{Z}/e_k\mathbb{Z}.$$

In particular,

$$l = \frac{e_1\cdots e_n}{c_{n-1}} = \mathrm{lcm}(e_1,\dots,e_n). \tag{5.5}$$

This suggests to identify the free part of $H^2(\Gamma,\mathbb{Z})$ with the subgroup $\frac{1}{l}\mathbb{Z}$ of $\mathbb{Q}$, so that the homomorphism α becomes the natural inclusion $\mathbb{Z}\hookrightarrow \frac{1}{l}\mathbb{Z}$ (cf. [25], Lemma 4.2).

Consider the isomorphism

$$c_1^G\colon\ \mathrm{Pic}(X;G) \to H^2(\Gamma,\mathbb{Z}).$$

Recall that for any element L of $\mathrm{Pic}(X;G)$ we can assign the degree of the vector bundle L^G on X/G and the Seifert numbers q_j/e_j, $j = 1,\dots,n$. Let D_j be the reduced divisor with support equal to the fibre over the branch point p_j of $p\colon X \to Y = \mathbb{P}^1$. Obviously it is a G-invariant divisor and hence $L_j = \mathcal{O}_X(D_i)$ is a G-linearized line bundle. It follows from the Hurwitz formula that we have an isomorphism of G-linearized line bundles

$$K_X \cong p^*(K_{\mathbb{P}^1}) \otimes \bigotimes_{j=1}^{n} L_j^{e_j-1}. \tag{5.6}$$

The Seifert numbers of K_X are equal to $\big((e_i-1)/e_1,\dots,(e_n-1)/e_n\big)$. Thus the image of $c_1(L_j)$ in the group $\bigoplus_{j=1}^n e_i^{-1}\mathbb{Z}/\mathbb{Z}$ is equal to $(0,\dots,0,1/e_j,0,\dots,0)$. This shows that any G-linearized line bundle is isomorphic to the bundle

$$L \cong p^*\big(\mathcal{O}_{\mathbb{P}^1}(m)\big) \otimes \bigotimes_{j=1}^{n} L_j^{q_j},$$

where q_j/e_j are the Seifert numbers of L and m is an integer. Since $L_j^{q_j}$ is a subsheaf of $\mathcal{O}_X(e_jD_j) \cong p^*\mathcal{O}_{\mathbb{P}^1}(1)$ and also contains $\mathcal{O}_X$ we easily see that $(L_j^{q_j})^G$ is a subsheaf of $\mathcal{O}_{\mathbb{P}^1}(1)$ and contains $\mathcal{O}_{\mathbb{P}^1}$. This easily implies that $(L_j^{q_j})^G \cong \mathcal{O}_{\mathbb{P}^1}$ (see [16, p. 39]). The projection formula shows that

$$L^G \cong \mathcal{O}_{\mathbb{P}^1}(m).$$

Comparing the degrees we find

$$\deg L = |G|m + \sum_{j=1}^{n} q_j|G|/e_j = |G|\Big(m + \sum_{j=1}^{n}\frac{q_j}{e_j}\Big).$$

The number

$$\mathrm{vdeg}(L) := m + \sum_{j=1}^{n} \frac{q_j}{e_j} = \deg L/|G|$$

is called the *Chern number* (or the *virtual degree*) of a G-linearized line bundle L. For example,

$$\mathrm{vdeg}(K_X) = \mu(\Gamma) := n - 2 - \sum_{j=1}^{n} \frac{1}{e_j}.$$

($= 1/(2\pi) \times$ the area of the polygon with angles π/e_i on the hyperbolic plane).

For any G-linearized vector bundle E its determinant $\det E$ is a G-linearized line bundle. We set

$$\mathrm{vdeg}(E) = \mathrm{vdeg}(\det E).$$

Grothendieck shows in [19] that

$$\mathrm{vdeg}(E) = \deg(E^G) + \sum_{j=1}^{n}\sum_{i=1}^{r} \frac{q_j^{(i)}}{e_j}.$$

Thus we obtain that the condition for an indecomposable E to be isomorphic to E_ρ is that $\deg(\det E) = |G| \,\mathrm{vdeg}(\det E) = 0$. Note that $\det(E)^G \neq \det(E^G)$ since the sum of the coordinates of the corresponding Seifert vector $(q_j^{(1)}/e_j, \ldots, q_j^{(r)}/e_j)$ differs from the Seifert invariant of E at p_j by an integer.

The forgetful homomorphism $\mathrm{Pic}(X;G) \to \mathrm{Pic}(X)^G$ composed with the degree map corresponds to the restriction map $H^2(\Gamma,\mathbb{Z}) \to H^2(\Gamma',\mathbb{Z}) = H^2(X,\mathbb{Z})$. It sends K_X to $2g-2$. On the other hand, the composition $\mathrm{Pic}(Y) \to \mathrm{Pic}(X;G) \to \mathrm{Pic}(X)$ is the multiplication by $|G|$. This implies that

$$c_1^G(K_X) - s\gamma_0 \in \mathrm{Tors}(H^2(\Gamma,\mathbb{Z})).$$

where

$$s = l\frac{2g-2}{|G|} = \mathrm{lcm}(e_1,\ldots,e_n)\Big(n - 2 - \sum_{i=1}^{n}\frac{1}{e_i}\Big). \tag{5.7}$$

In fact, one can show that a generator γ_0 can be changed in such a way that we have the equality

$$c_1^G(K_X) = s\gamma_0.$$

The generator γ_0 is equal to $c_1^G(L_0)$, where L_0 is G-linearized line bundle with $\mathrm{vdeg}(L_0) = \mu_\Gamma/s$ and the Seifert invariants $\frac{q_i}{e_i}$, where $sq_i \equiv 1 \mod e_i$ (see [14] or [25, Corollary 4.4]). In the case when the surfcae V_Γ is a hypersurface in $\mathbb{C}^3$ with equation $f(x,y,z) = 0$, where $f(x,y,z)$ is a quasi-homogeneous polynomial of degree d with positive integer weights w_1, w_2, w_3 we have (see [4, 13, 37])

$$s = d - w_1 - w_2 - w_3. \tag{5.8}$$

Let Γ' be a surface subgroup of Γ of genus g with $G = \Gamma/\Gamma'$. We apply the previous discussion by taking $C = X$. The class γ of the extension (2.5) defining our group Π is equal to $c_1^G(K_C)$. Recall that the natural homomorphism of the cohomology groups $H^2(\Gamma,\mathbb{Z}) \to H^2(\Gamma,\mathbb{Z}/m\mathbb{Z})$ is defined in terms of the extension classes as follows. Given an extension $1 \to \mathbb{Z} \to A \to \Gamma \to 1$ with the class e, its image in $H^2(\Gamma,\mathbb{Z}/m\mathbb{Z})$ is the class of the extension $1 \to \mathbb{Z}/m\mathbb{Z} \to A/mA \to \Gamma \to 1$. The exact sequence

$$0 \to \mathbb{Z} \to \mathbb{Z} \to \mathbb{Z}/m\mathbb{Z} \to 0$$

together with the fact that $H^3(\Gamma,\mathbb{Z}) = 0$ ([25, p. 3788]) gives an isomorphism

$$H^2(\Gamma,\mathbb{Z})/mH^2(\Gamma,\mathbb{Z}) \cong H^2(\Gamma,\mathbb{Z}/m\mathbb{Z}). \tag{5.9}$$

This shows that the image of the class e is equal to zero in $H^2(\Gamma,\mathbb{Z}/m\mathbb{Z})$ if and only if e is divisible by m in $H^2(\Gamma,\mathbb{Z})$. In particular, we obtain that $\gamma = s\gamma_0$ is mapped to zero in $H^2(\Gamma,\mathbb{Z}/m\mathbb{Z})$ if and only if m divides s.

On the hand, let us consider the image of γ under the composition of $H^2(\Gamma,\mathbb{Z}) \to H^2(\Gamma,\mathbb{Z}/m\mathbb{Z})$ and the restriction map $H^2(\Gamma,\mathbb{Z}/m\mathbb{Z}) \to H^2(\Gamma',\mathbb{Z}/m\mathbb{Z}) = \mathbb{Z}/m\mathbb{Z}$. The image of γ is equal to $2g-2 \bmod m$. So the extension

$$1 \to \langle c\rangle/\langle c^m\rangle \to \Pi'/\langle c^m\rangle \to \Gamma' \to 1 \tag{5.10}$$

splits if and only if m divides $2g-2$.

Recall that the group Π' is generated by c, a_i, b_i, $i = 1,\dots,g$ with defining relation $[a_1,b_1]\cdots[a_g,b_g] = c^{2-2g}$. If $m \mid 2-2g$ we can define an explicit distinguished splitting by sending the standard generators x_i, y_i of $\Gamma' = \pi_1(C)$ to $a_i \bmod \langle c^m\rangle$, $b_i \bmod \langle c^m\rangle$. We call such a splitting *canonical splitting*. Its image is a subgroup of $\Pi'/\langle c^m\rangle$ which is normal in $\Pi/\langle c^m\rangle$. The quotient group is a cyclic central extension $m.G$ of G. The extension

$$1 \to \langle c\rangle/\langle c^m\rangle \to \Pi/\langle c^m\rangle \to \Gamma \to 1 \tag{5.11}$$

can be reconstructed from $m.G$ by using an isomorphism

$$\Pi/\langle c^m\rangle \cong m.G \times_G \Gamma, \tag{5.12}$$

where $\Gamma \to G$ is the quotient map $\Gamma \to \Gamma/\Gamma'$ and $m.G \to G$ is the defined by the extension. In other words, the class of the extension (5.11) is the image of the extension $m.G$ under the inflation map $H^2(G,\mathbb{Z}/m\mathbb{Z}) \to H^2(\Gamma,\mathbb{Z}/m\mathbb{Z})$.

6. Admissible unitary representations of Π

As we have explained in the Introduction, the first side of the McKay correspondence must be a class of linear representations of $\Pi = \pi_1(V_\Gamma^*)$. Since any representation of a finite group is conjugate to a unitary representation, it is natural to deal with unitary representations of Π. Any unitary representation of Γ defines via the composition with $\Pi \to \Gamma$ a unitary representation of Π.

The proof of the next lemma was kindly explained to me by Daniel Allcock.

Lemma 6.1. *Let $\rho\colon \Pi \to \mathrm{GL}(r)$ be an irreducible linear representation. Then $\rho(c)$ is of finite order. There exists a surface subgroup Γ' of Γ of genus g such that*

$$\rho(c)^{2g-2} = 1.$$

PROOF. It follows from the presentation (2.6) of Π that $c^a \in [\Pi,\Pi]$ for some $a > 0$. Since the commutator of $\mathrm{GL}(r)$ is equal to $\mathrm{SL}(r)$ we obtain that $\rho(c)^a \in \mathrm{SL}(r)$. Since ρ is irreducible, $\rho(c)^a$ belongs to the center of $\mathrm{SL}(r)$, hence $\rho(c)^{ar} = 1$. This proves that $\rho(c)$ is of finite order.

Let Γ_1 be any surface subgroup of Γ of some genus g'. Since $\Gamma_1/[\Gamma_1,\Gamma_1] \cong \mathbb{Z}^{2g'}$, there exists a subgroup Γ_2 of γ_1 of index divisible by the order n of $\rho(c)$. The normal subgroup $\Gamma' = \bigcap_{g\in G} g\Gamma_2 g^{-1}$ of Γ is a surface subgroup of Γ of index divisible by n. By Hurwitz's formula, n divides $2g-2$, where g is the genus of Γ'. □

Since any unitary representation decomposes into the direct sum of irreducible representations, it follows from the previous lemma that the image of the center of Π is always a finite cyclic group.

Definition 6.2. A unitary linear representation $\rho : \Pi \to \mathrm{U}(r)$ is called *admissible* if its irreducible summands remain irreducible after restriction to any subgroup of finite index. The order of the image of the center of Π will be called the *level* of the representation. Two admissible representation ρ and ρ' are called *equivalent* if there exists a subgroup Π' of Π of finite index such that the restriction of ρ and ρ' to Π' are isomorphic.

It is known that an irreducible representation is admissible if and only if its image in $\mathrm{GL}(r)$ is Zariski-dense.

Remark 6.3. A linear representation of Π that factors through an irreducible representation of a finite group is an irreducible unitary representation of Π with $\rho(c)$ of finite order. The reason why we would like to ignore such representation is that there are "too many" of these representations. For example, it is known that the group Γ of signature $(0; 2a, 2b, c)$ surjects to all but finitely many symmetric groups S_n [27].

Lemma 6.4. *Let ρ be an admissible representation of Π of level m. Assume that the extension*

$$1 \to \langle c\rangle/\langle c^k\rangle \to \Pi/\langle c^k\rangle \to \Gamma \to 1$$

splits for some $k \mid m$. Then ρ is equivalent to an admissible representation of level dividing m/k.

PROOF. Let $r\colon \Gamma \to \Pi/\langle c^n\rangle$ be a splitting and $\widetilde{\Gamma}$ be the pre-image of $r(\Gamma)$ in $\Pi/\langle c^m\rangle$. It defines an extension

$$0 \to \mathbb{Z}/(m/k)\mathbb{Z} \to \widetilde{\Gamma} \to \Gamma \to 1.$$

This extension is isomorphic to the extension

$$0 \to \langle c\rangle/\langle c^{m/k}\rangle \to \Pi/\langle c^{m/k}\rangle \to \Gamma \to 1.$$

The representation ρ factors through $\Pi/\langle c^m\rangle$ and then restricts to $\widetilde{\Gamma}$ to define a representation of $\widetilde{\Gamma}$. Its composition with the projection $\Pi \to \Pi/\langle c^{m/k}\rangle$ defines an admissible representation $\rho'\colon \Pi \to \mathrm{U}(r)$ of level dividing m/k. Let us show that ρ and ρ' are equivalent.

We have $\Pi/\langle c^m\rangle \cong \widetilde{\Gamma} \times \langle c^k\rangle/\langle c^m\rangle$. Let $\widetilde{\Gamma}'$ be the pre-image of $\widetilde{\Gamma}$ in Π. It is a subgroup of finite index of Π that fits in the extension

$$1 \to \langle c^m\rangle \to \widetilde{\Gamma}' \to \Gamma \to 1.$$

It is clear that ρ and ρ' define the same representation of $\widetilde{\Gamma}'$. □

Proposition 6.5. *Assume that s $(= \mathrm{lcm}(e_1, \ldots, e_n)\mu(\Gamma))$ is coprime to the order of $\mathrm{Tors}(H^2(\Gamma, \mathbb{Z}))$ $(= e_1 \cdots e_n/\mathrm{lcm}(e_1, \ldots, e_n))$. Every admissible representation of dimension r is equivalent to an admissible representation of level m dividing r and coprime to s.*

PROOF. Since ρ is irreducible and c belongs to the center of Π, the image of c is a diagonal matrix $[\alpha, \ldots, \alpha]$, where $\alpha^m = 1$. Let $f\colon \Pi \to \mathrm{U}(1)$ be the composition of ρ and the determinant map $\mathrm{U}(r) \to \mathrm{U}(1)$. It factors through $\Pi^{\mathrm{ab}} = \Pi/[\Pi, \Pi]$

and $f(c) = \alpha^r$. The group Π^{ab} is easy to compute since we know its presentation. It is equal to the abelian group defined by the square matrix

$$\begin{pmatrix} 1 & e_1 & 0 & 0 & \dots & 0 \\ 1 & 0 & e_2 & 0 & \dots & 0 \\ \vdots & \vdots & \vdots & \vdots & & \vdots \\ 1 & 0 & 0 & 0 & \dots & e_n \\ 1 & 1 & 1 & 1 & \dots & 1 \end{pmatrix}.$$

Its order is equal to the determinant of the matrix and we find

$$|\Pi^{\mathrm{ab}}| = e_1 \cdots e_n \Big(n - 2 - \sum_{i=1}^{n} \frac{1}{e_i} \Big).$$

Note that this number coincides with the determinant of the intersection matrix of the exceptional curve $\mathbf{E}_0$. It follows from [31] that the group Π^{ab} is isomorphic to the group of connected components of the local Picard group of the singularity $(V_\Gamma, \{0\})$ (see [31]).

Comparing this with the order of Γ^{ab} given in (5.4), we obtain

$$|\Pi^{\mathrm{ab}}| = s|\Gamma^{\mathrm{ab}}|.$$

This shows that the kernel of the natural map $\Pi^{\mathrm{ab}} \to \Gamma^{\mathrm{ab}}$ induced by the surjection $\Pi \to \Gamma$ is isomorphic to $\mathbb{Z}/s\mathbb{Z}$. Thus the image $\bar{c}$ of c in Π^{ab} satisfies $\bar{c}^s = 1$, hence $f(c)^s = (\alpha)^{rs} = 1$. Since m is the order of α in $\mathbb{C}^*$, we get $m \mid rs$. It remains to show that ρ is equivalent to an admissible representation of level m coprime to s.

Assume $k = (m, s) > 1$. Recall from the previous section that Π fits in the extension defined by the class $\gamma = s\gamma_0$, where γ_0 projects to a generator of $H^2(\Gamma, \mathbb{Z})/\operatorname{Tors}$. By the assumption $\big(s, \big|\operatorname{Tors}(H^2(\Gamma, \mathbb{Z}))\big|\big) = 1$, hence the multiplication by k is an automorphism of $\operatorname{Tors}\big(H^2(\Gamma, \mathbb{Z})\big)$. Thus γ is divisible by k in $H^2(\Gamma, \mathbb{Z})$. Applying (5.9) (with m replaced by k) we obtain that the extension

$$1 \to \langle c \rangle / \langle c^n \rangle \to \Pi / \langle c^n \rangle \to \Gamma \to 1$$

splits. It remains to apply Lemma 6.4. □

Note that the proof also shows that we may replace the assumption on s with the assumption that m is coprime to $\big|\operatorname{Tors}\big(H^2(\Gamma, \mathbb{Z})\big)\big|$.

Theorem 6.6. *Let Γ' be a surface subgroup of Γ of genus g with finite quotient G and $C = \mathbb{D}/\Gamma'$. Assume m divides $2g-2$. There is a natural bijective correspondence between admissible r-dimensional representations of Π of level dividing m and $m.G$-linearized stable rank r vector bundles on C, where the image of the class of the extension $m.G$ under the inflation map $H^2(G, \mathbb{Z}/m\mathbb{Z}) \to H^2(\Gamma, \mathbb{Z}/m\mathbb{Z})$ is equal to the class of extension* (5.11).

PROOF. Let $\rho\colon \Pi \to \mathrm{U}(r)$ be an admissible representation of level m. Fix a surface subgroup Γ' of genus g such and m divides $2g-2$. As we explained at the end of Section 5, the extension $m.\Gamma$ defined by (5.11) is equal to the image of an extension $m.G$ under the inflation map $H^2(G, Z/m\mathbb{Z}) \to H^2(\Gamma, \mathbb{Z}/m\mathbb{Z})$. Now the representation $\rho\colon \Pi \to \mathrm{U}(r)$ factors through $\Pi/\langle c^m \rangle = m.\Gamma$. Since $m.\Gamma$ is defined by the extension $m.G$, the group $m.\Gamma$ is isomorphic to the fundamental group of the DM-curve $[\mathbb{D}/m.G]$. This defines a $m.G$-linearized $m.G$-stable vector bundle over C. Conversely, such a vector bundle E defines a unitary irreducible representation of $m.\Gamma \cong \Pi/\langle c^m \rangle$ and hence an irreducible representation ρ of Π of level dividing m. Suppose the restriction of ρ to some subgroup H of finite index is reducible. Then we find a surface subgroup Γ'' of finite index in both H and Γ'. The pull-back E' of E to the cover $\pi\colon C' = \mathbb{D}/\Gamma' \to C$ is defined by a reducible unitary representation, hence E is not stable but semi-stable. Let $K = \mathrm{Aut}(C'/C)$ and F be a unique maximal semi-stable subbundle of E (see [39], p. 15). It is a K-invariant subbundle of E' of degree 0. The subbundle F^K of $E^K \cong \mathcal{E}$ is of degree 0 and rank equal to rank of F. Hence it defines a destabilizing subbundle of E, contradicting its stability. This proves that ρ is an admissible representation of Π. It is easy to see that the construction establishes a bijective correspondence between the two sets. □

Remark 6.7. We know that $|G|$ is divisible by $l = \mathrm{lcm}(e_1, \ldots, e_n)$. In fact, $|G|$ must be divisible by $2l$ if l is even and the number of e_i's such that l/e_i is odd. This condition is sufficient for the existence of a not necessarily normal torsion-free subgroup of finite index [15]. It is not known what is the minimal index of a surface subgroup of Γ. However, if $n = 3$, accepted the extended Riemann hypothesis, in most cases the group Γ' can be found such that $G = \Gamma/\Gamma'$ is a simple group isomorphic to $\mathrm{PSL}_2(\mathbb{F}_q)$ [24]. Here, "in most cases" means that, fixing two of the numbers e_is, the values of the remaining e_j for which this assertion is not true belongs to a set of density 0.

Example 6.8. Let Γ be the triangle group with $(e_1, e_2, e_3) = (2, 3, 7)$. In this case C can be chosen to be the Klein curve of genus 3 isomorphic to the modular curve $X(7)$. The group $G = L_2(7)$ is the simple group of order 168, and $\mathrm{SL}(2, \mathbb{F}_7)$ is its non-split double extension. The order of $M(G)$ is equal to 2. The number s is equal to 1. Thus the possible levels of an admissible representation of Π are $1, 2, 4$. In particular, the dimension r of an admissible representation of level $m > 1$ must be even. It is known (see [11]) that there are exactly two non-isomorphic $\mathrm{SL}(2, \mathbb{F}_7)$-linearized stable bundles on C of rank 2. Each of them defines an admissible representation of level 2. Also it is known there are exactly two non-isomorphic $\mathrm{SL}(2, \mathbb{F}_7)$-linearized stable bundles on C of rank 3 which arise from a G-linearized stable bundle. Each of them defines an admissible representation of level 1.

7. The McKay correspondence

Let $\rho : \Pi \to \mathrm{U}(r)$ be an admissible representation of Π of level m and Γ' be a surface subgroup of genus g with m dividing $2 - 2g$. Let E_ρ be a $m.G$-linearized stable vector bundle on $C = \mathbb{D}/\Gamma'$ constructed in Theorem 6.6. Let $f : \overline{V(C)} \to C$ be the projectivization of the tangent bundle of C. Let $V(C)^*$ be the complement of the sections S_0 and S_∞. It is a $\mathbb{C}^*$-bundle over C. We know that extension (5.11) splits. The canonical splitting defines a surjection

$$\pi_1(V(C)^*) = \Pi' \to \Pi'/\langle c^m \rangle = \Gamma' \times \mathbb{Z}/mZ \to \mathbb{Z}/m\mathbb{Z}.$$

Let $X^* \to V(C)^*$ be the corresponding cyclic cover. It extends to a cyclic cover $p\colon X \to \overline{V(C)}$ of degree m ramified over $S_0 \cup S_\infty$. The group $m.G$ acts on X with the center $\mathbb{Z}/m\mathbb{Z}$ acting as the group of deck transformations. The surface X is a minimal ruled surface over C isomorphic to the projectivization of a G-invariant mth root of the tangent bundle of C (it does not admit a G-linearization unless m divides s). Let $\tilde{E}_\rho$ be the pull-back of E_ρ on X. It is a $m.G$-linearized vector bundle. The restriction of the vector bundle $E'_\rho = (\tilde{E}_\rho)^{\mathbb{Z}/m\mathbb{Z}}$ to $V(C)^*$ coincides with the pull-back of E_ρ on $V(C)^*$ but differs on the whole $\overline{V(C)}$.

Now we have constructed an extension E'_ρ of the pull-back of E_ρ on $V(C)^*$ to the whole $\overline{V(C)}$. Let $\pi : \overline{V(C)} \setminus S_0 \to (\overline{V(C)} \setminus S_0)/G$ be the projection to the quotient. We know that the quotient has only cyclic quotient singularities of types A_{e_i-1} lying on the image of the curve S_∞. The minimal resolution of the quotient is the surface $\widetilde{V}_\Gamma = \overline{V}_\Gamma \setminus \mathbf{E}_0$. Now we are in the situation of the ordinary McKay correspondence. The standard procedure is to take the coherent sheaf $\pi_*^G \mathcal{E}'_\rho$, where $\mathcal{E}'_\rho$ is the sheaf of sections of E'_ρ. Its pull-back on the resolution modulo the torsion subsheaf is a locally free sheaf $\mathcal{F}_\rho$ on $\widetilde{V}_\Gamma$ (see [18]). Equivalently, one can resolve the singularities of the quotient via the G-Hilbert scheme $H = \mathrm{Hilb}_G(\overline{V(C)} \setminus S_0)$ (see [36]). It comes with a universal G-equivariant family $\alpha\colon \mathcal{Z} \to H$ whose fibres are 0-dimensional subschemes Z of $\overline{V(C)} \setminus S_0$ with $H^0(Z, \mathcal{O}_Z)$ isomorphic to the regular representation of G. Let $\beta\colon \mathcal{Z} \to \overline{V(C)} \setminus S_0$ be the natural projection. Then $\alpha_*^G(\beta^* \mathcal{E}'_\rho)$ is a locally free sheaf isomorphic to $\mathcal{F}_\rho$.

Thus we have defined a correspondence

$$\{\text{admissible representations of } \Pi\} \xrightarrow{\rho \mapsto \mathcal{F}_\rho} \{\text{vector bundles on } \widetilde{V}_\Gamma\}. \tag{7.1}$$

Let $K_0(\Pi)$ be the Grothendieck group based on the set of admissible representations of Π and $K_0(\widetilde{V}_\Gamma)$ be the Grothendieck group based on vector bundles on the surface $\widetilde{V}_\Gamma$.

Conjecture 7.1. *The correspondence* (7.1) *defines an isomorphism of groups*

$$K_0(\Pi) \cong K_0(\widetilde{V}_\Gamma).$$

Note that the group $K_0(\widetilde{V}_\Gamma)$ is easy to compute. We know that the surface $\overline{V}_\Gamma$ is isomorphic to the blow-up of $\mathbb{P}^2$ at collinear points $p_1, \ldots, p_n$. The pre-image of the line containing these points is the union of the curve $\mathbf{E}_\infty$ and n exceptional curves F_i of the first kind contained in the fibres of the projection to $\mathbb{P}^1$. Since we know how K_0-groups change under blow-ups we immediately obtain the following.

Proposition 7.2. (i) *The map*

$$\phi\colon K_0(\widetilde{V}_\Gamma) \to \mathbb{Z} \oplus \mathrm{Pic}(\widetilde{V}_\Gamma), E \mapsto (\mathrm{rank}\, E, c_1(E))$$

is an isomorphism of abelian groups.

(ii) *The group* $\mathrm{Pic}(\widetilde{V}_\Gamma)$ *is a free abelian group generated by the classes* ω_i *which are dual to the divisor classes of irreducible components of* $\mathbf{E}_\infty$ *with respect to the intersection pairing* $\mathrm{Pic}(\widetilde{V}_\Gamma) \times \mathrm{Pic}(\widetilde{V}_\Gamma)^\infty$, *where* $\mathrm{Pic}(\widetilde{V}_\Gamma)^\infty$ *is the subgroup of divisors supported on* $\mathbf{E}_\infty$.

Note that in the ordinary McKay correspondence we have an analogous statement for the minimal resolution of $\mathbb{C}^2/\Pi$ ([18, Proposition 2]). Moreover, the images in $K_0(\widetilde{V}_\Gamma)$ of the elements of $K_0(\Pi)$ corresponding to irreducible representations ρ_i are equal to (d_i, ω_i), where $d_i = \dim \rho_i$. If we identify ω_i with the negatives of fundamental weights of the root system defined by the intersection matrix of $\mathbf{E}_0$, then the numbers d_i are the coefficients of the maximal root. In the hyperbolic case I do not know what are the elements in $K_0(\Pi)$ whose images in $K_0(\widetilde{V}_\Gamma)$ are projected to the elements ω_i of $\mathrm{Pic}(\widetilde{V}_\Gamma)$.

8. Weighted projective lines

Here we briefly comment on a certain relation of our work with the canonical algebras of C. Ringel and weighted projective lines of W. Geigle and H. Lenzing. There is enormous literature on this subject, so we refer only to [7] and [26], where further references can be found.

Let Γ be a Fuchsian group of signature $(0; e_1, \dots, e_n)$. The algebra of automorphic forms $A([\Gamma, \Gamma])$ is generated by n elements $f_1, \dots, f_n$ with relations of the form

$$f_1^{e_1} + a_i f_i^{e_i} + f_n^{e_n} = 0, \ i = 2, \dots, n-1,$$

where $a_2 = 1, a_i \in \mathbb{C}^*, i = 3, \dots, n-1$. This result belongs to H. Poincaré who called the functions $f_1, \dots, f_n$ the *Halphen functions* ([12; 34, p. 237]). This result implies easily that the étale cover $U \to V_\Gamma^*$ corresponding to the commutator group $[\Pi, \Pi]$ is isomorphic to the punctured affine surface $U_\Gamma^* = U_\Gamma \setminus \{0\}$, where U is an affine complete intersection surface in $\mathbb{C}^n$ given by the equations

$$z_1^{e_1} + a_i z_i^{e_i} + z_n^{e_n} = 0, \quad i = 2, \dots, n-1.$$

This result was proven by J. Milnor [30] ($n = 3$) and W. Neumann [32] ($n \geq 3$) independently of Poincaré's result.

The ring R of regular functions on the surface U_Γ is a finitely generated algebra with action of the group

$$(\lambda_1, \dots, \lambda_n) \in \mathbb{C}^{*n} : \lambda_1^{e_1} = \dots = \lambda_n^{e_n}.$$

This is the diagonalizable algebraic group $D(M)$, where M is an abelian group with presentation $\{g_1, \dots, g_n : e_1 g_1 = \dots = e_n g_n\}$ and D denotes the dual algebraic group. Using the theory of elementary divisors for abelian groups we find that

$$M \cong H^2(\Gamma, \mathbb{Z}).$$

The action of $D(M)$ is equivalent to the grading

$$R = \bigoplus_{m \in M} R_m$$

of the algebra R by the group M. A *weighted projective line* is the ring R together with the above grading. In notation of section 5, let $\gamma = c_1^G(K_X) \in H^2(\Gamma, \mathbb{Z})$. It is equal to $s\gamma_0$, where γ_0 generates the free part of $H^2(\Gamma, \mathbb{Z})$. The quotient group $M/\gamma M$ is isomorphic to $\Pi^{\mathrm{ab}} = \Pi/[\Pi, \Pi]$. The corresponding finite group $D(M/\gamma M)$ acts on the affine surface U_Γ with quotient isomorphic to our surface V_Γ. The action is free on U_Γ^*. So, in this sense the surface U_Γ^* is a maximal unramified abelian cover of the surface V_Γ^*.

Let $\mathbb{X}$ be a weighted projective line as above. Lenzing and others consider the category $\operatorname{coh}\mathbb{X}$ of coherent sheaves on $\mathbb{X}$. By definition, it is the category of M-graded R-modules modulo the Serre subcategory of finite length modules. Let R' be the same ring but with the grading defined by the free subgroup $\gamma\mathbb{Z}$ of M. For any M-graded R-module P, the coherent sheaf $\mathcal{F} = P^\sim$ restricted to U_Γ^* is a Π^{ab}-sheaf that descents to a coherent sheaf on $V_\Gamma^* = U_\Gamma/\Pi^{\mathrm{ab}}$. It corresponds to $\mathbb{Z}$-graded $A(\Gamma)$-module $P^{\Pi^{\mathrm{ab}}}$. Admissible unitary representations of Π correspond to coherent sheaves on $\mathbb{X}$ which define locally free sheaves on V_Γ^* isomorphic to the pre-image of a vector bundle on the DM-stack $\mathbb{D}/\Pi$ defined by an admissible unitary representation of Π.

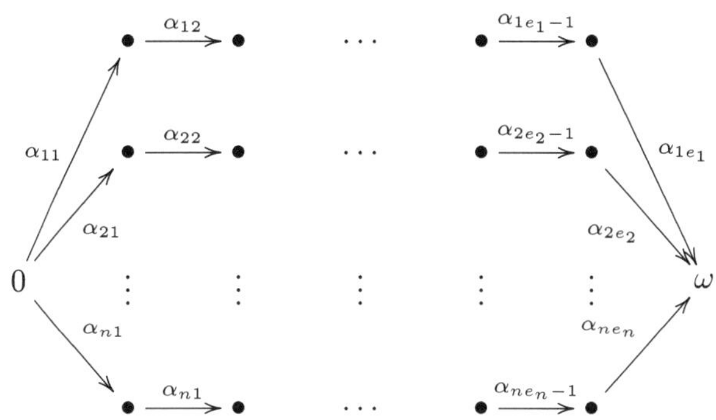

FIGURE 3

The Grothendieck group $K_0(\text{coh } \mathbb{X})$ was computed by Geigle and Lenzing. It is isomorphic to $K_0(\widetilde{V}_\Gamma)$. The intersection matrix of the curve $\mathbf{E}_\infty$ is the negative of a symmetric Cartan matrix. It defines a Kac–Moody Lie algebra $\mathfrak{g}$ and associated loop algebra $\mathcal{L}\mathfrak{g}$ with root lattice $K_0(\text{coh}\,\mathbb{X})$. Crowley-Boevey proves that indecomposable coherent sheaves on $\mathbb{X}$ correspond to positive roots, a unique indecomposable sheaf for a real root and infinitely many for an imaginary root. It would be very interesting to determine which roots correspond to admissible unitary representations of Π.

Note that the category coh $\mathbb{X}$ is derived equivalent to the category of modules over the canonical algebra $C(a, e_1, \ldots, e_n)$ of Ringel. The canonical algebra is the path algebra with relations of the following quiver obtained from our graph describing the curve $\mathbf{E}_\infty$ by adding one vertex joined to the extreme vertices of all arms of the following graph (we borrowed the picture from [26]).

Here the path algebra is taken with the relations

$$a_i\alpha_{ie_i}\cdots\alpha_{i2}\alpha_{i1} + \alpha_{1e_1}\cdots\alpha_{12}\alpha_{11} + \alpha_{ne_n}\cdots\alpha_{n2}\alpha_{n1} = 0, \quad i = 2, \ldots, n-1.$$

References

1. M. Artin and J.-L. Verdier, *Reflexive modules over rational double points*, Math. Ann. **270** (1985), no. 1, 79–82.
2. K. Behrend and B. Noohi, *Uniformization of Deligne–Mumford curves*, J. Reine Angew. Math. **599** (2006), 111–153.
3. T. Bridgeland, A. King, and M. Reid, *The McKay correspondence as an equivalence of derived categories*, J. Amer. Math. Soc. **14** (2001), no. 3, 535–554.
4. E. Brieskorn, A. Pratoussevitch, and F. Rothenhäusler, *The combinatorial geometry of singularities and Arnold's series E, Z, Q*, Mosc. Math. J. **3** (2003), 273–333.
5. J.-L. Brylinski, *A correspondence dual to McKay's*, unpublished, available at `arXiv:alg-geom/9612003`.
6. K. Corlette and C. Simson, *On the classification of rank 2 representations of quasiprojective fundamental groups*, available at `arXiv:math.AG/0702287`.
7. W. Crawley-Boevey, *Quiver algebras, weighted projective lines, and the Deligne–Simpson problem*, Proceedings of the International Congress of Mathematicians (Madrid, 2006), Volume II (M. Sanz-Solé, J. Soria, J. L. Varona, and J. Verdera, eds.), European Math. Soc., Zürich, 2007, pp. 117–129.
8. ———, *On the exceptional fibres of Kleinian singularities*, Amer. J. Math. **122** (2000), no. 5, 1027–1037.
9. P. Deligne, *Équations différentielles à points singuliers réguliers*, Lecture Notes in Math., vol. 163, Springer, Berlin, 1970.

10. I. V. Dolgačev, *Conic quotient singularities of complex surfaces*, Funkcional. Anal. i Priložen. **8** (1974), no. 2, 75–76 (Russian); English transl., Functional Anal. Appl. **8** (1975), no. 2, 160–161.
11. I. V. Dolgačv, *Automorphic forms, and quasihomogeneous singularities*, Funkcional. Anal. i Priložen. **9** (1975), no. 2, 67–68 (Russian); English transl., Functional Anal. Appl. **9** (1975), no. 2, 149–151.
12. I. V. Dolgachev, *On algebraic properties of algebras of automorphic forms*, Modular Functions in Analysis and Number Theory (T. A. Metzger, ed.), Lecture Notes Math. Statist., vol. 5, Univ. Pittsburgh, Pittsburgh, PA, 1983, pp. 20–29.
13. ______, *On the link space of a Gorenstein quasihomogeneous surface singularity*, Math. Ann. **265** (1983), no. 4, 529–540.
14. I. Dolgachev, *Invariant stable bundles over modular curves* $X(p)$, Recent Progress in Algebra (Taejon/Seoul, 1997) (S. G. Hahn, H. C. Myung, and E. Zelmanov, eds.), Contemp. Math., vol. 224, Amer. Math. Soc., Providence, RI, 1999, pp. 65–99.
15. A. L. Edmonds, J. H. Ewing, and R. S. Kulkarni, *Torsion free subgroups of Fuchsian groups and tessellations of surfaces*, Invent. Math. **69** (1982), no. 3, 331–346.
16. G. Ellingsrud and K. Lønsted, *Equivariant K-theory for curves*, Duke Math. J. **51** (1984), no. 1, 37–46.
17. H. Esnault and H. Knörrer, *Reflexive modules over rational double points*, Math. Ann. **272** (1985), no. 4, 545–548.
18. G. Gonzalez-Sprinberg and J.-L. Verdier, *Construction géométrique de la correspondance de McKay*, Ann. Sci. École Norm. Sup. (4) **16** (1983), no. 3, 409–449.
19. A. Grothendieck, *Sur le mémoire de Weil: généralisation des fonctions abéliennes*, Séminaire Bourbaki, Vol. 4, Exp. No. 141, Soc. Math. France, Paris, 1995, pp. 57–71.
20. Y. Ito and I. Nakamura, *Hilbert schemes and simple singularities*, New Trends in Algebraic Geometry (Warwick, 1996) (Hulek. K., F. Catanese, C. Peters, and M. Reid, eds.), London Math. Soc. Lecture Note Ser., vol. 264, Cambridge Univ. Press, Cambridge, 1999, pp. 151–233.
21. M. Kapranov and É. Vasserot, *Kleinian singularities, derived categories and Hall algebras*, Math. Ann. **316** (2000), no. 3, 565–576.
22. H. Knörrer, *Group representations and the resolution of rational double points*, Finite Groups — Coming of Age (Montréal, 1982) (J. McKay, ed.), Contemp. Math., vol. 45, Amer. Math. Soc., Providence, RI, 1985, pp. 175–222.
23. B. Kostant, *On finite subgroups of* SU(2)*, simple Lie algebras, and the McKay correspondence*, Proc. Nat. Acad. Sci. U.S.A. **81** (1984), no. 16, Phys. Sci., 5275–5274.
24. R. S. Kulkarni, *Normal subgroups of Fuchsian groups*, Quart. J. Math. Oxford Ser. (2) **36** (1985), no. 143, 325–344.
25. J. B. Lee, K. B. Lee, and F. Raymond, *Seifert manifolds with fiber spherical space forms*, Trans. Amer. Math. Soc. **348** (1996), no. 9, 3763–3798.
26. H. Lenzing and J. de la Peña, *Extended canonical algebras and Fuchsian singularities*, available at arXiv:math/0611532.
27. M. W. Liebeck and A. Shalev, *Fuchsian groups, coverings of Riemann surfaces, subgroup growth, random quotients and random walks*, J. Algebra **276** (2004), no. 2, 552–601.
28. J. McKay, *Graphs, singularities, and finite groups*, The Santa Cruz Conference on Finite Groups (Santa Cruz, CA, 1979) (B. Cooperstein and G. Mason, eds.), Proc. Sympos. Pure Math., vol. 37, Amer. Math. Soc., Providence, RI, 1980, pp. 183–186.
29. ______, *Cartan matrices, finite groups of quaternions, and Kleinian singularities*, Proc. Amer. Math. Soc. **81** (1981), no. 1, 153–154.
30. J. Milnor, *On the* 3*-dimensional Brieskorn manifolds* $M(p,q,r)$, Knots, Groups, and 3-Manifolds (L. P. Neuwirth, ed.), Ann. of Math. Stud., vol. 84, Princeton Univ. Press, Princeton, NJ, 1975, pp. 175–225.
31. D. Mumford, *The topology of normal singularities of an algebraic surface and a criterion for simplicity*, Inst. Hautes Études Sci. Publ. Math. **9** (1961), 5–22.
32. W. Neumann, *Brieskorn complete intersections and automorphic forms*, Invent. Math. **42** (1977), 285–293.
33. M. S. Narasimhan and C. S. Seshadri, *Stable and unitary vector bundles on a compact Riemann surface*, Ann. of Math. (2) **82** (1965), 540–567.
34. H. Poincaré, *Mémoire sur les fonctions fuchsiennes*, Acta Math. **1** (1882), no. 1, 193–294.

35. H. Pinkham, *Normal surface singularities with C^* action*, Math. Ann. **227** (1977), no. 2, 183–193.
36. M. Reid, *La correspondance de McKay*, Séminaire Bourbaki, Vol. 1999/2000, Astérisque, vol. 276, Soc. Math. France, Paris, 2002, pp. 53–72.
37. I. Sčerbak, *Algebras of automorphic forms with three generators*, Funkcional. Anal. i Priložen. **12** (1978), no. 2, 93–94 (Russian); English transl., Functional Anal. Appl. **12** (1978), no. 2, 156–158.
38. P. Scott, *The geometries of* 3-*manifolds*, Bull. London Math. Soc. **15** (1983), no. 5, 401–487.
39. C. S. Seshadri, *Fibrés vectoriels sur les courbes algébriques*, Astérisque, vol. 96, Soc. Math. France, Paris, 1982.
40. T. A. Springer, *Poincaré series of binary polyhedral groups and McKay's correspondence*, Math. Ann. **278** (1987), no. 1-4, 99–116.
41. R. Steinberg, *Finite subgroups of* SU_2*, Dynkin diagrams and affine Coxeter elements*, Pacific J. Math. **118** (1985), no. 2, 587–598.
42. Ph. Wagreich, *Algebras of automorphic forms with few generators*, Trans. Amer. Math. Soc. **262** (1980), no. 2, 367–389.
43. ———, *The structure of quasihomogeneous singularities*, Singularities. Part 2 (Arcata, CA, 1981) (P. Orlik, ed.), Proc. Sympos. Pure Math., vol. 40, Amer. Math. Soc., Providence, RI, 1983, pp. 593–611.

Department of Mathematics, University of Michigan, 530 Church Street, Ann Arbor, MI 48109-1043, USA

E-mail address: idolga@umich.edu

Centre de Recherches Mathématiques
CRM Proceedings and Lecture Notes
Volume **47**, 2009

Arithmetic Groups and the Affine E_8 Dynkin Diagram

John F. Duncan

ABSTRACT. Several decades ago, John McKay suggested a correspondence between nodes of the affine E_8 Dynkin diagram and certain conjugacy classes in the Monster group. Thanks to Monstrous Moonshine, this correspondence can be recast as an assignment of discrete subgroups of $PSL_2(\mathbb{R})$ to nodes of the affine E_8 Dynkin diagram. The purpose of this article is to give an explanation for this latter correspondence using elementary properties of the group $PSL_2(\mathbb{R})$. We also obtain a super analogue of McKay's observation, in which conjugacy classes of the Monster are replaced by conjugacy classes of Conway's group — the automorphism group of the Leech lattice.

1. Introduction

Several decades ago, John McKay suggested the following assignment of conjugacy classes in the Monster group to nodes of the affine E_8 Dynkin diagram (cf. [2, Section 14]).

(1.1) (1A) — 2A — 3A — 4A — 5A — 6A, with 6A joined to 3C and to 4B, and 4B joined to 2B

This has become known as *McKay's Monstrous* E_8 *observation.* Since elements of the Monster group determine principal moduli for genus zero subgroups of $PSL_2(\mathbb{R})$ via Monstrous Moonshine (cf. [1,5]), the assignment (1.1) entails a correspondence between nodes of the affine E_8 Dynkin diagram and certain discrete subgroups[1] of $PSL_2(\mathbb{R})$ that are commensurable with $PSL_2(\mathbb{Z})$.

(1.2) (1) — 2+ — 3+ — 4+ — 5+ — 6+, with 6+ joined to $3 \parallel 3$ and to $4 \parallel 2+$, and $4 \parallel 2+$ joined to 2

2000 *Mathematics Subject Classification.* 11F06.
This is the final form of the paper.

[1]The notation in (1.2) for subgroups of $PSL_2(\mathbb{R})$ follows [4,5]; we write $n+$ for $\Gamma_0(n)+$, and $n \parallel h+$ for $\Gamma_0(n \parallel h)+$, for example.

In this article we furnish a prescription for recovering this latter correspondence (1.2), that is given solely in terms of elementary properties of the group $\mathrm{PSL}_2(\mathbb{R})$.

For the purposes of this article, we will say that a subgroup $\Gamma < \mathrm{PSL}_2(\mathbb{R})$ is *arithmetic* if it is commensurable with $\mathrm{PSL}_2(\mathbb{Z})$ in the strong sense; viz. if the intersection of Γ with $\mathrm{PSL}_2(\mathbb{Z})$ has finite index in each of Γ and $\mathrm{PSL}_2(\mathbb{Z})$.

In Section 2 we set up a general context for studying arithmetic subgroups of $\mathrm{PSL}_2(\mathbb{R})$. The approach here is closely modeled on that described in [3]. In Section 3.2 we furnish arithmetic conditions that are satisfied only by the arithmetic groups appearing in (1.2). Then in Section 3.3 we show how the arithmetic properties of these groups can be used to recover also the edges of the affine E_8 Dynkin diagram (1.2).

When considering the correspondence (1.2), it is hard not to be reminded of *the McKay correspondence*, which may be regarded as giving a prescription for recovering the extended Dynkin diagrams from representations of finite subgroups of $\mathrm{SU}(2)$ (cf. [8]). In particular, the McKay correspondence indicates how to recover the affine E_8 Dynkin diagram from the category of representations of the binary icosahedral group, $2.\mathrm{Alt}_5 \subset \mathrm{SU}(2)$. Our prescription for recovering the edges of (1.2) is reminiscent of this procedure.

In Section 4 we demonstrate how to obtain a kind of "super analogue" of McKay's Monstrous E_8 observation, in which conjugacy classes of Conway's group Co_0 take on the rôle played by conjugacy classes of the Monster in (1.1).

2. Arithmetic groups

Let us agree to say that $\Gamma < \mathrm{PSL}_2(\mathbb{R})$ is an *arithmetic* subgroup of $\mathrm{PSL}_2(\mathbb{R})$ if the intersection $\mathrm{PSL}_2(\mathbb{Z}) \cap \Gamma$ has finite index both in $\mathrm{PSL}_2(\mathbb{Z})$ and in Γ. In order to study the arithmetic subgroups of $\mathrm{PSL}_2(\mathbb{R})$ we will adopt the approach of Conway [3], whereby one analyzes these groups in terms of their actions on lattices. We review this approach in the present section. Actually, our exposition will employ a slightly different language to that of [3]; the story is nonetheless the same. As a small supplement to the ideas of [3], we include, in Section 2.10, a description of how to identify the cusps of certain arithmetic subgroups of $\mathrm{PSL}_2(\mathbb{R})$, in terms of such a group's action on lattices.

2.1. Orientation. Let $\mathbf{k}$ be an ordered field. Write $\mathbf{k}^\times$ for the multiplicative group of non-zero elements of $\mathbf{k}$, and write $\mathbf{k}^+$ for its subgroup of *positive elements.* Note that an ordered field is necessarily of characteristic 0, so there is a natural embedding $\mathbb{Q} \hookrightarrow \mathbf{k}$.

Let V be a vector space of dimension n over $\mathbf{k}$. We may consider the *nth exterior power* $\bigwedge^n V$. This vector space has a distinguished point; viz. the origin, and we may consider the complement $(\bigwedge^n V)^\times = \bigwedge^n V \setminus \{0\}$. The set $(\bigwedge^n V)^\times$ is naturally a $\mathbf{k}^\times$-torsor. Upon picking any non-zero element, w say, of $\bigwedge^n V$, we arrive at a decomposition of $(\bigwedge^n V)^\times$ into two disjoint $\mathbf{k}^+$-torsors: $(\bigwedge^n V)^\times = \mathbf{k}^+ \cdot w \cup \mathbf{k}^+ \cdot (-w)$. We define an *orientation* for V to be a choice of $\mathbf{k}^+$-torsor in $\bigwedge^n V$. An *oriented vector space over* $\mathbf{k}$ is a pair (V, W) where V is a vector space over $\mathbf{k}$, and W is a $\mathbf{k}^+$-torsor in $\bigwedge^n V$.

Consider the canonical linear map $\Delta \colon V^n \to \bigwedge^n V$ given by

$$(v_1, \ldots, v_n) \mapsto \Delta(v_1, \ldots, v_n) := v_1 \wedge \cdots \wedge v_n. \tag{2.1}$$

The set $\{v_1,\dots,v_n\}$ underlying the n-tuple $(v_1,\dots,v_n)$ is a linearly independent subset of V if and only if $\Delta(v_1,\dots,v_n)\neq 0$. We say that an n-tuple $(v_1,\dots,v_n)$ is an *ordered basis* for V if $\Delta(v_1,\dots,v_n)\neq 0$. We write $\mathcal{B}$ for the set of ordered bases for V.

$$\mathcal{B} := \Delta^{-1}\big((\textstyle\bigwedge^n V)^\times\big) \subset V^n. \tag{2.2}$$

For (V,W) an oriented vector space, we say that an ordered set $(v_1,\dots,v_n)\in V^n$ is an *oriented basis* for V if $\Delta(v_1,\dots,v_n)\in W$. Let us write $\mathcal{B}^+$ for the set of oriented bases in V.

$$\mathcal{B}^+ := \Delta^{-1}(W) \subset V^n. \tag{2.3}$$

Let $\operatorname{End}(V)$ denote the $\mathbf{k}$-algebra of $\mathbf{k}$-linear transformations of V. We may regard $\operatorname{End}(V)$ as acting on V from the right

$$V\times \operatorname{End}(V)\to V\colon (v,A)\mapsto v\cdot A. \tag{2.4}$$

This action extends naturally to an (right) action of $\operatorname{End}(V)$ on V^n.

$$(v_1,\dots,v_n)\cdot A := (v_1\cdot A,\dots,v_n\cdot A) \tag{2.5}$$

As is usual, we take $\operatorname{GL}(V)$ to be the complement of $\det^{-1}(0)$ in $\operatorname{End}(V)$, where $\det\colon \operatorname{End}(V)\to\mathbf{k}$ is the unique map such that

$$\Delta(\mathbf{v}\cdot A) = \det(A)\Delta(\mathbf{v}) \tag{2.6}$$

for all $\mathbf{v}\in V^n$ and $A\in\operatorname{End}(V)$. We take $\operatorname{SL}(V)$ to be the kernel of the restriction $\det\colon \operatorname{GL}(V)\to\mathbf{k}^\times$, and we set $\operatorname{GL}^+(V)$ to be the preimage of $\mathbf{k}^+$ under this map.

$$\operatorname{GL}(V) = \det{}^{-1}(\mathbf{k}^\times) \tag{2.7}$$

$$\operatorname{GL}^+(V) = \det{}^{-1}(\mathbf{k}^+) \tag{2.8}$$

$$\operatorname{SL}(V) = \det{}^{-1}(1) \tag{2.9}$$

Observe that the sets $\mathcal{B}$ and $\mathcal{B}^+$ are naturally torsors for $\operatorname{GL}(V)$ and $\operatorname{GL}^+(V)$, respectively.

For any choice of ordered basis $\mathbf{v}=(v_1,\dots,v_n)\in\mathcal{B}$ there is a corresponding isomorphism $\phi_{\mathbf{v}}\colon \operatorname{GL}(V)\to\operatorname{GL}_n(\mathbf{k})$ given by setting $\phi_{\mathbf{v}}(A)$ to be the matrix $(a^i_j)\in \operatorname{GL}_n(\mathbf{k})$ such that $v_j\cdot A=\sum_i v_i a^i_j$ for $1\leq j\leq n$. Thus we obtain a right action of $\operatorname{GL}_n(\mathbf{k})$ on V^n subject to a choice of ordered basis $\mathbf{v}\in\mathcal{B}$. Observe that there is also a canonical (left) action of $\operatorname{GL}_n(\mathbf{k})$ on V^n given by setting

$$(a^i_j)\cdot(v_1,\dots,v_n) := (v'_1,\dots,v'_n) \tag{2.10}$$

where $v'_j=\sum_i a^i_j v_i$. Furthermore, this action preserves the subset $\mathcal{B}\subset V^n$; indeed, $\operatorname{GL}_n(\mathbf{k})$ acts simply transitively on $\mathcal{B}$, so that $\mathcal{B}$ is naturally a (left) $\operatorname{GL}_n(\mathbf{k})$-torsor. The set $\mathcal{B}^+$, consisting of oriented bases for V, is naturally a (left) torsor for $\operatorname{GL}^+_n(\mathbf{k})$.

2.2. Lattices. Let $V=(V,W)$ be an oriented vector space of dimension n over $\mathbf{k}$, as in Section 2.1. Write $\mathcal{B}^+$ for the subset of V^n consisting of oriented bases for V.

A *lattice* in V is an additive subgroup, L say, of V, such that L is equivalent to $\mathbb{Z}^n$ as a $\mathbb{Z}$-module, and such that $\operatorname{Span}_{\mathbf{k}} L$ (the $\mathbf{k}$-linear span of L) is V. Let us write $\mathcal{L}$ for the set of all lattices in V. Suppose that $L\in\mathcal{L}$, and let $\{v_1,\dots,v_n\}\subset V$ be a set of generators for L as a $\mathbb{Z}$-module, so that

$$L=\mathbb{Z}v_1+\cdots+\mathbb{Z}v_n. \tag{2.11}$$

Then $\{v_1, \ldots, v_n\}$ is a linearly independent subset of V, for if $\sum a_i v_i = 0$ for some $a_i \in \mathbf{k}$, then the $\mathbf{k}$-linear span of L is vector space of dimension less than n, contradicting the property that $\operatorname{Span}_{\mathbf{k}} L = V$. Conversely, if $\{v_1, \ldots, v_n\}$ is a linearly independent subset of V then the additive subgroup of V generated by the v_i is a lattice in V. We conclude that there is a natural surjective map from the set of bases for V to the set of lattices in V.

$$\{v_1, \ldots, v_n\} \mapsto \mathbb{Z}v_1 + \cdots + \mathbb{Z}v_n \tag{2.12}$$

Consider the (pre)composition of this map with the natural map $\mathcal{B}^+ \to \mathcal{P}(V)$, sending an n-tuple in $\mathcal{B}^+$ to it's underlying set. Since any basis for V can be made an oriented basis by equipping it with a suitable ordering, we conclude that this composition furnishes a (natural) surjective map from $\mathcal{B}^+$ to the set of lattices in V.

$$\mathcal{B}^+ \to \mathcal{L}\colon (v_1, \ldots, v_n) \mapsto \mathbb{Z}v_1 + \cdots + \mathbb{Z}v_n \tag{2.13}$$

We would like to know when two oriented bases for V determine the same lattice. Recall that $\mathcal{B}^+$ is naturally a $\mathrm{GL}_n^+(\mathbf{k})$ torsor (cf. Section 2.1). If

$$\mathbb{Z}v_1 + \cdots + \mathbb{Z}v_n = \mathbb{Z}v_1' + \cdots + \mathbb{Z}v_n' \tag{2.14}$$

then there are some integers m_j^i and n_j^i such that $v_j' = \sum_i m_j^i v_i$ and $v_j = \sum_i n_j^i v_i'$ for $1 \leq j \leq n$, and hence the matrices (m_j^i) and (n_j^i) are mutually inverse. In other words, if two oriented bases determine the same lattice then they are related, via left-multiplication, by an element of $\mathrm{SL}_n(\mathbb{Z})$. Conversely, if $A \in \mathrm{SL}_n(\mathbb{Z})$ and $(v_i) \in \mathcal{B}^+$, then we have the equality (2.14) once we set $(v_i') = (A \cdot v_i)$.

We have shown that the set $\mathcal{L}$ is naturally isomorphic to the orbit space $\mathrm{SL}_n(\mathbb{Z})\backslash\mathcal{B}^+$, where $\mathcal{B}^+$ is naturally a torsor for $\mathrm{GL}_n^+(\mathbf{k})$. Given a choice of oriented basis $\mathbf{v} \in \mathcal{B}^+$, we obtain an identification of $\mathcal{B}^+$ with $\mathrm{GL}_n^+(\mathbf{k})$

$$\mathrm{GL}_n^+(\mathbf{k}) \leftrightarrow \mathcal{B}^+ : A \leftrightarrow A \cdot \mathbf{v} \tag{2.15}$$

and hence, an identification of $\mathcal{L}$ with $\mathrm{SL}_n(\mathbb{Z})\backslash \mathrm{GL}_n^+(\mathbf{k})$.

From now on we will identify $\mathcal{L}$ with the orbit space $\mathrm{SL}_n(\mathbb{Z})\backslash\mathcal{B}^+$

2.3. Projective lattices. Let V, $\mathcal{B}^+$, and $\mathcal{L}$ be as is Section 2.2. For simplicity of exposition, let us assume that $n = \dim V$ is even — the case that n is odd requires one to occasionally replace $\mathbf{k}^\times$ with $\mathbf{k}^+$.

There is a natural embedding (of groups) $\mathbf{k}^\times \hookrightarrow \mathrm{GL}_n^+(\mathbf{k})$. The image of $\mathbf{k}^\times$ under this map is central (indeed, this $\mathbf{k}^\times$ is the center of $\mathrm{GL}_n^+(\mathbf{k})$), so there is a natural action of $\mathbf{k}^\times$ on the space of lattices, $\mathcal{L} = \mathrm{SL}_n(\mathbb{Z})\backslash\mathcal{B}^+$.

$$\alpha \cdot \mathrm{SL}_n(\mathbb{Z})(v_1, \ldots, v_n) := \mathrm{SL}_n(\mathbb{Z})(\alpha v_1, \ldots, \alpha v_n) \tag{2.16}$$

We wish to consider the quotient space $\mathrm{P}\mathcal{L} = \mathbf{k}^\times\backslash\mathcal{L}$. That is, we would like to regard two lattices as equivalent if one is a non-zero scalar multiple of the other; we call the elements of $\mathrm{P}\mathcal{L}$ the *projective lattices* in V.

If X is a set equipped with an action of $\mathbf{k}^\times$, we write $\mathrm{P}X$ for the orbit space $\mathbf{k}^\times\backslash X$. We write $x \mapsto [x]$ for the natural map $X \to \mathrm{P}X$.

Just as $\mathcal{L}$ is naturally identified with the orbit space $\mathrm{SL}_n(\mathbb{Z})\backslash\mathcal{B}^+$. The set $\mathrm{P}\mathcal{L}$ is naturally identified with the orbit space $\mathrm{PSL}_n(\mathbb{Z})\backslash\mathrm{P}\mathcal{B}^+$, where

$$\mathrm{PSL}_n(\mathbb{Z}) = \mathrm{SL}_n(\mathbb{Z})/(\mathbf{k}^\times \cap \mathbb{Z}), \tag{2.17}$$

$$\mathrm{P}\mathcal{B}^+ = \mathbf{k}^\times\backslash\mathcal{B}^+. \tag{2.18}$$

We call $\mathrm{P}\mathcal{B}^+$ the set of *projective oriented bases* for V. It is naturally a torsor for $\mathrm{PGL}_n^+(\mathbf{k}) = \mathrm{GL}_n^+(\mathbf{k})/\mathbf{k}^\times$, so that after choosing an element $[\mathbf{v}] \in \mathrm{P}\mathcal{B}^+$ we obtain an identification of $\mathrm{PGL}_n^+(\mathbf{k})$ with $\mathrm{P}\mathcal{B}^+$ by setting

$$\mathrm{PGL}_n^+(\mathbf{k}) \leftrightarrow \mathrm{P}\mathcal{B}^+ \colon [A] \leftrightarrow [A]\cdot[\mathbf{v}] := [A\mathbf{v}], \tag{2.19}$$

and hence, also, an identification of $\mathrm{PSL}_n(\mathbb{Z})\backslash \mathrm{PGL}_n^+(\mathbf{k})$ with $\mathrm{P}\mathcal{L}$.

2.4. Commensurable lattices. Let V, $\mathcal{B}^+$, and $\mathcal{L}$ be as in Section 2.2. Let us chose a lattice $L_1 \in \mathcal{L}$, and let $\mathbf{v}_1 \in \mathcal{B}^+$ satisfy $L_1 = \mathrm{SL}_n(\mathbb{Z})\mathbf{v}_1$. We say that a lattice L in V is *commensurable* with L_1 if the intersection $L \cap L_1$ has finite index both in L and in L_1. Let V_1 be the $\mathbb{Q}$-linear span of L_1 in V, so that V_1 is a vector space of dimension n over $\mathbb{Q}$ that is contained in V, and the $\mathbf{k}$-linear span of V_1 is V.

The following result is straightforward.

Proposition 2.1. *The lattices in V that are commensurable with L_1 are exactly the additive subgroups of V_1 that are equivalent to $\mathbb{Z}^n$ as $\mathbb{Z}$-modules.*

The rational vector space V_1, determined by L_1, inherits an orientation from V (in the sense of Section 2.1) which we denote by W_1. We write $\mathcal{B}_1^+$ for the set of oriented bases for V_1 (cf. Section 2.2), and we write $\mathcal{L}_1$ for the orbit space $\mathrm{SL}_n(\mathbb{Z})\backslash\mathcal{B}_1^+$, which is naturally identified with the set of lattices in V_1 — in other words, by Proposition 2.1, $\mathcal{L}_1$ is the subset of $\mathcal{L}$ consisting of the lattices in V that are commensurable with L_1. Our choice $\mathbf{v}_1 \in \mathcal{B}_1^+$ allows us to construct an identification of $\mathcal{L}_1$ with $\mathrm{SL}_n(\mathbb{Z})\backslash \mathrm{GL}_n^+(\mathbb{Q})$ (cf. Section 2.2).

We will require to distinguish the lattices in V_1 only up to their orbits under $\mathbb{Q}^\times$. Similar to Section 2.3, we write $\mathrm{P}\mathcal{L}_1$ for the orbit space $\mathbb{Q}^\times\backslash\mathcal{L}_1$. Then $\mathrm{P}\mathcal{L}_1$ is naturally identified with $\mathrm{PSL}_n(\mathbb{Z})\backslash\mathrm{P}\mathcal{B}_1^+$ for $\mathrm{P}\mathcal{B}_1^+ = \mathbb{Q}^\times\backslash\mathcal{B}_1^+$. The choice $\mathbf{v}_1 \in \mathcal{B}_1^+$ allows us to identify $\mathrm{P}\mathcal{B}_1^+$ with $\mathrm{PSL}_n(\mathbb{Z})\backslash \mathrm{PGL}_n^+(\mathbb{Q})$ (cf. Section 2.3).

We will continue to write $x \mapsto [x]$ for the projection maps $X \to \mathrm{P}X$. It should be clear from the context whether we are taking orbits with respect to actions by $\mathbb{Q}^\times$ or $\mathbf{k}^\times$.

2.5. Hyperdistance. We continue in the setting of Section 2.4. In particular, we continue to assume that $n = \dim V$ is even.

There is a canonically defined integer valued function on $M_n(\mathbb{Q})$, which we call the *rational projective determinant*, and denote Pdet, which is defined as follows. If $A = (a^i_j) \in M_n(\mathbb{Q})$ is non-zero, then there is a smallest positive rational number $\alpha_A \in \mathbb{Q}^+$ such that $\alpha_A a^i_j \in \mathbb{Z}$ for all $1 \le i,j \le n$; explicitly, if we write each a^i_j as a quotient $a^i_j = b^i_j/c^i_j$ for some $b^i_j, c^i_j \in \mathbb{Z}$, then $\alpha_A = \gcd\{c^i_j\}/\operatorname{lcm}\{b^i_j\}$. We set

$$\mathrm{Pdet}(A) := \begin{cases} \det(\alpha_A A) = \alpha_A^n \det(A), & \text{if } A \neq 0; \\ 0, & \text{if } A = 0. \end{cases} \tag{2.20}$$

Evidently, $\mathrm{Pdet}(\alpha A) = \mathrm{Pdet}(A)$ for any $\alpha \in \mathbb{Q}^\times$ (so long as n is even), so that the projective determinant induces a well defined function,

$$\mathrm{Pdet}\colon \mathrm{P}M_n(\mathbb{Q}) \to \mathbb{Z}, \tag{2.21}$$

which we also call the *rational projective determinant*, where $\mathrm{P}M_n(\mathbb{Q}) = \mathbb{Q}^\times\backslash M_n(\mathbb{Q})$ is the monoid of *rational projective matrices*.

Lemma 2.2. *If* $A \in \mathrm{SL}_n(\mathbb{Z})$ *then* $\mathrm{Pdet}(AX) = \mathrm{Pdet}(X) = \mathrm{Pdet}(XA)$ *for any* $X \in M_n(\mathbb{Q})$.

PROOF. We have $\mathrm{Pdet}(AX) = \alpha_{AX}^n \det(AX)$ by definition, and $\det(AX) = \det(X)$ for $A \in \mathrm{SL}_n(\mathbb{Z})$, so it suffices to show that $\alpha_{AX} = \alpha_X$ for $A \in \mathrm{SL}_n(\mathbb{Z})$. Since both A and its inverse have integer entries, αAX belongs to $M_n(\mathbb{Z})$ if and only if αX does, for any $\alpha \in \mathbb{Q}^+$. Thus the sets $\{\alpha \in \mathbb{Q}^+ \mid \alpha AX \in M_n(\mathbb{Z})\}$ and $\{\alpha \in \mathbb{Q}^+ \mid \alpha X \in M_n(\mathbb{Z})\}$ coincide, and thus their minimal elements coincide. □

We will be most interested in the restriction of Pdet to the group $\mathrm{PGL}_n^+(\mathbb{Q})$, where it takes only positive integer values. The following result is an immediate consequence of Lemma 2.2.

Proposition 2.3. *The projective determinant induces a well-defined positive integer valued function on the orbit space* $\mathrm{PSL}_n(\mathbb{Z})\backslash \mathrm{PGL}_n^+(\mathbb{Q})$

$$\mathrm{Pdet}\colon \mathrm{PSL}_n(\mathbb{Z})\backslash \mathrm{PGL}_n^+(\mathbb{Q}) \to \mathbb{Z}_{>0}. \tag{2.22}$$

This function is invariant under the natural right action of $\mathrm{PSL}_n(\mathbb{Z})$.

In particular then, the projective determinant furnishes us with a natural method for comparing projective lattices in V_1. For suppose given a pair of projective lattices $L, L' \in \mathrm{P}\mathcal{L}_1$. We regard $\mathrm{P}\mathcal{L}_1$ as identified with $\mathrm{PSL}_n(\mathbb{Z})\backslash \mathrm{P}\mathcal{B}_1^+$, so that we have

$$L = \mathrm{PSL}_n(\mathbb{Z}) \cdot [\mathbf{v}], \quad L' = \mathrm{PSL}_n(\mathbb{Z}) \cdot [\mathbf{v}'], \tag{2.23}$$

for some $[\mathbf{v}], [\mathbf{v}'] \in \mathrm{P}\mathcal{B}_1^+$. The set $\mathrm{P}\mathcal{B}_1^+$ is a $\mathrm{PGL}_n^+(\mathbb{Q})$-torsor, so there is a unique $g \in \mathrm{PGL}_n^+(\mathbb{Q})$ such that $g \cdot [\mathbf{v}] = [\mathbf{v}']$. We define a positive integer $\delta(L, L')$ by setting

$$\delta(L, L') = \mathrm{Pdet}(g). \tag{2.24}$$

Proposition 2.4. *The function* $\delta\colon \mathrm{P}\mathcal{L} \times \mathrm{P}\mathcal{L} \to \mathbb{Z}_{>0}$ *given by* (2.24) *is well-defined.*

PROOF. We should check that the value of $\delta(L, L')$ does not depend upon the choice of representatives (viz. $[\mathbf{v}]$ and $[\mathbf{v}']$), for the orbits L and L'. So let $L, L' \in \mathrm{P}\mathcal{L}_1$ be as in (2.23), with $g \in \mathrm{PGL}_n^+(\mathbb{Q})$ satisfying $g \cdot [\mathbf{v}] = [\mathbf{v}']$, and suppose that $L = \mathrm{PSL}_n(\mathbb{Z}) \cdot [\mathbf{w}]$ and $L' = \mathrm{PSL}_n(\mathbb{Z}) \cdot [\mathbf{w}']$. Then there are some $h, h' \in \mathrm{PSL}_n(\mathbb{Z})$ such that $[\mathbf{w}] = h \cdot [\mathbf{v}]$ and $[\mathbf{w}'] = h' \cdot [\mathbf{v}']$, and $h'gh^{-1}$ is the unique element of $\mathrm{PGL}_n^+(\mathbb{Q})$ satisfying $(h'gh^{-1}) \cdot [\mathbf{w}] = [\mathbf{w}']$. We require to show that $\mathrm{Pdet}(g) = \mathrm{Pdet}(h'gh^{-1})$, but this is guaranteed by Proposition 2.3. □

In the case that $n = 2$ the function δ acquires a special property.

Proposition 2.5. *For* $n = 2$, *the function* $\delta : \mathrm{P}\mathcal{L} \times \mathrm{P}\mathcal{L} \to \mathbb{Z}_{>0}$ *is symmetric.*

PROOF. If $X \in M_n(\mathbb{Z})$ and X is invertible, then $\det(X)X^{-1} \in M_n(\mathbb{Z})$. Applying this same rule with $\det(X)X^{-1}$ in place of X, we see that

$$\frac{\det(X)^n}{\det(X)} \frac{X}{\det(X)} \in M_n(\mathbb{Z}) \tag{2.25}$$

Thus, in the special case that $n = 2$, we have that an invertible matrix X lies in $M_n(\mathbb{Z})$ if and only if $\det(X)X^{-1}$ does. Consequently, for any $A \in \mathrm{GL}_2^+(\mathbb{Q})$ and any $\alpha \in \mathbb{Q}^+$ we have that αA belongs to $M_2(\mathbb{Z})$ if and only if $\alpha \det(A) A^{-1}$ does, so that if α_A is the minimal positive rational for which $\alpha_A A$ belongs to $M_2(\mathbb{Z})$, then $\alpha_{A^{-1}} = \det(A)\alpha_A$ is the minimal positive rational that has this property for A^{-1}. It is easy to compute now that $\mathrm{Pdet}(A) = \alpha_A^2 \det(A)$ and $\mathrm{Pdet}(A^{-1}) = \alpha_{A^{-1}}^2 \det(A^{-1})$ coincide. This verifies the claim, since if $\delta(L, L') = \mathrm{Pdet}(g)$ then $\delta(L', L) = \mathrm{Pdet}(g^{-1})$. □

Remark. It is easy to find $g \in \mathrm{PGL}_n^+(\mathbb{Q})$ such that $\mathrm{Pdet}(g) \neq \mathrm{Pdet}(g^{-1})$ when $n > 2$.

Following Conway [3], we call the function δ, when defined on pairs of projective lattices in a rational vector space of dimension 2, *hyperdistance*. The logarithm of δ is in fact a metric on $\mathrm{P}\mathcal{L}$ for $n = 2$.

From now on we will restrict attention to the case that $n = \dim V = 2$.

2.6. Names. We continue in the setting of Section 2.5. In particular, we assume that $n = \dim V = 2$, and we retain our choice of lattice $L_1 \in \mathcal{L}$, and generating set $\mathbf{v}_1 \in \mathcal{B}_1^+$. We regard the set $\mathrm{P}\mathcal{L}_1$, of projective lattices in V_1, as identified with the coset space $\mathrm{PSL}_2(\mathbb{Z})\backslash \mathrm{PGL}_2^+(\mathbb{Q})$ via this choice.

$$\mathrm{PSL}_2(\mathbb{Z})\backslash \mathrm{PGL}_2^+(\mathbb{Q}) \leftrightarrow \mathrm{P}\mathcal{L}_1 \colon \mathrm{PSL}_2(\mathbb{Z})[A] \leftrightarrow \mathrm{PSL}_2(\mathbb{Z}) \cdot [A\mathbf{v}_1] \tag{2.26}$$

Of course, $\mathrm{PGL}_2^+(\mathbb{Q})$ acts naturally, from the right, on $\mathrm{PSL}_2(\mathbb{Z})\backslash \mathrm{PGL}_2^+(\mathbb{Q})$, so the identification (2.26) also allows us to define a right action of $\mathrm{PGL}_2^+(\mathbb{Q})$ on $\mathrm{P}\mathcal{L}_1$. Evidently, hyperdistance (cf. Section 2.5) is invariant with respect to this $\mathrm{PGL}_2^+(\mathbb{Q})$-action.

We will now introduce *names* for the elements of the coset space

$$\mathrm{PSL}_2(\mathbb{Z})\backslash \mathrm{PGL}_2^+(\mathbb{Q}), \tag{2.27}$$

and hence also for the projective lattices in V_1. Consider the natural map $\mathrm{GL}_2^+(\mathbb{Q}) \to \mathrm{PSL}_2(\mathbb{Z})\backslash \mathrm{PGL}_2^+(\mathbb{Q})$. Our strategy (following [3]) is to try and define a canonical preimage in $\mathrm{GL}_2^+(\mathbb{Q})$ for each coset $\mathrm{PSL}_2(\mathbb{Z})g$ in the image of this map.

Let $g \in \mathrm{PGL}_2^+(\mathbb{Q})$. We will write

$$\begin{pmatrix} a & b \\ c & d \end{pmatrix} \mapsto \begin{bmatrix} a & b \\ c & d \end{bmatrix} \tag{2.28}$$

for the canonical map $M_2(\mathbb{Q}) \to \mathrm{P}M_2(\mathbb{Q})$. Then

$$g = \begin{bmatrix} a & b \\ c & d \end{bmatrix} \tag{2.29}$$

for some $a, b, c, d \in \mathbb{Q}$ such that $ad - bc > 0$. Choose integers $s, t \in \mathbb{Z}$ such that $sa + tc = 0$, and observe that then $sb + td \neq 0$, by linear independence of the columns of an invertible matrix. We may assume that s, t have no common factors, and thus there are integers $m, n \in \mathbb{Z}$ such that $mt - sn = 1$. In other words, there is an $h \in \mathrm{PSL}_2(\mathbb{Z})$ such that

$$h \cdot g = g' = \begin{bmatrix} a' & b' \\ 0 & d' \end{bmatrix} \tag{2.30}$$

for some $a', b', d' \in \mathbb{Q}$, with $d' \neq 0$. We have

$$\begin{bmatrix} a' & b' \\ 0 & d' \end{bmatrix} = \begin{bmatrix} a'' & b'' \\ 0 & 1 \end{bmatrix} \tag{2.31}$$

for $a'' = a'/d' > 0$ and $b'' = b'/d'$. There is a unique integer, N say, such that $0 \leq b'' + N < 1$. Left-multiplying by

$$\begin{bmatrix} 1 & N \\ 0 & 1 \end{bmatrix}, \tag{2.32}$$

we arrive at an element g''' of $\mathrm{PSL}_2(\mathbb{Z})g$ of the form

$$g''' = \begin{bmatrix} a''' & b''' \\ 0 & 1 \end{bmatrix} \tag{2.33}$$

where $a''' > 0$ and $0 \leq b''' < 1$. Let us write $\mathcal{M}$ for the set of matrices in $M_2(\mathbb{Q})$ of the form

$$\begin{pmatrix} M & b \\ 0 & 1 \end{pmatrix} \tag{2.34}$$

where $M > 0$ and $0 \leq b < 1$. We have established the following result.

Proposition 2.6. *The assignment*

$$\begin{pmatrix} M & b \\ 0 & 1 \end{pmatrix} \mapsto \mathrm{PSL}_2(\mathbb{Z}) \begin{bmatrix} M & b \\ 0 & 1 \end{bmatrix} \tag{2.35}$$

defines a bijective correspondence between the elements of $\mathcal{M}$ and the elements of $\mathrm{PSL}_2(\mathbb{Z})\backslash \mathrm{PGL}_2^+(\mathbb{Q})$.

Given $M \in \mathbb{Q}^+$ and $b \in \mathbb{Q} \cap [0,1)$, let us write $g_{M,b}$ for the projective matrix

$$g_{M,b} := \begin{bmatrix} M & b \\ 0 & 1 \end{bmatrix}. \tag{2.36}$$

Let us write $L_{M,b}$ for the corresponding projective lattice with respect to the identification (2.26); i.e., we set

$$L_{M,b} := \mathrm{PSL}_2(\mathbb{Z})g_{M,b}. \tag{2.37}$$

We will typically abbreviate $g_{M,0}$ to g_M, and $L_{M,0}$ to L_M. Observe that, under this notational convention, we have

$$L_1 = L_{1,0} = \mathrm{PSL}_2(\mathbb{Z})g_1 = \mathrm{PSL}_2(\mathbb{Z}) \leftrightarrow \mathrm{PSL}_2(\mathbb{Z})[\mathbf{v}_1] \tag{2.38}$$

in agreement with our notation L_1 for the lattice we picked in Section 2.4.

Observe that to compute the hyperdistance $\delta(L_{M,b}, L_1)$, from an arbitrary lattice $L_{M,b}$ to the distinguished lattice L_1, is the same as computing the projective determinant of the matrix $g_{M,b}$. More generally, computing the hyperdistance between $L_{M',b'}$ and $L_{M,b}$ is the same as computing the projective determinant of $g_{M',b'}g_{M,b}^{-1}$.

$$\begin{aligned} (g_{M',b'}g_{M,b}^{-1}) \cdot (g_{M,b} \cdot [\mathbf{v}_0]) &= g_{M',b'} \cdot [\mathbf{v}_0] \\ \delta(L_{M',b'}, L_{M,b}) &= \mathrm{Pdet}(g_{M',b'}g_{M,b}^{-1}) \end{aligned} \tag{2.39}$$

Our names $L_{M,b}$ for the projective lattices in V_1 are not particularly canonical. We could, for example, have chosen to seek coset representatives for $\mathrm{PSL}_2(\mathbb{Z})\backslash \mathrm{PSL}_2^+(\mathbb{Q})$ with vanishing top-right entry, unital top-left entry, positive bottom-right entry, and bottom-left entry in $\mathbb{Q} \cap [0,1)$. That is, we could have sought representatives in the form $\bar{g}_{b,M}$ where

$$\bar{g}_{b,m} = \begin{bmatrix} 1 & 0 \\ b & M \end{bmatrix}. \tag{2.40}$$

As it turns out, the matrices $\bar{g}_{b,M}$, for $b \in \mathbb{Q} \cap [0,1)$ and $M \in \mathbb{Q}^+$, also furnish a complete and irredundant list of coset representatives for $\mathrm{PSL}_2(\mathbb{Z})$ in $\mathrm{PSL}_2(\mathbb{Q})$. We will write $\bar{L}_{b,M}$ for the projective lattice corresponding to $\mathrm{PSL}_2(\mathbb{Z})\bar{g}_{b,M}$. Of course, every projective lattice $L_{M,b}$ can be written as $\bar{L}_{b',M'}$ for some b', M'. The correspondence is such that

$$L_{M,0} = \bar{L}_{0,1/M}, \quad L_{M,f/g} = \bar{L}_{f'/g,1/g^2M}, \tag{2.41}$$

where, for $0 < f < g$ and $\gcd\{f,g\} = 1$, the integer f' is uniquely determined by the conditions that $0 < f' < g$ and $ff' \equiv 1 \pmod g$.

When $L_{M,b} = \bar{L}_{b',M'}$ we (following [3]) call $\bar{L}_{b',M'}$ the *reverse name* for $L_{M,b}$. The reverse names for projective lattices will be useful in Section 2.11.

2.7. Stabilizers. Let us continue with the notation and conventions of Section 2.6. In particular, we consider the $\mathbb{Q}^\times$-orbits of lattices in V that are commensurable with our distinguished lattice L_1 — the set $\mathrm{P}\mathcal{L}_1$, of projective lattices in V_1 — and these orbits are in natural correspondence with the $\mathrm{PSL}_2(\mathbb{Z})\backslash \mathrm{PGL}_2^+(\mathbb{Q})$. Each projective lattice $L \in \mathrm{P}\mathcal{L}_1$ arises as $L = L_{M,b}$ for some $M \in \mathbb{Q}^+$ and $b \in \mathbb{Q} \cap [0,1)$.

From now on, let us write G for the group $\mathrm{PGL}_2^+(\mathbb{Q})$ regarded as a group with right-action on the space $\mathrm{P}\mathcal{L}_1$ of projective lattices in V_1.

$$\begin{gathered} \mathrm{P}\mathcal{L}_1 \leftrightarrow \mathrm{PSL}_2(\mathbb{Z})\backslash \mathrm{PGL}_2^+(\mathbb{Q}) \circlearrowleft \mathrm{PGL}_2^+(\mathbb{Q}) =: G \\ \mathrm{PSL}_2(\mathbb{Z}) \cdot (g \cdot [\mathbf{v}_1]) \leftrightarrow \mathrm{PSL}_2(\mathbb{Z})g \end{gathered} \tag{2.42}$$

For any $L \in \mathrm{P}\mathcal{L}_1$, we may ask for the subgroup of G that fixes L; i.e., the group $\mathrm{Fix}_G(L)$. We will write G_L for $\mathrm{Fix}_G(L)$, and $G_{M,b}$ for G_L if $L = L_{M,b}$. We may also consider the group that fixes several lattices,

$$G_{(L^1,\dots,L^k)} = \bigcap_i G_{L^i} = \{g \in G \mid L^i \cdot g = L^i, 1 \le i \le k\}, \tag{2.43}$$

or the group that stabilizes a set of lattices,

$$G_{\{L^1,\dots,L^k\}} = \left\{g \in G \;\middle|\; \{L^1 \cdot g, \dots, L^k \cdot g\} = \{L^1, \dots, L^k\}\right\}. \tag{2.44}$$

Let us compute some examples of the groups $G_{(L,\dots)}$, $G_{\{L,\dots\}}$. Note that, by definition,

$$(\mathrm{PSL}_2(\mathbb{Z}) \cdot [\mathbf{v}_1]) \cdot g = \mathrm{PSL}_2(\mathbb{Z})g \cdot [\mathbf{v}_1] \tag{2.45}$$

for $g \in G$. The identity (2.45) makes it clear that the group $G_1 = \mathrm{Fix}_G(L_1)$ is none other than the familiar *modular group*, $\mathrm{PSL}_2(\mathbb{Z})$. Evidently, G acts transitively on $\mathrm{P}\mathcal{L}_1$, so the lattice L_M, for $M \in \mathbb{Q}^+$, is fixed by a conjugate of $G_1 \cong \mathrm{PSL}_2(\mathbb{Z})$; viz.

$$G_M = g_M^{-1} G_1 g_M = \left\{ \begin{bmatrix} a & b/M \\ cM & d \end{bmatrix} \;\middle|\; a,b,c,d \in \mathbb{Z}, ad - bc = 1 \right\}. \tag{2.46}$$

Thus we see that the intersection $G_{(1,N)} = G_1 \cap G_N$, for N a positive integer, is the *Hecke group of level* N, usually denoted $\Gamma_0(N)$.

For N a prime power, $N = p^a$ say, the group $G_{\{1,N\}}$ that preserves the set $\{L_1, L_N\}$ is the group obtained from $G_{(1,N)}$ by adjoining a *Fricke involution*:

$$G_{\{1,N\}} = \left\langle G_{(1,N)}, \begin{bmatrix} 0 & -1 \\ N & 0 \end{bmatrix} \right\rangle. \tag{2.47}$$

This is the group denoted $\Gamma_0(N) + N$ (or just $\Gamma_0(N)+$, since N is a prime power) in [5]. Notice that the subgroup of G stabilizing the set $\{L_1, L_{p^a}\}$ is the same as the subgroup of G stabilizing the set $\{L_1, L_p, \ldots, L_{p^a}\}$, containing the lattices associated to all powers of p that divide p^a, and this latter set is precisely the set of lattices L for which the product of hyperdistances $\delta(L_1, L)\delta(L, L_{p^a})$ coincides with the hyperdistance between L_1 and L_{p^a}. Given L' and L'' in $P\mathcal{L}_1$, let us write $(L', L'')+$ for the set of lattices L satisfying $\delta(L', L)\delta(L, L'') = \delta(L', L'')$.

$$(L', L'')+ = \{L \in P\mathcal{L}_1 \mid \delta(L', L)\delta(L, L'') = \delta(L', L'')\} \tag{2.48}$$

Following [3], we call the set $(L', L'')+$ the (L', L'')-*thread*. The sets $(1,6)+$ and $(1,4)+$ appear in (2.49).

(2.49) 

In terms of threads, we have $G_{\{1,N\}} = G_{(1,N)+}$ when N is a prime power.

More generally, for arbitrary $N \in \mathbb{Z}_{>0}$, the group $G_{(1,N)+}$ is the group obtained by adjoining to $G_{(1,N)}$, the sets $W_e = W_e(N)$, where $e \in \mathbb{Z}_{>0}$ is a divisor of N, and

$$W_e = \left\{ \begin{bmatrix} ae & b \\ cN & de \end{bmatrix} \;\middle|\; a, b, c, d \in \mathbb{Z},\ ade^2 - bcN = e \right\}. \tag{2.50}$$

Note that $ade^2 - bcN = e$ implies $\gcd\{e, N/e\} = 1$, so the set W_e is non-empty only when e is an *exact divisor* of N. The sets W_e are exactly the cosets of $G_{(1,N)} = W_1$ in $G_{(1,N)+}$. (We will sometimes write W_e for a certain element in $W_e(N)$ — the validity of the statement in which this W_e appears should be independent of the choice that is made.)

Suppose given $h, n \in \mathbb{Z}_{>0}$ such that $h \mid n$. Then

$$G_{(h,n)} = \left\{ \begin{bmatrix} a & b/h \\ cn & d \end{bmatrix} \;\middle|\; a, b, c, d \in \mathbb{Z}, ad - bcn/h = 1 \right\}. \tag{2.51}$$

This group is conjugate to $G_{(1,n/h)}$ since $G_{(h,n)} = g_h^{-1} G_{(1,n/h)} g_h$ (cf. (2.36)). The group $G_{(h,n)+}$ is evidently obtained from $G_{(1,n/h)+}$ via conjugation by g_h, and so $G_{(h,n)+}$ consists of $G_{(h,n)}$ together with the (other co)sets $g_h^{-1} W_e(n/h) g_h$ for e an exact divisor of n/h. The significance of the groups $G_{(h,n)+}$ is demonstrated by the following result.

Theorem 2.7 (Atkin – Lehner). *For N a positive integer, the normalizer of $G_{(1,N)}$ in $\mathrm{PSL}_2(\mathbb{R})$ is the group $G_{(h,n)+}$, where $n = N/h$, and h is the largest divisor of 24 such that $h^2 | N$.*

Remark. A beautiful proof of the Atkin – Lehner theorem appears in [3].

Recall that the action of G on $\mathrm{P}\mathcal{L}_1$ preserves hyperdistance (cf. Section 2.5). It follows that for any $L \in \mathrm{P}\mathcal{L}_1$, the group G_L acts by permutations on the set $\mathrm{HC}_N(L)$, of lattices at hyperdistance N from L. We call $\mathrm{HC}_N(L)$ the *hypercircle of hyperradius N about L*

$$\mathrm{HC}_N(L) = \{L' \in \mathrm{P}\mathcal{L}_1 \mid \delta(L, L') = N\}. \tag{2.52}$$

In fact, the action of G_L on $\mathrm{HC}_N(L)$ is transitive.

Proposition 2.8. *For any positive integer N, the group $G_L = \mathrm{Fix}_G(L)$ acts transitively on the set of lattices that are hyperdistant N from L.*

We will furnish a proof of Proposition 2.8 in Section 2.11.

We may consider the subgroup of G that fixes all the lattices in $\mathrm{HC}_N(L)$ for given $N \in \mathbb{Z}_{>0}$ and $L \in \mathrm{P}\mathcal{L}_1$. When L is the distinguished lattice $L = L_1$, this is just the group $\Gamma(N)$—the *principal congruence group of level* N.

$$1 \to \Gamma(N) \to \mathrm{PSL}_2(\mathbb{Z}) \to \mathrm{PSL}_2(\mathbb{Z}/N) \to 1 \tag{2.53}$$

Recall that a group $H < G_1 \cong \mathrm{PSL}_2(\mathbb{Z})$ is called a *congruence group* if H contains $\Gamma(N)$ for some N. A congruence group H is said to have *level* N if N is the smallest positive integer such that $\Gamma(N) < H$. We will apply these definitions also to arithmetic groups: we will say that a subgroup $\Gamma < \mathrm{PSL}_2(\mathbb{R})$ that is commensurable with G_1 is a *congruence group* if Γ contains $\Gamma(N)$ for some positive integer N, and we will say that such a group Γ has *level* N, if N is the minimal positive integer for which $\Gamma(N) < \Gamma$.

Remark. See Lemma 2.19 for an explicit description of $\mathrm{HC}_N(1)$ in the case that N is a prime power.

2.8. Trees. Let p be a prime, and consider the set $(\mathrm{P}\mathcal{L}_1)_p \subset \mathrm{P}\mathcal{L}_1$ consisting of (projective) lattices that are hyperdistant a power of p from L_1

$$(\mathrm{P}\mathcal{L}_1)_p = \{L \in \mathrm{P}\mathcal{L}_1 \mid \delta(L_1, L) \in p^{\mathbb{Z}}\}. \tag{2.54}$$

Following Conway [3], we call $(\mathrm{P}\mathcal{L}_1)_p$ the *p-adic tree* in $\mathrm{P}\mathcal{L}_1$.

For $L, L' \in \mathrm{P}\mathcal{L}_1$ we say L and L' are *p-adically equivalent*, and write $L \sim_p L'$, if $p \nmid \delta(L, L')$. It is easy to check that the p-adic equivalence class of any projective lattice $L \in \mathrm{P}\mathcal{L}_1$ has a unique representative in $(\mathrm{P}\mathcal{L}_1)_p$.

Proposition 2.9. *For any $L \in \mathrm{P}\mathcal{L}_1$, and any prime p, there is a unique $L' \in \mathrm{P}\mathcal{L}_1$ such that $L \sim_p L'$ and $L' \in (\mathrm{P}\mathcal{L}_1)_p$.*

Consequently, we obtain a map $\pi_p \colon \mathrm{P}\mathcal{L}_1 \to (\mathrm{P}\mathcal{L}_1)_p$, called the *$p$-adic projection*, by setting $\pi_p(L) = L'$ when L and L' are as in Proposition 2.9.

As explained in [3], the set $(\mathrm{P}\mathcal{L}_1)_p$ has a natural tree structure; viz. if we regard the elements of $(\mathrm{P}\mathcal{L}_1)_p$ as vertices of a graph, with edges joining just those lattices that are hyperdistant p from each other, we obtain a graph with the property that there is a unique shortest path between any two vertices. That is, we obtain a tree. The p-adic tree $(\mathrm{P}\mathcal{L}_1)_p$ has infinitely many nodes, and each node is $p+1$ valent.

A finite subset $S \subset \mathrm{P}\mathcal{L}_1$ is called a *cell* if, for each prime p, the subtree of $(\mathrm{P}\mathcal{L}_1)_p$ generated by the set $\pi_p(S) \subset (\mathrm{P}\mathcal{L}_1)_p$ is either a point, or two points joined by an edge.

The methods of [3] illustrate the utility of the trees $(\mathrm{P}\mathcal{L}_1)_p$ and the projections π_p. The following result is a prime example of this.

Proposition 2.10 ([3]). *If Γ is a subgroup of $\mathrm{PGL}_2^+(\mathbb{Q})$ that is commensurable with $G_1 \cong \mathrm{PSL}_2(\mathbb{Z})$, then Γ stabilizes a cell in $\mathrm{P}\mathcal{L}_1$.*

It is well known that a subgroup of $\mathrm{PGL}_2^+(\mathbb{R})$ that is commensurable with $\mathrm{PSL}_2(\mathbb{Z})$ is contained in $\mathrm{PGL}_2^+(\mathbb{Q})$ (cf. Proposition 2.14). For any cell $S \subset \mathrm{P}\mathcal{L}_1$ we can find lattices L and L' in S that maximize $\delta(L, L')$. After conjugation by an element of $\mathrm{PGL}_2^+(\mathbb{Q})$ we may assume that $L = L_1$. Since $G_1 \cong \mathrm{PSL}_2(\mathbb{Z})$ acts transitively on the lattices of any given hyperdistance from L_1 (cf. Proposition 2.8), we may conjugate the pair (L, L') to (L_1, L_N) for some N. These observations, together with Proposition 2.10, quickly imply the following result, known as Helling's theorem.

Theorem 2.11 ([7]). *The maximal arithmetic subgroups of* $\mathrm{PSL}_2(\mathbb{R})$ *are the conjugates of* $G_{(1,N)+}$ *for square-free* N.

2.9. Characters. Let N be a positive integer, and let h be the largest divisor of 24 such that $h^2|N$. Observe that $G_{(h,N/h)}$ contains $G_{(1,N)}$. According to [5] (cf. [4]), there is a canonically defined subgroup of index h in $G_{(h,N/h)}$ that contains $G_{(1,N)}$, and it may be realized as the kernel of a (non-canonically defined) homomorphism

$$\lambda\colon G_{(h,N/h)} \to \mathbb{Z}/h. \tag{2.55}$$

We will write $G^{(h)}_{(h,N/h)}$ for this group that arises as $\ker(\lambda)$. This group is denoted $\Gamma_0(n|h)$ in [5].

The cases that $N = 9$ and $N = 8$ will be of particular relevance in Section 3.

2.9.1. *Example*: $N = 9$. If $N = 9$ then $h = 3$ and $N/h = 3$. Evidently, $G_{(3,3)} = G_3$. Define generators x and y for $G_3/G_{(1,9)}$ (recall from Theorem 2.7 that $G_{(1,9)}$ is normal in G_3) by setting

$$x = G_{(1,9)}T^{1/3} = G_{(1,9)}\begin{bmatrix} 1 & \frac{1}{3} \\ 0 & 1 \end{bmatrix}, \tag{2.56}$$

$$y = G_{(1,9)}(T^3)^t = G_{(1,9)}\begin{bmatrix} 1 & 0 \\ 3 & 1 \end{bmatrix}. \tag{2.57}$$

A suitable map $\lambda\colon G_3/G_{(1,9)} \to \mathbb{Z}/3$ may be defined by assigning $\lambda(y) = \sigma$ and $\lambda(x) = \sigma^{-1}$, where σ is a generator for $\mathbb{Z}/3$ (cf. [4]).

Let us also analyze the situation in terms of the 3-adic tree (cf. Section 2.8). The group G_3 preserves the set $\mathrm{HC}_3(3)$ — the hypercircle of hyperradius 3 about L_3. There are exactly 4 lattices in $\mathrm{HC}_3(3)$; they appear in the diagram (2.58), which displays the smallest subtree of the 3-adic tree $(\mathrm{P}\mathcal{L}_1)_3$ that contains $\mathrm{HC}_3(3)$.

(2.58)

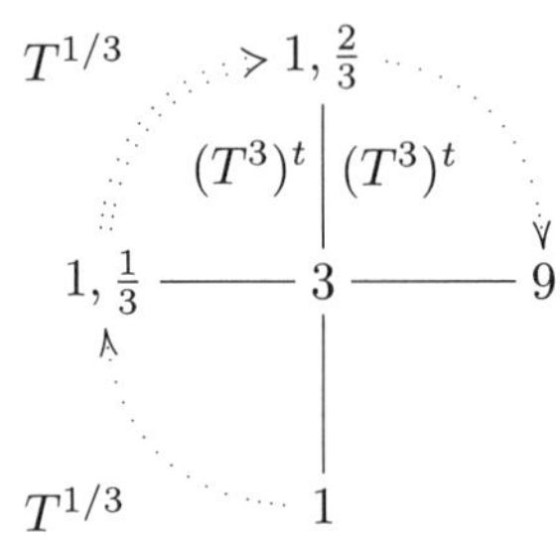

Observe that $G_{1,9}$ fixes every element of $\mathrm{HC}_3(3)$. Thus, there is a natural map $G_3/G_{1,9} \to \mathrm{Sym}(\mathrm{HC}_3(3))$. Diagram (2.58) also displays the 3-cycles in $\mathrm{Sym}(\mathrm{HC}_3(3))$ generated by (the images in $\mathrm{Sym}(\mathrm{HC}_3(3))$ of) x and y. Evidently, the image of $G_3/G_{1,9}$ in $\mathrm{Sym}(\mathrm{HC}_3(3))$ is just $\mathrm{Alt}(\mathrm{HC}_3(3))$ — a copy of the alternating group on 4 letters. This group has order 12. It contains 8 elements of order 3, and 3 elements of order 2. Any non-trivial homomorphism $\mathrm{Alt}(\mathrm{HC}_3(3)) \to \mathbb{Z}/3$ must be trivial on elements of order 2, and non-trivial on the elements of order 3.

We conclude that $G_3^{(3)}$ consists of all the elements of G_3 that induce permutations of order 2 (or 1) on the set $\mathrm{HC}_3(3)$, and the permutations of order 2 occurring are just those that arise as a product of 2 disjoint transpositions

$$G_3^{(3)} = \langle G_{(1,9)}, (T^3)^t T^{1/3}, T^{-1/3}(T^3)^t T^{-1/3} \rangle. \tag{2.59}$$

The group $G_3^{(3)}$ is denoted $\Gamma_0(3|3)$ in [5].

2.9.2. *Example:* $N = 8$. If $N = 8$ then $h = 2$ and $N/h = 4$. The group $G_{(2,4)}$ contains $G_{(1,8)}$ normally by Theorem 2.7. Consider the set $S = \mathrm{HC}_2(2) \cup \mathrm{HC}_2(4)$.

(2.60)

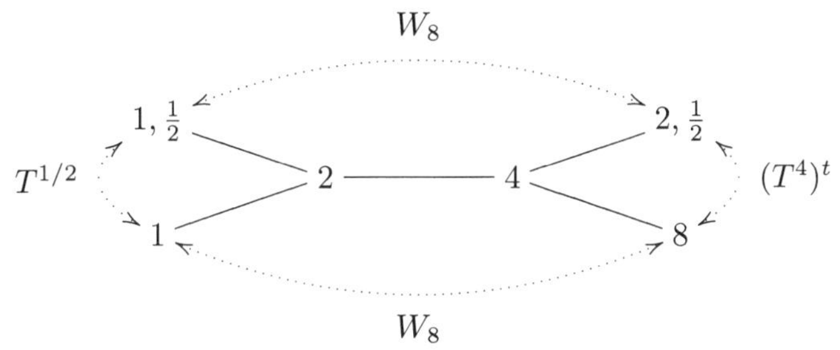

Diagram (2.60) displays the smallest subtree of the 2-adic tree (cf. Section 2.8) containing S, together with the actions on this tree induced by some elements of $G \cong \mathrm{PGL}_2^+(\mathbb{Q})$ (cf. Section 2.7). The symbol W_8 in (2.60) denotes an (arbitrary) element of $g_2^{-1}W_2(2)g_2$ (cf. Section 2.7).

The group $G_{(2,4)}$ acts by permutations on S; the subgroup $G_{(1,8)}$ is exactly the subgroup that fixes every element of S. We see from (2.60) that $G_{(2,4)}$ is generated by $G_{(1,8)}$ together with $T^{1/2}$ and $(T^4)^t$. The group $G_{\{2,4\}}$ — the full normalizer of $G_{(1,8)}$ — is obtained by adjoining W_8 to $G_{(2,4)}$.

We thus obtain generators x, y for $G_{(2,4)}/G_{(1,8)}$ by setting $x = G_{(1,8)}T^{1/2}$ and $y = G_{(1,8)}(T^4)^t$, say. We obtain a suitable map $\lambda\colon G_{(2,4)}/G_{(1,8)} \to \mathbb{Z}/2$ by setting $\lambda(x) = \lambda(y) = \sigma$, where σ is the non-trivial element of $\mathbb{Z}/2$

$$G_{(2,4)}^{(2)} = \langle G_{(1,8)}, (T^4)^t T^{1/2} \rangle. \tag{2.61}$$

We can also carry out this story with $G_{\{2,4\}}$ in place of $G_{(2,4)}$; that is, we can adjoin the set $g_2^{-1}W_2(2)g_2$ to (2.61), just as we adjoin this same set to $G_{(2,4)}$ in order to recover $G_{(2,4)}$ (cf. (2.50)). We set

$$G_{\{2,4\}}^{(2)} = \langle G_{(1,8)}, (T^4)^t T^{1/2}, W_8 \rangle. \tag{2.62}$$

Evidently, $G_{\{2,4\}}^{(2)}$ may be characterized as the kernel of the following composition of natural mappings: $G_{\{2,4\}} \to \mathrm{Sym}(S) \to \mathbb{Z}/2$. This group is denoted $\Gamma_0(4|2)+$ in [5].

2.10. Cusps. From now on we will restrict to the case that $\mathbf{k} = \mathbb{R}$.

There is an obvious embedding $\mathrm{PSL}_2(\mathbb{R}) \hookrightarrow \mathrm{PGL}_2^+(\mathbb{R})$ arising from the embedding of groups $\mathrm{SL}_2(\mathbb{R}) \hookrightarrow \mathrm{GL}_2^+(\mathbb{R})$. Observe that this map

$$\mathrm{PSL}_2(\mathbb{R}) \hookrightarrow \mathrm{PGL}_2^+(\mathbb{R}) \tag{2.63}$$

is in fact surjective, so that $\mathrm{PGL}_2^+(\mathbb{R}) = \mathrm{PSL}_2(\mathbb{R})$. There is also an embedding $\mathrm{GL}_2^+(\mathbb{Q}) \hookrightarrow \mathrm{GL}_2^+(\mathbb{R})$ coming the the embedding of fields $\mathbb{Q} \hookrightarrow \mathbb{R}$. The map $\mathrm{PGL}_2^+(\mathbb{Q}) \to \mathrm{PGL}_2^+(\mathbb{R})$ sending a $\mathbb{Q}^\times$-orbit in $\mathrm{GL}_2^+(\mathbb{Q})$ to the $\mathbb{R}^\times$-orbit of its image in $\mathrm{GL}_2^+(\mathbb{R})$ is readily checked to be an embedding; we conclude that $\mathrm{PGL}_2^+(\mathbb{Q})$ embeds naturally in $\mathrm{PSL}_2(\mathbb{R})$.

Our choice $\mathbf{v}_1 \in \mathcal{B}_1^+$ allows us to identify $PV^\times = \mathbb{R}^\times \backslash V^\times$ with the real projective line $\mathbb{P}^1(\mathbb{R}) = \mathbb{R} \cup \{\infty\}$; viz.

$$(2.64) \qquad [\alpha u_1 + \beta v_1] \leftrightarrow \begin{cases} \alpha/\beta, & \text{if } \beta \neq 0; \\ \infty, & \text{if } \beta = 0; \end{cases}$$

where $\mathbf{v}_1 = (u_1, v_1)$. The resulting left action of $\mathrm{PGL}_2(\mathbb{R})$ on $PV^\times$ is given by

$$(2.65) \qquad \begin{bmatrix} a & b \\ c & d \end{bmatrix} \cdot [\alpha u_1 + \beta v_1] = [(a\alpha + b\beta)u_1 + (c\alpha + d\beta)v_1].$$

In terms of the identification $PV^\times \leftrightarrow \mathbb{R} \cup \{\infty\}$ this translates to the familiar prescription

$$(2.66) \qquad \begin{bmatrix} a & b \\ c & d \end{bmatrix} \cdot \alpha = \frac{a\alpha + b}{c\alpha + d}.$$

Evidently, it is natural to regard the group $\mathrm{PGL}_2^+(\mathbb{R})$ as acting both from the right on projective lattices $\mathrm{P}\mathcal{L}$, and from the left on the projective line $\mathbb{P}^1(\mathbb{R})$. Similarly, we may regard our group $G \cong \mathrm{PGL}_2^+(\mathbb{Q})$ (see Section 2.7) as acting both from the right on the projective lattices $\mathrm{P}\mathcal{L}_1$ in V_1, and from the left on the rational projective line $\mathbb{P}^1(\mathbb{Q})$. Also, there is a natural embedding $\mathbb{P}^1(\mathbb{Q}) \hookrightarrow \mathbb{P}^1(\mathbb{R})$.

Recall that T typically denotes the element of $\mathrm{PSL}_2(\mathbb{R}) = \mathrm{PGL}_2^+(\mathbb{R})$ represented by the upper-triangular unipotent matrix with 1 in the top right-hand corner. We convene to write T^A, for $A \in \mathbb{R}$, for the element of $\mathrm{PGL}_2^+(\mathbb{R})$ represented by the upper-triangular unipotent matrix with A in the top right-hand corner.

$$(2.67) \qquad T^A := \begin{bmatrix} 1 & A \\ 0 & 1 \end{bmatrix} \in \mathrm{PGL}_2^+(\mathbb{R})$$

Evidently, $T^A T^B = T^{A+B}$ for $A, B \in \mathbb{R}$, so that the assignment $A \mapsto T^A$ defines an embedding (of groups) $\mathbb{R} \hookrightarrow \mathrm{PGL}_2^+(\mathbb{R})$. Observe that the stabilizer of $\infty \in \mathbb{P}^1(\mathbb{R})$ in $\mathrm{PGL}_2^+(\mathbb{R})$ (let us write $\mathrm{Fix}(\infty)$ for this group) contains the image of $\mathbb{R}$ under this embedding.

$$(2.68) \qquad \mathrm{Fix}(\infty) = \left\{ \begin{bmatrix} a & b \\ 0 & 1 \end{bmatrix} \,\middle|\, a, b \in \mathbb{R}, a > 0 \right\}$$

In fact, the image of $\mathbb{R}$ here is a normal subgroup of $\mathrm{Fix}(\infty)$, and $\mathrm{Fix}(\infty)$ may be described as the split extension $\mathrm{Fix}(\infty) \cong \mathbb{R} \rtimes \mathbb{R}^+$, corresponding to the natural action of (the multiplicative group) $\mathbb{R}^+$ on (the additive group) $\mathbb{R}$. Given a subgroup $H < \mathrm{PGL}_2^+(\mathbb{R})$ let us write $\mathrm{Fix}_H(\infty)$ for the intersection $\mathrm{Fix}(\infty) \cap H$. We have

$$(2.69) \qquad \mathrm{Fix}_G(\infty) \cong \mathbb{Q} \rtimes \mathbb{Q}^+.$$

Observe that $\mathrm{Fix}_G(\infty)$ acts transitively on the subset $\mathbb{P}^1(\mathbb{Q}) \subset \mathbb{P}^1(\mathbb{R})$, and no element of $\mathrm{P}\mathcal{L}_1$ is fixed by any non-trivial element of the subgroup $\mathbb{Q}^+ < \mathrm{Fix}_G(\infty)$. Observe also that the intersection $\mathrm{Fix}_{G_1}(\infty) = \mathrm{Fix}_G(\infty) \cap G_1$ is contained in the image (under $A \mapsto T^A$) of $\mathbb{Q}$; indeed, $\mathrm{Fix}_{G_1}(\infty) = \langle T \rangle$. This property is shared by the subgroups of $\mathrm{PSL}_2(\mathbb{R})$ that are commensurable with $G_1 \cong \mathrm{PSL}_2(\mathbb{Z})$ (cf. Section 2.7).

Proposition 2.12. *Suppose $H < \mathrm{PSL}_2(\mathbb{R})$ is commensurable with G_1. Then $\mathrm{Fix}_H(\infty)$ is contained in the image of $\mathbb{Q}$ under $A \mapsto T^A$.*

PROOF. Let $h \in \mathrm{Fix}_H(\infty)$. Choose $a, b \in \mathbb{R}$ such that

$$h = \begin{bmatrix} a & b \\ 0 & 1 \end{bmatrix}. \tag{2.70}$$

If H is commensurable with G_1 then $h^n \in G_1$ for some $n \in \mathbb{Z}$. This implies that $a^n = 1$, and hence $a = 1$, since a must be positive. Given that $a = 1$, we have $nb \in \mathbb{Z}$, so that $b \in \mathbb{Q}$, and $h = T^b$ lies in the image of $\mathbb{Q}$ in $\mathrm{Fix}_G(\infty)$, as required. □

Recall that an element $\gamma \in \mathrm{PSL}_2(\mathbb{R})$ is called *parabolic* if there is a unique fixed point for its action on $\mathbb{P}^1(\mathbb{R})$. Equivalently, $\gamma \in \mathrm{PSL}_2(\mathbb{R})$ is parabolic if it is not the identity, and if $\gamma = [A]$ for some $A \in \mathrm{SL}_2(\mathbb{R})$ with $\mathrm{tr}(A) = 2$. For any subgroup $H < \mathrm{PGL}_2^+(\mathbb{R})$ we may ask for the orbits of H on $\mathbb{P}^1(\mathbb{R})$ that are comprised of points that are fixed by some parabolic element of H. (If $h \cdot s = s$ for some parabolic $h \in H$ and some $s \in \mathbb{P}^1(\mathbb{R})$ then every $s' \in H \cdot s$ is fixed by some parabolic element of H.) Such an orbit for the action of H on $\mathbb{P}^1(\mathbb{R})$ is called a *cusp* of H. We write $\mathcal{C}(H)$ for the set of cusps of H, and we define

$$C(H) := \bigcup_{H \cdot s \in \mathcal{C}(H)} H \cdot s, \tag{2.71}$$

so that $C(H)$ consists of all the points in $\mathbb{P}^1(\mathbb{R})$ which are fixed by some parabolic element of H.

Proposition 2.13. *The assignment $H \mapsto C(H)$ is constant on commensurability classes.*

PROOF. It suffices to show that $C(H) = C(H')$ whenever H' is a subgroup of finite index in H. Certainly, $C(H') \subset C(H)$ in this case, so let $s \in C(H)$. Then there is some parabolic $h \in H$ such that $h \cdot s = s$. Since H' has finite index in H, there is some $n > 0$ such that $h^n \in H'$, but h^n is parabolic whenever h is (since $A^2 = \mathrm{tr}(A)A - 1$ for $A \in \mathrm{SL}_2(\mathbb{R})$), so there is a parabolic element of H' that fixes s, and $C(H) \subset C(H')$, as required. □

For $H < \mathrm{PGL}_2^+(\mathbb{R})$ set $c(H)$ to be the number of cusps of H.

$$c(H) := |\mathcal{C}(H)|. \tag{2.72}$$

It is easy to check that $C(G) = \mathbb{P}^1(\mathbb{Q})$, and it is clear that G acts transitively on $\mathbb{P}^1(\mathbb{Q})$, so we have $c(G) = 1$. In fact, these statements remain valid with $G_1 \cong \mathrm{PSL}_2(\mathbb{Z})$ in place of G.

$$c(G_1) = 1, \quad C(G_1) = \mathbb{P}^1(\mathbb{Q}), \quad \mathcal{C}(G_1) = \{G_1 \cdot \infty\}. \tag{2.73}$$

By Proposition 2.13, we have $C(H) = \mathbb{P}^1(\mathbb{Q})$ for any subgroup of $\mathrm{PGL}_2^+(\mathbb{R})$ that is commensurable with G_1. This has the following useful consequence.

Proposition 2.14. *If $H < \mathrm{PSL}_2(\mathbb{R})$ is commensurable with $G_1 \cong \mathrm{PSL}_2(\mathbb{Z})$ then H is contained in $G \cong \mathrm{PGL}_2^+(\mathbb{Q})$.*

PROOF. We see from Proposition 2.13 that the action of such a group H on $\mathbb{P}^1(\mathbb{R})$ stabilizes $\mathbb{P}^1(\mathbb{Q})$. Any element of H can be written as a Möbius transformation, and in particular, is determined by its action on 0, 1 and ∞, for example. Since these points are mapped to points of $\mathbb{P}^1(\mathbb{Q})$, any element of H can be represented by a rational matrix. □

Let $H < G$ such that $C(H) = \mathbb{P}^1(\mathbb{Q})$. Then $c(H^g) = c(H)$ for any $g \in G$, since G acts transitively on $\mathbb{P}^1(\mathbb{Q})$. This shows, for example, that $c(G_L) = 1$ for any $L \in \mathrm{P}\mathcal{L}_1$.

We may ask what the action of a group on $P\mathcal{L}_1$ says about its cusps. The following lemma is useful in this regard.

Lemma 2.15. *Let H be a group equipped with a transitive left-action on a set L, and a transitive right-action on a set R. Let $l \in L$ and $r \in R$, and write H_x for the stabilizer of x in H, for x in L or R. Then the H_r-orbits on L are in bijective correspondence with the H_l-orbits on R.*

$$L/H_r \leftrightarrow H_l\backslash R \tag{2.74}$$

PROOF. Since the H-actions are transitive, we have bijections $L \leftrightarrow H_l\backslash H$ and $H/H_r \leftrightarrow R$, so both sides of (2.74) are in bijection with $H_l\backslash H/H_r$. □

Note that the correspondence in (2.74) is given explicitly by

$$l \cdot hH_r \leftrightarrow H_l h \cdot r. \tag{2.75}$$

Lemma 2.15 has the following immediate consequence.

Proposition 2.16. *Let H be a subgroup of G with a single cusp, let $\mathcal{O}$ be an orbit for the action of H on $P\mathcal{L}_1$, and let H_0 be the subgroup of H fixing some $L_0 \in \mathcal{O}$. Then the cusps of H_0 are in natural bijective correspondence with the orbits of $\mathrm{Fix}_H(\infty)$ on $\mathcal{O}$.*

Suppose now that H is a subgroup of $\mathrm{PSL}_2(\mathbb{R})$ commensurable with G_1, so that H is contained in $G \cong \mathrm{PGL}_2^+(\mathbb{Q})$, by Proposition 2.14, and suppose that H has a single cusp (i.e., acts transitively on $\mathbb{P}^1(\mathbb{Q})$). For a subgroup H_0 of H, we have $\mathcal{C}(H_0) = H_0\backslash\mathbb{P}^1(\mathbb{Q})$ for the cusps of H_0. If H_0 has finite index in H then we may define a function

$$w_H\colon \mathcal{C}(H_0) \to \mathbb{Q}^+ \tag{2.76}$$

in the following way. For $H_0 \cdot x \in H_0\backslash\mathbb{P}^1(\mathbb{Q})$ let $g \in H$ such that $x = g \cdot \infty$, and consider the intersection $H_0^g \cap \mathrm{Fix}_H(\infty)$. Since H is commensurable with G_1, the group $\mathrm{Fix}_H(\infty)$ is an infinite cyclic group, by Proposition 2.12. Since H_0 has finite index in H, the intersection $H_0^g \cap \mathrm{Fix}_H(\infty)$ is also an infinite cyclic group. Define $w_H(H_0 \cdot x) = A \in \mathbb{Q}^+$ just when T^A generates $H_0^g \cap \mathrm{Fix}_H(\infty)$.

Proposition 2.17. *The function w_H, of* (2.76)*, is well-defined.*

PROOF. If $g, g' \in H$ satisfy $g \cdot \infty = g' \cdot \infty = x$ then $g^{-1}g' = T^B$ for some $B \in \mathbb{Q}$, since $\mathrm{Fix}_H(\infty)$ is contained in the image of $\mathbb{Q}$ in $\mathrm{Fix}_G(\infty)$, by Proposition 2.12. We have $H_0^{g'} = (H_0^g)^{T^B}$, so that $H_0^{g'} \cap \mathrm{Fix}_H(\infty)$ and $H_0^g \cap \mathrm{Fix}_H(\infty)$ coincide, since T^B centralizes $\mathrm{Fix}_H(\infty)$. This shows that the value of $w_H(H_0 \cdot x)$ doesn't depend upon the choice of $g \in H$ mapping ∞ to x. Suppose that $H_0 \cdot x = H_0 \cdot x'$. Then $x' = h \cdot x$ for some $h \in H_0$, and $g \cdot \infty = x$ implies $hg \cdot \infty = x'$. We have $H_0^{hg} = H_0^g$, so the value of w_H at the cusp $H_0 \cdot x$ is independent of the choice of representative point $x \in \mathbb{P}^1(\mathbb{Q})$. □

For H and H_0 as above, and $C \in \mathcal{C}(H_0)$, we call $w_H(C)$ the *width of C relative to H*. If $H = G_1$ then w_H recovers the usual notion of width for cusps of finite index subgroups of the modular group. Observe that if $H' < \mathrm{PSL}_2(\mathbb{R})$ is commensurable with G_1 and contains H, then H_0 is a subgroup of finite index in H', and the functions $w_{H'}$ and w_H coincide on $\mathcal{C}(H_0)$. Consequently, for a group $H_0 < \mathrm{PSL}_2(\mathbb{R})$ commensurable with G_1, to know all the widths of a cusp $H_0 \cdot x \in H_0 \backslash \mathbb{P}^1(\mathbb{Q})$, it suffices to know the values $w_H(H_0 \cdot x)$ for each supergroup $H > H'$ that is maximal subject to being commensurable with G_1. The subgroups of $\mathrm{PSL}_2(\mathbb{R})$ that are maximal subject to being commensurable with G_1 are determined by Proposition 2.10.

Even though the functions w_H and $w_{H'}$ in general don't coincide for $H_0 < H \cap H'$, the particular value $w_H(H_0 \cdot \infty)$ is easily checked to be independent of the choice of supergroup H. Thus we can speak unambiguously of the *width of* H_0 *at* ∞, to be denoted $w(H_0 \cdot \infty)$, whenever H_0 is commensurable with G_1.

Proposition 2.18. *Suppose H and H_0 are as in Proposition* 2.16, *and suppose that H is commensurable with G_1. Then*

$$\frac{w_H(H_0 h \cdot \infty)}{A} = \#(L_0 \cdot h \operatorname{Fix}_H(\infty)). \tag{2.77}$$

for all $h \in H$, where $A \in \mathbb{Q}^+$ is the width of H at ∞.

PROOF. Observe that the size of the orbit $L_0 \cdot h \operatorname{Fix}_H(\infty)$ is just the smallest $n \in \mathbb{Z}_{>0}$ such that $T^{nA} \in \operatorname{Fix}_H(L_0 \cdot h)$. Observe also that T^B, for $b \in \mathbb{Q}$, belongs to $\operatorname{Fix}_H(L_0 \cdot h)$ if and only if $T^B \in H_0^h$. For $T^B \in \operatorname{Fix}_H(L_0 \cdot h)$ if and only if $hT^B = h_0 h$ for some $h_0 \in H_0$, and this occurs if and only if $T^B \in h^{-1} H_0 h$. Thus if $B = w_H(H_0 h \cdot \infty)$, so that B is the smallest positive rational such that $T^B \in H_0^h$, then $B = nA$ where n is $\#(L_0 \cdot h \operatorname{Fix}_H(\infty))$. This completes the proof. □

Consider the case that $H = G_1$ is the modular group, and $\mathcal{O} = \mathrm{HC}_N(1)$ is the hypercircle of hyperradius N about L_1 for some positive integer N (cf. Section 2.7), and take $L_0 = L_N$, so that $H_0 = G_{(1,N)}$. By Proposition 2.16, the cusps of $G_{(1,N)}$ are in natural correspondence with the orbits of $\operatorname{Fix}_G(\infty) \cap G_1 = \langle T \rangle$ on $\mathrm{HC}_N(1)$, and by Proposition 2.18, the width of each cusp is just the cardinality of the corresponding orbit of $\langle T \rangle$.

It turns out that the orbit structure of $\langle T \rangle$ on $\mathrm{HC}_N(1)$ is not difficult to describe. We will not give the full analysis here (since none of it would be new), but we will furnish the following first step, which will be of use in Section 2.11.

Lemma 2.19. *Suppose $N = p^n$ for some positive integer n. Then $\mathrm{HC}_N(1)$ consists of the lattices of the form $p^{n-2a}, k/p^a$ where a and k satisfy one of the following conditions*:

(1) $a = 0$ *and* $k = 0$;
(2) $0 < a < n$ *and* $0 < k < p^a$ *and* $\gcd\{k, p\} = 1$;
(3) $a = n$ *and* $0 \le k < p^n$.

In other words, we have

$$\mathrm{HC}_{p^n}(1) = \{p^n\} \cup \left\{ \frac{p^{n-a}}{p^a}, \frac{k}{p^a} \;\middle|\; 0 < a < n, 0 < k < p^a, \gcd\{k,p\} = 1 \right\} \cup \left\{ \frac{1}{p^n}, \frac{k}{p^n} \;\middle|\; 0 \le k < p^n \right\}. \tag{2.78}$$

2.11. Orbits. In this section we use Lemma 2.19 to furnish a proof of Proposition 2.8. Since $G \cong \mathrm{PSL}_2^+(\mathbb{Q})$ acts transitively on $\mathrm{P}\mathcal{L}_1$, we have verified Proposition 2.8 as soon as we show that $G_1 \cong \mathrm{PSL}_2(\mathbb{Z})$ acts transitively on the hypercircles $\mathrm{HC}_N(1)$ for all N. Actually, a stronger result is true.

Proposition 2.20. *If M and N are positive integers with* $\gcd\{M,N\}=1$, *then the group $G_{(1,M)}$ acts transitively on* $\mathrm{HC}_N(1)$.

PROOF. Consider first the case that $N=p^n$ is a prime power. Then M is a positive integer such that $p\nmid M$. The group $G_{(1,M)}$ contains both T and $(T^M)^t$.

$$T=\begin{bmatrix}1&1\\0&1\end{bmatrix},\quad (T^M)^t=\begin{bmatrix}1&0\\M&1\end{bmatrix}. \tag{2.79}$$

Recall from Lemma 2.19 that $\mathrm{HC}_{p^n}(1)$ consists of the lattices $L_{A,b}$ with $A=p^{n-a}/p^a$ and $b=k/p^a$ for some $0\leq a\leq n$ and some $0\leq k<p^a$, with $\gcd\{k,p^a\}=1$ in case $a<n$. Recall also, from Section 2.6, that every projective lattice $L_{A,f/g}$ (for coprime positive integers f,g with $0\leq f<g$) has a reverse name $\bar{L}_{f'/g,1/g^2A}$, where f' is the unique positive integer less than g such that $ff'\equiv 1 \pmod g$. It is easy to check then that $L_{A,b}$ belongs to $\mathrm{HC}_{p^n}(1)$ just when $\bar{L}_{b,A}$ does. The precise correspondence is as follows.

$$\begin{aligned}L_{p^n,0}&=\bar{L}_{0,1/p^n};\\ L_{p^{n-a}/p^a,k/p^a}&=\bar{L}_{k'/p^a,1/p^n}, && p\nmid k,\ 0<a<n;\\ L_{1/p^n,k/p^a}&=\bar{L}_{k'/p^a,p^{n-a}/p^a}, && p\nmid k,\ 0<a\leq n;\\ L_{1/p^n,0}&=\bar{L}_{0,p^n}.\end{aligned} \tag{2.80}$$

In the above, k' denotes the unique positive integer less than p^a such that $kk'\equiv 1 \pmod{p^a}$. The reverse labels $\bar{L}_{b,A}$ are convenient for describing the action of T^t. For example, we have $\bar{L}_{b,A}\cdot(T^M)^t=\bar{L}_{b+AM,A}$. In particular,

$$\bar{L}_{0,1/p^n}\cdot(T^M)^t=\bar{L}_{M/p^n,1/p^n}. \tag{2.81}$$

Since M is coprime to p, we see that there is some power of T^t in $G_{(1,M)}$ that induces the permutation

$$\bar{L}_{k/p^n,1/p^n}\mapsto\bar{L}_{(k+1)/p^n,1/p^n} \tag{2.82}$$

on the lattices $\bar{L}_{b,A}$ in $\mathrm{HC}_{p^n}(1)$ with $A=1/p^n$ and $b=k/p^n$ for some k satisfying $0\leq k<p^n$. Recall that $T\in G_{(1,M)}$ induces a cyclic permutation on the lattices $L_{A,b}$ in $\mathrm{HC}_{p^n}(1)$ for the very same A and b—viz. the permutation obtained by removing the bars in (2.82). Comparing with (2.80) we see that the $G_{(1,M)}$ orbit containing $L_{p^n,0}$ contains every lattice in $\mathrm{HC}_{p^n}(1)$; that is, $G_{(1,M)}$ acts transitively on $\mathrm{HC}_{p^n}(1)$ if $p\nmid M$.

More generally, consider the action of $G_{(1,M)}$ on $\mathrm{HC}_N(1)$ for N coprime to M. Then the elements of $\mathrm{HC}_N(1)$ are in natural correspondence with the elements of the Cartesian product

$$\mathrm{HC}_{p_1^{a_1}}(1)\times\cdots\times\mathrm{HC}_{p_k^{a_k}}(1) \tag{2.83}$$

for $N=p_1^{a_1}\dots p_k^{a_k}$ a prime decomposition of N (cf. Section 2.8). The factors in (2.83) are sets of mutually coprime order acted on transitively by $G_{(1,M)}$. It follows that $G_{(1,M)}$ acts transitively on their product. This completes the proof. □

3. Diagrams

In this section we give our prescription for recovering McKay's Monstrous E_8 observation (1.2)—at least, its reformulation in terms of discrete subgroups of $\mathrm{PSL}_2(\mathbb{R})$ (cf. Section 1)—using elementary properties of the group $\mathrm{PSL}_2(\mathbb{R})$.

3.1. Setting. We adopt the setting of Section 2 (more particularly, of Section 2.10), with $\mathbf{k} = \mathbb{R}$, so that V is an oriented real vector space of dimension 2, and $\mathbf{v}_1 = (u_1, v_1)$ is an ordered basis for V, and $V_1 \subset V$ is the rational vector space generated by $\{u_1, v_1\}$. The choice $\mathbf{v}_1 \in \mathcal{B}^+$ (cf. Section 2.2) entails well-defined actions of the groups $\mathrm{PSL}_2(\mathbb{R})$, $\mathrm{PGL}_2^+(\mathbb{Q})$, and $\mathrm{PSL}_2(\mathbb{Z})$ on V, and on various objects related to V (cf. Section 2.7, Section 2.10).

3.2. Vertices. If Γ is a group and Γ' is a finite index subgroup of Γ, we write $[\Gamma : \Gamma']$ for the index of Γ' in Γ.

There is a well known formula for the index of $G_{(1,N)}$ (a.k.a. $\Gamma_0(N)$) in $G_1 \cong \mathrm{PSL}_2(\mathbb{Z})$. As demonstrated in [3] it is easy to recover this formula by considering the projections $\pi_p(L_N)$. For if $N = p_1^{a_1} \cdots p_k^{a_k}$ is a prime decomposition of N, then, by Proposition 2.8, the index of $G_{(1,N)}$ in G_1 is the product over i of the number of lattices at hyperdistance $p_i^{a_i}$ from the distinguished lattice L_1 (cf. Section 2.6), since $\pi_p(L_N) = L_{p_i^{a_i}}$. The cardinality of $\mathrm{HC}_{p_i^{a_i}}(L_1)$ is $(p_i+1)p_i^{a_i-1}$ (by Lemma 2.19, for example). We thus obtain the following expression,

$$[G_1 : G_{(1,N)}] = \prod_i (p_i+1)p_i^{a_i-1}, \tag{3.1}$$

which easily implies the following lemma.

Lemma 3.1. *If the index of $G_{(1,n)}$ in G_1 does not exceed 12, then n belongs to the following set:*

$$\{1, 2, 3, 4, 5, 6, 7, 8, 9, 11\} \tag{3.2}$$

Given a group $\Gamma < \mathrm{PSL}_2(\mathbb{R})$ that is commensurable with $G_1 \cong \mathrm{PSL}_2(\mathbb{Z})$, let us write $I_\Gamma^{G_1}$ for the index of $\Gamma \cap G_1$ in G_1, and $I_{G_1}^\Gamma$ for the index of $\Gamma \cap G_1$ in Γ.

(3.3)

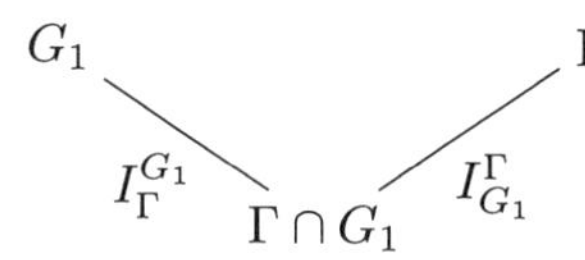

Proposition 3.2. *Suppose $\Gamma < \mathrm{PSL}_2(\mathbb{R})$ satisfies the following conditions:*

(1) *Γ is arithmetic;*
(2) *Γ has width 1 at ∞ (cf. Section 2.10);*
(3) *there is some N such that Γ contains and normalizes $G_{(1,N)}$, and the quotient $\Gamma/G_{(1,N)}$ is a group of exponent 2;*
(4) *$I_\Gamma^{G_1} \leq 12$ and $I_\Gamma^{G_1}/I_{G_1}^\Gamma \leq 3$.*

Then Γ is one of the groups in $\mathfrak{E}$, where

$$\mathfrak{E} = \left\{G_1, G_{(1,2)}, G_{\{1,2\}}, G_{\{1,3\}}, G_{\{1,4\}}, G_{\{1,5\}}, G_{\{1,2,3,6\}}, G_{\{2,4\}}^{(2)}, G_3^{(3)}\right\}. \tag{3.4}$$

Remark. Since $G_{(1,N)}$ and its normalizer are both commensurable with G_1 for any N, condition (3) implies condition (1).

Remark. We may also write $G_{(1,N)+}$ for $G_{\{1,N\}}$ when $N \in \{2,3,4,5\}$. We may write $G_{(1,6)+}$ for $G_{\{1,2,3,6\}}$. The notation in (3.4) has been chosen because it suggests, more strongly, how we can determine the correct valence for each Γ in $\mathfrak{E}$ as a vertex in the affine E$_8$ Dynkin diagram (cf. Section 3.3.3).

Proof. By the Atkin–Lehner theorem (Theorem 2.7), the intersection of the normalizer of $G_{(1,N)}$ with G_1 is $G_{(1,N/h)}$ where h is the largest divisor of 24 such that $h^2|N$. By condition (3) then, we have $G_{(1,N)} < \Gamma \cap G_1 < G_{(1,N/h)}$, for some N, with h as in the previous sentence. Equivalently, we have

$$G_{(1,hn)} < \Gamma \cap G_1 < G_{(1,n)} \tag{3.5}$$

for some n and some h, where h is a divisor of $\gcd\{n,24\}$ such that neither $4h$ nor $9h$ divide n.

By Lemma 3.1, the only possibilities for the n in (3.5) are those in the set (3.2) if the first inequality of condition (4) is to be satisfied; we will consider these 10 cases separately.

Let us agree to write Γ_1 for the intersection $\Gamma \cap G_1$.

- *Case* $n = 1$. The inequality (3.5) reduces to $G_1 < \Gamma_1 < G_1$ in this case, so $\Gamma_1 = G_1$. The normalizer of G_1 is G_1, so $\Gamma = \Gamma_1 = G_1$ in this case. Conditions (1) through (4) are satisfied when $\Gamma = G_1$.
- *Case* $n = 2$. If $n = 2$ then $h \in \{1, 2\}$.

If $h = 1$ then $\Gamma_1 = G_{(1,2)}$. The normalizer of $G_{(1,2)}$ is $G_{\{1,2\}}$. The former group has index 2 in the latter, so Γ is one of $G_{(1,2)}$ or $G_{\{1,2\}}$. The conditions (1) through (4) are satisfied in both cases.

If $h = 2$ then Γ is assumed to normalize and contain $G_{(1,4)}$, so that $G_{(1,4)} < \Gamma_1 < G_{(1,2)}$, and $\Gamma < G_2$.

(3.6)

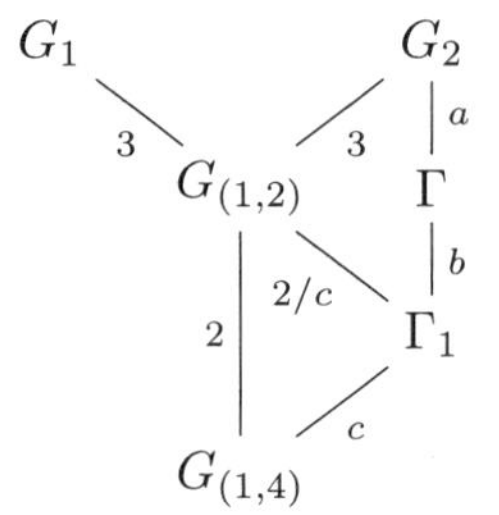

The indices of these containments are as in (3.6)), for some a, b, c. Evidently, $abc = 6$. In order for $\Gamma/G_{(1,4)}$ to have exponent 2 (cf. condition (3)), it must be that $a = 3$ or $a = 6$.

If $a = 6$ then $\Gamma = \Gamma_1 = G_{(1,4)}$, and $I_{\Gamma}^{G_1}/I_{G_1}^{\Gamma} = I_{\Gamma}^{G_1} = 6$, in violation of condition (4). So $a = 3$, and $bc = 2$. If $c = 2$ then $b = 1$ and $\Gamma = \Gamma_1 = G_{(1,2)}$, and we have seen already that this group satisfies conditions (1) through (4). If $b = 2$ and $c = 1$, then Γ is a subgroup of index 3 in G_2 whose intersection with G_1 is exactly $G_{(1,4)}$. The group G_2 acts by permutations on $\mathrm{HC}_2(2)$ — the hypercircle of hyperradius 2 about L_2 — and $G_{(1,4)}$ is exactly the subgroup stabilizing L_4. It follows that the natural map $G_2 \to \mathrm{Sym}(\mathrm{HC}_2(2))$ is surjective, and the image of Γ under this map is the group generated by one transposition.

(3.7)

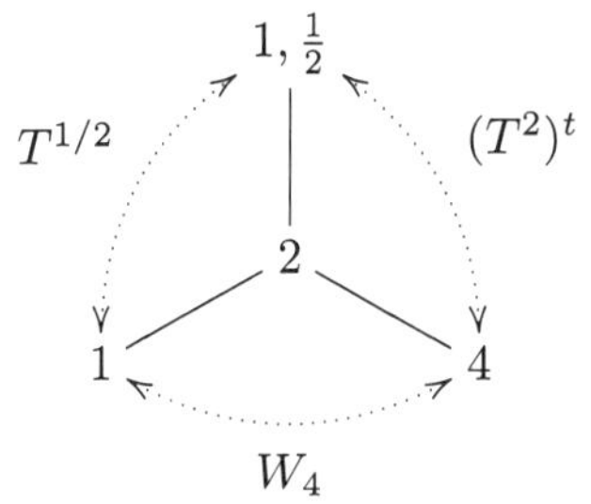

Diagram (3.7) displays the smallest subtree of the 2-adic tree (cf. Section 2.8) containing $\mathrm{HC}_2(2)$, together with elements of G_2 that give rise to each of the transpositions in $\mathrm{Sym}(\mathrm{HC}_2(2))$. (We may take $W_4 = g^{1/2}Sg^2$.) Our group Γ does not contain $T^{1/2}$ by condition (2), and it does not contain $(T^2)^t$ since this element lies in G_1 but not in $\Gamma_1 = G_{(1,4)}$. The remaining possibility is that $\Gamma = \langle G_{(1,4)}, W_4\rangle = G_{\{1,4\}}$. This group satisfies conditions (1) through (4).

- *Case* $n = 3$. If $n = 3$ then $h \in \{1, 3\}$.

If $h = 1$ then $\Gamma_1 = G_{(1,3)}$ and $\Gamma < G_{\{1,3\}}$. The group $G_{(1,3)}$ has index 2 in $G_{\{1,3\}}$, so either $\Gamma = G_{(1,3)}$ or $\Gamma = G_{\{1,3\}}$. In the former case the second inequality of condition (4) is violated. In the latter case conditions (1) through (4) are all satisfied.

If $h = 3$ then Γ is assumed to normalize and contain $G_{(1,9)}$, so that $G_{(1,9)} < \Gamma_1 < G_{(1,3)}$, and $\Gamma < G_3$.

(3.8)

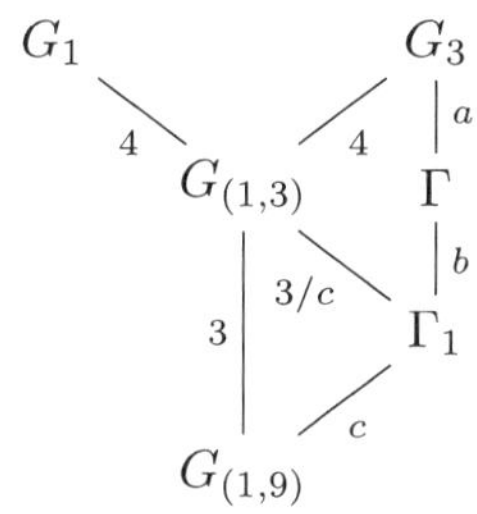

The indices of these containments are as in (3.8), for some a, b, c. Evidently, $abc = 12$. In order for $\Gamma/G_{(1,9)}$ to have exponent 2 (cf. condition (3)), it must be that $3|a$, and in particular, we must have $c = 1$; i.e., $\Gamma_1 = G_{(1,9)}$. In order that the second inequality of condition (4) be satisfied, we require that $b \geq 4$. Consequently, Γ is a subgroup of index 3 in G_3 whose intersection with G_1 is exactly the group $G_{(1,9)}$.

The group G_3 acts by permutations on $\mathrm{HC}_3(3)$, and $G_{(1,9)}$ is exactly the kernel of the corresponding map $G_3 \to \mathrm{Sym}(\mathrm{HC}_3(3))$ (cf. Section 2.9.1). The image of G_3 under this map is a copy of the alternating group on 4 symbols. The image of Γ in $\mathrm{Sym}(\mathrm{HC}_3(3))$ must be a subgroup of order 4 in this Alt_4; such a subgroup is unique — it is the kernel of any non-trivial homomorphism $\mathrm{Alt}_4 \to \mathbb{Z}/3$. We conclude that $\Gamma = G_3^{(3)}$ (cf. Section 2.9.1). This group satisfies conditions (1) through (4).

- *Case* $n = 4$. If $n = 4$ then $h \in \{2, 4\}$.

If $h = 2$ then $G_{(1,8)} < \Gamma_1 < G_{(1,4)}$ and $\Gamma < G_{\{2,4\}}$.

(3.9)

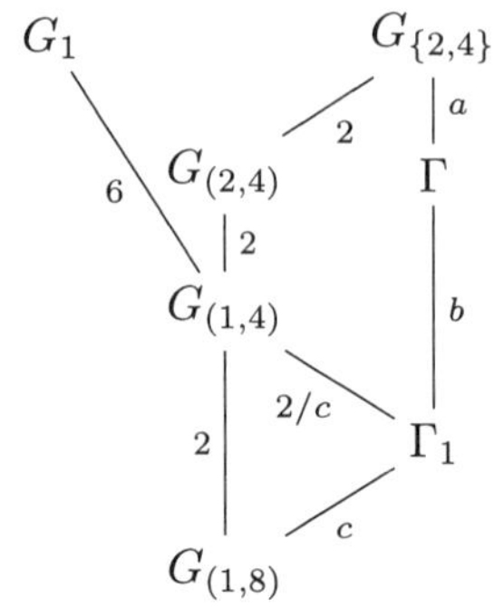

We have $abc = 8$ for a, b, c as in (3.9). The group $G_{\{2,4\}}$ does not have width 1 at ∞, so $a \geq 2$. On the other hand, by condition (4) we have $I_{\Gamma}^{G_1}/I_{G_1}^{\Gamma} = 12/bc \leq 3$, so $bc \geq 4$, so we conclude $a = 2$ and $bc = 4$.

The group $G_{\{2,4\}}$ acts by permutations on the set $S = \mathrm{HC}_2(2) \cup \mathrm{HC}_2(4)$, and the image of $G_{\{2,4\}}$ in $\mathrm{Sym}(S)$ is a group of the shape

$$(\mathbb{Z}/2 \times \mathbb{Z}/2) \rtimes \mathbb{Z}/2 \tag{3.10}$$

(i.e., a dihedral group of order 8) where the central factors $\mathbb{Z}/2$ are generated by $T^{1/2}$ and $(T^4)^t$, and a non-central $\mathbb{Z}/2$ is generated by the Atkin – Lehner involution W_8 (cf. Section 2.9.2). The group $G_{(1,8)}$ is the kernel of the natural map $G_{\{2,4\}} \to \mathrm{Sym}(S)$. Consequently, the image of Γ in $\mathrm{Sym}(S)$ is a subgroup of order four in (3.10). By condition (2) it does not contain $T^{1/2}$. By condition (3) it is not cyclic. There is exactly one possibility: $\Gamma = \langle G_{(1,8)}, (T^4)^t T^{1/2}, W_8 \rangle$; i.e., $\Gamma = G_{\{2,4\}}^{(2)}$ (cf. Section 2.9.2). This group satisfies all the required properties.

If $h = 4$ then $G_{(1,16)} < \Gamma_1 < G_{(1,4)}$ and $\Gamma < G_4$.

(3.11)

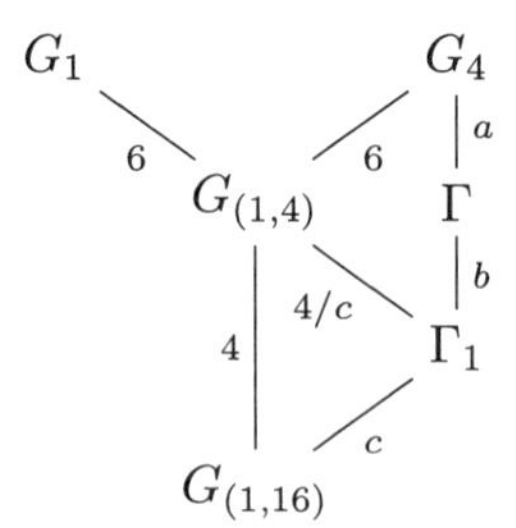

We have $abc = 24$ for a, b, c as in (3.11). By condition (3), we have $3 \mid a$. By condition (4) we have $I_{\Gamma}^{G_1}/I_{G_1}^{\Gamma} = 24/bc \leq 3$, so $bc \geq 8$, and this implies $a = 3$ and $bc = 8$.

(3.12)

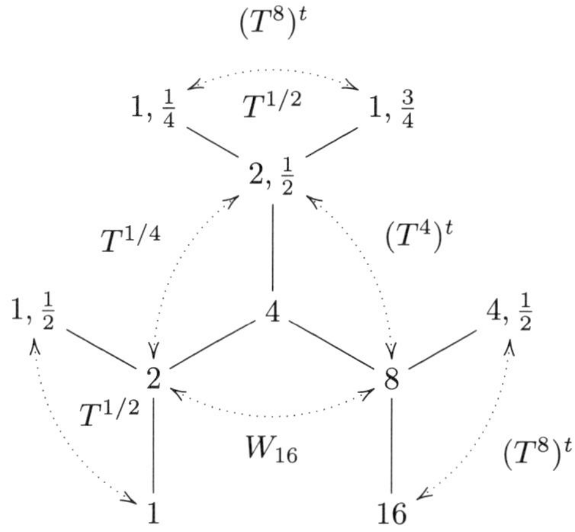

The diagram in (3.12) shows the part of the 2-adic tree containing HC$_4(4)$; the group G_4 acts as automorphisms of this tree, and the subgroup $G_{(1,16)}$ fixes every node. We see from (3.12) that the quotient $G_4/G_{(1,16)}$ has the structure $(\mathbb{Z}/2 \times \mathbb{Z}/2) \rtimes \mathrm{Sym}_3$, with the elementary abelian subgroup generated by $T^{1/2}$ and $(T^8)^t$, and the symmetric group acting on it generated by $T^{1/4}$ and $(T^4)^t$ (cf. (3.7)). If Γ is a subgroup of G_4 of order 8 containing $G_{(1,16)}$, then the image of Γ in $G_4/G_{(1,16)}$ contains the subgroup $\mathbb{Z}/2 \times \mathbb{Z}/2$. It follows that Γ contains $T^{1/2}$, but this violates condition (2).

• *Case* $n = 5$. If $n = 5$ then $h = 1$ and $\Gamma_1 = G_{(1,5)}$, and $G_{(1,5)} < \Gamma < G_{\{1,5\}}$. Similar to the case that $N = nh = 3$, the only possibility satisfying conditions (1) through (4) is $\Gamma = G_{\{1,5\}}$.

• *Case* $n = 6$. If $n = 6$ then $h \in \{1, 2, 3\}$.

If $h = 1$ then $\Gamma_1 = G_{(1,6)}$, and $\Gamma < G_{\{1,2,3,6\}}$. The group $G_{(1,6)}$ has index 12 in G_1, and index 4 in $G_{\{1,2,3,6\}}$, so in order to satisfy condition (4), it must be that $\Gamma = G_{\{1,2,3,6\}}$. This group satisfies all the required properties.

If $h = 2$ then $G_{(1,12)} < \Gamma_1 < G_{(1,6)}$, and $\Gamma < G_{\{2,6\}}$. The group $G_{(1,6)}$ has index 12 in G_1, so it must be that $\Gamma_1 = G_{(1,6)}$, by the first inequality of condition (4).

(3.13)

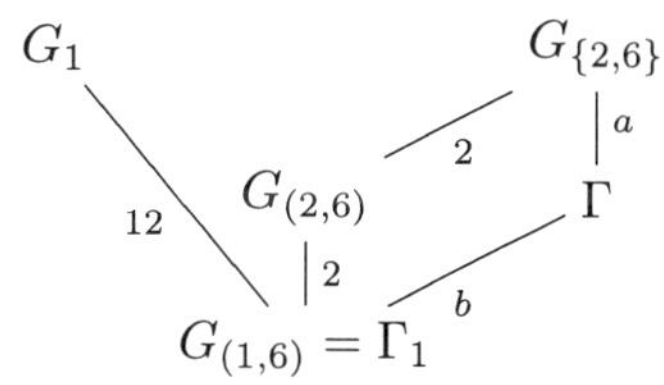

We have $ab = 4$ for a, b as in (3.13). For condition (4) we require $b \geq 4$, so that $\Gamma = G_{\{2,6\}}$, but this group does not satisfy condition (2). Diagram (3.14) shows the cell $\{L_2, L_6\}$ together with sufficiently many adjacent nodes (in the 2 and 3-adic trees) so as to allow us to display the action of $G_{\{2,6\}}/G_{(1,6)}$ (and verify the indices in (3.13)).

(3.14)
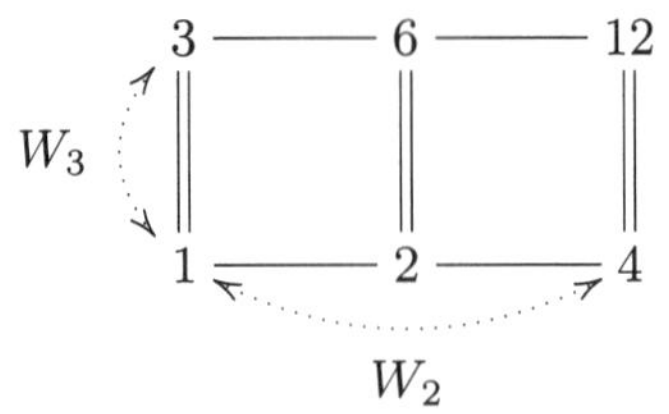

If $h = 3$ then $G_{(1,18)} < \Gamma_1 < G_{(1,6)}$ and $\Gamma < G_{\{3,6\}}$. Again, we must have $\Gamma_1 = G_{(1,6)}$, by condition (4).

(3.15)
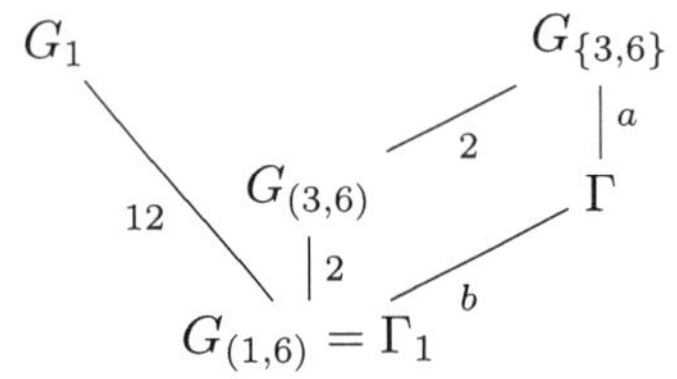

Considering indices (cf. (3.15))) we see that Γ must coincide with $G_{\{3,6\}}$ if condition (4) is to be satisfied, but this group fails to satisfy condition (2). Diagram (3.16) is the analogue of (3.14) for $h = 3$.

(3.16)
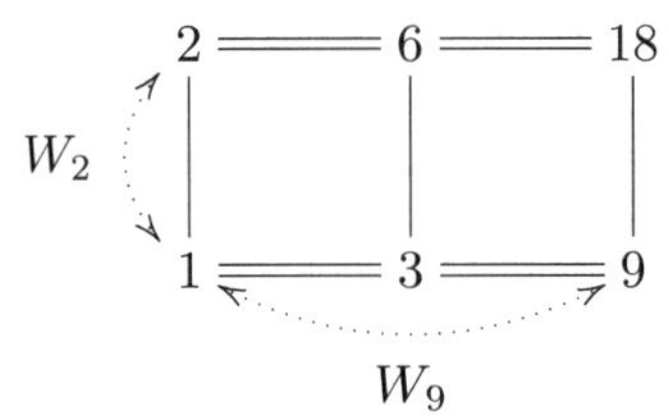

• *Case* $n = 7$. If $n = 7$ then $h = 1$, so we have $\Gamma_1 = G_{(1,7)}$ and $\Gamma < G_{\{1,7\}}$. Then $G_{(1,7)}$ has index 8 in G_1, and index 2 in $G_{\{1,7\}}$, so there is no possibility for Γ that satisfies condition (4).

• *Case* $n = 8$. If $n = 8$ then $h \in \{4, 8\}$. Since $G_{(1,8)}$ has index 12 in G_1, we have $\Gamma_1 = G_{(1,8)}$. The group Γ is assumed to normalize $G_{(1,8h)}$, and the quotient $\Gamma/G_{(1,8h)}$ should have exponent 2, by condition (3). Observe that $G_{(1,8)}/G_{(1,8h)}$ is a cyclic group of order h—we may take the image of $(T^8)^t$ as a generator. Since $h > 2$, there are no groups in $\mathfrak{C}$ that arise for $n = 8$.

• *Case* $n = 9$. For $n = 9$ we have $h \in \{3, 9\}$. Since $G_{(1,9)}$ has index 12 in G_1, we have $\Gamma_1 = G_{(1,9)}$. The group Γ is assumed to normalize $G_{(1,9h)}$, and, by condition (3), it is required that $\Gamma/G_{(1,9h)}$ have exponent 2, but $G_{(1,9h)}$ has index h in Γ_1. We conclude that no elements of $\mathfrak{C}$ arise for $n = 9$.

• *Case* $n = 11$. Similar to the case that $n = 7$, we must have $\Gamma_1 = G_{(1,11)}$ and $G_{(1,11)} < \Gamma < G_{\{1,11\}}$. But $G_{(1,11)}$ has index 12 in G_1, and index 2 in $G_{\{1,11\}}$, so there is no possibility for Γ that satisfies condition (4).

We have accounted for all the groups appearing in the set $\mathfrak{C}$, and we have verified that no other subgroups of $\mathrm{PSL}_2(\mathbb{R})$ satisfy conditions (1) through (4). □

3.3. Edges. The groups in $\mathfrak{C}$ are exactly those that label the vertices in the diagram (1.2). We now seek to reconstruct the edges of the diagram (1.2), by examination of the properties of the groups in $\mathfrak{C}$.

3.3.1. *Thread.* By condition (3) of Proposition 3.2, each group $\Gamma \in \mathfrak{E}$ satisfies

$$G_{(1,N)} < \Gamma < N_G\big(G_{(1,N)}\big) = G_{(h,N/h)+} \tag{3.17}$$

for some positive integer N (where h is the largest divisor of 24 such that h^2 divides N — cf. Theorem 2.7). For $\Gamma \in \mathfrak{E}$ let N_Γ be the minimal N with this property. (The group $\Gamma = G_{(1,2)}$ satisfies (3.17) for $N = 2$ and $N = 4$.) Now let a_Γ be the largest divisor of 24 such that a_Γ^2 divides N_Γ, and $G_{(a,N/a)}$ contains $\Gamma \cap G_{(a,N/a)}$ to index a, for $a = a_\Gamma$ and $N = N_\Gamma$.

(3.18)

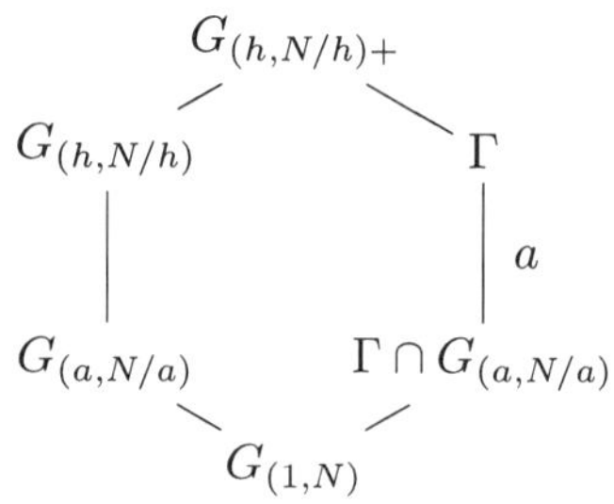

Lemma 3.3. *If $\Gamma \neq G_{\{1,4\}}$ then a_Γ is the largest divisor of* 24 *such that a_Γ^2 divides N_Γ; that is, $a = a_\Gamma = h$ in* (3.18). *If $\Gamma = G_{\{1,4\}}$ then $a_\Gamma = 1$.*

For each $\Gamma \in \mathfrak{E}$ define a subgroup $\Gamma_0 < \Gamma$ by setting $\Gamma_0 = \Gamma \cap G_{(a,N/a)}$, where $a = a_\Gamma$ and $N = N_\Gamma$. If we agree to write also $G^{(1)}_{(1,N)}$ for $G_{(1,N)}$, and $G^{(a)}_{(a,a)}$ for $G^{(a)}_a$, then we have

$$\Gamma_0 = G^{(a)}_{(a,N/a)} \tag{3.19}$$

for $a = a_\Gamma$ and $N = N_\Gamma$, for each $\Gamma \in \mathfrak{E}$.

The groups Γ_0, for $\Gamma \in \mathfrak{E}$, may be recovered in a more geometric way as follows.

Relax for a moment the assumption that $\Gamma \in \mathfrak{E}$, and suppose, more generally, that Γ is an arithmetic subgroup of $\mathrm{PSL}_2(\mathbb{R})$ such that Γ contains and normalizes some $G_{(1,N)}$. Then Γ normalizes the set $(1,N)+$ and the normalizer of Γ stabilizes some subset of $(1,N)+$.

Reinstate now the assumption that $\Gamma \in \mathfrak{E}$, and define S_Γ to be the largest subset of $(1,N)+$ that is stabilized by $N_G(\Gamma)$. For each $\Gamma \in \mathfrak{E}$ except for $\Gamma = G_{\{1,4\}}$, the normalizer $N_G(\Gamma)$, of Γ, is a maximal discrete subgroup of $\mathrm{PSL}_2(\mathbb{R})$, and therefore, by Proposition 2.10, stabilizes a unique cell in $\mathrm{P}\mathcal{L}_1$. The set S_Γ then recovers this cell. In the case that $\Gamma = G_{\{1,4\}}$, we have $N_G(\Gamma) = \Gamma$, while the group G_2 properly contains Γ. We have $S_\Gamma = \{L_1, L_2, L_4\}$ (cf. (2.49)) for $\Gamma = G_{\{1,4\}}$.

For $\Gamma \in \mathfrak{E}$ consider the subgroup of $N_G(\Gamma)$ fixing the elements of S_Γ point-wise; this group is exactly Γ_0, for each $\Gamma \in \mathfrak{E}$.

The following lemma is easily checked — by inspection of the first two lines of Table 2, for example.

Lemma 3.4. *For each $\Gamma \in \mathfrak{E}$, the subgroup Γ_0 is contained normally in Γ, and the quotient Γ/Γ_0 is a group of exponent* 2.

The group Γ_0 may be regarded as the group we obtain from Γ by removing its Atkin – Lehner involutions.

3.3.2. *Level.* Let Γ be an arithmetic subgroup of $\mathrm{PSL}_2(\mathbb{R})$, and suppose that Γ is a congruence group (cf. Section 2.7). Then Γ has a well-defined level; viz. the minimal positive integer N such that $\Gamma(N) < \Gamma$. Let us write $\mathrm{lev}(\Gamma)$ for the level of an arithmetic congruence group Γ. For $\Gamma \in \mathfrak{E}$ we define the *normalized level* of Γ, to be denoted $\mathrm{lev}_0(\Gamma)$, by setting

$$\mathrm{lev}_0(\Gamma) = \frac{\mathrm{lev}(\Gamma)}{a_\Gamma} \tag{3.20}$$

where a_Γ is as in Section 3.3.1.

The values a_Γ and the normalized levels $\mathrm{lev}_0(\Gamma)$ are displayed in Table 1.

3.3.3. *Valency.* By Lemma 3.4, the group Γ_0 is normal in Γ for each $\Gamma \in \mathfrak{E}$, and the quotient Γ/Γ_0 has exponent 2. For $\Gamma \in \mathfrak{E}$ we define the *valency* of Γ, to be denoted $\mathrm{val}(\Gamma)$, by setting $\mathrm{val}(\Gamma) = m + 1$, where m satisfies

$$\Gamma/\Gamma_0 \cong (\mathbb{Z}/2)^m. \tag{3.21}$$

The value $\mathrm{val}(\Gamma)$ encodes the number of Atkin–Lehner involutions in Γ. The valencies of the groups in $\mathfrak{E}$ are recorded in Table 2.

3.3.4. *Faithfulness.* Recall that in the classical McKay Correspondence for the binary icosahedral group $2.\mathrm{Alt}_5$, the nodes in the affine E_8 Dynkin diagram are labeled by irreducible representations of $2.\mathrm{Alt}_5$. In particular, four of the nodes correspond to faithful representations of $2.\mathrm{Alt}_5$, and the remaining five nodes correspond to representations that factor through Alt_5.

With this in mind, we will say that a group $\Gamma \in \mathfrak{E}$ is *faithful* if Γ is "not far" from $G_{\{1,2\}}$, in the sense that $[\Gamma, \Gamma \cap G_{\{1,2\}}] \leq 2$. We set $\mathfrak{E}_1$ to be the subset of groups in $\mathfrak{E}$ that are faithful, and we set $\mathfrak{E}_0 = \mathfrak{E} \setminus \mathfrak{E}_1$.

$$\mathfrak{E}_1 = \{G_{\{1,2\}}, G_{\{1,4\}}, G_{\{1,2,3,6\}}, G_{(1,2)}\} \tag{3.22}$$

$$\mathfrak{E}_0 = \{G_1, G_{\{1,3\}}, G_{\{1,5\}}, G_3^{(3)}, G_{\{2,4\}}^{(2)}\} \tag{3.23}$$

3.3.5. *Prescription.* The following proposition is easily checked.

Proposition 3.5. *There is a unique graph with vertex set $\mathfrak{E}$ satisfying the following properties:*

- *The valence of $\Gamma \in \mathfrak{E}$ is $\mathrm{val}(\Gamma)$.*
- *The identity $2\,\mathrm{lev}_0(\Gamma) = \sum_{\Gamma' \in \mathrm{adj}(\Gamma)} \mathrm{lev}_0(\Gamma')$ holds for all $\Gamma \in \mathfrak{E}$, where $\mathrm{adj}(\Gamma)$ denotes the set of vertices that are adjacent to Γ.*
- *If $\Gamma \in \mathfrak{E}_1$ then $\mathrm{adj}(\Gamma) \subset \mathfrak{E}_0$.*

TABLE 1. Γ, a_Γ and $\mathrm{lev}_0(\Gamma)$, for $\Gamma \in \mathfrak{E}$

G_1	$G_{\{1,2\}}$	$G_{\{1,3\}}$	$G_{\{1,4\}}$	$G_{\{1,5\}}$	$G_{\{1,2,3,6\}}$	$G_3^{(3)}$	$G_{\{2,4\}}^{(2)}$	$G_{(1,2)}$
1	1	1	1	1	1	3	2	1
1	2	3	4	5	6	3	4	2

TABLE 2. Γ, Γ_0 and $\mathrm{val}(\Gamma)$, for $\Gamma \in \mathfrak{E}$

G_1	$G_{\{1,2\}}$	$G_{\{1,3\}}$	$G_{\{1,4\}}$	$G_{\{1,5\}}$	$G_{\{1,2,3,6\}}$	$G_3^{(3)}$	$G_{\{2,4\}}^{(2)}$	$G_{(1,2)}$
G_1	$G_{(1,2)}$	$G_{(1,3)}$	$G_{(1,4)}$	$G_{(1,5)}$	$G_{(1,6)}$	$G_3^{(3)}$	$G_{(2,4)}^{(2)}$	$G_{(1,2)}$
1	2	2	2	2	3	1	2	1

The graph whose existence and uniqueness is guaranteed by Proposition 3.5 is displayed in (3.24).

(3.24)
$$
\begin{array}{l}
\qquad\qquad\qquad\qquad\qquad\qquad\qquad\qquad\quad G_3^{(3)} \\
\qquad\qquad\qquad\qquad\qquad\qquad\qquad\quad / \\
\textcircled{G_1} — G_{\{1,2\}} — G_{\{1,3\}} — G_{\{1,4\}} — G_{\{1,5\}} - G_{\{1,2,3,6\}} \\
\qquad\qquad\qquad\qquad\qquad\qquad\qquad\qquad\quad G_{\{2,4\}}^{(2)} \diagdown \\
\qquad\qquad\qquad\qquad\qquad\qquad\qquad\qquad\qquad\qquad G_{(1,2)}
\end{array}
$$

In terms of the more standard notation for the discrete groups of Monstrous Moonshine, this is exactly the diagram (1.2) we sought to recover.

4. Super analogue

For each $\Gamma \in \mathfrak{C}$, we define an arithmetic group ${}^s\Gamma < \mathrm{PSL}_2(\mathbb{R})$ as follows. Recall that for each $\Gamma \in \mathfrak{C}$, the corresponding group Γ_0 (cf. Section 3.3.1) can be written in the form $G^{(a)}_{(a,N/a)}$ for some $a = a_\Gamma$ and $N = N_\Gamma$. Define groups ${}^s\Gamma_0$ by setting

$$ {}^s\Gamma_0 = G^{(a)}_{(a,2N/a)} \tag{4.1} $$

when $\Gamma_0 = G^{(a)}_{(a,N/a)}$. We will arrive at the group ${}^s\Gamma$ by adjoining to ${}^s\Gamma_0$ the appropriate "scalings" of the Atkin–Lehner involutions of Γ; i.e., scalings of the cosets Γ/Γ_0. More precisely, a coset $W_e(N)$ (cf. Section 2.7) in Γ/Γ_0, for $N = N_\Gamma$ and e an exact divisor of N, is scaled (i.e., sent) to $W_{2e}(2N)$ or $W_e(2N)$ according as $2 \mid e$ or not.

$$ {}^sW_e(N) = \begin{cases} W_{2e}(2N), & \text{if } 2 \mid e; \\ W_e(2N), & \text{else.} \end{cases} \tag{4.2} $$

This handles the cosets of Γ_0 in Γ in the cases that $\Gamma \neq G^{(2)}_{\{2,4\}}$. For $\Gamma = G^{(2)}_{\{2,4\}}$ we apply the same idea, but the notation is slightly more involved: the unique non-trivial coset in Γ/Γ_0 is $g_2^{-1}W_2(2)g_2$ in this case, and we set ${}^2(g_2^{-1}W_2(2)g_2) = g_2^{-1}W_4(4)g_2$. Now we define ${}^s\Gamma$, for $\Gamma \in \mathfrak{C}$, by setting

$$ {}^s\Gamma = \bigcup_{X \in \Gamma/\Gamma_0} {}^sX. \tag{4.3} $$

It is straightforward to check that ${}^s\Gamma$ is an arithmetic subgroup of $\mathrm{PSL}_2(\mathbb{R})$ for each $\Gamma \in \mathfrak{C}$. Even more than this, each group ${}^s\Gamma$ also appears in Monstrous Moonshine. Define a set ${}^s\mathfrak{C}$ by setting

$$ {}^s\mathfrak{C} = \{{}^s\Gamma \mid \Gamma \in \mathfrak{C}\} \tag{4.4} $$

We may consider the diagram obtained from (1.2) by replacing the labels $\Gamma \in \mathfrak{C}$ with the corresponding groups ${}^s\Gamma \in {}^s\mathfrak{C}$. In the notation of [4,5], the labeling of the affine E8 Dynkin diagram thus obtained is displayed in (4.5).

(4.5)
$$
\begin{array}{l}
\qquad\qquad\qquad\qquad\qquad\qquad\qquad\qquad 6 \parallel 3 \\
\qquad\qquad\qquad\qquad\qquad\qquad\qquad\quad / \\
\textcircled{2} — 4+ — 6+6 — 8+ — 10+10 — 12+ \\
\qquad\qquad\qquad\qquad\qquad\qquad\qquad\quad \diagdown \\
\qquad\qquad\qquad\qquad\qquad\qquad\qquad\qquad 8 \parallel 2+ \\
\qquad\qquad\qquad\qquad\qquad\qquad\qquad\qquad\qquad \diagdown \\
\qquad\qquad\qquad\qquad\qquad\qquad\qquad\qquad\qquad\quad 4
\end{array}
$$

Each of the groups in (4.5) is attached to a unique conjugacy class of the Monster group, courtesy of Monstrous Moonshine. Each conjugacy class nZ in M so obtained has the property that if $g \in nZ$ then $g^{n/2} \in 2B$. Further, each of these conjugacy classes is a "lift" of some conjugacy class in the Conway group Co_0. More precisely, for each group ${}^s\Gamma$ in ${}^s\mathfrak{C}$ there is a Frame shape[2] $A_1^{\alpha_1}A_2^{\alpha_2}\cdots$ say, such that the eta product

$$\frac{\eta(A_1\tau)^{\alpha_1}\eta(A_2\tau)^{\alpha_2}\cdots}{\eta(2A_1\tau)^{\alpha_1}\eta(2A_2\tau)^{\alpha_2}\cdots} \tag{4.6}$$

furnishes a principal modulus (a.k.a. Hauptmodul) for ${}^s\Gamma$, and also encodes the eigenvalues of the elements of a unique conjugacy class of Co_0.

$$(4.7)\quad \textcircled{1^{24}} - 2^{24}/1^{24} - 3^{12}/1^{12} - 4^8/1^8 - 5^6/1^6 - 2^6 6^6/1^6 3^6 \begin{cases} - 3^8 \\ - 4^{12}/2^{12} - 1^8 2^8 \end{cases}$$

Diagram (4.7) shows the labeling obtained by replacing the groups in (4.5) with the corresponding Frame shapes of Co_0.

Evidently, (4.7) furnishes a reformulation of McKay's Monstrous E_8 observation, in which conjugacy classes of the Monster are replaced by conjugacy classes of Co_0.

It is amusing to observe that the highest root labeling of the affine E_8 Dynkin diagram can be read off directly from the Frame shapes in (4.7): consider the maximal A_i in $A_1^{\alpha_1}A_2^{\alpha_2}\ldots$ (i.e., the order of the corresponding class in Co_0). One can also predict the valency of the vertex corresponding to any Frame shape in (4.7): the valency of the vertex labeled $A_1^{\alpha_1}A_2^{\alpha_2}\cdots$ is the number of negative α_i plus 1.

A direct link between the classes in (4.7) and the principal moduli of the groups in (4.5) can be obtained by considering the McKay–Thompson series associated to conjugacy classes of Co_0 via the action of this group on a suitably defined vertex operator superalgebra (cf. [6]).

Acknowledgement. This article is dedicated to John McKay, with gratitude and respect. The author is grateful to Noam Elkies and Curtis McMullen for helpful conversations.

References

1. R. E. Borcherds, *Monstrous moonshine and monstrous Lie superalgebras*, Invent. Math. **109** (1992), no. 2, 405–444.
2. J. H. Conway, *A simple construction for the Fischer–Griess monster group*, Invent. Math. **79** (1985), no. 3, 513–540.
3. ———, *Understanding groups like* $\Gamma_0(N)$, Groups, Difference Sets, and the Monster (Columbus, OH, 1993) (K. T. Arasu, J. F. Dillon, K. Harada, S. Sehgal, and R. Solomon, eds.), Ohio State Univ. Math. Res. Inst. Publ., vol. 4, de Gruyter, Berlin, 1996, pp. 327–343.
4. J. Conway, J. McKay, and A. Sebbar, *On the discrete groups of Moonshine*, Proc. Amer. Math. Soc. **132** (2004), no. 8, 2233–2240.
5. J. H. Conway and S. P. Norton, *Monstrous moonshine*, Bull. London Math. Soc. **11** (1979), no. 3, 308–339.

[2] a generalization of the cycle notation for permutations, which encodes the eigenvalues of an orthogonal transformation of finite order that is writable over $\mathbb{Z}$.

6. J. F. Duncan, *Super-moonshine for Conway's largest sporadic group*, Duke Math. J. **139** (2007), no. 2, 255–315.
7. H. Helling, *Bestimmung der Kommensurabilitätsklasse der Hilbertschen Modulgruppe*, Math. Z. **92** (1966), 269–280.
8. J. McKay, *Graphs, singularities, and finite groups*, The Santa Cruz Conference on Finite Groups (Santa Cruz, CA, 1979) (B. Cooperstein and G. Mason, eds.), Proc. Sympos. Pure Math., vol. 37, Amer. Math. Soc., Providence, RI, 1980, pp. 183–186.

DEPARTMENT OF MATHEMATICS, HARVARD UNIVERSITY, ONE OXFORD STREET, CAMBRIDGE, MA 02138, USA
E-mail address: duncan@math.harvard.edu
URL: http://math.harvard.edu/~jfd/

Centre de Recherches Mathématiques
CRM Proceedings and Lecture Notes
Volume **47**, 2009

The Galois Action on Character Tables

Terry Gannon

To John, prophet and mentor

ABSTRACT. A geometric interpretation and generalization for the Galois action on finite group character tables is sketched. The generalization is a Galois action on the space $\mathrm{Map}_G(G^n, \overline{\mathbb{Q}})/\mathrm{S}_n$ for each finite G, where G acts by simultaneous conjugation on the n-tuples G^n and the symmetric group S_n permutes the components.

1. Introduction

Throughout this paper, let G be a finite group. Its finite-dimensional representations (over $\mathbb{C}$) are completely reducible into a direct sum of irreducible representations. Moreover, such a representation ρ is uniquely determined (up to equivalence) by its *character* $\mathrm{ch}_\rho(g) := \mathrm{Tr}\big(\rho(g)\big)$.

A character is a *class function*, i.e., is constant on conjugacy classes: $\mathrm{ch}_\rho(k^{-1}gk) = \mathrm{ch}_\rho(g)$. We collect the characters of irreducible representations, evaluated at each conjugacy class, into a square matrix called the *character table*. The character tables for the symmetric group S_3 and alternating group A_5 are given below ($\alpha = (1+\sqrt{5})/2$, $\alpha' = (1-\sqrt{5})/2$). The character table is an important group invariant, with rich structure. For example, each 1-dimensional representation is a symmetry permuting the rows, and dually, any central element is a symmetry permuting the columns.

The Galois symmetry of the character table isn't as well-known. Any element of the *cyclotomic field* $\mathbb{Q}[\xi_n]$ for $\xi_n := \exp[2\pi \mathrm{i}/n]$ can be written (in many ways) as a polynomial $p(\xi_n) = \sum_k a_k \xi_n^k$, $a_k \in \mathbb{Q}$; it is called a cyclotomic integer if all $a_k \in \mathbb{Z}$. The entries of the character table of G are cyclotomic integers, $\mathrm{ch}_\rho(g) \in \mathbb{Z}[\xi_n]$, where n can be taken to be the order (or even the exponent) of G. This can be seen by diagonalizing the matrix $\rho(g)$, which will have order dividing n. For example in the A_5 table, $\alpha = 1 + \xi_5^2 + \xi_5^3$ and $\alpha' = 1 + \xi_5 + \xi_5^4$ are both cyclotomic integers. In fact somewhat more is true: any (complex) representation of G can be realized (in

2000 *Mathematics Subject Classification.* Primary 20C15; Secondary 11R32.

This research is supported in part by NSERC.
This the final form of the paper.

TABLE 1. The character tables of S_3 and A_5

S_3	(1)	(12)	(123)
ch_1	1	1	1
ch_s	1	-1	1
ch_2	2	0	-1

A_5	(1)	(12)(34)	(123)	(12345)	(13524)
ch_1	1	1	1	1	1
ch_4	4	0	1	-1	-1
ch_5	5	1	-1	0	0
ch_3	3	-1	0	α	α'
$ch_{3'}$	3	-1	0	α'	α

many ways) by matrices with entries in $\mathbb{Q}[\xi_n]$ (this can be elegantly proved using induced representations — see, e.g., [4, Theorem 10.3]).

The Galois group $\mathcal{G}_n := \text{Gal}(\mathbb{Q}[\xi_n]/\mathbb{Q})$ is the multiplicative group $\mathbb{Z}_n^\times$ consisting of all classes mod n of integers coprime to n. The integer $l \in \mathbb{Z}_n^\times$ corresponds to the automorphism $\sigma_l \in \mathcal{G}_n$ taking ξ_n to ξ_n^l. This implies that $\sigma_l\big(p(\xi_n)\big) = p(\xi_n^l)$ for any polynomial p with rational coefficients.

The Galois group $\mathcal{G}_n \cong \mathbb{Z}_n^\times$ acts on the rows of the character table of G, by $\sigma.\,\text{ch}_\rho = \text{ch}_{\sigma\rho}$, where $\sigma\rho$ is defined by first realizing ρ as matrices over $\mathbb{Q}[\xi_n]$, and then evaluating $(\sigma\rho)(g) = \sigma\big(\rho(g)\big)$ entry-wise. That $\sigma\rho$ is a representation of G is clear because σ preserves addition and multiplication; that $\sigma\rho$ is independent of how ρ is realized over $\mathbb{Q}[\xi_n]$, is now clear by considering characters. For example, $l \equiv \pm 2 \pmod 5$ will permute the A_5-characters ch_3 and $ch_{3'}$. (Of course, $\sigma\rho$ can also be expressed in a basis-independent way, e.g., by having $\sigma \in \text{Gal}(\mathbb{C}/\mathbb{Q})$ act on $\rho \in \text{Hom}\big(G, \text{End}(\mathbb{C}^m)\big)$ by $\sigma \circ \rho \circ \sigma^{-1}$.)

The Galois group $\mathbb{Z}_n^\times$ acts on columns of character tables by $\sigma_l.g = g^l$ (l is only defined mod n, but $g^n = 1$ so this isn't a problem). That this takes conjugacy classes to conjugacy classes is clear: $(k^{-1}gk)^l = k^{-1}g^lk$.

These two actions are compatible, in the sense that σ_l applied to the number $\text{ch}_\rho(g) \in \mathbb{Q}[\xi_n]$ is given by both the row- and column-actions:

$$\sigma_l\big(\text{ch}_\rho(g)\big) = \text{ch}_{\sigma_l\rho}(g) = \text{ch}_\rho(g^l). \tag{1}$$

The first equality is clear by the definition of $\sigma\rho$: $\text{Tr}\left(\sigma\big(\rho(g)\big)\right) = \sigma\left(\text{Tr}\big(\rho(g)\big)\right)$. The second equality follows by diagonalizing $\rho(g)$ (which can be done over $\mathbb{Q}[\xi_n]$): its eigenvalues are nth roots of unity so will be raised to the lth power.

For example, the S_3-character values are all integers, and so are invariant under any Galois automorphism. This is equivalent to saying that g^l is conjugate to g, for any $g \in S_3$ and any $l \in \mathbb{Z}_6^\times$. Since g^l will likewise be conjugate to g in any symmetric group S_m, all character values for any S_m will necessarily be integers.

There are many theorems in group theory whose best (or only) known proof uses character table technology, including this Galois action. A famous example is Burnside's p^aq^b-theorem, which says that as long as three distinct primes don't divide the order of a group, that group will necessarily be solvable. This is the best possible result, in the sense that the alternating group A_5 is not solvable (much to the chagrin of quintic solvers) and only three primes divide $|A_5| = 60$.

We can formulate this Galois action more abstractly as follows. The characters are class functions, so lie in (in fact span) the space

$$\text{Map}_G(G, \mathbb{Q}[\xi_n]) \tag{2}$$

of G-equivariant maps, where G acts on itself by conjugation and fixes the scalars. The Galois group $\mathcal{G}_n$ acts on this space in two ways: it manifestly acts on the scalars, and it acts on G as above. The compatibility (1) of these two actions is clearly a basis-dependent statement; that such a basis (e.g., the irreducible characters) exists, is equivalent to the statement that the representation of $\mathcal{G}_n$ on the group algebra $\mathbb{Q}\langle g\rangle$, for any $g \in G$, is a subrepresentation of that on the field $\mathbb{Q}[\xi_n]$.

In any case, it can be asked why should the cyclotomic Galois group $\mathcal{G}_n$ act on finite group character tables. More precisely, these Galois groups act by definition on cyclotomic numbers; any other action that can't be reduced quite directly to this one can be termed "mysterious." In this sense, perhaps the action of $\mathcal{G}_n$ on G can be labeled mysterious. Can we relate it to more familiar Galois actions, perhaps as a special case of some general class? I think this is a question John McKay could ask. We'll suggest an answer in the following pages.

However, especially in this era of George Bush and Tony Blair, we've been taught that we shouldn't look too closely at motivations. Indeed, for an additional ulterior motive for this material, see [3, Chapter 6]. Certainly, the "explanation" we propose in Section 3 for this action is much more complicated than the standard proof sketched above. At the end of the day, we should judge a question by how intriguing the resulting picture is, which it helped shape. So withhold your judgment until you conclude Section 3!

2. A warm-up example

To help clarify what is wanted here, let's consider a toy model. It is a vaguely similar action, though is much simpler to "explain." What we are after is essentially its profinite version. This section can be skipped if the material is too unfamiliar; on the other hand, it may also be of independent interest.

In Moonshine and orbifold string theory, the following action of $\mathrm{SL}_2(\mathbb{Z})$ arises. Let G be any finite group. Consider all commuting pairs $(g,h) \in G \times G$. Identify simultaneous conjugates: $(g,h) \sim (k^{-1}gk, k^{-1}hk)$. For example, S_3 has 8 such equivalence classes, while A_5 has 22. On these equivalence classes of pairs, we get a (right) $\mathrm{SL}_2(\mathbb{Z})$ action by

$$(g,h)\begin{pmatrix} a & b \\ c & d \end{pmatrix} = (g^a h^c, g^b h^d). \tag{3}$$

Where does this come from? The answer is geometry. If we forget the commuting pairs and simultaneous conjugation, the $\mathrm{SL}_2(\mathbb{Z})$-action becomes a braid group $\mathcal{B}_3$-action. In particular, writing T, U for the $\mathrm{SL}_2(\mathbb{Z})$-generators $\left(\begin{smallmatrix} 1 & 1 \\ 0 & 1 \end{smallmatrix}\right)$ and $\left(\begin{smallmatrix} 1 & 0 \\ -1 & 1 \end{smallmatrix}\right)$, the actions

$$(g,h).T = (g, gh)\ , \quad (g,h).U = (gh^{-1}, h) \tag{4}$$

no longer satisfy $(TU)^6 = 1$, and so instead of $\mathrm{SL}_2(\mathbb{Z})$ they generate its extension $\langle T, U \mid TUT = UTU \rangle = \mathcal{B}_3$.

Now, one of the basic facts about the braid groups [1] is that they act on the free groups. In particular, $\mathcal{B}_3$ acts on $\mathcal{F}_3 = \langle x, y, z\rangle$ by

$$\sigma_1.x = xyx^{-1}, \quad \sigma_1.y = x, \quad \sigma_1.z = z, \quad \sigma_2.x = x, \quad \sigma_2.y = yzy^{-1}, \quad \sigma_2.z = y.$$

The easiest way to see this is through the realization of the braid group as the fundamental group of configuration space. Indeed, the pure braid group $\mathcal{P}_n$ is $\pi_1(C_n)$ where $C_n = \{(z_1, z_2, \ldots, z_n) \in \mathbb{C}^n \mid z_i \neq z_j\})$ and the free group $\mathcal{F}_n$ is

$\pi_1(\mathbb{C} - \{1, 2, \ldots, n\})$, so given the obvious bundle $C_{n+1} \to C_n$ (dropping a coordinate), there will be an action of the fundamental group $\mathcal{P}_n$ of the base on the fundamental group $\mathcal{F}_n$ of the fibre. The $\mathcal{B}_n$-action is obtained through symmetrization of C_n.

In any case, $\mathcal{B}_3$ acts on $\mathcal{F}_3$ (by group automorphisms), so it also acts on the *set* $\mathrm{Hom}(\mathcal{F}_3, G) \cong G^3$ (this latter action won't respect the $G \times G \times G$ group structure). Collapsing a triple $(g_1, g_2, g_3) \in G^3$ to the pair $(g_1 g_2^{-1}, g_2 g_3^{-1})$ recovers (4). This $\mathcal{B}_3$-action becomes a well-defined $\mathrm{SL}_2(\mathbb{Z})$-action if we either restrict to a commuting pair, or we identify simultaneous conjugates.

Similarly, we get a $\mathcal{B}_n$-action on $G^n \cong \mathrm{Hom}(\mathcal{F}_n, G)$ and hence on G^{n-1}. The role of $\mathrm{SL}_2(\mathbb{Z})$ will now be served by $\mathcal{B}_n/\langle z \rangle$ (n even) or $\mathcal{B}_n/\langle z^2 \rangle$ (n odd), where z is the generator of the center of $\mathcal{B}_n$. This 'explains' the $\mathrm{SL}_2(\mathbb{Z})$-action (3) in the sense that it embeds it in a larger context, showing that it belongs to an infinite family of related actions, hence generalizing it. We want to do the same to the Galois action on character tables — to embed it in an infinite family of other Galois actions. In this geometric interpretation of Galois on character tables, $\mathcal{B}_n$ is replaced by $\mathrm{Gal}(\overline{\mathbb{Q}}/\mathbb{Q})$ and $\mathcal{F}_n$ by its profinite completion.

Of course there is never a unique "explanation," nor need there even be a best one. For a different explanation, (3) can also be obtained from the following construction (see, e.g., [3, 11] for the background). Given a Hopf algebra, the Drinfel′d double produces a quasi-triangular Hopf algebra (i.e., a Hopf algebra co-commutative up to an isomorphism). Given such a quasi-triangular Hopf algebra, Reshetikhin – Turaev obtain from this a modular category. These are interesting because modular categories supply knot invariants on arbitrary 3-manifolds. In a modular category (or what is essentially the same thing, a 3-dimensional topological field theory or 2-dimensional rational conformal field theory), we get representations of any surface mapping class group. Now apply this construction to the group algebra $\mathbb{C}G$ (a Hopf algebra). Fix any compact genus g surface Σ_g, and consider the set of all group homomorphisms $\pi_1(\Sigma_g) \to G$, where we identify two maps if they are conjugates by G. The mapping class group $\Gamma_{g,0}$ acts naturally on $\pi_1(\Sigma_g)$ and hence on $\mathrm{Hom}(\pi_1(\Sigma_g), G)$, and in genus $g = 1$ this action reduces to that of (3) (recall that π_1 of a torus is $\mathbb{Z}^2$ and $\Gamma_{1,0}$ is $\mathrm{SL}_2(\mathbb{Z})$).

The second explanation certainly should be regarded as an interesting generalization of (3). Perhaps one advantage though of the first explanation is that it is both more intrinsically group-theoretic and more elementary, not relying on relatively heavy machinery. In any case, the interpretation we'll suggest shortly for the Galois action on character tables is related to this first explanation.

3. The geometric meaning of Galois

Seeking a geometric interpretation of a Galois action leads naturally to geometric Galois. Classical Galois theory concerns field extensions L over a base field K; geometric Galois concerns unramified coverings $f \colon Y \to X$ of a space X (see, e.g., [6] for a gentle account). The role of the algebraic closure $\overline{K}$ is played by the universal covering space $\widetilde{X}$, and the Galois group $\mathrm{Gal}(\overline{K}/L)$ by the fundamental group $\pi_1(Y)$. The action of Galois automorphisms on $\alpha \in L$ corresponds to that of deck transformations on the fibre $f^{-1}(p)$ above $p \in X$ (a *deck* or *covering transformation* is a homeomorphism $\gamma \colon Y \to Y$ satisfying $f \circ \gamma = \gamma$; we identify the deck transformation with the permutation it induces on the fibre $f^{-1}(p)$). The deck transformations form a group analogous to Gal(L/K). A *Galois covering* is one whose deck transformation group is transitive on $f^{-1}(p)$; if the degree n of the covering is finite, this is the same as saying the order of that group is also n.

In complete analogy with classical Galois theory, geometric Galois establishes a bijection between subgroups H of $\pi_1(X)$ and coverings $Y = \widetilde{X}/H$, where $H \cong \pi_1(Y)$. Y is a Galois covering iff H is normal in $\pi_1(X)$, in which case the deck transformation group is isomorphic to $\pi_1(X)/H$.

For example, consider the covering (spiral staircase) $Y_n\colon x_1^n x_2 = 1$ of the punctured complex plane $X\colon x_1x_2 = 1$. Here, $\pi_1(X) \cong \mathbb{Z}$ and $\pi_1(Y_n)$ can be naturally identified with the subgroup $n\mathbb{Z}$. The deck transformations are $(x_1, x_2) \mapsto (\xi_n^k x_1, x_2)$, and form the group $\mathbb{Z}_n \cong \mathbb{Z}/n\mathbb{Z}$.

Geometric Galois would have been understood shortly after Poincaré defined the fundamental group — it can be understood in terms of function fields. But it took Grothendieck much later to explain how to simultaneously combine geometric Galois with the Galois theory of number fields. Let X be an algebraic variety which can be defined over $\mathbb{Q}$, i.e., defined by polynomials $p_i(x_1, \ldots, x_n)$ with rational coefficients. Fix (if it exists) a rational basepoint $p \in X(\mathbb{Q})$. Choose any finite index normal subgroup N of the topological fundamental group $\pi_1(X)$. Then there will be a covering $f\colon X_N \to X$ with $\pi_1(X_N) = N$. The generalized Riemann existence theorem ensures that X_N will be an algebraic variety over the algebraic closure $\overline{\mathbb{Q}}$ (i.e., the coefficients of the defining polynomials will be algebraic numbers). In particular, the fibre $f^{-1}(p)$ will all have coordinates in $\overline{\mathbb{Q}}$.

The absolute Galois group $\Gamma_{\mathbb{Q}} = \text{Gal}(\overline{\mathbb{Q}}/\mathbb{Q})$ acts on the points in $f^{-1}(p)$ component-wise. It acts on varieties over $\overline{\mathbb{Q}}$, equivalently the finite index subgroups N, coefficientwise on the defining polynomials. Grothendieck wrote that an automorphism $\sigma \in \Gamma_{\mathbb{Q}}$ acts on the deck transformations by

$$\gamma_N \mapsto \sigma \circ \gamma_{\sigma^{-1}N} \circ \sigma^{-1}, \tag{5}$$

where as always we identify a deck transformation with the associated permutation of $f^{-1}(p)$.

This isn't quite well-defined yet because σ may mix up the deck transformations for different coverings — when $\sigma N \neq N$ we must explain how $\gamma_{\sigma^{-1}N}$ is related to γ_N. This isn't hard, because for each σ the deck transformation groups $\pi_1(X)/(\sigma N)$ are all naturally isomorphic (the isomorphism being given by the appropriate Galois automorphism). A more elegant solution to this well-definedness issue will be given shortly. But an example should help clarify the basic idea of (5). Return to the covering Y_n of the punctured plane X. Take $p = (1,1)$, so $f^{-1}(p) = \{(\xi_n^j, 1)\}$. The deck transformation γ_k, for $k \in \mathbb{Z}_n$, sends $(\xi_n^j, 1)$ to $(\xi_n^{j+k}, 1)$. Since Y_n is also defined over $\mathbb{Q}$, any $\sigma \in \Gamma_{\mathbb{Q}}$ fixes it (hence also fixes $N = n\mathbb{Z}$). Note also that the $\Gamma_{\mathbb{Q}}$-action here factors through to that of the cyclotomic group $\mathcal{G}_n \cong \mathbb{Z}_n^\times$: then (5) says that $l \in \mathbb{Z}_n^\times$ acts on the deck transformation γ_k by

$$(\xi_n^j, 1) \mapsto (\xi_n^{l(l^{-1}j+k)}, 1) = (\xi_n^{j+lk}, 1),$$

i.e., $\mathbb{Z}_n^\times$ acts on $\mathbb{Z}_n$ by multiplication.

There is a cleaner way to formulate this, which in addition makes (5) well-defined. All of these normal subgroups N can be handled simultaneously, by using the inverse-limit (algebra's way to integrate): the inverse-limit $\varprojlim$ of the deck transformation groups $\pi_1(X)/N$ is called the *profinite completion* $\widehat{\pi_1(X)}$ of $\pi_1(X)$

and the *algebraic fundamental group* of X. Grothendieck's (5) defines "componentwise" the $\Gamma_{\mathbb{Q}}$-action on $\widehat{\pi_1(X)}$.

For example, return to the coverings Y_n of the punctured plane X. The algebraic fundamental group here is $\varprojlim \mathbb{Z}_n = \widehat{\mathbb{Z}}$, the direct product over all primes p of the p-adic integers $\widehat{\mathbb{Z}}_p$. It can be realized as the collection of all sequences $\hat{k} = (k_1, k_2, \dots) \in \prod_{n=1}^{\infty} \mathbb{Z}_n$ with the property that $k_n \equiv k_m \pmod{n}$ whenever n divides m. (Any other profinite completion is constructed similarly.) Then $\sigma \in \Gamma_{\mathbb{Q}}$ acts on $\widehat{\mathbb{Z}}$ through the *cyclotomic character* χ_{cyclo} defining the projection of $\Gamma_{\mathbb{Q}}$ onto the abelianization $\Gamma_{\mathbb{Q}}/[\Gamma_{\mathbb{Q}}, \Gamma_{\mathbb{Q}}]$. By the Kronecker – Weber theorem, the abelianization is simply $\text{Gal}(\bigcup_n \mathbb{Q}[\xi_n]/\mathbb{Q}) = \varprojlim \mathcal{G}_n = \widehat{\mathbb{Z}}^{\times}$, the invertible elements in $\widehat{\mathbb{Z}}$, i.e., all sequences $\hat{l} = (l_1, l_2, \dots) \in \widehat{\mathbb{Z}}$ such that each l_n is coprime to n. The action of $\sigma \in \Gamma_{\mathbb{Q}}$ on $\widehat{\mathbb{Z}}$ is given by multiplication by $\chi_{\text{cyclo}}(\sigma)$.

Section 2 tells us how to proceed. We get an action of $\widehat{\mathbb{Z}}^{\times}$, or if you prefer $\Gamma_{\mathbb{Q}}$, on the (continuous) homomorphisms $\text{Hom}(\widehat{\mathbb{Z}}, G) \cong G$, namely $\hat{l}$ sends $g \in G$ to g^{l_n} where n is the order of g. This is precisely the Galois action on columns of the character table! The more straightforward action on rows can be recovered from this by a duality argument.

In other words, we should replace the definition (2) of the space of class functions, with

$$\text{Map}_G(\text{Hom}(\widehat{\mathbb{Z}}, G), \overline{\mathbb{Q}}). \tag{6}$$

This expression is superior to (2) in that the two Galois actions on the class functions are both manifest in (6): $\Gamma_{\mathbb{Q}}$ acts on $\widehat{\mathbb{Z}}$ much as it acts on $\overline{\mathbb{Q}}$.

But why should we limit ourselves to X being the punctured plane? Take X for instance to be $\mathbb{C}$ with n punctures. This can always be taken to be an algebraic variety over $\mathbb{Q}$ — e.g., the polynomial $x_1 x_2^2 (x_2 - 1)^2 = x_2^2 - x_2 + 1$ realizes the plane with two punctures. For n punctures, Grothendieck's action (5) becomes a homomorphism $\text{Gr}_G^n \colon \Gamma_{\mathbb{Q}} \to \text{Aut}(G)$, for any finite group G with n generators (more precisely, this action depends not only on the abstract group G but also on a choice of n generators). It can be reclothed into an action of $\Gamma_{\mathbb{Q}}$ on $\widehat{\mathcal{F}_n}$ (again by group automorphisms), and hence an action of $\Gamma_{\mathbb{Q}}$ on the *set* G^n, for any finite group G. A minor miracle is that, by Belyi's theorem (see, e.g., [9]), these are all faithful actions for $n \geq 2$ (as G varies)! This faithfulness is very surprising, considering how large the kernel is for $n = 1$.

The homomorphism Gr_G^n is a generalization to $n > 1$ of the fact that $g \mapsto g^l$ is an automorphism of cyclic G (but not of general G). This $\Gamma_{\mathbb{Q}}$-action *stabilizes*, in the sense that $\text{Gr}_G^n = \text{Gr}_G^m$ whenever G has at most m and n generators, so we can unambiguously write Gr_G.

The $\Gamma_{\mathbb{Q}}$-action on G^n can be described more explicitly as follows:

$$\sigma.(g_1, \dots, g_n) = \big(\text{Gr}_H(g_1), \dots, \text{Gr}_H(g_n)\big) \tag{7}$$

where $H = \langle g_1, \dots, g_n \rangle$. In particular, σ commutes both with simultaneous conjugation and with permutations of the components: $\sigma.(k^{-1} g_1 k, \dots, k^{-1} g_n k) = k^{-1}\big(\sigma.(g_1, \dots, g_n)\big)k$ for any $k \in G$, and $\pi\big(\sigma.(g_1, \dots, g_n)\big) = \sigma.(g_{\pi 1}, \dots, g_{\pi n})$ for any permutation $\pi \in \text{S}_n$. Therefore (6) generalizes to

$$\text{Map}_G(\text{Hom}(\widehat{\mathcal{F}_n}, G), \overline{\mathbb{Q}})/\text{S}_n \ , \tag{8}$$

which as a set is just $\mathrm{Map}_G(G^n, \overline{\mathbb{Q}})/\mathrm{S}_n$, where the symmetric group S_n acts by permuting the n-tuples. Again we have two manifest (and for $n \geq 2$, faithful!) $\Gamma_{\mathbb{Q}}$-actions.

The most obvious question is, are there any group-theoretic consequences of this action, when $n \geq 2$? Is there perhaps some generalization of character tables which realize this action, i.e., some generalization of character which can supply a natural basis for the space (8)? The obvious answer is very classical: the n-characters (essentially the coefficients of Frobenius' group determinant [5]) span the space (8) (with a little help). However, these are always $\mathbb{Q}[\xi_{|G|}]$-valued, so compatibility (1) will fail for them for $n > 1$. In this sense the n-characters are too close to the usual characters; there should be (from this Galois perspective) a better basis of (8).

There are plenty of other more recent multi-variable generalizations of character tables. For instance, note that the subspace of n-tuples $(g_1, \ldots, g_n)$ of mutually commuting elements will be mapped to itself by $\Gamma_{\mathbb{Q}}$; this suggests considering the spaces of Section 2, or the generalizations of characters associated with elliptic cohomology (see, e.g., [7]). It may also be interesting to study this Galois action specifically for the reductive groups over finite fields, where we have a more geometric alternative to the character table, namely character sheaves [8] (I thank Clifton Cunningham for this interesting suggestion).

The case of the twice-punctured plane is especially accessible, because of its connection to dessins d'enfants [9] and also to modular curves [10]: its coverings are precisely the algebraic curves defined over $\overline{\mathbb{Q}}$, and they are isomorphic to $(\mathbb{H} \cup \mathbb{Q} \cup \{\mathrm{i}\infty\})/\Gamma$ for finite index subgroups of $\mathrm{SL}_2(\mathbb{Z})$. For example, explicit genus-0 calculations can be found in [2].

For example, Galois coverings with deck transformation group isomorphic to the dihedral group D_n (of order $2n$) are given by $4x_1^n x_2 = 2x_1^n - x_1^{2n} - 1$. The fibre above $x_2 = \frac{1}{2}$ are the $2n$ points $x_1 \in \{\xi_{4n}^j \mid j = \pm 1, \pm 3, \ldots, \pm(2n-1)\}$. The deck transformations are parametrized by pairs (ϵ, k) where $k \in \mathbb{Z}_n$ and $\epsilon = \pm 1$; it acts on the fibre by $(\epsilon, k).\xi_{4n}^j = \xi_{4n}^{\epsilon j + 4k}$. These coverings are also defined over $\mathbb{Q}$, and the Grothendieck action (5) again factors through the cyclotomic character: $\mathrm{Gr}_{\mathrm{D}_n}(\sigma)$ sends (ϵ, k) to $(\epsilon, \chi_{\mathrm{cyclo}}(\sigma).k)$ where $\hat{l}.k = l_n k$. Note that this $\widehat{\mathbb{Z}}^\times$-action differs from the $\widehat{\mathbb{Z}}$-action $g \mapsto g^{\hat{l}}$ coming from the once-punctured plane: for the latter, $(-1, k)$ would be fixed by any $\hat{l}$, since it has order 2. Nevertheless the dihedral groups are close enough to the cyclic ones that the cyclotomic character arises again. However as mentioned earlier, for any $\sigma \in \Gamma_{\mathbb{Q}}$, there will be a finite group G (with two generators) such that σ acts nontrivially on G.

This short paper has tried to identify a McKay-esque question: where does the $g \mapsto g^l$ symmetry of character tables really come from? We sketched a geometric source for it, relating it to more familiar Galois actions, thus fitting the $g \mapsto g^l$ action into the first window of an infinite tower. We propose that these form a natural generalization of $g \mapsto g^l$. The big question now is, do these other Galois actions on finite groups play any role in group theory? This question, and several others begged by this interpretation, will be addressed in future work.

References

1. J. S. Birman, *Braids, links, and mapping class groups*, Ann of Math. Stud., vol. 82, Princeton Univ. Press, Princeton, NJ, 1974.

2. J.-M. Couveignes and L. Granboulan, *Dessins from a geometric point of view*, The Grothendieck Theory of Dessins d'Enfant (Luminy, 1993) (L. Schneps, ed.), London Math. Soc. Lecture Note Ser., vol. 200, Cambridge Univ. Press, Cambridge, 1994, pp. 79 – 113.
3. T. Gannon, *Moonshine beyond the Monster*, Cambridge Monogr. Math. Phys., Cambridge Univ. Press, Cambridge, 2006.
4. I. M. Isaacs, *Character theory of finite groups*, Dover Publ., New York, 1994.
5. K. W. Johnson, *The Dedekind – Frobenius group determinant: new life in an old problem*, Groups St. Andrews 1997 in Bath. II (Bath, 1997) (C. M. Campbell, E. F. Robertson, N. Ruskuc, and G. C. Smith, eds.), London Math. Soc. Lecture Note Ser., vol. 261, Cambridge Univ. Press, Cambridge, 1999, pp. 417 – 428.
6. M. Kuga, *Galois' dream: group theory and differential equations*, Birkhäuser, Boston, MA, 1993.
7. N. J. Kuhn, *Character rings in algebraic topology*, Advances in Homotopy Theory (Cortona, 1988) (S. M. Salamon, B. Steer, and W. A. Sutherland, eds.), London Math. Soc. Lecture Note Ser., vol. 139, Cambridge Univ. Press, Cambridge, 1989, pp. 111 – 126.
8. G. Lusztig, *Character sheaves and generalizations*, The Unity of Mathematics (Cambridge, MA, 2003) (P. Etingof, V. Retakh, and I. M. Singer, eds.), Progr. Math., vol. 244, Birkhäuser, Boston, MA, 2006, pp. 443 – 455.
9. L. Schneps, *Dessins d'enfants on the Riemann sphere*, The Grothendieck Theory of Dessins d'Enfant (Luminy, 1993) (L. Schneps, ed.), London Math. Soc. Lecture Note Ser., vol. 200, Cambridge Univ. Press, Cambridge, 1994, pp. 47 – 77.
10. A. J. Scholl, *Modular forms on noncongruence subgroups*, Séminaire de Théorie des Nombres, Paris 1985 – 86 (Paris, 1985 – 86) (C. Goldstein, ed.), Progr. Math., vol. 71, Birkhäuser, Boston, MA, 1987, pp. 199 – 206.
11. V. G. Turaev, *Quantum invariants of knots and 3-manifolds*, de Gruyter Stud. Math., vol. 18, de Gruyter, Berlin, 1994.

Department of Mathematical Sciences, University of Alberta, Edmonton, AB T6G 2G1, Canada
E-mail address: tgannon@math.ualberta.ca

Centre de Recherches Mathématiques
CRM Proceedings and Lecture Notes
Volume **47**, 2009

Hecke Operators in Equivariant Elliptic Cohomology and Generalized Moonshine

Nora Ganter

Dedicated to John McKay

1. Introduction

1.1. Background on generalized Moonshine. We recall Norton's generalized Moonshine conjecture. Let g, h be a pair of commuting elements in the Fischer – Griess Monster group M. We denote the centralizer of g in M by $C_{\mathrm{M}}(g)$.

Conjecture 1.1 (Norton [28]). *It is possible to associate with each such pair a modular function $f(g,h;\tau)$ with the following properties:*

(a) *Up to a constant factor (which will be a root of unity), there is an equality*

$$f(g^a h^c, g^b h^d; \tau) = f\left(g, h, \frac{a\tau + b}{c\tau + d}\right)$$

whenever $ad - bc = 1$,

(b) *For any group element g and nonzero rational number l the coefficient of $q^l = \mathrm{e}^{2\pi i l \tau}$ in $f(g,h;\tau)$ is, as a function of h, a character of a central extension*[1] *of $C_{\mathrm{M}}(g)$. Note that nonzero characters can occur for nonintegral l, but that generalized characters are not needed.*

(c) *Conjugation of g and h leaves the function unchanged.*

(d) *Unless $f(g,h;\tau)$ is a constant function, its invariance group will be a modular group of genus zero, commensurable with the standard modular group* $\mathrm{SL}_2(\mathbb{Z})$.

Norton further explains: "It is the simultaneous action of $\mathrm{SL}_2(\mathbb{Z})$ on $\langle g, h\rangle$ and τ that seems to suggest an explanation of the Moonshine phenomenon; but this only gives a clue to the actions of the elements of $\mathrm{SL}_2(\mathbb{Z})$ on τ, so the reason why the full invariance group should have genus zero remains obscure."

2000 *Mathematics Subject Classification.* Primary 55N34; Secondary 11F32, 20D08, 58J28.

The author was supported by NSF-grant DMS-0504539. During her work on this paper, the author enjoyed the hospitality of Concordia University and of the Mathematical Sciences Research Institute.

This is the final form of the paper.

[1]For a discussion of how characters of central extensions of a group correspond to twisted characters of this group, see for example [36].

Such simultaneous actions are well known to topologists: they arise in the context of equivariant elliptic cohomology. There are more indications of a connection between Moonshine and elliptic cohomology:

Generalized Moonshine is supposed to specialize to "classical" Moonshine for pairs of the form $(1, g)$. This part is often referred to as the "untwisted sector". In this situation, part (b) of Norton's conjecture involves only integral powers of q. Moreover, no central extension is needed, i.e., the coefficients are honest characters of $M = C_{\mathrm{M}}(1)$.

In this context, the Moonshine functions are usually denoted

$$j_{\langle g\rangle}(\tau) = f(1, g; \tau) = q^{-1} + a_1 q + a_2 q^2 + \cdots ,$$

since they are generalizations of the j-function

$$j_{\langle 1\rangle} = j - 744 = q^{-1} + 196884q + 21493760q^2 + \cdots$$

and depend only on the cyclic subgroup generated by g.

We recall from [28] the notions of *replicable function* and *twisted Hecke operator.* Let f be a Laurent series in q of the form

$$f(q) = q^{-1} + a_1 q + a_2 q^2 + \cdots . \tag{1}$$

Then the nth *Faber polynomial* Φ_n of f is the unique polynomial of degree n for which the Laurent series $\Phi_n(f)$ has the form

$$\Phi_n(f) = q^{-n} + b_1 q + b_2 q^2 + \cdots .$$

The series f is called *replicable*, if there exist *replicates* $f^{(a)}$ of f, which satisfy the equalities

$$\Phi_n(f) = \sum_{\substack{ad=n\\ 0\le b<d}} f^{(a)}\left(\frac{a\tau + b}{d}\right) \tag{2}$$

for all $n \in \mathbb{N}$. One shows by induction that the $f^{(a)}$ are uniquely determined if they exist: assume we know that $f^{(a)}$ is defined for $a < n$. By solving the equality (2) for $f^{(n)}(n\tau)$, we arrive at the question whether a series in q, which we already defined, is indeed a series in q^n.

A replicable series f is called *completely replicable* if all the replicates of f are again replicable and their replicates are given by

$$\left(f^{(a)}\right)^{(b)} = f^{(ab)}.$$

Complete replicability was proved to be equivalent to the genus zero property (part (d) of the conjecture) in [9, 27].

If f is replicable, then the nth *twisted Hecke operator* acting on f is defined as

$$\widehat{T}_n(f) := \frac{1}{n}\Phi_n(f). \tag{3}$$

The classical Moonshine functions are known to be completely replicable, with the replicates given by the Adams operations

$$j^{(a)}_{\langle h\rangle}(q) = j_{\langle h^a\rangle}(q).$$

The collection of all classical Moonshine functions is a *McKay – Thompson series*, i.e., it is a series of the form (1) with coefficients in the representation ring $R(M)$. A function f satisfying condition (b) of Norton's conjecture is often referred to as a generalized McKay-Thompson series. The following definition of replicability for

McKay – Thompson series can be found in [28]. It is stricter than asking for the individual functions $j_{\langle g\rangle}$ to be replicable but seems to be the correct notion in the context of Moonshine.

Definition 1.2. A McKay – Thompson series f is called replicable, if the formula (2) holds when $f^{(a)}$ is the ath Adams operation applied to f.

We will rephrase Definition 1.2 in terms of power operations in elliptic cohomology in Section 6.4 below.

1.2. Background on equivariant elliptic cohomology and statement of results. Adams operations were originally defined in the context of (equivariant) K-theory. Hecke operators in (equivariant) elliptic cohomology are a generalization[2] of the Adams operations [1], and it seems a natural question to ask for a "topological" explanation for the relationship between Hecke and Adams operations occurring in Moonshine. We will start by explaining the geometry behind the generalized Moonshine conjecture and its relationship to equivariant elliptic cohomology. This discussion is not specific to the Monster and can be formulated for any finite group G. We will translate parts (a) and (c) of Norton's conjecture to
"*f is an element of the twisted equivariant elliptic cohomology of the one point space, in the sense of Ginzburg, Kapranov and Vasserot*: $f \in \mathrm{Ell}_G^{\alpha}(\mathrm{pt})$."

Here α is a cocycle representing an element of

$$\mathrm{H}^3(BG;\mathbb{R}/\mathbb{Z}) = \mathrm{H}^4(BG,\mathbb{Z}).$$

In other words,
"*f is a section of the line bundle $\mathcal{L}^{\alpha}$ over the moduli stack of principal G-bundles over (complex) elliptic curves.*"

Here $\mathcal{L}^{\alpha}$ is a line bundle associated to α (cf. Section 2.2).

In this context, Ginzburg, Kapranov and Vasserot have defined the action of correspondences. In particular, one can consider the action of the Hecke correspondence. In Section 6.1, I will spell out this geometric definition of the Hecke operators: they act on such sections via pull-backs along isogenies.

We return to Norton's conjecture. One easily translates part (b) to
"*f is an element of the (twisted) Devoto equivariant Tate K-theory of the one point space.*"

Devoto's equivariant Tate K-theory

$$K_{\mathrm{Dev},g}(X) = \bigoplus_{[g]} K_{C_g}(X^g)[\![q^{1/|g|}]\!]$$

was introduced to study characteristic classes of bundles over orbifold loop spaces. For our statement to be true as stated, we will need to invert q. *Twisted* here means that we will have to replace the centralizers with the appropriate central extensions and $q^{1/|g|}$ with $q^{1/(h|g|)}$, where h is the order of $\alpha|_{B\langle g\rangle}$. The equivariant Witten genus takes values here. In [19], I defined power operations P_n in an appropriate subring of Devoto equivariant Tate K-theory. If a function $f(g,h;\tau)$ satisfies condition (a) of Norton's conjecture, this implies that f is in this subring. My definitions in [19] were guided by the work of Dijkgraaf, Moore, Verlinde and Verlinde on orbifold genera of symmetric powers. The results of [19], together with [11],

[2]More precisely, they are the chromatic level 2 analogue of the Adams operations (cf. also [18]).

make $K_{\mathrm{Dev},G}$ an equivariant theory with power operations and a Hopkins–Kuhn–Ravenel character theory. In this situation, we have the following combinatorial definition of Hecke operators (cf. [1, 18]):

$$T_n(f) = \frac{1}{n} \sum_{\substack{[\sigma,\rho] \\ \text{transitive}}} P_n(f)(\sigma, \rho),$$

where the sum runs over all conjugacy classes of pairs of commuting elements (σ, ρ) of the symmetric group Σ_n with the property that $\langle \sigma, \rho \rangle$ acts transitively on $\{1, \ldots, n\}$.

The main result of this paper can be summarized as follows:

Theorem 1.3. *The following three definitions of Hecke operators result in exactly the same formula for functions $f(g, h; \tau)$:*

(1) *The geometric definition using isogenies, when f is an element of (twisted) Ginzburg–Kapranov–Vasserot equivariant elliptic cohomology;*

(2) *The combinatorial definition, using the equivariant power operations of* [19], *when f is an element of Devoto's equivariant Tate K-theory on which P_n is defined;*

(3) *for $g = 1$, the definition of twisted Hecke operators acting on McKay–Thompson series.*

The formula is

$$T_n(f)(g, h; \tau) = \sum_{\substack{ad=n \\ 0 \le b < d}} f\left(g^d, g^{-b}h^a; \frac{a\tau + b}{d}\right).$$

Similarities between equivariant elliptic cohomology and generalized Moonshine have been studied in the past by Baker, Devoto, and Thomas (cf., e.g., [5, 11]). Baker's work includes a definition of twisted Hecke operators in elliptic cohomology, using Adams operations. The theory of Hecke operators in elliptic cohomology was developed further in Ando's work, which brought the theory of isogenies of formal groups into the picture [2].

Acknowledgements. It is a pleasure to dedicate this paper to John McKay, who introduced me to the subject of generalized Moonshine. Many thanks go to Jorge Devoto, from whom I learned a lot about the connections between Moonshine and elliptic cohomology, to Jack Morava and Dan Freed for their very helpful email correspondence, to Alex Ghitza for plenty of help and comments on earlier versions of the paper, to Michael Hopkins, Matthew Ando, and Christian Haesemeyer for long and helful conversations about the subject.

2. Principal bundles over complex elliptic curves

Let G be a finite group, let $\mathcal{C}(G)$ denote the set of all pairs of commuting elements in G, and consider the right action of G on $\mathcal{C}(G)$ given by simultaneous conjugation

$$(g, h) \mapsto (s^{-1}gs, s^{-1}hs).$$

We define the *generalized centralizer* group $C_G(g, h)$ of the pair (g, h) by

$$C_G(g, h) = \mathrm{Stab}_G(g, h).$$

The *generalized conjugacy class* $[g,h]_G$ is defined as the orbit of (g,h) under G. The set of all conjugacy classes of pairs of commuting elements in G is denoted $\overline{\mathcal{C}}(G)$.

Let E be an elliptic curve over $\mathbb{C}$. Assume that we have picked an isomorphism

$$E \cong \mathbb{C}/\langle \tau, 1\rangle. \tag{4}$$

Throughout the paper, we will identify $\langle \tau, 1\rangle$ with $\mathbb{Z}^2$ via the isomorphism

$$\begin{aligned} \mathbb{Z}^2 &\to \langle \tau, 1\rangle \\ (1,0) &\mapsto -1 \\ (0,1) &\mapsto \tau. \end{aligned} \tag{5}$$

Further, we will view $\mathbb{C}$ with the action of the additive subgroup $\langle \tau, 1\rangle \cong \mathbb{Z}^2$ as a contractible free $\mathbb{Z}^2$-space $E\mathbb{Z}^2$, and E as a classifying space

$$E = B\mathbb{Z}^2 = E\mathbb{Z}^2/\mathbb{Z}^2.$$

Lemma 2.1. *The set of isomorphism classes of principal G-bundles over E is*

$$\mathrm{Prin}_G(E) \cong [B\mathbb{Z}^2, BG] \cong \mathrm{Hom}(\mathbb{Z}^2, G)/\,\mathrm{Inn}(G) = \overline{\mathcal{C}}(G),$$

the set of all conjugacy classes of pairs of commuting elements of G.

Such a pair of commuting elements may also be viewed as encoding the monodromy of the corresponding principal bundle along the two standard circles of the torus.

Under the isomorphism (5), the map

$$\begin{aligned} \varphi\colon \langle \tau', 1\rangle &\to \langle \tau, 1\rangle \\ \tau' &\mapsto a\tau + b \\ 1 &\mapsto c\tau + d \end{aligned}$$

corresponds to the endomorphism of $\mathbb{Z}^2$ given by the matrix

$$\varphi = \begin{pmatrix} d & -b \\ -c & a \end{pmatrix}, \tag{6}$$

and

$$\begin{aligned} \varphi^\vee\colon \langle \tau, 1\rangle &\to \langle \tau', 1\rangle \\ \tau &\mapsto d\tau' - b \\ 1 &\mapsto -c\tau' + a \end{aligned}$$

becomes the pseudo-inverse of this matrix,

$$\varphi^\vee = \begin{pmatrix} a & b \\ c & d \end{pmatrix}.$$

The map $\phi := B\varphi$ is an isogeny

$$\phi\colon \mathbb{C}/\langle \tau', 1\rangle \to \mathbb{C}/\langle \tau, 1\rangle$$

with dual isogeny $\phi^\vee = B(\varphi^\vee)$. Any isogeny of complex elliptic curves can be obtained in this way. If $ad-bc = 1$, then ϕ becomes an isomorphism with $\phi^{-1} = \phi^\vee$. It is the isomorphism obtained when comparing two different choices for (4).

Proposition 2.2. *The isogeny $\phi^\vee$ pulls back the principal bundle over $\mathbb{C}/\langle \tau', 1\rangle$ which is classified by the pair (g,h) to the bundle over $\mathbb{C}/\langle \tau, 1\rangle$ which is classified by $(g^a h^c, g^b h^d)$.*

PROOF. After identifying both $\langle\tau,1\rangle$ and $\langle\tau',1\rangle$ with $\mathbb{Z}^2$ as in (5), the map $\varphi^\vee$ sends $(1,0)$ to (a,c) and $(0,1)$ to (b,d). Recall that (g,h) stands for the map from $\mathbb{Z}^2$ to G sending $(1,0)$ to g and $(0,1)$ to h. If we precompose this map with $\varphi^\vee$, we obtain

$$(g,h)\circ\varphi^\vee = (g^a h^c, g^b h^d). \qquad \square$$

2.1. The moduli space of principal G-bundles. Let

$$(s,\gamma)\in G\times \mathrm{SL}_2(\mathbb{Z})$$

act on

$$\mathcal{C}(G)\times\mathfrak{H}$$

from the right by

$$(g,h;\tau)\mapsto \left(s^{-1}(g^a h^c, g^b h^d)s; \gamma^{-1}(\tau)\right),$$

where

$$\gamma = \begin{pmatrix} a & b\\ c & d\end{pmatrix} \quad\text{and}\quad \gamma^{-1}(\tau) = \frac{d\tau - b}{-c\tau + a}.$$

Then the quotient

$$(\overline{\mathcal{C}}(G)\times\mathfrak{H})/\,\mathrm{SL}_2(\mathbb{Z})$$

is the coarse moduli space X_G of isomorphism classes of principal G-bundles over complex elliptic curves. Apart from the roots of unity, conditions (a) and (c) of Norton's conjecture are precisely the coherence data satisfied by holomorphic functions on X_G (compare also [11]).

In order to understand the roots of unity, we need to consider the moduli stack

$$\mathcal{M}_G := (\mathcal{C}(G)\times\mathfrak{H})/\!\!/(G\times\mathrm{SL}_2(\mathbb{Z})).$$

Write

$$S := \begin{pmatrix} 0 & 1\\ -1 & 0\end{pmatrix} \quad\text{and}\quad T := \begin{pmatrix} 1 & 1\\ 0 & 1\end{pmatrix}.$$

The groupoid $\mathcal{C}(G)/\!\!/G$ is equivalent to the category of principal G-bundles over the torus and isomorphisms between them. To show this, consider the pointed circle $(\mathbb{S}^1,*)$ and the torus

$$(\mathbb{S}^1,*)\times(\mathbb{S}^1,*).$$

Let (P,p) be a principal G-bundle over this torus together with a choice of basepoint over $(*,*)$. Let (g,h) be the commuting pair given by the monodromy of P around the two circles, starting at p. Note that for any two pointed principal bundles (P,p) and (P',p') with the same monodromy, there is a unique basepoint-preserving isomorphism from P to P'.

For any commuting pair (g,h) in G, fix once and for all a pointed principal G-bundle $(P_{g,h},p)$ over the torus with this monodromy. A (not necessarily basepoint-preserving) morphism from (P,p) to (P',p') is a pair (f,s), where $s\in G$, and $f\colon (P,ps)\to P'$ is a basepoint-preserving isomorphism. For any $s\in G$, the pointed bundle $(P_{g,h},ps)$ has monodromy $(s^{-1}gs, s^{-1}hs)$, and there is a unique basepoint-preserving isomorphism

$$f\colon (P_{g,h},ps)\to(P_{s^{-1}gs,s^{-1}hs},p').$$

Note that the automorphism group of (g,h) in $\mathcal{C}(G)/\!\!/G$ is given by the generalized centralizer group $C_G(g,h)$. The automorphisms of $(g,h;\tau)$ in $\mathcal{M}_G$ might be more complicated: The automorphisms of $P_{g,h}$ over E are parametrized by $C_G(g,h)$, but now there might also be automorphisms of E which pull back $P_{g,h}$

to a bundle isomorphic to $P_{g,h}$. For instance, $(1, ST)$ is an automorphism of $(g, g^{-1}; (\sqrt{3}\mathrm{i}-1)/2)$ whenever g has order three, and (s, S) is an automorphism of $(g, h; \mathrm{i})$ whenever $s(g,h)s^{-1} = (h^{-1}, g)$.

2.2. Line bundles over $\mathcal{M}_G$. The purpose of this section is to construct a homomorphism from $\mathrm{H}^4(BG,\mathbb{Z})$ to the Picard group of line bundles over the moduli stack of principal G-bundles.

2.2.1. *The cocycle α.* Fix, once and for all, a cellular decomposition of BG. Let

$$i_{(2)} \colon BG^{(2)} \to BG$$

denote the inclusion of the 2-skeleton in BG.

Definition 2.3 (compare [36]). A cocycle $\alpha \in \mathrm{Z}^3(BG;\mathbb{R}/\mathbb{Z})$ is called *normalized*, if

$$i_{(2)}^*(\alpha) = 0.$$

Lemma 2.4. *Any cocycle $\alpha \in \mathrm{Z}^3(BG;\mathbb{R}/\mathbb{Z})$ is cohomologous to a normalized one.*

PROOF. The statement follows from the commuting diagram

$$\begin{array}{ccc} \mathrm{C}^2(BG;\mathbb{R}/\mathbb{Z}) & \xrightarrow{i_{(2)}^*} & \mathrm{C}^2(BG^{(2)};\mathbb{R}/\mathbb{Z}) \\ \downarrow \delta & & \downarrow \delta \\ \mathrm{Z}^3(BG;\mathbb{R}/\mathbb{Z}) & \xrightarrow{i_{(2)}^*} & \mathrm{Z}^3(BG^{(2)};\mathbb{R}/\mathbb{Z}). \end{array}$$

Here C^* are the groups of cochains and Z^* the groups of cocycles. The upper horizontal arrow is surjective, since $\mathbb{R}/\mathbb{Z}$ is an injective abelian group. □

We fix a normalized cocycle $\alpha \in \mathrm{Z}^3(BG;\mathbb{R}/\mathbb{Z})$. Recall that we have

$$\mathrm{H}^4(BG;\mathbb{Z}) \cong \mathrm{H}^3(BG;\mathbb{R}/\mathbb{Z}).$$

In the case that G is the Monster, a candidate for $[\alpha] \in \mathrm{H}^4(BG;\mathbb{Z})$ was described in [29].

2.3. Construction and properties of the Freed – Quinn line bundle $\mathcal{L}^\alpha$. Inspired by the work of Dijkgraaf, Vafa, Verlinde and Verlinde [12], Freed and Quinn use α to define a (possibly degenerate) line bundle $L^\alpha = \{L^\alpha_{g,h}\}$ on $\overline{\mathcal{C}}(G)$. They proceed to describe the $\mathrm{SL}_2(\mathbb{Z})$-action on this line bundle. Their discussion can be interpreted as constructing an $\mathrm{SL}_2(\mathbb{Z})$-equivariant line bundle

$$\begin{array}{c} \mathfrak{H} \times L^\alpha \\ \downarrow \\ \mathfrak{H} \times \overline{\mathcal{C}}(G), \end{array}$$

and thus a line bundle $\mathcal{L}^\alpha$ on the moduli stack $\mathcal{M}_G$. We briefly recall their construction, which simplifies when α is normalized.

Construction 2.5. Let P be a principal G bundle over $\mathbb{T}^2$, and consider the groupoid $\mathcal{C}_P$ whose objects are cellular classifying maps $f\colon P \to EG$ for P, and whose morphisms from f to f' are homotopy classes rel boundary of cellular G-homotopies

$$f\colon [0,1] \times P \to EG$$

from f to f'. We define a functor $\mathcal{F}$ from $\mathcal{C}_P$ to the category of metrized complex lines as follows: For every f, one sets[3] $\mathcal{F}(f) = \mathbb{C}$. Further $\mathcal{F}(f \xrightarrow{h} f')$ is multiplication with

$$\mathrm{e}^{2\pi \mathrm{i}(\widehat{h}^*\alpha)(x)},$$

where x is a 3-cycle representing the fundamental class in $[0,1] \times \mathbb{T}^2$, and $\bar{h} = h/G$. It can be shown that this is independent of the choice of representatives h and x and that $\mathcal{F}$ has no holonomy. The line L_P^α is then defined as the space of invariant sections of $\mathcal{F}$.

Let ψ be an automorphism of P (not necessarily covering the identity on $\mathbb{T}^2$). Then the action of ψ on L_P^α is given by multipication with the *Chern – Simons* invariant of the glued mapping cylinder

$$\mathrm{cyl}_\psi(P) := ([0,1] \times P)/\sim, \quad \text{where } \big(0, \psi(p)\big) \sim (1, p).$$

This invariant is computed by choosing a cellular classifying map $h\colon \mathrm{cyl}_\psi(P) \to EG$ and a 3-cycle x representing the fundamental class of $\mathrm{cyl}_{\bar{\psi}}(\mathbb{T}^2)$ and then evaluating $\bar{h}^*(\alpha)$ at x. More generally, let $\gamma \in \mathrm{SL}_2(\mathbb{Z})$ be an automorphism of $\mathbb{T}^2$. Then the induced isomorphism γ_* of metrized integration lines is defined to be the identity

$$\mathrm{id}_{\mathbb{C}}\colon \mathcal{F}(f) = \mathbb{C} \to \mathbb{C} = \mathcal{F}(f \circ \gamma). \tag{7}$$

Note that $\mathbb{T}^2$ can be replaced with any Riemann surface. We will often drop α from the notation and write L_P for L_P^α. If P is a principal bundle over $\mathbb{T}^2$ classified by (g, h), we will sometimes write $L_{g,h}$ for L_P.

Proposition 2.6. *Let $\phi\colon E' \to E$ be an isogeny of degree n, and let P be a principal G-bundle over E. Then there is an isometry of metrized lines,*

$$\phi_*\colon L_{\phi^*(P)} \to L_P^{\otimes n}.$$

The construction of ϕ_* is natural in E.

PROOF. Write $\mathbb{I}$ for $[0,1]$. Let A be the kernel of ϕ, and let $C_3^A(\mathbb{I} \times E')$ denote the group of A-invariant singular 3-chains in $\mathbb{I} \times E'$. Since A acts fixed point free on $\mathbb{I} \times E'$, every element $x \in C_3^A(\mathbb{I} \times E')$ is of the form

$$x = \sum \lambda_i \Big(\sum_{a \in A} a\Big) s_i,$$

where the λ_i are integers and the s_i are singular simplices in $\mathbb{I} \times E'$. Hence the map $\mathrm{id}_{\mathbb{I}} \times \phi_*$ takes values in $nC_3(\mathbb{I} \times E)$. If $x \in C_3^A(\mathbb{I} \times E')$ represents the relative fundamental class of $\mathbb{I} \times E'$, then $(1/n)\phi_*(x)$ represents the relative fundamental class of $\mathbb{I} \times E$. Let $\mathcal{C}_P$ and $\mathcal{C}_{\phi^*P}$ be the groupoids in Construction 2.5. Precomposition with ϕ gives a map of groupoids

$$\iota\colon \mathcal{C}_P \to \mathcal{C}_{\phi^*P}.$$

[3]The metrized integration line $I_{\mathbb{T}^2, \bar{f}^*\alpha}$ of [15] is canonically trivialized if α is normalized and f is cellular.

Let $\mathcal{F}_{P,\alpha}$ and $\mathcal{F}_{\phi^*P,\alpha}$ be the functors in Construction 2.5. The discussion above gives rise to a natural isomorphism

$$\mathcal{F}_{P',\alpha} \circ \iota \cong \left(\mathcal{F}_{P,\alpha}\right)^{\otimes n}.$$

It remains to observe that the invariant sections of $\mathcal{F}_{P',\alpha}$ and $\mathcal{F}_{P',\alpha} \circ \iota$ are canonically isomorphic. □

For later reference, we note that the definition of the isomorphism in Proposition 2.6 used only the structure of ϕ as a principal A-bundle. Note also that the action of A on L_{ϕ^*P} is trivial.

Lemma 2.7. *Let $\alpha = 0$. Then the line bundle L^α over $\overline{\mathcal{C}}(G)$ is canonically trivial, and the* $\mathrm{SL}_2(\mathbb{Z})$*-action preserves this trivialization.*

PROOF. Let $\alpha = 0$, then the line of the previous construction is canonically trivialized. Explicitly, such a trivialization is given by picking an arbitrary object $f \in \mathrm{ob}(\mathcal{C}_P)$ and evaluating invariant sections of $\mathcal{F}$ at f. Since our choice of trivialization was independent of the choice of f, it follows that

$$\gamma_* = \mathrm{id}_{\mathbb{C}} \colon L_P^\alpha \to L_{\gamma^*P}^\alpha.$$ □

Lemma 2.8. *Let $a\colon H \to G$ be a map of finite groups. Then a induces a map of groupoids*

$$\mathcal{C}(a)\colon \mathcal{C}(H)/\!\!/H \to \mathcal{C}(G)/\!\!/G,$$

which pulls back L^α to $L^{a^*(\alpha)}$ as an $\mathrm{SL}_2(\mathbb{Z})$*-equivariant bundle.*

PROOF. Fix a cellular classifying map

$$F\colon EH[G] \to EG,$$

and write Ba for the induced map $\overline{F}$ of classifying spaces. Let P be a principal H-bundle over $\mathbb{T}^2$, let $\mathcal{C}_P$ be the groupoid in [15, p. 6], and similarly, $C_{P[G]}$. We have an equivalence of groupoids

$$\begin{aligned} \iota\colon \mathcal{C}_P &\to \mathcal{C}_{P[G]} \\ f &\mapsto F \circ f[G] \\ h &\mapsto F \circ h[G]. \end{aligned}$$

Here $f\colon P \to EH$ ranges over the classifying maps of P, and h ranges over the homotopies between such classifying maps. Let $\mathcal{F}_{P,Ba^*(\alpha)}$ be the functor defined in [15, p. 6], and note that

$$\mathcal{F}_{P,Ba^*(\alpha)} = \mathcal{F}_{P[G],\alpha} \circ \iota.$$

It follows that the corresponding metrized lines of invariant sections of the two functors are the same:

$$L_P^{Ba^*(\alpha)} = L_{P[G]}^\alpha.$$

Let now F' denote a different choice for F, and let ι' be the corresponding equivalence of groupoids. Then there exists a cellular homotopy between F and F' and hence a natural transformation between ι and ι', yielding a canonical isomorphism between the spaces of invariant sections of $\iota^*\mathcal{F}$ and $(\iota')^*\mathcal{F}$. The construction of the proof is compatible with the functoriality condition in the following sense: Let $\psi\colon P \to P'$ be a map of principal H-bundles covering an orientation-preserving diffeomorphism of the base (for the proof of this lemma it is enough if this isomorphism is the identity of the torus). Then the induced isometries ψ_* and $\psi[G]_*$ of the above lines agree. □

Corollary 2.9. *Let G be a finite group, and let α be a normalized 3-cocycle on BG with values in $\mathbb{R}/\mathbb{Z}$. Then the line $L^{\alpha}_{1,1}$ is canonically trivialized and carries a trivial $\mathrm{SL}_2(\mathbb{Z})$-action.*

PROOF. By the previous lemma, it is enough to consider the case where $G = 1$ is the trivial group. Let $\mathrm{id}_{\mathbb{T}^2}: P \to \mathbb{T}^2$ denote the trivial principal bundle over $\mathbb{T}^2$. Since trivial bundles can be classified by constant maps, we have an equivalence of groupoids $\iota\colon \mathcal{C}_P^{(2)} \to \mathcal{C}_P$, where $\mathcal{C}_P^{(2)}$ is the subgroupoid whose objects are classifying maps $f\colon \mathbb{T}^2 \to B1^{(2)}$ and whose morphisms are homotopy classes relative boundary of homotopies $h\colon [0,1] \times \mathbb{T}^2 \to B1^{(2)}$. Since α is normalized, it follows that $\iota^*(L^{\alpha}_P)$, and hence L^{α}_P is canonically trivialized. □

Corollary 2.10. *Let $P_{g,h}$ be a principal bundle over the torus which is classified by $[g,h]_G$. Let $n = |g| \cdot |h|$. Then $(L^{\alpha}_{[g,h]})^{\otimes n}$ is canonically trivial.*

PROOF. The trivialization is given by Proposition 2.6 applied to the n-fold covering which wraps the first circle around itself $|g|$ times and the second one around itself $|h|$ times. This covering pulls back $P_{[g,h]}$ to a trivial bundle. □

Lemma 2.11. *Let p be a map from the torus to the circle, and let P be the pull-back along p of a principal G-bundle Q over the pointed circle. Then p induces a trivialization of L^{α}_P.*

PROOF. Let f be a classifying map for Q. Then $f \circ p$ is a classifying map for P and hence yields a trivialization of L^{α}_P. If f' is a different classifying map for Q, then there is a cellular homotopy $h\colon \mathbb{I} \times \mathbb{S}^1 \to EG$, and $\bar{h} \circ (\mathrm{id}_{\mathbb{I}} \times p)^*(\alpha) = 0$. Hence our trivialization is independent of our choice of f. □

For every $g \in G$, we fix, once and for all, a principal G-bundle P_g over the pointed circle with monodromy g. We will follow the convention to trivialize lines of the form $L^{\alpha}_{g,1}$ or $L^{\alpha}_{1,g}$ using the projection to the first (respectively second) generating circle of the torus.

Let d be a natural number, and let f_g be a classifying map for P_g. Precomposing f_g with the degree d self-map of the circle gives a classifying map for P_{g^d}. This yields isomorphisms of trivialized lines

$$i_d\colon L_{1,g^d} \cong \mathbb{C} = \mathbb{C} \cong L_{1,g},$$

which are compatible in the sense that $i_{ad} = i_d \circ i_a$ and $i_{k|g|+1} = \mathrm{id}_{L_{1,g}}$. We will occasionally use this isomorphism to identify these lines and write $L_{\langle g\rangle} := L_{1,g}$. Here $\langle g\rangle$ stands for the cyclic subgroup generated by g. Let

$$\gamma = \begin{pmatrix} a & b \\ nc & d \end{pmatrix} \in \Gamma_0(n).$$

Then $ad \equiv 1 \bmod n$, and $\langle g^d\rangle = \langle g\rangle$. Hence

$$\gamma_*^{-1}\colon L_{1,g} \to L_{1,g^d}$$

is an automorphism of $L_{\langle g\rangle}$. We caution the reader that this action of $\Gamma_0(n)$ on $L_{\langle g\rangle}$ is not independent of the choice of generator g. Instead, replacing g by g^l amounts to pulling back $\alpha|_{B\langle g\rangle}$ along the map $(-)^l$. We will see in the proof of Lemma 2.19 below that restriction along this map raises the action to the lth power.

Lemma 2.12. *Let $h = |\alpha|_{B\langle g\rangle}|$. Then T^n acts as multiplication by an hth root of unity on the line $L^\alpha_{g,1}$. Similarly, any element $\gamma \in \Gamma_0(n)$ acts as multiplication by a hth root of unity on the line $L^\alpha_{\langle g\rangle}$.*

We will see in Corollary 2.15 below that T^n actually acts by multiplication with a primitive h^{th} root of 1.

PROOF. Let ϕ be the n-fold covering map from the torus to itself which has degree n on the first circle and is the identity on the second. Then ϕ pulls back $P_{g,1}$ to the trivial bundle, and we have

$$T^n \circ \phi = \phi \circ T.$$

Hence ϕ induces an n-fold covering of the glued cylinders of Construction 2.5:

$$\phi\colon \operatorname{cyl}_T(P_{1,1}) \to \operatorname{cyl}_{T^n}(P_{g,1}).$$

Just as in the proof of Proposition 2.6, we may conclude that the Chern–Simons invariant of $\operatorname{cyl}_T(P_{1,1})$ is equal to the nth power of the Chern–Simons invariant of $\operatorname{cyl}_{T^n}(P_{g,1})$. The first gives the action of T on $L^\alpha_{1,1}$. By Corollary 2.9, this action is 1. The Chern–Simons invariant of $\operatorname{cyl}_{T^n}(P_{g,1})$ gives the action of T^n on $L^\alpha_{g,1}$. It follows that this action is an nth root of 1. From this point of view, it is also clear why its order divides the order of α. Fix g, and let $\gamma \in \Gamma_0(n)$ be as above. The same argument, using the degree n map of the second circle, shows that for any two classifying maps f and f' of P_{1,g^d}, the corresponding trivializations of L_{1,g^d} differ by an nth root of unity. Writing f_g and f_{g^d} for the classifying maps used to trivialize $L_{1,g}$ and L_{1,g^d}, the claim follows when we take $f = f_{g^d}$ and $f' = f_g \circ \gamma^{-1}$. □

2.4. Sections of L^α. The following definitions and facts are taken from [15, 5]. We let E be the space of sections of L^α over $\overline{\mathcal{C}}(G)$. In the notation of [15], we have

$$E = \mathcal{L}^2(\overline{\mathcal{C}}_{\mathbb{S}^1\times\mathbb{S}^1}, \overline{\mathcal{L}}_{\mathbb{S}^1\times\mathbb{S}^1}).$$

Further, Freed and Quinn define a line bundle $\overline{\mathcal{L}}_{[0,1]\times\mathbb{S}^1}$ on the set $\overline{\mathcal{C}'}_{[0,1]\times\mathbb{S}^1}$ of isomorphism classes of principal G-bundles over $[0,1]\times\mathbb{S}^1$ with basepoints chosen over the basepoints of the boundary. They prove that the sections of this line bundle

$$A := \mathcal{L}^2(\overline{\mathcal{C}'}_{[0,1]\times\mathbb{S}^1}, \overline{\mathcal{L}}_{[0,1]\times\mathbb{S}^1})$$

form a coalgebra, whose underlying vector space is the direct sum of complex lines

$$A \cong \bigoplus_{g,h\in G} L_{g,h}.$$

If we view A as the space of complex valued linear functions on its dual algebra

$$A^* \cong \bigoplus_{g,h\in G} L^*_{g,h},$$

then $E \subseteq A$ is identified with the subspace of central sections.

The algebra A^* is semisimple, and E possesses an orthonormal basis of character functions $\{\chi_\lambda\}$, where λ runs through the irreducible representations of A^*. The support of χ_λ is

$$\operatorname{supp}(\chi_\lambda) = \{(g,h) \mid g \in [g_0], h \in C_g\}$$

for a conjugacy class $[g_0]$ depending on λ. For fixed g, the restriction of χ_λ to (the units of)

$$\bigoplus_{h \in C_g} L^*_{g,h}$$

is a character of the central extension $\widehat{C}_g$ of C_g defined by these units. Thus

$$E \cong \mathbb{C}\{\chi_\lambda\} \cong \bigoplus_{[g] \subseteq G} R(\widehat{C}_g) \otimes \mathbb{C}.$$

Following Freed and Quinn, we write $e_{g,1} \in L_{g,1}$ and $e^*_{g,1}$ for the elements of unit length corresponding to the trivialization we picked above and note that $e^*_{g,1}$ is the unit of the group $\widehat{C}_g$.

2.4.1. *Sections of $\mathcal{L}^\alpha$.* By construction, a section of $\mathcal{L}^\alpha$ is equivalent to an $\mathrm{SL}_2(\mathbb{Z})$-equivariant map

$$f \colon \mathfrak{H} \to E.$$

Fix an element $g \in G$ and consider the $[g]$-component of f,

$$f_g \colon \mathfrak{H} \to R(\widehat{C}_g) \otimes \mathbb{C}.$$

Lemma 2.13. *The element $T^n \in \mathrm{SL}_2(\mathbb{Z})$ acts as multiplication by a root of unity on the summand $R(\widehat{C}_g) \otimes \mathbb{C}$. Its order divides n and the order of α. More precisely T^n acts by multiplication with $\mathrm{e}^{2\pi \mathrm{i} \alpha(x)}$, where x is the 3-cycle*

$$x = \sum_{k=0}^{n} (g, g^k, g) \in \mathrm{Z}_3(B\langle g \rangle).$$

PROOF. The action of T on $A^*|_{[g]}$ is given by multiplication with the element $e^*_{g,g} \in L^*_{g,g}$ defined in [15, (5.5)]. Hence it is enough to understand the action of T^n on $L^\alpha_{g,1}$. We saw in Lemma 2.12 that this action is given by multiplication with a root of unity whose order divides n and the order of α. To determine it explicitly, we triangulate the fundamental domain $\mathbb{I} \times \mathbb{I}$ of $\mathbb{T}^2 = \mathbb{R}/\mathbb{Z}$ by choosing as 1-simplices s_1 from $(0,0)$ to $(0,1)$, s_2 from $(0,0)$ to $(1,0)$, s_3 from $(0,1)$ to $(1,1)$, s_4 from $(1,0)$ to $(1,1)$ and s_5 from $(0,1)$ to $(1,0)$. Then T pulls back the

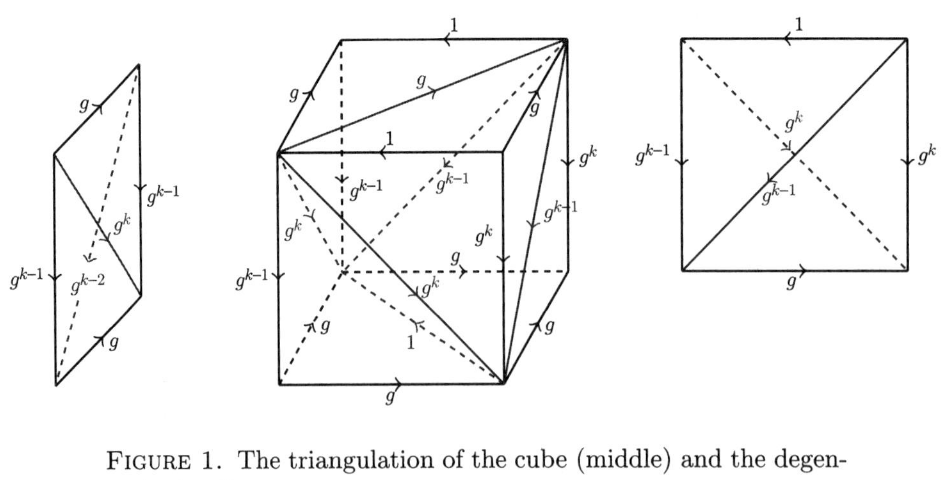

FIGURE 1. The triangulation of the cube (middle) and the degenerate simplices.

corresponding triangulation of $\mathbb{T}^2$ to the one corresponding to the following choice of 1-simplices on $\mathbb{I}^2$: a simplex t_1 from $(0,1)$ to $(0,0)$, t_2 from $(0,0)$ to $(1,0)$, t_3 from $(0,1)$ to $(1,1)$, t_4 from $(1,1)$ to $(1,0)$, and t_5 from $(0,0)$ to $(1,1)$. Cutting open the pointed bundle $P_{g,g}$ along the two standard circles of $\mathbb{T}^2$ gives a bundle over $\mathbb{I}^2$ with basepoints over the corners and parallel transports g along s_1, s_2, s_3 and s_4, and 1 along s_5. The same procedure for $T^*P_{g,g} = P_{g,1}$ gives parallel transports g along t_2, t_3, and t_5, and 1 along t_1 and t_4. We triangulate the faces of the cube $\mathbb{I}^3$ by choosing the first of the above triangulations on $\mathbb{I}^2 \times \{0\}$ and the second one on $\mathbb{I}^2 \times \{1\}$ and adding the 1-simplices r_1 from $(1,1,1)$ to $(0,1,0)$, r_2 from $(1,1,1)$ to $(1,0,0)$, r_3 from $(0,0,1)$ to $(0,1,0)$, r_4 from $(0,0,1)$ to $(1,0,0)$, r_5 from $(1,1,1)$ to $(1,1,0)$, r_6 from $(1,0,1)$ to $(1,0,0)$, r_7 from $(0,1,1)$ to $(0,1,0)$, and r_8 from $(0,0,1)$ to $(0,0,0)$. Note that there exists a bundle Q_k over $\mathbb{I}^3$, pointed over the corners, such that the parallel transports along r_5 and r_7 are g^{k-1} and those along r_6 and r_8 are g^k. The cube $\mathbb{I}^3$ is divided into five 3-simplices whose edges are the 1-simplices specified above, and the cycles defined by pushing forward these 3-simplices along a classifying map of Q_k are

$$\pm(g^{k-1},g,1), \quad \pm(g^{k-1},1,g), \quad \pm(1,g,g^{k-1}), \quad \pm(g,1,g^{k-1}) \quad \text{and} \quad \pm(g,g^{k-1},1).$$

We add two more degenerate 3-simplices, one to the face $\{0\}\times\mathbb{I}^2$ giving the cocycle $\pm(1,g^{k-1},g)$ and one along the face $\mathbb{I}\times\{0\}\times\mathbb{I}$ giving the cocycle $\pm(g,g^{k-2},g)$. This allows us to glue the first circle of the torus back together via

$$(0,x,y) \sim (1,x,y).$$

This gives a triangulation of the mapping cylinder of $T|_{\mathbb{S}^1\times\mathbb{I}}$. Further, we are able to glue Q_k to Q_{k+1} along the ends of the cylinder in such a way that their triangulations fit together. Since $Q_n = Q_0$, we may join Q_0 through Q_{n-1} together to get a bundle over the mapping cylinder of T^n. Since α is normalized, it vanishes on cycles containing the identity. Hence T^n acts as claimed. $\square$

Lemma 2.14. *The cycle $\sum_{k=0}^{n-1}(g,g^k,g)$ is a generator of the cyclic group*

$$\mathrm{H}_3(B\langle g\rangle) \cong \mathbb{Z}/n\mathbb{Z}.$$

PROOF. We will use the terminology and notation of [7]. Let $\mathbb{Z} \leftarrow F_\bullet$ be the standard resolution of $\mathbb{Z}$ over $\mathbb{Z}G$, and consider the commuting diagram of $\mathbb{Z}\langle g\rangle$-modules

$$\begin{array}{ccccccccc}
\mathbb{Z} & \xleftarrow{\ \varepsilon\ } & \mathbb{Z}\langle g\rangle & \xleftarrow{\ g-1\ } & \mathbb{Z}\langle g\rangle & \xleftarrow{\ \mathrm{norm}\ } & \mathbb{Z}\langle g\rangle & \xleftarrow{\ g-1\ } & \mathbb{Z}\langle g\rangle \\
 & & \downarrow{\scriptstyle f_0} & & \downarrow{\scriptstyle f_1} & & \downarrow{\scriptstyle f_2} & & \downarrow{\scriptstyle f_3} \\
\mathbb{Z} & \longleftarrow & F_0 & \xleftarrow{\ \partial_1\ } & F_1 & \xleftarrow{\ \partial_2\ } & F_2 & \xleftarrow{\ \partial_3\ } & F_3,
\end{array}$$

where

$$\mathrm{norm}(1) = 1 + g + g^2 + \cdots + g^{n-1},$$

and in bar notation, the maps f_i are given by

$$f_0(1) = 1, \quad f_1(1) = [g], \quad f_2(1) = \sum_{k=0}^{n-1}[g^k|g], \quad \text{and} \quad f_3(1) = \sum_{k=0}^{n-1}[g|g^k|g].$$

These are the first maps of an augmentation preserving map of chain complexes $f_\bullet$ from the periodic resolution to the standard resolution. On coinvariants, $f_\bullet$ induces an isomorphism of complexes of the form

$$\mathbb{Z} \xleftarrow{=} \mathbb{Z} \xleftarrow{0} \mathbb{Z} \xleftarrow{n} \mathbb{Z} \xleftarrow{0} \mathbb{Z} \xleftarrow{n} \cdots,$$

where $1 \in \mathbb{Z}\langle g\rangle$ becomes $1 \in \mathbb{Z}$. □

Corollary 2.15. *The order of the action T^n on the g-twisted sector*

$$\bigoplus_{h \in C_g} L_{g,h}$$

equals the order of $\alpha|_{\langle g\rangle}$. In particular, it divides the order of α and that of g.

Following [8], we will write h for the order of $\alpha|_{\langle g\rangle}$ and n for the order of g, and set $N = nh$.

Corollary 2.16. *The map f_g factors through*

$$\begin{aligned} \mathfrak{H} &\to \mathbb{C}^\times \\ \tau &\mapsto \mathrm{e}^{2\pi \mathrm{i}\tau/N}. \end{aligned}$$

Corollary 2.17. *There is a Fourier expansion*

$$f_g \in R(\widehat{C}_g) \otimes \mathbb{C}[\![q^{1/N}]\!]$$

of f_g, where $q = \mathrm{e}^{2\pi\mathrm{i}\tau}$. If T^n acts by multiplication with $\mathrm{e}^{2\pi \mathrm{i}s/h}$ on $L_{g,1}$, then f_g has the form

$$f_g(\tau) = \sum_k q^{k/n + s/N}.$$

Proof. The first part follows immediately from the previous corollary. The second part is proved by a comparison of coefficients of the two sides of the equation

$$f(q^{1/N}\mathrm{e}^{2\pi\mathrm{i}/h}) = f(\tau + n) = f(\tau)\mathrm{e}^{2\pi\mathrm{i}s/h} = f(q^{1/N})\mathrm{e}^{2\pi\mathrm{i}s/h}.$$ □

In [14] such a fraction s/N turns up as the conformal weight of the g-twisted sector. If $g = 1$, only integral powers of q occur, and the central extension of the centralizer is trivial:

$$\widehat{C}_1 = U(1) \times G.$$

Proposition 2.18. *The group $\Gamma_0(n)$ acts on $L_{\langle g\rangle}$ by multiplication with h^{th} roots of 1. The group $\Gamma_0(N)$ acts trivially on $L_{\langle g\rangle}$.*

Proof. The first claim follows immediately from Lemmas 2.8 and 2.12. Let $\gamma \in \Gamma_0(N)$, and let $p_h \colon \mathbb{T}^2 \to \mathbb{T}^2$ be the degree h map of the first circle. Then we have

$$p_h \circ \gamma = \widehat{\gamma} \circ p_h$$

with $\widehat{\gamma} \in \Gamma_0(n)$. As in the proof of Lemma 2.12, it follows that $\gamma_* = (\widehat{\gamma}_*)^h$. But we just saw that $\widehat{\gamma}_*$ acts as an hth root of unity. □

Lemma 2.19. *If $[\alpha]$ is an element of order h in $\mathrm{H}^3(B\langle g\rangle, \mathbb{R}/\mathbb{Z})$, then the restriction $[\alpha|_{B\langle g^h\rangle}]$ has order 1 in $\mathrm{H}^3(B\langle g^h\rangle, \mathbb{R}/\mathbb{Z})$.*

Proof. There is a $\mathbb{Z}\langle g^h\rangle$-equivariant, augmentation-preserving map between the periodic resolutions, which is the inclusion of $\mathbb{Z}\langle g^h\rangle$ in $\mathbb{Z}\langle g\rangle$ in even degrees and sends 1 to $1 + g + \cdots + g^{h-1}$ in odd degrees. On coinvariants, this map becomes the identity in even degrees and multiplication with h in odd degrees. □

An alternative proof for this lemma is given as follows: Let T^t denote the transpose of T. Note that we can use the degree h map of the second circle to conjugate $(T^t)^n$ into $(T^t)^{n/h}$. It follows that the action of $(T^t)^{n/h}$ on $L_{\langle g^h\rangle}$ is the hth power of the action of $(T^t)^n$ on $L_{\langle g\rangle}$.

Corollary 2.20. *The group $\Gamma_0(n/h)$ acts trivially on $L_{\langle g^h\rangle} = L_{\langle g\rangle}^{\otimes h}$.*

Definition 2.21 (cf. [8]). Let $\Gamma_0(n|h)$ be the subgroup of $\mathrm{SL}_2(\mathbb{R})$ consisting of matrices of the form

$$\begin{pmatrix} a & b/h \\ nc & d \end{pmatrix},$$

where a, b, c and d are integers.

Note that $\Gamma_0(n|h)$ is conjugate to $\Gamma_0(n/h)$ via conjugation with

$$\begin{pmatrix} h & 0 \\ 0 & 1 \end{pmatrix}^{-1} \quad \text{or with} \quad \begin{pmatrix} 1 & 0 \\ 0 & h \end{pmatrix}.$$

Remark 2.22. The conjecture in [8] suggests that $\Gamma_0(n|h)$ acts by h^{th} roots of 1 on $L_{\langle g\rangle}$. While the previous corollary does not tell us how to define this action when h does not divide b, it shows that in order for it to be consistent with the rest of the picture, it needs to be an action by h^{th} roots of unity.

3. Abelian groups and level structures

Let $E = \mathbb{C}/\langle\tau, 1\rangle$. Since any classifying map $f\colon \mathbb{Z}^2 \to G$ factors through an abelian subgroup of G, it is often enough to study principal bundles over E with abelian structure group. These also are important for the definition of Hecke operators.

3.1. Isogenies. Let A be a finite abelian group. Our first observation is that principal A-bundles with connected total space are almost the same as isogenies with kernel A. More precisely, let $\phi\colon E' \to E$ be an isogeny with kernel A. Then A acts on E' by addition, making ϕ the structure map of a principal A-bundle over E.

Proposition 3.1. *Let $f\colon \mathbb{Z}^2 \to A$ be a group homomorphism. Then f is surjective if and only if the principal A-bundle $\phi\colon P \to E$ over E classified by Bf is isomorphic to an isogeny with kernel isomorphic to A, viewed as a principal A-bundle in the way just described. This in turn is the case if and only if P is connected.*

Proof. Let $f\colon \mathbb{Z}^2 \to A$ be surjective and let n be the order of A. Write Λ for $\ker(f)$. Then Λ is an index n subgroup of $\langle\tau, 1\rangle$, and therefore a rank two sublattice. The short exact sequence

$$0 \to \Lambda \to \mathbb{Z}^2 \to A \to 0$$

gives rise to a fibre sequence of classifying spaces

$$B\Lambda \to B\mathbb{Z}^2 \xrightarrow{Bf} BA,$$

whose first map is the principal A-bundle classified by Bf. It is constructed as the quotient map

$$B\Lambda := E\mathbb{Z}^2/\Lambda = \mathbb{C}/\Lambda \to \mathbb{C}/\langle\tau, 1\rangle = B\mathbb{Z}^2, \tag{8}$$

with the A action induced by the action of $\langle\tau,1\rangle\cong\mathbb{Z}^2$ on $\mathbb{C}\cong E\mathbb{Z}^2$. Hence it is an isogeny whose kernel is isomorphic to A. Moreover, any such isogeny into E is of the form (8) for some sublattice $\Lambda\subseteq\langle\tau,1\rangle$, and therefore classified by the surjective map

$$\mathbb{Z}^2\cong\langle\tau,1\rangle\twoheadrightarrow\langle\tau,1\rangle/\Lambda.$$

Let now P be connected. We will see in Lemma 4.1 that this implies that P can be viewed as a complex elliptic curve and ϕ as an isogeny. More precisely, P is the quotient of $\mathbb{C}$ by an index n subgroup $\Lambda\subseteq\langle\tau,1\rangle$. Thus ϕ is an isogeny with kernel $\langle\tau,1\rangle/\Lambda$, and Proposition A.7 implies that A is isomorphic to $\langle\tau,1\rangle/\Lambda$. □

Remark 3.2. The (base-preserving) automorphisms of a principal A-bundle over E are given by multiplication with elements of A:

$$\mathrm{Aut}(P_{a_1,a_2})\cong C_A(a_1,a_2)=A.$$

Isogenies, at the other hand, come equipped with a choice of basepoint over zero and therefore have no automorphisms over E.

If f is not surjective, we can write $f=i\circ f'$, where $f'\colon\mathbb{Z}^2\to B$ is surjective, and $i\colon B\to A$ is injective. Then the A-bundle classified by Bf is a collection of isogenies into E with kernel B.

3.2. Duals. Instead of considering isogenies into E, it is often more natural to consider the dual picture, which involves isogenies with source E and the corresponding level structures on E. Let $\widehat{A}:=\mathrm{Hom}(A,\mathbb{S}^1)$ be the Pontrjagin dual of A. Since A is finite and abelian, we have $\widehat{\widehat{A}}=A$. We further know

$$\widehat{\mathbb{Z}}=\mathbb{S}^1\quad\text{and}\quad\widehat{\mathbb{S}^1}=\mathbb{Z}$$

Let $i_{\langle\omega_1,\omega_2\rangle}$ denote the isomorphism from E to $\mathbb{S}^1\times\mathbb{S}^1$ corresponding to the choice of basis (ω_1,ω_2). While we had identified

$$B\mathbb{Z}^2=\mathbb{S}^1\times\mathbb{S}^1$$

with E in the non-standard way, using $i_{\langle-1,\tau\rangle}$, we will follow the convention to identify

$$\widehat{\mathbb{Z}^2}\cong\mathbb{S}^1\times\mathbb{S}^1$$

with E in the standard way, using $i_{\langle\tau,1\rangle}$.

Proposition 3.3 (compare [20, (1.4.4)]**).** *These identifications yield an isomorphism*

$$\mathrm{Prin}_A(E)\cong\mathrm{Hom}(\widehat{A},E),$$

where $\mathrm{Prin}_A(E)$ *denotes the set of isomorphism classes of principal A-bundles over E. Let $\phi\colon E'\to E$ be an isogeny. Then pullback along $\phi^\vee$ on the left-hand side becomes composition with ϕ on the right-hand side.*

PROOF. The isomorphism is given by

$$\mathrm{Prin}_A(E)\cong\mathrm{Hom}(\mathbb{Z}^2,A)\cong\mathrm{Hom}(\widehat{A},\widehat{\mathbb{Z}^2})\cong\mathrm{Hom}(\widehat{A},E).$$

Here the first isomorphism uses $i_{\langle-1,\tau\rangle}$ to identify $\mathbb{S}^1\times\mathbb{S}^1$ with E, while the last one uses $i_{\langle\tau,1\rangle}$. Let $\varphi\colon\mathbb{Z}^2\to\mathbb{Z}^2$ be as in (6) (on page 177), let $\varphi^\vee$ denote its pseudo-inverse, and write $\phi=B\varphi$. Then ϕ is viewed as a map from E' to E via the identifications $i_{\langle -1,\tau' \rangle}$ and $i_{\langle -1,\tau \rangle}$. The first isomorphism sends $(\phi^\vee)^*$ to $(\varphi^\vee)^*$. The second isomorphism maps $(\varphi^\vee)^*$ to $\widehat{\phi^\vee}_*$. We have a commutative diagram

$$\text{(9)} \qquad \begin{array}{ccccc} E' & \xrightarrow{i_{\langle \tau',1 \rangle}} & \mathbb{S}^1 \times \mathbb{S}^1 & \xrightarrow{\cong} & \widehat{\mathbb{Z}^2} \\ \downarrow{\scriptstyle \phi} & & \downarrow{\scriptstyle B(\varphi^\vee)^t} & & \downarrow{\scriptstyle \widehat{\varphi^\vee}} \\ E & \xrightarrow{i_{\langle \tau,1 \rangle}} & \mathbb{S}^1 \times \mathbb{S}^1 & \xrightarrow{\cong} & \widehat{\mathbb{Z}^2}, \end{array}$$

where $(\varphi^\vee)^t$ is the transpose of $\varphi^\vee$. This completes the proof. $\square$

Lemma 3.4. *Let $f \in \mathrm{Hom}(\widehat{A}, E)$, and let ξ_f be the principal A-bundle over E corresponding to f. Let g be an automorphism of A. Then the pullback $g^*(\xi_f)$ of ξ_f along g is the principal A-bundle corresponding to $f \circ \widehat{g^{-1}} \in \mathrm{Hom}(\widehat{A}, E)$,*

$$g^*(\xi_f) \cong \xi_{f \circ \widehat{g^{-1}}}.$$

PROOF. Let $\widehat{f}\colon \mathbb{Z}^2 \to A$ be the dual of f. Then ξ_f is classified by $B\widehat{f}$, and $\xi_{f \circ \widehat{g^{-1}}}$ is classified by

$$B(\widehat{f \circ \widehat{g^{-1}}}) = B(g^{-1} \circ \widehat{f}) = Bg^{-1} \circ B\widehat{f}.$$

The proposition now follows from Corollary A.5. $\square$

A homomorphism $f\colon \mathbb{Z}^2 \to A$ is surjective, if and only if its dual homomorphism $\widehat{f}\colon \widehat{A} \to E$ is injective.

Proposition 3.5. *Let $f \in \mathrm{Inj}(\widehat{A}, E)$, and let ξ_f be the principal A-bundle over E classified by $B\widehat{f}$. Further, let $\Psi_f\colon E \to E'$ be the isogeny with kernel f. Then ξ_f is isomorphic to the principal bundle defined by the dual isogeny $\Psi_f^\vee\colon E' \to E$.*

PROOF. Consider the short exact sequence

$$0 \to \widehat{A} \xrightarrow{f} E \xrightarrow{\Psi_f} E' \to 0.$$

Its Pontrjagin dual is a short exact sequence of the form

$$0 \to \ker(\widehat{f}) \to \mathbb{Z}^2 \xrightarrow{\widehat{f}} A \to 0.$$

Write ψ for the inclusion of $\ker(\widehat{f})$ in $\mathbb{Z}^2$. Then ψ is such that under the isomorphisms $i_{\langle \tau,1 \rangle}$ and $i_{\langle \tau',1 \rangle}$, we have $\Psi_f = \widehat{\psi}$. It follows from the proof of Proposition 3.1 that under the isomorphisms $i_{\langle -1,\tau \rangle}$ and $i_{\langle -1,\tau' \rangle}$, we have $\xi_f = B\psi$. Finally, we consider the commutative diagram (9) with $\xi_f^\vee$ in the role of ϕ and ψ in the role of $\varphi^\vee$ to obtain $\xi_f^\vee = \Psi_f$. $\square$

Corollary 3.6. *There is a one-to-one correspondence*

$$\mathrm{Inj}(\widehat{A}, E) \to \{\textit{principal } A\textit{-bundles over } E \textit{ with connected total space}\}/\cong$$

sending f to $\Psi_f^\vee$.

The decomposition

$$\mathrm{Hom}(\widehat{A}, E) \cong \coprod_{B \subseteq A} \mathrm{Inj}(\widehat{B}, E)$$

is invariant under the action of $\mathrm{SL}_2(\mathbb{Z})$ on $A \times A$. Hence it induces a decomposition of the coarse moduli space

$$X_A = \coprod_{B \subseteq A} X_B^{\mathrm{inj}},$$

where X_B^{inj} is the moduli space of elliptic curves with level $\widehat{B}$-structures, i.e., of isogenies $\phi\colon E \to E'$ together with a choice of isomorphism from their kernel to $\widehat{B}$.

Note that

$$\mathrm{Inj}(\widehat{A}, E)$$

is invariant under the action of $\mathrm{Aut}(A)$ on $\mathrm{Hom}(\widehat{A}, E)$, and that

$$\mathrm{Inj}(\widehat{A}, E)/\,\mathrm{Aut}(A)$$

corresponds to the set of all isogenies out of E which allow a choice of isomorphism from $\widehat{A}$ to their kernels. Since no choice of this isomorphism is specified, and $\widehat{A}$ is non-canonically isomorphic to A, this condition is equivalent to allowing an isomorphism from A to their kernels.

Definition 3.7. We will write $\mathcal{M}_A^{\mathrm{conn}}$ for the component of $\mathcal{M}_A$ parametrizing bundles with connected total space and

$$\widetilde{\mathcal{M}}_0(A) := \mathcal{M}_A^{\mathrm{conn}} /\!\!/ \mathrm{Aut}(A)$$

for the quotient stack of $\mathcal{M}_A^{\mathrm{conn}}$ by the action of $\mathrm{Aut}(A)$ on the fibers. We will write $X_0(A)$ for the corresponding coarse moduli space.

Note that by Remark 3.2 the stack $\widetilde{\mathcal{M}}_0(A)$ is not the same as the moduli-stack $\mathcal{M}_0(A)$ of isogenies allowing an isomorphism from A to their kernel. However, the forgetful map

$$\mathcal{M}_0(A) \to \widetilde{\mathcal{M}}_0(A)$$

induces an isomorphism of coarse moduli spaces.

Definition 3.8. The Fricke involution W_A is an involution of $\mathcal{M}_0(A)$. It sends the isogeny ϕ to its dual isogeny

$$\phi^\vee : E' \to E.$$

If the kernel of ϕ is isomorphic to $\widehat{A}$, then the kernel of $\phi^\vee$ is (also) isomorphic to A.

3.3. Cyclic groups. Let now A be the cyclic group $\mathbb{Z}/n\mathbb{Z}$. An injective map from $\mathbb{Z}/n\mathbb{Z}$ to E is the same as a choice of point of exact order n on E. Such a choice is called a $\Gamma_1(n)$-level structure on E. Therefore, the space of principal $\mathbb{Z}/n\mathbb{Z}$ bundles on E is non-canonically isomorphic to

$$E[n] \cong \mathrm{Hom}(\mathbb{Z}/n\mathbb{Z}, E) \cong \coprod_{d|n} \mathrm{Level}_{\Gamma_1(d)}(E),$$

and the moduli space $X_{\mathbb{Z}/n\mathbb{Z}}$ gets (non-canonically) identified with

$$X_{\mathbb{Z}/n\mathbb{Z}} \cong \coprod_{d|n} X_1(d).$$

Note that

$$\mathrm{Inj}(\mathbb{Z}/n\mathbb{Z}, E)/\,\mathrm{Aut}(\mathbb{Z}/n\mathbb{Z}) \cong \mathrm{Level}_{\Gamma_0(n)}(E).$$

Therefore, we get a (canonical) isomorphism

$$X_{\mathbb{Z}/n\mathbb{Z}}^{\mathrm{inj}}/\,\mathrm{Aut}(\mathbb{Z}/n\mathbb{Z}) \cong X_0(n).$$

The Fricke involution $W_n := W_A$ takes a cyclic subgroup $C \subseteq E$ of order n to the cyclic subgroup

$$E[n]/C \subseteq E/C.$$

Let $g \in \mathbb{Z}/n\mathbb{Z}$ be a generator. Then the triple $(1, g; \tau)$ corresponds to the cyclic subgroup $\langle 1/n \rangle$ of $\mathbb{C}/\langle \tau, 1 \rangle$. Therefore W_n sends it to the cyclic subgroup of $\mathbb{C}/\langle \tau, 1/n \rangle$ generated by τ/n. This again corresponds to the subgroup generated by $1/n$ in

$$\mathbb{C}/\langle -1/n\tau, 1 \rangle.$$

Thus

$$W_n(1, g; \tau) = \left(1, g; \frac{-1}{n\tau}\right).$$

4. Symmetric groups and coverings

The symmetric groups play a central role in the definition of power operations and will turn out to be important in the theory of replicability. Recall the one-to-one correspondence

$$\left\{\begin{matrix}\text{Isom. classes of principal} \\ \Sigma_n\text{-bundles over } X\end{matrix}\right\} \longrightarrow \left\{\begin{matrix}\text{Isom. classes of } n\text{-fold} \\ \text{covers of } X\end{matrix}\right\}$$
$$P \longmapsto P \times_{\Sigma_n} \underline{n},$$

where $\underline{n}$ is a fixed set with n elements.

Lemma 4.1. *Let $\pi\colon P \to E$ be an n-fold covering of a complex elliptic curve E with connected total space. Then P can be viewed as a complex elliptic curve and π as an isogeny.*

PROOF. By the theory of covering spaces, the universal cover of E,

$$\mathbb{C} \to \mathbb{C}/\langle \tau, 1 \rangle \cong E,$$

factors through π. More precisely, if we choose a basepoint of P over zero, this identifies P with the quotient of $\mathbb{C}$ by an index n subgroup $\Lambda \subseteq \langle \tau, 1 \rangle$. Thus π becomes an isogeny. □

It follows that every Σ_n-bundle with connected total space is induced from a principal bundle with abelian structure group: Let A be an abelian group of order n, and pick a bijection between A and $\underline{n}$. Then the action of A on itself by multiplication defines a map

$$i\colon A \hookrightarrow \Sigma_n.$$

Note that a different choice of bijection leads to a different map which is conjugate to i by an element of Σ_n. In particular, this applies to precomposition of i with any automorphism of A.

Lemma 4.2. *Let*

$$\phi\colon Y \to E$$

be an isogeny together with an isomorphism from $\ker(\phi)$ *to A. Consider the n-fold covering map underlying ϕ, and let ξ_Y be the corresponding principal Σ_n-bundle. Now consider ϕ as principal A-bundle. Then the isomorphism class of the Σ_n-bundle $\phi[\Sigma_n]$ associated to ϕ via i is equal to that of ξ_Y.*

Note that (in particular) the isomorphism class of $\phi[\Sigma_n]$ is independent of the choice of i and of the choice of isomorphism $\ker(\phi) \cong A$.

PROOF. By Lemma A.6 we have

$$\phi[\Sigma_n] \times_{\Sigma_n} \underline{n} \cong \phi \times_A \underline{n},$$

which is the n-fold covering underlying ϕ. Thus ξ_Y and $\phi[\Sigma_n]$ correspond to the same covering, which proves the claim. □

Take an n-fold covering P of E with connected total space, and view it as a principal A-bundle. Note that any deck transformation can be realized as $(-)+a$ with $a \in A$. Hence the automorphisms of P as an n-fold covering are the same as the automorphisms of P as principal A-bundle. For future reference, we note that their number is $n = |A|$.

We now direct our attention to coverings whose total spaces are not connected. Recall that if P and P' are Σ_n- and Σ_m-principal bundles with associated covering spaces Y and Y' respectively, then the associated covering space of

$$(P \times P') \times_{\Sigma_n \times \Sigma_m} \Sigma_{n+m}$$

is the disjoint union $Y \cup Y'$. Let now Y be the disjoint union of connected components

$$Y = Y_1 \cup \cdots \cup Y_k.$$

Then the Y_i can be viewed as having abelian structure groups $A_1, \ldots, A_k$ of orders $n_1, \ldots, n_k$ with $n = \sum n_i$. Each of the Y_i has an associated Σ_{n_i}-principal bundle

$$P_i = Y_i \times_{A_i} \Sigma_{n_i},$$

and P is the bundle

$$P = \left(\prod_i P_i\right) \times_{\Sigma_1 \times \cdots \times \Sigma_k} \Sigma_n.$$

Note that each A_i is a quotient of $\mathbb{Z}^2$, making the disjoint union

$$A_1 \cup \cdots \cup A_k \cong \underline{n},$$

into an $\mathbb{Z}^2$-set, whose orbits are the A_i. The $\mathbb{Z}^2$-action is transitive if and only if Y is connected. We obtain a map

$$\mathbb{Z}^2 \to \Sigma_n,$$

well defined up to conjugation in Σ_n, and this is the pair of commuting elements which classifies P. More precisely, we have the following:

Proposition 4.3. *The decomposition of covers into connected components defines an equivalence*

$$e\colon \mathcal{M}_{\Sigma_n} \xrightarrow{\cong} \coprod_{n=\sum |A| N_A} \prod_{[A]} \widetilde{\mathcal{M}}_0(A) \wr \Sigma_{N_A}$$

where the product as well as the sum under the coproduct is over all isomorphism classes of finite abelian groups, and all products are taken over the target maps $\widetilde{\mathcal{M}}_A \to \mathcal{M}_1$.

Example 4.4. Let $n = 9$, and consider the decomposition $9 = |\{1\}| + |\mathbb{Z}/4\mathbb{Z}| \cdot 2$. The corresponding summand

$$\mathcal{M}_1 \times_{\mathcal{M}_1} \widetilde{\mathcal{M}}_{\mathbb{Z}/4\mathbb{Z}} \wr \Sigma_2$$

parametrizes ninefold covers of elliptic curves that decompose into three connected components, two of which are fourfold covers corresponding to isogenies with kernel $\mathbb{Z}/4\mathbb{Z}$ and one of which is a onefold cover.

PROOF OF PROPOSITION 4.3. We start by considering bundles over a torus. Fix a partition

$$n = \sum_T |T| N_T,$$

where the T are elements of a fixed system of representatives of the set of isomorphism classes of finite transitive $\mathbb{Z}^2$ sets. Such decompositions classify conjugacy classes $[\sigma,\rho]_{\Sigma_n}$ of pairs of commuting elements in Σ_n, and hence they classify isomorphism classes of principal Σ_n-bundles over $\mathbb{T}^2$. For each T, let

$$A_T := \mathbb{Z}^2 / \operatorname{Stab}_{\mathbb{Z}^2}(T)$$

denote the corresponding abelian group. Note that the isomorphism class of T is uniquely determined by the surjective map $f_T\colon \mathbb{Z}^2 \to A$, that for any automorphism g of A, the map $g \circ f_T$ determines the same isomorphism class $[T]$, and that we have

$$N_A = \sum_{A_T = A} N_T.$$

Now the n-fold cover of $\mathbb{T}^2$ corresponding to $[\sigma,\rho]_{\Sigma_n}$ decomposes into connected components, N_T of which are isomorphic to the principal A_T-bundle with connected total space that is classified by f_T. The automorphism group of this n-fold cover over $\mathbb{T}^2$ is isomorphic to

$$C_{\Sigma_n}(\sigma,\rho) = \prod_T A_T^{N_T} \wr \Sigma_{N_T}.$$

For each $[A]$, pick, once and for all, an ordering of the classes $[T]$ with $A_T \cong A$. Then, for each pair $[\sigma,\rho]_{\Sigma_n}$ and for each orbit-type $[T]$ occuring in its orbit decomposition, pick an ordering of the orbits isomorphic to T. In this way, we assign to $[\sigma,\rho]_{\Sigma_n}$ a point

$$e(\sigma,\rho) \in \prod_A \big(\operatorname{Surj}(\mathbb{Z}^2, A)/\operatorname{Aut}(A)\big)^{N_A}.$$

Our choices also determine an isomorphism

$$\operatorname{Aut}(e(\sigma,\rho)) \cong C_{\Sigma_n}(\sigma,\rho),$$

which we use to continue e to automorphisms of $[\sigma,\rho]$. Let now $\gamma \in \mathrm{SL}_2(\mathbb{Z})$. Then γ acts from the right on (σ,ρ) and on $\operatorname{Surj}(\mathbb{Z}^2, A)$, for each A. The transitive $\mathbb{Z}^2$-set $T \cdot \gamma$ classified by $\gamma \circ f_T$ might be different from T, but we have

$$A_{T\cdot\gamma} = A_T.$$

The action of γ on (σ,ρ) is compatible with the action of (σ,ρ) on the connected components of the corresponding bundle. However, we might have numbered the components belonging to the same group A differently for $(\sigma,\rho)\cdot\gamma$ than we did for (σ,ρ), with the result that they differ by an element $\sigma_\gamma \in \Sigma_{N_A}$. To extend the map e to the translation groupoid

$$(\overline{\mathcal{C}}(\Sigma_n) \times \mathfrak{H}) \rtimes \mathrm{SL}_2(\mathbb{Z}),$$

we set

$$e\big((\sigma,\rho;\tau),\gamma\big) := \prod_A \big(((\sigma,\rho)_A;\tau),(\gamma \wr \sigma_\gamma)\big).$$

Here $(\sigma,\rho)_A$ stands for the collection of the (σ,ρ)-orbits that belong to A.

By construction, e is fully faithful and essentially surjective. Different choices of orderings would result in a map e', which would differ from e by a natural transformation on its target. Hence, as a map of orbifolds, e is independent of these choices. $\square$

Definition 4.5. We will write $\widetilde{\mathcal{H}}_n$ for the moduli stack of all n-fold covers with connected fibre,

$$\widetilde{\mathcal{H}}_n = \coprod_{|A|=n} \widetilde{\mathcal{M}}_0(A).$$

Here the disjoint union runs over all isomorphism classes of abelian groups A of order n.

Equivalently, $\widetilde{\mathcal{H}}_n$ is the substack of $\mathcal{M}_{\Sigma_n}$ defined as

$$\widetilde{\mathcal{H}}_n := (\{f : \mathbb{Z}^2 \to \Sigma_n \mid f \text{ makes } \underline{n} \text{ into a transitive } \mathbb{Z}^2\text{-set}\} \times \mathfrak{H})/\!\!/ \big(\Sigma_n \times \mathrm{SL}_2(\mathbb{Z})\big).$$

The following lemma provides an alternative proof for the fact that the automorphism group of an n-fold covering with connected total space is isomorphic to the kernel A of the corresponding isogeny.

Lemma 4.6. *Let (σ, ρ) be a pair of commuting elements of Σ_n which acts transitively on $\{1, \ldots, n\}$. Write $A := \langle \sigma, \rho \rangle$. Then A has order n, the principal Σ_n-bundle classified by (σ, ρ) corresponds to an isogeny with kernel A, and we have*

$$C_{\Sigma_n}(\sigma, \rho) = \langle \sigma, \rho \rangle = A.$$

PROOF. If (σ, ρ) acts transitively on $\{1, \ldots, n\}$, then the corresponding n-fold covering has connected total space and hence is an isogeny. Write A' for its kernel, then the bundle $P_{(\sigma,\rho)}$ is induced by a principal A'-bundle. Hence the classifying map of $P_{(\sigma,\rho)}$ factors through A', yielding an isomorphism to A:

$$\mathbb{Z}^2 \to A' \xrightarrow{\cong} \langle \sigma, \rho \rangle \to \Sigma_n.$$

To determine the centralizer of $\langle \sigma, \rho \rangle$, let $\pi \in C_{\Sigma_n}(\sigma, \rho)$. Because of the transitivity of the action of A, we can write $\pi(1)$ in the form $\pi(1) = \sigma^a \rho^b(1)$. Further, for $1 \leq x \leq n$, we have c and d such that $x = \sigma^c \rho^d(1)$. It follows that

$$\pi(x) = \pi(\sigma^c \rho^d(1)) = \sigma^c \rho^d(\pi(1)) = \sigma^c \rho^d(\sigma^a \rho^b(1)) = \sigma^a \rho^b(\sigma^c \rho^d(1)) = \sigma^a \rho^b(x).$$

Hence $\pi = \sigma^a \rho^b$. $\square$

5. Hopkins – Kuhn – Ravenel character theory

Let $\mathcal{A}(G)$ be the category having as objects the abelian subgroups of G and with morphisms from B to A being the G-equivariant maps from G/B to G/A. Then $\mathcal{A}(G)$ is a full subcategory of the standard orbit category. The results of the previous sections can be summarized as follows:

$$\mathcal{M}_G \simeq \varinjlim_{\mathcal{A}(G)} \mathcal{M}_A. \tag{10}$$

Here $\varinjlim$ stands for a weak 2-colimit. Roughly speaking, that means that we take the disjoint union of all the $\mathcal{M}_A$ and instead of quotienting by the equivalence relation defined by maps in $\mathcal{A}(G)$, as we would for a classical colimit, we add in extra isomorphisms.

In fact, one can replace $\mathcal{A}$ by the full subcategory whose objects are abelian groups which are generated by two elements. The equivalence (10) is an analogue of the generalized Artin theorem [22, Theorem A].

5.1. Products. Let G and H be finite groups. Then we have an isomorphism of moduli stacks

$$\mathcal{M}_{G\times H} \xrightarrow{\cong} \mathcal{M}_G \times_{\mathcal{M}_1} \mathcal{M}_H,$$

which pulls back $\mathcal{L}^\alpha \otimes \mathcal{L}^\beta$ to $\mathcal{L}^{\alpha\times\beta}$. This yields external products

$$\begin{aligned}\boxtimes\colon \operatorname{Ell}_G^\alpha(\mathrm{pt}) \otimes \operatorname{Ell}_H^\beta(\mathrm{pt}) &\to \operatorname{Ell}_{G\times H}^{\alpha\times\beta}(\mathrm{pt}),\\ (f_G \boxtimes f_H)(g_1, h_1, g_2, h_2; \tau) &= f_G(g_1, g_2)\cdot f_H(h_1, h_2).\end{aligned}$$

In the case that $G = H$, one composes with pull-back along the diagonal of $G \times G$ to obtain internal products

$$\otimes\colon \operatorname{Ell}_G^\alpha(\mathrm{pt}) \otimes \operatorname{Ell}_G^\beta(\mathrm{pt}) \to \operatorname{Ell}_G^{\alpha+\beta}(\mathrm{pt}).$$

5.2. Change of groups. Let $a\colon H \to G$ be a map of finite groups. We write

$$\overline{\mathcal{C}}(a)\colon \overline{\mathcal{C}}(H) \to \overline{\mathcal{C}}(G)$$

for the map of moduli sets induced by a, and note that there is a canonical isomorphism of line bundles

$$\overline{\mathcal{C}}(a)^*(L^\alpha) \cong L^{a^*(\alpha)}.$$

For a section f of L^α over $\overline{\mathcal{C}}_G$, we define the section $\operatorname{res}|_a(f)$ of $L^{a^*(\alpha)}$ over $\overline{\mathcal{C}}(H)$ by

$$\operatorname{res}|_a(f) := f \circ \overline{\mathcal{C}}(a).$$

We recall the measure μ on $\overline{\mathcal{C}}(G)$ defined in [15, (2.1)]:

$$\mu([P_{g,h}]) = \frac{1}{|\operatorname{Aut}(P_{g,h})|} = \frac{1}{|C_G(g,h)|}.$$

If f is a section of $L^{a^*(\alpha)}$ over $\overline{\mathcal{C}}(H)$, we define a section $\operatorname{ind}|_a(f)$ of L^α over $\overline{\mathcal{C}}(G)$ by

$$\operatorname{ind}|_a(f)\cdot\mu = \int_{\text{fiber}} f\cdot \mathrm{d}\mu = \sum_{fiber} f([h_1, h_1]_H)\frac{1}{|C_H(h_1, h_2)|},$$

where at the point $[g_1, g_2]_G$, the sum is over all H-conjugacy classes $[h_1, h_2]_H$ which satisfy $[h_1, h_2]_G = [g_1, g_2]_G$. We have

$$\begin{aligned}\operatorname{ind}|_a(f)([g_1, g_2]_G) &= |C_G(g_1, g_2)|\cdot \sum_{[h_1,h_2]_H\in\text{fiber}} \frac{1}{|C_H(h_1, h_2)|}\cdot f([h_1, h_1]_H)\\ &= |C_G(g_1, g_2)|\cdot \sum_{a(h_1,h_2)\sim_G(g_1,g_2)} \frac{1}{|H|}\cdot f([h_1, h_1]_H).\end{aligned}$$

Two special cases are important: in the case that a is an inclusion of groups, we obtain the formula of [22, Theorem D],

$$\operatorname{ind}|_H^G(f) = \frac{1}{|H|} \sum_{\substack{s\in G|\\ s^{-1}(g_1,g_2)s\in H^2}} f(s^{-1}g_1 s, s^{-1}g_2 s).$$

In the case that a is the unique map from G to the trivial group, we get

$$\operatorname{ind}|_G^1(f)(1,1) = \frac{1}{|G|} \sum_{gh=hg} f(g,h),$$

where the sum is over all pairs of commuting elements of G.

Definition 5.1. We define the inner product on $\mathrm{Ell}_G^0(\mathrm{pt})$ as the composite of the (internal) product with $\operatorname{ind}|_G^1$,

$$\langle -,- \rangle_G : \mathrm{Ell}_G^0(\mathrm{pt}) \otimes \mathrm{Ell}_G^0(\mathrm{pt}) \to \mathrm{Ell}_G^0(\mathrm{pt}) \to \mathrm{Ell}_1^0(\mathrm{pt}).$$

We have

$$\langle f_1, f_2 \rangle_G = \frac{1}{|G|} \sum_{gh=hg} f_1(g,h) \cdot f_2(g,h).$$

All of these constructions extend to $\mathrm{SL}_2(\mathbb{Z})$-equivariant maps from $\mathfrak{H}$ to $\bigoplus L_{g,h}$.

6. Hecke operators and power operations

6.1. The Hecke correspondence. Let $\widetilde{\mathcal{H}}_{n,G} \subseteq \mathcal{M}_{G\times\Sigma_n}$ be defined by

$$\widetilde{\mathcal{H}}_{n,G} := \mathcal{M}_G \times_{\mathcal{M}_1} \widetilde{\mathcal{H}}_n,$$

where $\widetilde{\mathcal{H}}_n$ is as in Definition 4.5. Explicitly, we have

$$\widetilde{\mathcal{H}}_{n,G} = \left\{ \begin{array}{ccc} P' & \to & P \\ \downarrow & & \downarrow \\ E' & \xrightarrow{\phi} & E \end{array} \quad \begin{array}{c} P' = \phi^*(P) \\ \deg(\phi) = n \end{array} \right\} /\!\!/ \cong,$$

where P and P' are principal G-bundles and ϕ is an n-fold covering with connected total space. Consider the diagram

$$\begin{array}{ccc} & \widetilde{\mathcal{H}}_{n,G} & \\ {\scriptstyle s}\swarrow & & \searrow{\scriptstyle t} \\ \mathcal{M}_G & & \mathcal{M}_G, \end{array} \tag{11}$$

where the map s returns $P' \to E'$, and t returns $P \to E$. Note that s is only defined canonically up to canonical natural isomorphism in $\mathcal{M}_1$. The map t is induced by the map of groups

$$\mathrm{id} \times p \colon G \times \Sigma_n \to G,$$

where p is the unique map from Σ_n to the trivial group.

Note that this is not the usual Hecke correspondence; like $\widetilde{\mathcal{M}_0}(A)$, the stack $\widetilde{\mathcal{H}}_n$ has more automorphisms than $\mathcal{H}_n$ but induces the correct coarse moduli space. Therefore, in the case that G is the trivial group, the diagram of maps between coarse moduli spaces induced by (11) is the classical Hecke correspondence.

Definition 6.1. We define the trace map $t_!$ as the restriction of $\operatorname{ind}|_{\mathrm{id}\times p}$ to $\widetilde{\mathcal{H}}_{n,G}$.

Proposition 2.6 yields an isomorphism

$$s^*\mathcal{L}^\alpha \cong t^*(\mathcal{L}^\alpha)^{\otimes n}$$

over $\widetilde{\mathcal{H}}_{n,G}$.

Definition 6.2. The nth Hecke operator

$$T_n\colon \Gamma(\mathcal{L}^\alpha) \to \Gamma(\mathcal{L}^{n\alpha})$$

is defined as the composition

$$T_n = t_! \circ s^*.$$

Proposition 6.3. *Let f be a section of $\mathcal{L}^\alpha$. Then $T_n(f)$ is the following section of $(\mathcal{L}^\alpha)^{\otimes n}$:*

$$T_n(f)(g,h;\tau) = \frac{1}{n} \sum_{\substack{ad=n\\ 0\le b<d}} (\phi_{a,b,d})_* f\left(g^d, g^{-b}h^a; \frac{a\tau+b}{d}\right),$$

where

$$\phi_{a,b,d}\colon \mathbb{C}/\langle \tau', 1\rangle \to \mathbb{C}/\langle \tau, 1\rangle$$

sends $\tau' = (a\tau+b)/d$ to $a\tau+b$ and 1 to d, and

$$(\phi_{a,b,d})_*\colon L^\alpha_{g^d, g^{-b}h^a} \to (L^\alpha_{g,h})^{\otimes n}$$

is the induced isomorphism of line bundles of Proposition 2.6.

Proof. Let P be a principal G-bundle over $E \cong \mathbb{C}/\langle\tau,1\rangle$, let $E' \cong \mathbb{C}/\langle\tau',1\rangle$, where $\tau' = (a\tau+b)/d$ and let $\phi = \phi_{a,b,d}$. Then $\phi = B\varphi$, where

$$\begin{aligned} \varphi\colon \mathbb{Z}^2 &\to \mathbb{Z}^2\\ (1,0) &\mapsto (d,0)\\ (0,1) &\mapsto (-b,a). \end{aligned}$$

Let P be the principal bundle over E classified by the pair (g,h). Then the principal bundle Φ^*P over E' is classified by the pair $(g^d, g^{-b}h^a)$. □

Example 6.4. On the untwisted sector $g = 1$, the formula of Proposition 6.3 specializes to the formula for the twisted Hecke operators of classical Moonshine in (3).

Example 6.5. Let $G = \mathbb{Z}/l\mathbb{Z}$. Then

$$(e,f) \in (\mathbb{Z}/l\mathbb{Z})^2$$

corresponds to the point $(e\tau+f)/l$ of $E[l]$. Let ω_1, ω_2 and ϕ be as above. We have

$$\frac{e\tau+f}{l} = de\frac{\omega_1}{l} - be\frac{\omega_2}{l} + af\frac{\omega_2}{l}.$$

Hence the image of $(e\tau+f)/l$ under ϕ is the point of $E'[l]$ corresponding to $(de, af-be)$.

The action of the twisted Hecke operators on the q-expansions of the untwisted sectors is computed as follows:

$$\begin{aligned}\widehat{T}_n f(\tau) &= \frac{1}{n}\sum_{\substack{ad=n\\0\le b<d}} f^{(a)}\left(\frac{a\tau+b}{d}\right)\\ &= \frac{1}{n}\sum_{\substack{ad=n\\0\le b<d}}\sum_{m\in\mathbb{Z}} c^{(a)}(m)q^{am/d}\zeta_d^{bm}\\ &= \frac{1}{n}\sum_{ad=n}\sum_{m\in d\mathbb{Z}} d\cdot c^{(a)}(m)q^{am/d}\\ &= \sum_{m\in\mathbb{Z}}\sum_{a\mid(m,n)}\frac{1}{a}\cdot c^{(a)}\left(\frac{nm}{a^2}\right)q^m.\end{aligned}$$

6.2. Power operations in K-theory. In [4], Atiyah developed the theory of cohomology operations in equivariant K-theory (also known as *equivariant power operations*). In this section, we briefly recall the special cases of Atiyah's definitions that will serve as a model for our definitions in $\mathrm{Ell}^{\alpha}_G(\mathrm{pt})$. Recall that the coefficient ring of equivariant K-theory is the representation ring $K_G(\mathrm{pt}) = R(G)$. Atiyah's definitions, applied to representations, are given as follows:

$$\begin{aligned}P_n\colon R(G) &\to R(G\times\Sigma_n)\\ [V] &\mapsto [V^{\otimes n}],\end{aligned}$$

is called the nth *power operation.*[4] For finite groups G and H, and $h\in H$, Atiyah and Segal defined the map

$$\begin{aligned}\mathrm{Tr}(h\mid -)\colon R(G\times H) &\to R(G)\otimes_{\mathbb{Z}}\mathbb{Z}[\zeta]\\ W &\mapsto \sum W_\eta\otimes\eta.\end{aligned}$$

Here the sum runs over all eigenvalues η of the action of h on W, the space W_η is the eigenspace corresponding to η, and ζ is a primitive $|h|^{\mathrm{th}}$ root of unity.

The nth *symmetric power* of V is given by

$$\begin{aligned}\mathrm{sym}_n(V) &= \big(P_n(V)\big)^{\Sigma_n}\\ &= \frac{1}{n!}\sum_{\sigma\in\Sigma_n}\mathrm{Tr}(\sigma\mid P_n(V))\\ &= \sum_{[\sigma]_{\Sigma_n}}\frac{1}{|C_\sigma|}\cdot\mathrm{Tr}(\sigma\mid P_n(V)).\end{aligned}$$

Here $(-)^{\Sigma_n}$ picks out the Σ_n-invariant part of a $G\times\Sigma_n$-representation. All equalities are in $R(G)$, and the second equality can be proved using the splitting principle. Similarly, the nth *exterior power* of V is given by picking out the summand on

[4]Atiyah's power operations actually take values in $R(G\wr\Sigma_n)$ rather than $R(G\times\Sigma_n)$, but for our purposes, it is enough to pull them back via the diagonal map of G^n.

which Σ_n acts by the sign representation:

$$\begin{aligned}\lambda_n(V) &= \mathrm{Hom}_{G\times\Sigma_n}\big(\mathbb{C}[G]\otimes \mathrm{sgn}, P_n(V)\big)\\ &= \frac{1}{n!}\sum_{\sigma\in\Sigma_n}\mathrm{sgn}(\sigma)\cdot \mathrm{Tr}(\sigma \mid P_n(V)).\end{aligned}$$

To define the nth Adams operation, let $c_n \in \Sigma_n$ be a long cycle, and set

$$\psi_n(V) = \mathrm{Tr}(c_n \mid P_n(V)).$$

Again, one can use the splitting principle to compare this definition with others in the literature. The λ_i make $R(G)$ into a Λ-ring. Writing $\Lambda_t = \sum \lambda_n t^n$ for the total exterior power, and similarly S_t for the total symmetric power, one has the well known equations

$$\Lambda_{-t}(x) = \exp\Big(-\sum\frac{\psi_n(x)}{n}t^n\Big) \quad\text{and}\quad S_t(x) = \exp\Big(\sum\frac{\psi_n(x)}{n}t^n\Big). \tag{12}$$

Here t is a dummy variable. In [18, 9], I explained that, in the context of Atiyah's definitions, the equations (12) are generating functions, which encode the cycle decomposition of the elements of the symmetric groups. I also explained that for pairs of commuting elements of the symmetric group, there is an analogous argument using the decomposition of $\underline{n}$ into (σ,ρ)-orbits, i.e., into transitive $\mathbb{Z}$-sets.

To make the analogy with our setup even more explicit, we will view class functions on G as functions on G/G, where G acts on itself by conjugation, and note that G/G is the coarse moduli space of principal G-bundles on a circle. In this setup, the nth Adams operation is given by pull-back along the degree n self-map of the circle. The group of deck transformations of such a map has order n.

The nth symmetric power is the weighted sum of pull-backs along all isomorphism classes of n-fold coverings of $\mathbb{S}^1$. The weights are one over the order of the automorphism groups.

From this point of view, the generating function equation for the total symmetric power S_t in (12) encodes the fact that every covering of the circle decomposes into the disjoint union of covers with connected total space. More precisely, the analogue of Proposition 4.3 for covers of the circle is

$$\coprod_{k\geqslant 0}\mathrm{Cov}_k(\mathbb{S}^1)t^k = \coprod_{m\geqslant 0}\Big(\coprod_{n\geqslant 1}\mathrm{Cov}_n^{\mathrm{conn}}(\mathbb{S}^1)t^n\Big)\wr\Sigma_m.$$

Here $\mathrm{Cov}_k(\mathbb{S}^1)$ stands for the groupoid of k-fold covers of $\mathbb{S}^1$ and $\mathrm{Cov}_k^{\mathrm{conn}}(\mathbb{S}^1)$ for the groupoid of those with connected total space. The second equation in (12) is the weighted sum of the points of both sides of this decomposition.

6.3. Symmetric and exterior powers on Ell_G^α. In [18], I explained how Atiyah's theory can be carried over to Ell_G. The idea is simple: wherever Atiyah uses elements of the group, one replaces them by pairs of commuting elements. Then the Hecke operators take the role of the Adams operations. In this section, I will define the analogues of the symmetric and exterior powers. We will denote them by $\mathrm{sym}^{(2)}$ and $\lambda_n^{(2)}$. Depending on the taste of the reader, the 2 can stand for "pairs of commuting elements," for the fact that the torus is made from two circles, or for "chromatic level 2."

In order to define symmetric and exterior powers on Ell_G^α, we need to extend the "Hecke correspondence" (11) to all of $\mathcal{M}_{G\times\Sigma_n}$. We set

$$t := \mathcal{M}_{\mathrm{id}\times p}\colon \mathcal{M}_{G\times\Sigma_n} \to \mathcal{M}_G.$$

In order to define s, we recall the equivalence e of Proposition 4.3. For an abelian group A, let

$$s_A\colon \mathcal{M}_G \times_{\mathcal{M}_1} \widetilde{\mathcal{M}}_0(A) \to \mathcal{M}_G$$

be the source map. We define s to be e composed with the map which on the component indexed by the decomposition $\sum N_A|A| = n$ is given by

$$\prod_A s_A \wr \Sigma_{N_A}.$$

Over this component, we have

$$s^*\Big(\prod_A \mathcal{L}^\alpha \wr \Sigma_{N_A}\Big) \cong t^*(\mathcal{L}^\alpha)^{\otimes n}.$$

Definition 6.6. We define the nth symmetric power on sections f of $\mathcal{L}^\alpha$ by

$$\mathrm{sym}_n^{(2)}(f) := t_! \circ s^*\Big(\coprod_{n=\sum|A|N_A} \prod_A f \wr \Sigma_{N_A}\Big).$$

Here $\mathrm{sym}_n^{(2)}(f)$ is a section of $(\mathcal{L}^\alpha)^{\otimes n}$. We define the total symetric power $\mathrm{Sym}_q^{(2)}(f)$ by

$$\mathrm{Sym}_q^{(2)}(f) := \sum_{n\geqslant 0} \mathrm{sym}_n^{(2)}(f)\cdot q^n.$$

Here q is a dummy variable.

From now on, t will again denote our dummy variable (as opposed to the target map). The total symmetric power $\mathrm{Sym}_t^{(2)}$ takes values in the graded ring

$$\bigoplus_{n\geqslant 0} \mathrm{Ell}_G^{n\alpha}(\mathrm{pt})t^n.$$

In the case that $\alpha = 0$, this is the ring of formal power series in t. The series $\mathrm{Sym}_t^{(2)}(f)$ has a multiplicative inverse in $\bigoplus_{n\geqslant 0}\mathrm{Ell}_G^{-n\alpha}(\mathrm{pt})t^n$, which we will refer to as the total (alternating) exterior power, denoted

$$\Lambda_{-t}^{(2)}(f) = \sum_{n\geqslant 0} \lambda_n^{(2)}(f)(-t)^n.$$

The $\lambda_n^{(2)}$ make $\bigoplus \mathrm{Ell}_G^{-n\alpha}(\mathrm{pt})$ a Λ-ring.

Definition 6.7. Let (σ,ρ) be a commuting pair in Σ_n, and assume that (σ,ρ) induces the decomposition $n = \sum|A|N_A$. Let $s_{(\sigma,\rho)}$ denote the restriction of s to the component of $\mathcal{M}_{\Sigma_n}$ corresponding to this decomposition. We define the power operation $\psi_{(\sigma,\rho)}$ by

$$\psi_{(\sigma,\rho)}(f) := s^*_{(\sigma,\rho)}(f).$$

Definition 6.8. Let (σ,ρ) be a pair of commuting elements in Σ_n. We define the signature $\mathrm{sgn}(\sigma,\rho)$ of the pair by

$$\mathrm{sgn}(\sigma,\rho) := (-1)^{\#\text{even orbits of }(\sigma,\rho)}.$$

We have

$$\mathrm{sym}_n^{(2)}(f) = \frac{1}{n!} \sum_{\sigma\rho=\rho\sigma} \psi_{(\sigma,\rho)}(f)$$

and

$$\lambda_n^{(2)}(f) = \frac{1}{n!} \sum_{\sigma\rho=\rho\sigma} \mathrm{sgn}(\sigma,\rho) \cdot \psi_{(\sigma,\rho)}(f).$$

The proof of [18, Proposition 9.1] goes through to give the generating function equations

$$\mathrm{Sym}_t^{(2)}(f) = \exp\Big[\sum_{k\geqslant 0} T_k(f) \cdot t^k\Big]$$

and

$$\Lambda_t^{(2)}(f) = \exp\Big[-\sum_{k\geqslant 0} T_k(f) \cdot t^k\Big].$$

Remark 6.9. Another proof for the first equation and a detailed discussion of the generating function argument in terms of covers over elliptic curves can be found in [31]. In the case that G is the trivial group and $f = 1$, the statement is a special case of Roth's Lemma 2.9. His discussion is more general: it also applies to (simply branched) coverings of elliptic curves by curves of higher genus.

6.4. Replicability. Using these definitions, we can rephrase Definition 1.2 as follows (compare [6, (8.2)]):

Definition 6.10. A McKay-Thompson series $f(q)$ is called *replicable*, if (over the untwisted sector) it satisfies

$$(f(t) - f(q)) = t^{-1}\Lambda_{-t}^{(2)}(f(q)). \tag{13}$$

A few words of explanation are in order. The left-hand side of (13) equals

$$t^{-1} - f(q) + \sum_{n\geqslant 1} a_n t^n,$$

while the right-hand side is

$$t^{-1} - f(q) + \sum_{n\geqslant 1} \lambda_{n+1}^{(2)}(-t)^n.$$

Hence replicability means that for every $n \geqslant 1$, the function $\lambda_{n+1}^{(2)}(f)$ is a constant, namely $(-1)^{n+1}$ times the nth coefficient of the q-expansion of f.

7. The Witten genus

Let X be a compact, differentiable, spin manifold of dimension $2d$, let p_1 denote the first Pontrjagin class, and assume that $(p_1/2)(X) = 0$. Let $T_{\mathbb{C}}$ denote the complexification of the tangent bundle of X, and set $\overline{T}_{\mathbb{C}} := T_{\mathbb{C}} - \mathbb{C}^{2k}$. Then the Witten genus of X is defined as

$$\Phi_W(X) = \widehat{A}\Big(X, \bigotimes_{k=1}^{\infty} \mathrm{Sym}_{q^k}(\overline{T}_{\mathbb{C}})\Big).$$

Here $\mathrm{Sym}_q = \mathrm{Sym}_q^{(1)}$ stands for the total symmetric power in K-theory, and $\widehat{A}$ is the $\widehat{A}$-genus. If the dimension of X is a multiple of 24, the condition $p_1/2 = 0$ (often called "the vanishing of the *anomaly*") implies that

$$\frac{\Phi_W(X)}{\Delta^{2d/24}} = q^{-2d/24} \cdot \widehat{A}\Big(X, \bigotimes_{k=1}^{\infty} \mathrm{Sym}_{q^k}(T_{\mathbb{C}})\Big) \tag{14}$$

becomes a modular function. The Witten genus is linked to Moonshine by Hirzebruch's prize question [21, p. 86], which was answered affirmatively by Hopkins and Mahowald [26]:

> *Is there a 24-dimensional manifold X as above such that $\widehat{A}(X) = 1$ and $\widehat{A}(X, T_{\mathbb{C}}) = 0$?*

For such an X, the expression (14) becomes $j - 744$. Hirzebruch explains that if one could find an action of a finite group G on X by diffeomorphisms, this action would lift to the tangent bundle and its symmetric powers and make (14) a McKay–Thompson series. In the case that G is the Monster, one could hope for this series to be that of classical Moonshine. In fact, assume that the action of G lifts to the spin structure and that $\big((p_1/2)(X)\big)_G = 0$. Then (14) is the $g = 1$ part of a generalized McKay–Thompson series, $\phi_{W,G}(X)$, called the *equivariant Witten genus* of X [19]. Note that the equivariant genus in [19] differs from the one usually found in the literature by a normalization factor, so that my Φ_W there is actually given by the right-hand side of (14).

However, the setup of [19] describes only the $\alpha = 0$ case. Also, the condition that the equivariant class

$$\left(\frac{p_1}{2}\right)_G \in \mathrm{H}^4(X \times_G EG; \mathbb{Z})$$

should be zero turns out to be very restrictive. The following conjecture was explained to me by Matthew Ando, as was much of the next section:

Conjecture 7.1. *Let X be a compact, differentiable, G-equivariant spin manifold of dimension $24d$, and assume that $(p_1/2)(X) = 0$ and*

$$\left(\frac{p_1}{2}\right)_G (X) = \pi^*(\alpha),$$

where π is the map from X to a point, and α is an element of $\mathrm{H}^4(BG;\mathbb{Z})$. Then the equivariant Witten genus of X is a generalized Thompson series whose transformation behavior under $\mathrm{SL}_2(\mathbb{Z})$ makes it a section of $\mathcal{L}^{-\alpha}$.

The idea to interpret non-trivial anomaly as giving a section of a line bundle goes back to Witten and can be found in the work of Freed and Witten [16], Liu [24], and Ando [3]. The next section outlines how this conjecture could follow from the conjectural behavior of the Thom isomorphism in [20].

7.1. Twisted Thom isomorphisms. The Witten genus is closely related to the theory of Thom isomorphisms and transfer maps in elliptic cohomology. Assume that $V - TX$ is a spin bundle and that we are given a lift of the action of G to the spin structure. Let X^{V-TX} denote the equivariant Thom spectrum of the virtual vector bundle $V - TX$ [23]. Then $\mathrm{Ell}^0_G(X^{V-TX})$ is an invertible Ell^0_G module sheaf.

Pulling back first along the relative zero section $X^{-TX} \to X^{V-TX}$ and then along the Pontrjagin–Thom collapse $\mathbb{S}^0 \to X^{-TX}$, we obtain a map

$$\mathrm{Ell}_G^0(X^{V-TX}) \to \mathrm{Ell}_G^0(X^{-TX}) \to \mathrm{Ell}_G^0(\mathbb{S}^0) = \mathscr{O}_{\mathcal{M}_G}.$$

If the equivariant characteristic class

$$\left(\frac{p_1}{2}\right)_G (V - TX) \in \mathrm{H}^4(EG \times_G X; \mathbb{Z})$$

equals zero, the Thom isomorphism is a trivialization

$$\mathrm{Ell}_G^0(X) \xrightarrow{\cong} \mathrm{Ell}_G^0(X^{V-TX}).$$

Composing it with the map above, gives the *transfer-* (or *Gysin-*) map along $\pi\colon X_+ \to \mathbb{S}^0$,

$$\pi_!^V\colon \mathrm{Ell}_G^0(X) \to \mathrm{Ell}_G^0(\mathbb{S}^0),$$

and the Witten genus of X twisted with V is

$$\phi_{W,G}(V) = \pi_!^V(1).$$

If V is trivial, we drop it from the notation and have

$$\phi_{W,G}(X) = \pi_!(1).$$

Assume now that the weaker condition

$$\left(\frac{p_1}{2}\right)_G (V - TX) = \pi^*(\alpha)$$

is satisfied, where

$$\alpha \in \mathrm{H}^4(BG; \mathbb{Z}),$$

and assume that, non-equivariantly, we have

$$\left(\frac{p_1}{2}\right)(V - TX) = 0 \in \mathrm{H}^4(X; \mathbb{Z}).$$

Then we get a twisted Thom isomorphism

$$\mathrm{Ell}_G^0(X) \otimes_{\mathscr{O}_{\mathcal{M}_G}} \mathcal{L}^\alpha \xrightarrow{\cong} \mathrm{Ell}_G^0(X^{V-TX}),$$

yielding a transfer map

$$\pi_!^V\colon \mathrm{Ell}_G^0(X) \to (\mathcal{L}^\alpha)^{-1}.$$

Hence the Witten genus $\phi_{W,G}(V)$ takes values in the global sections of $\mathcal{L}^{-\alpha}$.

7.2. A few words about physics. Principal bundles over Riemann surfaces turn up in string theory, when the target space of the theory is an orbifold quotient $X/\!\!/G$. In string theory with target space X, one considers spaces of maps from the circle (a *closed string*) or from a Riemann surface Σ (the *worldsheet*) to X. An orbifold map from Σ to $X/\!\!/G$ turns out to be the same as an equivalence class of G-equivariant maps from a principal G-bundle over Σ to X:

$$\mathrm{map}_{\mathrm{orb}}(\Sigma, X/\!\!/G) = \coprod_{P \to \Sigma \text{pbdl}} \mathrm{map}_G(P, X)/\sim$$

(cf., e.g., [25, 30, 33, 35]). The space

$$\mathcal{L}(X/\!\!/G) = \mathrm{map}_{\mathrm{orb}}(\mathbb{S}^1, X/\!\!/G)$$

is called the *orbifold loop space* of $X/\!\!/G$, and the space

$$\mathcal{L}^2(X/\!\!/G) = \mathrm{map}_{\mathrm{orb}}(\mathbb{T}^2, X/\!\!/G)$$

is the double (orbifold) loop space of $X/\!\!/G$. They decompose into components

$$\mathcal{L}_g(X/\!\!/G) := \mathrm{map}_G(P_g, X) \quad \text{and} \quad \mathcal{L}^2_{g,h}(X/\!\!/G) = \mathrm{map}_G(P_{g,h}, X),$$

often referred to as the *twisted sectors* and the *membrane twisted sectors* of the theory.

In the non-equivariant case, the Witten genus $\Phi_W(X)$ is the genus 1 partition function of the supersymmetric non-linear sigma model for the target space X [34].

In [10], de Fernex, Lupercio, Nevins and Uribe consider a supersymmetric string sigma model whose target space is an orbifold and show that its partition function on a two-dimensional torus is the orbifold Witten genus. The orbifold Witten genus is related to the equivariant Witten genus by

$$\begin{aligned}\Phi_{W,\mathrm{orb}}(X/\!\!/G) &= \frac{1}{|G|}\sum_{gh=hg} \Phi_{W,G}(X)(g,h)\\ &= \mathrm{ind}\,|^1_G\big(\Phi_{W,G}(X)\big).\end{aligned}$$

In [10], the (g,h)-summand of $\Phi_{W,\mathrm{orb}}$ is the contribution of the corresponding twisted sector to the partition sum, which is a phase integral over the torus with boundary conditions twisted by g and h. Hence the discussion of [10] fits the equivariant Witten genus (the same thing as in [19]) into the framework of non-linear sigma models.

The idea to study maps into an orbifold target space via principal bundles on the worldsheet already turned up in Segal's celebrated work [32]. In the same paper, Segal prominently announced the existence of a "*fairly simple and natural conformal field theory,*" whose automorphism group is the Monster. In [17], Frenkel, Lepowski and Meurman constructed a vertex algebra whose automorphism group is the Monster, the so-called *Moonshine module*. Vertex algebras are often thought of as the mathematical language of two-dimensional conformal field theory. In his famous paper [6], Borcherds proved that the Moonshine module yields the McKay–Thompson series of the classical Moonshine conjecture. It is expected and in many cases proved that generalized Moonshine is given by the twisted sectors of the Moonshine module (cf. [14]).

In [13], Dixon, Ginsparg, and Harvey translate the result of [17] into the language of string theory. According to them, the target space of the Monster CFT is the orbifold

$$\mathbb{T}^{24}_{\mathrm{Leech}}/\!\!/(\mathbb{Z}/2\mathbb{Z}).$$

Here $\mathbb{T}^{24}_{\mathrm{Leech}} = \mathbb{R}^{24}/\Lambda_{\mathrm{Leech}}$, where Λ_{Leech} is the Leech lattice, and the $\mathbb{Z}/2\mathbb{Z}$ action is induced by multiplication with -1 on $\mathbb{R}^{24}$. Its orbifold partition function is $j-744$. The automorphism group of the theory is the Monster. For a pair of commuting elements (g,h) in the Monster, [13] describes the generalized Moonshine function $f(g,h;\tau)$ as the phase integral over the torus with boundary conditions twisted by g and h.

Given these facts, one could hope to reinterpret Hirzebruch's question and ask whether the orbifold Witten genus of $\mathbb{T}^{24}_{\mathrm{Leech}}/\!\!/(\mathbb{Z}/2)$ was given by $j-744$. After all, the Witten genus is the partition function of the theory, and the partition function is $j-744$. Unfortunately, the two theories are different, and in particular, their notions of (orbifold) partition function do not agree.[5] For instance, as a consequence

[5]More precisely, we have the sigma model whose partition function is the (orbifold) Witten genus on one hand; this is the $N=\frac{1}{2}$ suspersymmetric non-linear sigma model with right moving

of the rigidity theorem, the contribution of the non-twisted sector of the Witten genus, $\Phi_W(\mathbb{T}^{24}_{\text{Leech}})$, is zero, whereas the analogous contribution to the partition function of the Moonshine module is $j - 720$.

When Witten defines the Witten genus in [37], he explains that different non-linear sigma models lead to different genera.

Question 7.2. Is it possible to describe the partition function of the "holomorphic conformal field theory" in [13] as a push-forward construction in equivariant elliptic cohomology, similar to the description of the Witten genus in [19], and can we identify the corresponding genus? How does the discussion in [10] need to be modified to supply the geometric side of this construction (in terms of orbifold loop spaces)? Ideally, one would want this genus to preserve power operations. In this case, the theory would have a Dijkgraaf – Moore – Verlinde – Verlinde formula.

7.3. The Witten genus and replicability? In [19], I calculated the total symmetric power in Devoto's equivariant Tate K-theory applied to a G-equivariant vector bundle V and observed that it equals Witten's exponential characteristic class:

$$\mathrm{Sym}_q^{(2)}(V) = \bigotimes_{k=1}^{\infty} \mathrm{Sym}_{q^k}^{(1)}(V).$$

Recall that we have

$$\mathrm{Sym}_q^{(2)}(V) = \Lambda_{-q}^{(2)}(-V).$$

Question 7.3. Can one describe replicability (13) in the context of Hirzebruch's prize question as asking for an equality about iterated total exterior powers at chromatic level 2?

Question 7.4. Is it possible to make sence of the notion of replicability for generalized McKay Thompson series? In this case, the coefficient $\lambda^{(2)}_{n+1}(f(q))$ on the right-hand side of (13) is a section of the line bundle $(\mathcal{L}^{\alpha})^{\otimes n+1}$. What does it mean for such a thing to be a constant? And how should we take the fractional powers of t occurring over the twisted sectors on the left-hand side into account?

Note that these two issues are related: The fractional powers in the q-expansions of the twisted sectors come from the fact that T does not act trivially on the lines $L_{g,h}$. Let L_N be the line of invariant sections of the functor $\mathcal{F}$ on $\mu_N \rtimes \mu_N$ which is given by $\mathcal{F}(\zeta) = \mathbb{C}$ on objects and such that

$$\mathcal{F}(\zeta, \xi) \colon \mathcal{F}(\zeta) \to \mathcal{F}(\zeta\xi)$$

is multiplication with ξ. Then there is a $\mathbb{Z}/N\mathbb{Z}$-equivariant ring isomorphism

$$\bigoplus_{m\geqslant -1} L_N^{\otimes m} q^m \to \mathbb{C}[\![q^{1/N}]\!] \qquad a_m q^m \mapsto a_m(\mathrm{e}^{2\pi \mathrm{i}/N}) q^{m/N}.$$

Here $1 \in \mathbb{Z}/N\mathbb{Z}$ acts on $q^{1/N}$ by multiplication with $\mathrm{e}^{2\pi\mathrm{i}/N}$.

---

fermions only, coupled to the spin connection of the manifold M decribed in [37, (26)]. On the other hand, we have the holomorphic CFT associated to the Frenkel-Lepowsky-Meurman module described in [13] .

Appendix A. Generalities on principal bundles

This appendix collects some well known facts about principal bundles used in the paper. Let G be a finite group.

Definition A.1. A principal G-bundle ξ consists of a map $\pi\colon P \to X$, where P is a right G-space such that G acts strictly transitively on the fibers of π. The space P is called the *total space* and X the *base space* of ξ.

For simplicity, we will always assume that the base space X is connected.

Definition A.2. Let $i\colon G \hookrightarrow H$ be an inclusion of groups, and let ξ be a principal G-bundle. The principal H-bundle $\xi[H]$ associated to ξ via i has total space

$$P \times_G H := (P \times H)/\sim,$$

where $(p, h) \sim (pg, i(g^{-1})h)$. The action of H on $P \times_G H$ is defined by

$$(p, h) \cdot h' := (p, hh').$$

The projection $\pi[H]$ is defined by $\pi[H](p, h) := \pi(p)$.

If $f\colon X \to BG$ classifies the G-bundle ξ, the associated H-bundle $\xi[H]$ is classified by the map $Bi \circ f$.

Definition A.3. Let $\xi = (\pi\colon P \to X)$ be a principal G-bundle with connected total space, and let $\alpha\colon H \to G$ be an isomorphism of groups. We define the pull-back of ξ along α to be the principal H-bundle $\alpha^*(\xi)$ consisting of the same map π, where the H action on the total space P is given by

$$p * h := p \cdot \alpha(h).$$

In the case $H = G$, we will refer to this pulled back action as "*the G-action on P twisted by α.*"

Lemma A.4. *The bundle $\alpha^*(\xi)$ is isomorphic to the G-bundle associated to ξ via the inclusion $\alpha^{-1}\colon G \to G$.*

PROOF. Consider the map

$$\begin{aligned} f\colon P &\to P \times_{\alpha^{-1}} G \\ p &\mapsto (p, 1). \end{aligned}$$

It is obviously fiber preserving and bijective. We have

$$f(p)\cdot g = (p,1)\cdot g = (p,g) \sim (p\alpha(g), \alpha^{-1}(\alpha(g))\cdot 1) = (p\alpha(g), 1) = f(p\alpha(g)) = f(p*g).$$

Therefore f is G-equivariant with respect to the G-action on P twisted by α. □

Corollary A.5. *Let ξ be classified by the map $f\colon X \to BG$. Then $\alpha^*(\xi)$ is classified by*

$$B\alpha^{-1} \circ f.$$

Lemma A.6. *Let $G \subseteq H$ be an inclusion of groups, let ξ be a principal H-bundle, and let F be a left G-space. Then*

$$\xi \times_H F = \xi[G] \times_G F.$$

Proposition A.7. *Let $\xi = (\pi\colon P \to X)$ be a principal G-bundle over X, and $\zeta = (\rho\colon Q \to X)$ be a principal H-bundle over X, and assume that both total spaces P and Q are connected. Assume further that there exists an isomorphism of covering spaces of X*

$$f\colon P \to Q.$$

Then there exists an isomorphism of groups

$$i\colon G \to H$$

making f into an isomorphism of principal G-bundles

$$f\colon \xi \xrightarrow{\cong} i^*\zeta.$$

Moreover, i is uniquely determined by f.

PROOF. Fix an element $p \in P$. We define the map $i\colon G \to H$ as follows: for $g \in G$ let $i(g)$ be the unique element of H mapping $f(p)$ to $f(pg)$. It is clear that i is a bijection. We need to show that i is a map of groups. Fix $g \in G$, and consider the deck transformations D_g of π sending p to pg and $D_{i(g)}$ of ρ sending q to $q \cdot i(g)$. Then $D_{i(g)}$ and $f \circ D_g \circ f^{-1}$ both are deck transformations of Q sending $f(p)$ to $f(pg)$. We may conclude that they are the same deck transformation. Now we have

$$\begin{aligned} f(p) \cdot i(hg) &= f\big(p(hg)\big) \\ &= f\left(\left(f^{-1}\big(f(ph)\big)\right) \cdot g\right) \\ &= f(ph) \cdot i(g) \\ &= \big(f(p) \cdot i(h)\big) \cdot i(g). \end{aligned}$$

Further, for every $g \in G$, we have

$$f \circ D_g = D_i(g) \circ f,$$

proving that f is an isomorphism of principal G-bundles. Let now $\beta\colon G \to H$ be an isomorphism making f into an isomorphism between the principal G-bundles ξ and $\beta^*(\zeta)$. Then we have for every $g \in G$,

$$f(pg) = f(p)\beta(g),$$

implying that $\beta(g) = i(g)$. □

References

1. M. Ando, *Operations in complex-oriented cohomology theories related to subgroups of formal groups*, Ph.D. thesis, MIT, 1992.
2. ———, *Power operations in elliptic cohomology and representations of loop groups*, Trans. Amer. Math. Soc. **352** (2000), no. 12, 5619–5666.
3. ———, *The sigma orientation for analytic circle-equivariant elliptic cohomology*, Geom. Topol. **7** (2003), 91–153.
4. M. F. Atiyah, *Power operations in K-theory*, Quart. J. Math. Oxford Ser. (2) **17** (1966), 165–193.
5. A. Baker and C. B. Thomas, *Classifying spaces, Virasoro equivariant bundles, elliptic cohomology and moonshine*, Technical Report 99/39, Glasgow University Mathematics Department, `http://www.maths.gla.ac.uk/~ajb/dvi-ps/ab-cbt.pdf`.
6. R. E. Borcherds, *Monstrous moonshine and monstrous Lie superalgebras*, Invent. Math. **109** (1992), no. 2, 405–444.
7. K. S. Brown, *Cohomology of groups*, Grad. Texts in Math., vol. 87, Springer, New York, 1994.
8. J. H. Conway and S. P. Norton, *Monstrous moonshine*, Bull. London Math. Soc. **11** (1979), no. 3, 308–339.

9. C. J. Cummins and T. Gannon, *Modular equations and the genus zero property of moonshine functions*, Invent. Math. **129** (1997), no. 3, 413–443.
10. T. de Fernex, E. Lupercio, T. Nevins, and B. Uribe, *A localization principle for orbifold theories*, Recent Developments in Algebraic Topology (San Miguel de Allende, 2003) (A. Ádem, J. González, and G. Pastor, eds.), Contemp. Math., vol. 407, Amer. Math. Soc., Providence, RI, 2006, pp. 113–133.
11. J. A. Devoto, *Equivariant elliptic homology and finite groups*, Michigan Math. J. **43** (1996), no. 1, 3–32.
12. R. Dijkgraaf, C. Vafa, E. Verlinde, and H. Verlinde, *The operator algebra of orbifold models*, Comm. Math. Phys. **123** (1989), no. 3, 485–526.
13. L. Dixon, P. Ginsparg, and J. Harvey, *Beauty and the beast: superconformal symmetry in a Monster module*, Comm. Math. Phys. **119** (1988), no. 2, 221–241.
14. C. Dong, H. Li, and G. Mason, *Modular-invariance of trace functions in orbifold theory and generalized Moonshine*, Comm. Math. Phys. **214** (2000), no. 1, 1–56.
15. D. S. Freed and F. Quinn, *Chern–Simons theory with finite gauge group*, Comm. Math. Phys. **156** (1993), no. 3, 435–472.
16. D. S. Freed and E. Witten, *Anomalies in string theory with D-branes*, Asia J. Math **3** (1999), no. 4, 819–851.
17. I. B. Frenkel, J. Lepowsky, and A. Meurman, *A moonshine module for the Monster*, Vertex Operators in Mathematics and Physics (Berkeley, CA, 1983) (J. Lepowsky, S. Mandelstam, and I. M. Singer, eds.), Math. Sci. Res. Inst. Publ., vol. 3, Springer, New York, 1985, pp. 231–273.
18. N. Ganter, *Orbifold genera, product formulas and power operations*, Adv. Math. **205** (2006), no. 1, 84–133.
19. ———, *Stringy power operations in Tate K-theory*, available at `arXiv:math/0701565`.
20. V. Ginzburg, M. Kapranov, and E. Vasserot, *Elliptic algebras and equivariant elliptic cohomology*, available at `arXiv:q-alg/9505012`.
21. F. Hirzebruch, T. Berger, and R. Jung, *Manifolds and modular forms*, with appendices by N.-P. Skoruppa and P. Baum, Aspects Math., vol. E20, Friedr. Vieweg & Sohn, Braunschweig, 1992.
22. M. J. Hopkins, N. J. Kuhn, and D. C. Ravenel, *Generalized group characters and complex oriented cohomology theories*, J. Amer. Math. Soc. **13** (2000), no. 3, 553–594.
23. L. G. Lewis Jr., J. P. May, M. Steinberger, and J. E. McClure, *Equivariant stable homotopy theory*, with by J. E. McClure, Lecture Notes in Math., vol. 1213, Springer, Berlin, 1986.
24. K. Liu, *On modular invariance and rigidity theorems*, J. Differential Geom. **41** (1995), no. 2, 343–396.
25. E. Lupercio, B. Uribe, and M. A. Xicotencatl, *Orbifold string topology*, available at `arXiv:math/0512658`.
26. M. Mahowald and M. Hopkins, *The structure of 24 dimensional manifolds having normal bundles which lift to $BO[8]$*, Recent Progress in Homotopy Theory (Baltimore, MD, 2000) (D. M. Davis, J. Morava, G. Nishida, W. S. Wilson, and N. Yagita, eds.), Contemp. Math., vol. 293, Amer. Math. Soc., Providence, RI, 2002, pp. 89–110.
27. Y. Martin, *On modular invariance of completely replicable functions*, Moonshine, the Monster, and Related Topics (South Hadley, MA, 1994) (C. Dong and G. Mason, eds.), Contemp. Math., vol. 193, Amer. Math. Soc., Providence, RI, 1996, pp. 263–286.
28. G. Mason, *Finite groups and modular functions*, with an appendix by S. P. Norton, The Arcata Conference on Representations of Finite Groups (Arcata, CA, 1986) (P. Fong, ed.), Proc. Sympos. Pure Math., vol. 47, Part 1, Amer. Math. Soc., Providence, RI, 1987, pp. 181–210.
29. ———, *Orbifold conformal field theory and cohomology of the monster*, available at `http://www.newton.cam.ac.uk/programmes/NST/Mason.pdf`.
30. I. Moerdijk, *Orbifolds as groupoids: an introduction*, Orbifolds in Mathematics and Physics (Madison, WI, 2001) (A. Adem, J. Morova, and Y. Ruan, eds.), Contemp. Math., vol. 310, Amer. Math. Soc., Providence, RI, 2002, pp. 205–222.
31. M. Roth, *Counting covers of an elliptic curve*, available at `http://www.mast.queensu.ca/~mikeroth/`.

32. G. Segal, *The definition of conformal field theory*, Topology, Geometry and Quantum Field Theory (Oxford, 2002) (U. Tillmann, ed.), London Math. Soc. Lecture Note Ser., vol. 308, Cambridge Univ. Press, Cambridge, 2004, pp. 421 – 577.
33. E. Sharpe, *Quotient stacks and string orbifolds*, available at arXiv:hep-th/0103008.
34. S. Stolz and P. Teichner, *What is an elliptic object?*, Topology, Geometry and Quantum Field Theory (Oxford, 2002) (U. Tillmann, ed.), London Math. Soc. Lecture Note Ser., vol. 308, Cambridge Univ. Press, Cambridge, 2004, pp. 247 – 343.
35. H. Tamanoi, *Infinite product decomposition of orbifold mapping spaces*, available at arXiv:0706.0932v1.
36. S. Willerton, *The twisted Drinfeld double of a finite group via gerbes and finite groupoids*, available at arXiv:math.QA/0503266.
37. E. Witten, *Elliptic genera and quantum field theory*, Comm. Math. Phys. **109** (1987), no. 4, 525 – 536.

Department of Mathematics, University of Illinois at Urbana-Champaign, 1409 W. Green Street, Urbana, IL 61801, USA

E-mail address: ganter@math.uiuc.edu

Centre de Recherches Mathématiques
CRM Proceedings and Lecture Notes
Volume **47**, 2009

Sculptural Presentation of the Icosahedral Rotation Group

George W. Hart

ABSTRACT. A 1.2 meter diameter wooden geometric sculpture was assembled by participants at the McKay conference. Its sixty components are in one-to-one correspondence with the elements of the icosahedral rotation group. We constructed it on the first day, so the sculpture served as a centerpiece in the common room until we disassembled it on the last day. It presents the beauty of symmetry in a visual form, analogous to the mathematical beauty presented in more abstract form in talks in the conference hall.

The conference *Groups and Symmetries: From the Neolithic Scots to John McKay* was a meeting of participants with a keen interest in groups and symmetry. John McKay invited me to participate in an artistic role, by contributing a mathematical construction to the event. As a sculptor of geometric forms, I enjoy creating sculptures that are assembled as a "barn raising" by a group of participants. Examples of such works can be seen on my web pages [2]. These activities are celebrations of mathematical beauty in a sculptural form.

For the McKay conference at the CRM in Montreal, I provided a set of laser-cut wooden parts that assemble into a 1.2 meter diameter hanging sculpture. The result has sixty icosahedrally arranged components and can be seen as a physical presentation of the icosahedral symmetry group. Shown in Figure 1, the form is composed of identical cartoon-like plywood "people." Each is planar. The geometry and the pattern of connections are unintuitive, as adjacent "people" do not touch. Their limbs weave in or out past their neighbor's limbs and connect to parts slightly further away. So it takes some pondering and practice before the participants become fully comfortable with the structure.

In designing such a sculpture, I begin with a symmetrically arranged set of planes in space, calculate their lines of intersection, and devise a shape that connects one plane to another at the lines of intersection. I use a sculpture editing program [3] for implementing the symmetry transformations, visualizing the form, keeping track of the various lengths and angles, and outputting the part shape in a file format for laser cutting. For this sculpture, the sixty "people" are in the planes of Catalan's pentagonal hexacontahedron, the dual to the snub dodecahedron. This

2000 *Mathematics Subject Classification.* 52B10.
This is the final form of the paper.

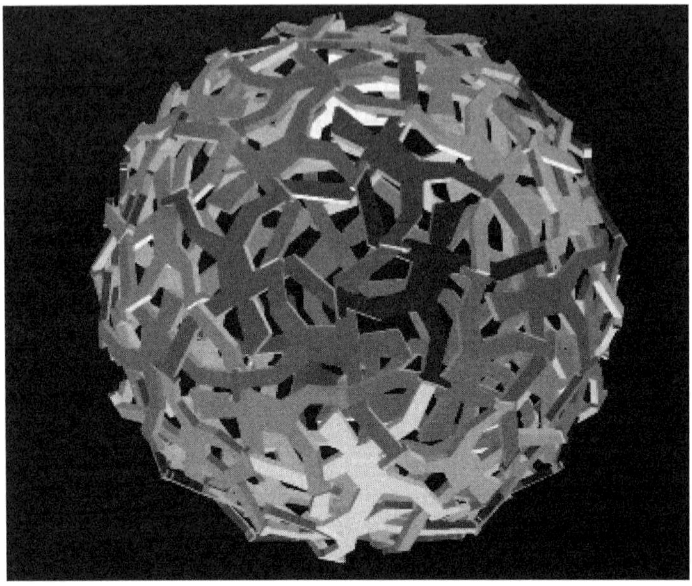

FIGURE 1. Computer model of sculpture's form, before fabricating components

is an isogonal polyhedron, meaning the symmetry group is transitive on the faces, so each component is surrounded by neighbors in an identical manner.

I began the assembly event by letting the participants explore the different ways that the parts might be connected. I did not give specific instructions except to say that the symmetry is icosahedral and transitive on the parts. Within a half hour the participants were able to try out and eliminate all the various ways of connecting the parts that were incompatible with these clues. For example, if seven parts seem to make a loop then the connection can not be correct, as the icosahedral group can not contain a subgroup of size seven. What remained was the correct method of assembly: every front arm connects to a front leg and every back arm connects to a back leg. Once that is known, the construction proceeds rapidly. Small groups of participants can make modules in parallel. Figure 2 shows a stage partway through the construction. The work is suspended from the ceiling so a half-dozen people can add to it from all sides at once.

At the wrists and feet, there is a hole for connecting parts together. They are neatly fastened with 120 aluminum connectors and 240 nylon thumbscrews. The screws are easily turned by hand, so no tools are required. The aluminum parts were drilled and tapped to accept the screws, bent to the proper dihedral angle, then spray painted to blend with the wood. Figure 3 shows a close-up of this hardware attached to one wooden arm.

After conceiving the design, I was able to cut the sixty Baltic birch plywood components on a laser cutter at the computer science and artificial intelligence laboratory at MIT. I am grateful to Erik Demaine and Martin Demaine for arranging this and for all their help. There we also test-fit the parts [1]. Laser cutting the components results in precise lengths and angles, ensuring that the parts fit exactly together while weaving in and out through each other in three dimensions. The disassembled parts are easily shipped from one venue to another. I have organized

FIGURE 2. Halfway through the construction

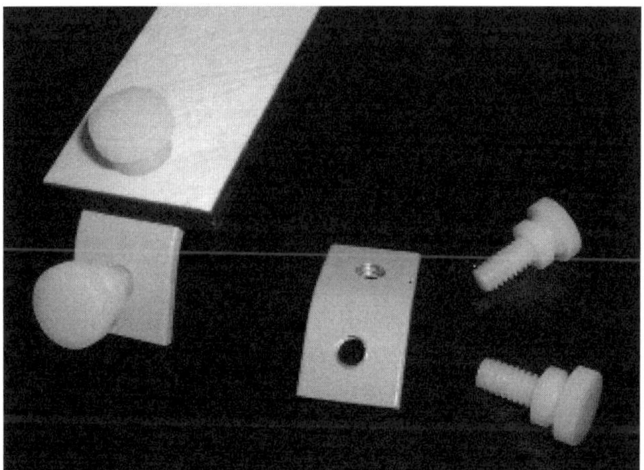

FIGURE 3. Aluminum connectors and nylon thumb screws

dozens of events like this at various colleges and conferences, using this same set of parts. About one hour is generally required for the assembly.

Figure 4 shows the Groups and Symmetries conference participants around the completed sculpture. We found it to be a unique bonding experience: an interdisciplinary activity combining the precision of geometric thinking, the beauty

FIGURE 4. Conference participants in group photo around sculpture

of wooden sculpture, the engineering of architectural structure, and the fun of an intricate assembly puzzle. I was honored to be able to present such an event at John McKay's celebration.

Acknowledgments. I thank John Harnad for arrangements which made my participation at the McKay conference possible. Portions of this material were presented at the Conference of the International Society of Art, Mathematics, and Architecture held at DePaul University, June 17 – 19, 2004.

References

1. E. Demaine, photos, available at `http://theory.lcs.mit.edu/~edemaine/photos/GeorgeHart_February2004`.
2. G. W. Hart, `http://www.georgehart.com`. web pages.
3. ______, *Symmetric sculpture*, J. Math. Arts **1** (2007), no. 1, 21 – 28.

COMPUTER SCIENCE DEPARTMENT, STONY BROOK UNIVERSITY, STONY BROOK, NY 11794, USA
E-mail address: `george@georgehart.com`

Centre de Recherches Mathématiques
CRM Proceedings and Lecture Notes
Volume **47**, 2009

Spherical Harmonics and the Icosahedron

Nigel Hitchin

Dedicated to John McKay

Abstract. We define a sextic invariant J on the seven-dimensional space of degree three spherical harmonics and show that J is positive if and only if the nodal set of the spherical harmonic contains the vertices of exactly two regular icosahedra. The proof uses the geometry of the Clebsch diagonal cubic surface, Atiyah's classification of vector bundles on an elliptic curve and a Fano threefold introduced by Mukai and Umemura.

1. Introduction

Spherical harmonics of degree l are the functions on the unit 2-sphere which satisfy $\Delta f = l(l+1)f$ for the Laplace–Beltrami operator Δ. They form an irreducible representation of SO(3) of dimension $2l+1$ and are the restrictions to the sphere of homogeneous polynomials $f(x_1, x_2, x_3)$ of degree l which solve Laplace's equation in $\mathbb{R}^3$. This paper concerns a curious relationship between the case $l = 3$ and the regular icosahedron.

We consider the zero set of f — the *nodal set* — and ask whether it contains the 12 vertices of a regular icosahedron. If, for example, $f = (5\cos^3\theta - 3\cos\theta)/2$ (the third-order Legendre polynomial) then its nodal set consists of the intersection of the sphere with the three planes $x_3 = 0, x_3 = \pm\sqrt{3/5}$ and it is easy to see that no three parallel planes can contain all the vertices of the icosahedron. On the other hand, if $\phi = (1+\sqrt{5})/2$, consider the standard icosahedron with vertices $(0, \pm\phi, \pm 1)$, $(\pm 1, 0, \pm\phi)$, $(\pm\phi, \pm 1, 0)$ (on the sphere of radius $\sqrt{\phi^2+1}$). They clearly lie on the nodal set of $f = x_1x_2x_3$, but then so do $(0, \pm 1, \pm\phi)$, $(\pm\phi, 0, \pm 1)$, $(\pm 1, \pm\phi, 0)$ so we have two icosahedra.

We define an invariant of f as follows: let

$$M_{ij}(f) = \int_{S^2} f^2(x)\left(x_ix_j - \frac{4}{15}\delta_{ij}\right) \mathrm{d}S$$

1991 *Mathematics Subject Classification.* Primary 33C55, 14H60; Secondary 14M12, 15A72.
Key words and phrases. Icosahedron, Clebsch cubic, spherical harmonic.
This is the final form of the paper.

and let $x^3 + a_1x^2 + a_2x + a_3$ be the characteristic polynomial of M. Put $J(f) = 3a_3 - 4a_1a_2$, then our result is:

Theorem. *Let f be a degree three spherical harmonic.*

- *If $J(f) > 0$ (resp. $J(f) < 0$) then the nodal set of f contains the vertices of two (resp. zero) regular icosahedra.*
- *When $J(f) = 0$ and f is of the form $f(x) = (u,x)((a,x)^2 - (x,x)(a,a)/5)$ with $(u,a) = 0$ then any regular icosahedron with a as a vertex lies on the nodal set, otherwise there is a unique such icosahedron.*

Our proof uses the geometry of the Clebsch diagonal cubic surface, vector bundles on an elliptic curve and a Fano threefold introduced by S. Mukai and H. Umemura.

2. Harmonic cubics

Let U be a 3-dimensional real vector space with positive definite inner product (a,b) and consider the 10-dimensional vector space $\operatorname{Sym}^3 U^*$ of degree 3 homogeneous polynomial functions on U. The inner product identifies U with U^* and induces an inner product on $\operatorname{Sym}^3 U^*$ which we normalize so that

$$(p(x), (a,x)^3) = p(a).$$

The special orthogonal group SO(3) acts on $\operatorname{Sym}^3 U^*$ and decomposes it into two orthogonal irreducible components: a 7-dimensional representation H of functions which satisfy Laplace's equation (the *harmonic cubics*), and a 3-dimensional representation consisting of cubics of the form $(a,x)(x,x)$ for $a \in U$. It is the space H which will concern us here. Restricting to the unit sphere these are eigenfunctions of the Laplace–Beltrami operator with eigenvalue 12.

If $p \in H$, then the inner product $p(a) = (p(x), (a,x)^3)$ only depends on the harmonic component of $(a,x)^3$. This is of the form

$$f(x) = (a,x)^3 - \lambda(a,x)(x,x)$$

and differentiating we find that it satisfies Laplace's equation only if $\lambda = \frac{3}{5}$. Thus

$$f_a(x) = (a,x)^3 - \tfrac{3}{5}(a,a)(a,x)(x,x) \tag{2.1}$$

is the harmonic part of $(a,x)^3$ and so

$$(p(x), f_a(x)) = p(a).$$

The function f_a is, up to a constant multiple, the unique harmonic cubic which is symmetric about the axis a — if $a = (0,0,1)$ then $f_a = r^3 P_3(\cos\theta)$ for the Legendre polynomial P_3.

3. The icosahedron

Let $\phi = (1+\sqrt{5})/2$ be the golden ratio. The standard model for a regular icosahedron of side length 2 has vertices $\pm a_1, \dots, \pm a_6$ given by

$$(0, \pm\phi, \pm 1), \quad (\pm 1, 0, \pm\phi), \quad (\pm\phi, \pm 1, 0). \tag{3.1}$$

It has 20 triangular faces, 30 edges and 12 vertices occurring in six opposite pairs.

The group of symmetries of the icosahedron is isomorphic to the alternating group A_5. The five objects it permutes are triples of orthogonal planes which pass through the vertices (see Fig. 1). In the standard model such a triple is provided

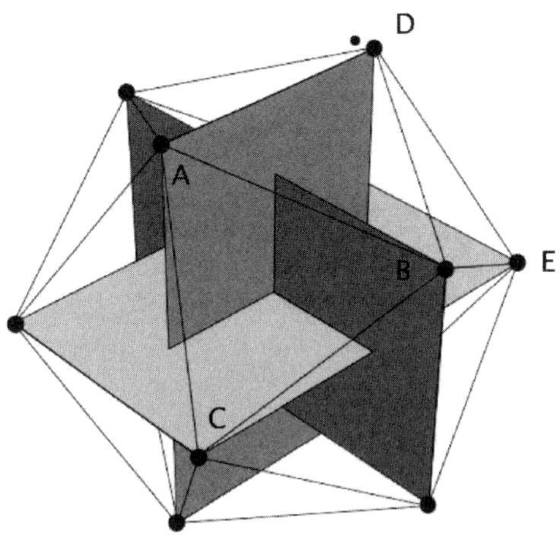

FIGURE 1. The icosahedron[1]

by the coordinate planes $x_1 = 0, x_2 = 0, x_3 = 0$ and clearly the zero set of the cubic $q(x) = x_1x_2x_3$ contains all the vertices of the standard icosahedron.

The first result provides the link between the two themes in the title.

Proposition 3.1. *The cubic polynomials which vanish on the vertices of a regular icosahedron form a four-dimensional space of harmonic cubics.*

PROOF. To prove that there is a 4-dimensional space of such cubic functions we adopt a geometric point of view, though a character argument for the icosahedral group can also be used. We use algebraic geometric results over the complex numbers and then specialize to the real situation. We shall still denote the vector space over the complex numbers by U, but recognize that it has a real structure, an antilinear involution. Then a homogeneous cubic polynomial p defines a cubic curve in the two-dimensional projective plane $\mathrm{P}(U)$. The image of the nodal set under the double covering $S^2 \to \mathrm{P}(U)^{\mathrm{r}}$ consists of the real points of the cubic curve. (We shall use the superscript r to denote real points).

The cubic curves passing through m points in the projective plane define the anticanonical system for the algebraic surface S obtained by blowing up these points. Take six points to be defined by the six axes through opposite vertices of a regular icosahedron. Clearly no three are collinear. Also, no conic passes through all six points because if it did, its transform by an element of A_5 would be another conic meeting it in at least 6 points and hence coinciding with it. The conic would then be invariant, and since U is an irreducible representation of A_5 this would be the null conic $(x, x) = 0$. But the axis of an icosahedron is not a null vector.

These facts mean that the anticanonical bundle K^* of S is ample and has a four-dimensional space V of sections. Furthermore, since $c_1^2(S) = 9 - 6 = 3$ these sections embed S in P^3 as a nonsingular cubic surface.

Now if the icosahedron is the standard model, then $q_1(x) = x_1x_2x_3$ defines a cubic which passes through the six points, as do its transforms $q_1, \ldots, q_5$ by the

[1]Figure 1 is an annotated version of *Image:Icosahedron-golden-rectangles.svg* in Wikipedia, `http://en.wikipedia.org/wiki/Icosahedron`.

action of the group A_5. These generate the four-dimensional irreducible permutation representation, hence the q_α span this space. The single relation is $\sum_\alpha q_\alpha = 0$. But q_1 and hence q_α satisfies Laplace's equation, so all linear combinations of such cubics are harmonic. □

4. The Clebsch cubic surface

In the course of Proposition 3.1 we encountered the cubic surface S obtained by blowing up $\mathrm{P}(U)$ at the six axes of an icosahedron. The rational map $\rho\colon \mathrm{P}(U) \to \mathrm{P}^3$ is described by $x \mapsto [q_1(x), \dots, q_5(x)]$ where $\sum_\alpha q_\alpha(x) = 0$. In fact $\sum_\alpha q_\alpha^3(x) = 0$ and this is the equation of the cubic surface. Indeed, it is invariant in the permutation representation of A_5 and since the elementary symmetric function σ_1 vanishes, there is a unique cubic invariant polynomial σ_3, or equivalently the sum of cubes above. This defines the *Clebsch diagonal cubic surface.* Figure 2 shows the plaster model acquired by J. J. Sylvester in 1886 for Oxford University (see [5]).

The cubic surface we have just described lies in the projective space $\mathrm{P}(V^*)$ where $V = H^0(S, K^*)$: given a point $u \in U$ the polynomials in V which vanish at u define a plane in $\mathrm{P}(V)$. Dually, $[p] \in \mathrm{P}(V)$ defines a plane in $\mathrm{P}(V^*)$ and its intersection with the surface $S \subset \mathrm{P}(V^*)$ blows down to the cubic curve $p = 0$ in $\mathrm{P}(U)$.

However, V has an A_5-invariant inner product which identifies it with its dual and so we may regard the Clebsch cubic as also lying in $\mathrm{P}(V)$. If $p(x) = \sum_\alpha y_\alpha q_\alpha(x)$

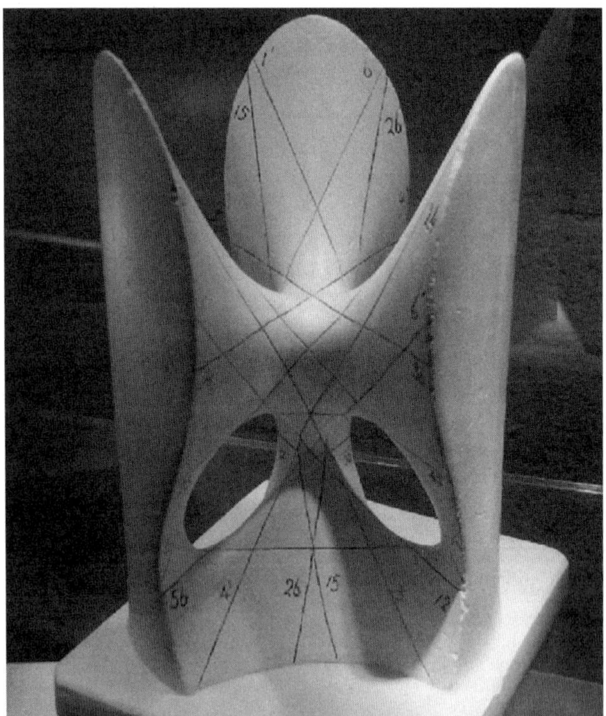

FIGURE 2. The Clebsch diagonal cubic surface

with $\sum_\alpha y_\alpha = 0$ then its equation is

$$\sum_1^5 y_\alpha^3 = 0.$$

The rational map $\rho\colon \mathrm{P}(U) \to S \subset \mathrm{P}(V)$ has a direct interpretation. Given $u \in U$ let p_u be the orthogonal projection of f_u (defined in (2.1)) onto V. If this projection is zero then $(p, f_u) = p(u) = 0$ for all $p \in V$, so that any cubic which vanishes at the a_i also vanishes at u. However, these cubics define a projective embedding of the blown-up plane, and so must separate points. It follows that $p_u = 0$ if and only if $[u] = [a_i]$ for some $1 \leq i \leq 6$.

Now a given p is orthogonal to f_u if and only if $p(u) = 0$, so $p_u \in V$ is orthogonal to the codimension one subspace of V consisting of cubics that vanish at u. This, however, is the definition of the Clebsch surface in $\mathrm{P}(V)$. Thus $u \mapsto [p_u]$ maps $\mathrm{P}(U) \setminus \{[a_1], \ldots, [a_6]\}$ to S. Since p_u is the orthogonal projection of f_u on V, it follows that $(p_u, q_\alpha) = (f_u, q_\alpha) = q_\alpha(u)$ from which we have an explicit form for the map:

$$\rho(u) = \sum_{\alpha=1}^5 q_\alpha(u) q_\alpha.$$

There remains the identification within $\mathrm{P}(V)$ of the six exceptional curves obtained by blowing up the points $[a_i] \in \mathrm{P}(U)$, and the extension of the map $u \mapsto [p_u]$. Consider $p_{u(t)}$ for $u(t) = a_1 + tv$ as $t \to 0$, where $(v, a_1) = 0$. We have

$$f_{u(t)} = f_{a_1} + 3t(v,x)\big((a_1,x)^2 - \tfrac{1}{5}(a_1,a_1)(x,x)\big) + \cdots .$$

Consider the cubic polynomial

$$h(x) = (v,x)\big((a_1,x)^2 - \tfrac{1}{5}(a_1,a_1)(x,x)\big) \tag{4.1}$$

where $(v, a_1) = 0$. We shall meet this type of harmonic cubic many times. The angle θ between two different axes of the icosahedron is given by

$$\cos\theta = \pm\frac{\phi}{1+\phi^2} = \pm\frac{1}{\sqrt{5}}$$

so $h(a_i) = 0$ if $i \neq 1$, but also $h(a_1) = 0$ since $(v, a_1) = 0$. Thus $q \in V$. As we saw above, $p_{a_1} = 0$ so f_{a_1} is orthogonal to V. Hence $p_{u(t)} = 3th + \cdots$ and $v \mapsto h$ defines a map from the line $\mathrm{P}(a_1^\perp)$, the projectivized tangent space to $\mathrm{P}(U)$ at $[a_1]$, to $\mathrm{P}(V)$. This is the exceptional curve E_1 obtained by blowing up $\mathrm{P}(U)$ at $[a_1]$, and is a line in the cubic surface $S \subset \mathrm{P}(V)$. Geometrically the cubic curve $h(x) = 0$ is the union of a line through $[a_1]$ and the unique conic C_1 passing through the five points $[a_2], \ldots, [a_6]$, so the line E_1 in $\mathrm{P}(V)$ consists of the pencil of lines through $[a_1]$ together with C_1.

For the six axes we get six disjoint lines — six of the 27 real lines shown on the model in Fig. 2. As is well known, the other 21 lines in the cubic surface are the six proper transforms $E_1', \ldots, E_6'$ of $C_1, \ldots, C_6$ together with the transforms E_{ij} of the 15 lines L_{ij} joining $[a_i]$ to $[a_j]$.

The line E_i' consists of the pencil of cubics in $\mathrm{P}(V)$ with a singularity at $[a_i]$. The polynomials $q_\alpha - q_\beta$, $1 \leq \alpha < \beta \leq 5$, clearly satisfy $\sum y_\alpha = 0 = \sum y_\alpha^3$ and so lie on the Clebsch cubic. There are 15 disjoint pairs $\{\alpha, \beta\}, \{\gamma, \delta\}$ and, as a, b vary,

$$a(q_\alpha - q_\beta) + b(q_\gamma - q_\delta)$$

describes 15 lines E_{ij} in S. Each $q_\alpha - q_\beta$ lies in three of these. On a general cubic surface points where three lines intersect are called Eckard points — the Clebsch cubic has the maximal number ten of these.

Remark 4.1. Under the composite map $S^2 \to \mathrm{P}(U)^{\mathrm{r}} \xrightarrow{\rho} S$, the real lines (apart from the blow-ups E_i) are recognizable in the geometry of the icosahedron. For a vertex a_1, the adjacent vertices $a_2, \dots, a_6$ lie on a circle in S^2 and this maps to $E_1' \subset S$. The points $a_1, a_2, -a_1, -a_2$ lie on a great circle and this maps to E_{12}. The three great circles E_{12}, E_{34}, E_{56} intersect in the midpoints of a pair of opposite triangular faces, which define the Eckard points. A description of these basic features can also be found in [6].

5. Skew forms

We now adopt a new viewpoint on our space H of harmonic cubics. If U is identified with the Lie algebra of SO(3), the action of a vector $u \in U$ on a homogeneous polynomial is via the differential operator

$$u \cdot f = \sum_{i,j,k} \epsilon_{ijk} x_i u_j \frac{\partial f}{\partial x_k}$$

This is skew-adjoint with respect to the inner product so each $u \in U$ defines a skew form

$$\omega_u(p, q) = (u \cdot p, q)$$

on the 7-dimensional space H. Since the Legendre polynomial f_u is symmetric about the axis u, clearly $u \cdot f_u = 0$. More generally we calculate (using the vector product $a \times b$)

$$u \cdot f_a(x) = 3(u \times a, x)\big((a, x)^2 - \tfrac{1}{5}(a, a)(x, x)\big)$$

which is precisely the harmonic cubic (4.1) we met before.

Any skew form on an odd-dimensional space is degenerate. In the case of ω_u we have:

Proposition 5.1. *The degeneracy subspace of ω_u on the space of harmonic cubics is spanned by f_u.*

PROOF. The degeneracy subspace is defined as the space of cubics $p \in H$ such that $\omega_u(p, q) = 0$ for all $q \in H$. Since

$$\omega_u(p, q) = (u \cdot p, q)$$

this holds if and only if $u \cdot p = 0$, i.e. if p is annihilated by the Lie algebra action of $u \in U \cong \mathfrak{so}(3)$. But f_u is the unique such function. □

Remark 5.1. Note that over the complex numbers this result still holds. The action of u on the irreducible representation H of SO(3) has a unique one-dimensional kernel whether u is semi-simple or nilpotent.

Proposition 5.2. *Let $\pm a_1, \dots, \pm a_6$ be the vertices of a regular icosahedron, and $X \subset H$ be the space spanned by the harmonic cubics f_{a_i}. Then ω_u restricted to X vanishes for all u.*

PROOF. We need to show that

$$\omega_u(f_a, f_b) = 0$$

for $a, b \in \{a_1, \dots a_6\}$ or equivalently that

$$(u \cdot f_a, f_b) = (u \cdot f_a)(b) = 0.$$

But $u \cdot f_a$ is given by (4.1) and, as we saw in Section 4, $(u \cdot f_a)(b) = 0$ for $a = a_i$, $b = a_j$. □

We call a subspace on which ω_u vanishes *isotropic*. Over the real numbers, we have a converse to Proposition 5.2.

Proposition 5.3. *Let $X \subset H$ be a 3-dimensional subspace which is isotropic for all ω_u. Then X is spanned by f_{a_i} where the a_i are the vertices of a regular icosahedron.*

PROOF. Over $\mathbb{C}$ this is the approach of S. Mukai [8], who described a family of Fano threefolds in terms of the subvariety of the Grassmannian $\mathrm{G}(3,7)$ of three-dimensional subspaces $X \subset \mathbb{C}^7$ on which a three-dimensional space of skew forms vanishes. Over $\mathrm{G}(3,7)$ we have the universal rank 3 bundle E, and each skew form defines a section of $\bigwedge^2 E^*$, also a rank three bundle. Thus $\omega_1, \omega_2, \omega_3$ define three sections of $\bigwedge^2 E^*$ and their vanishing gives a subvariety F of codimension 9. Since $\dim G(3,7) = 12$ this is three-dimensional. It contains, from Proposition 5.2, the three-dimensional space $\mathrm{SO}(3,\mathbb{C})/\mathrm{A}_5$ of (complex) icosahedra and Mukai shows that F is a smooth equivariant compactification of this quotient. The real points in this open orbit constitute the compact space $\mathrm{SO}(3)/\mathrm{A}_5$, the Poincaré sphere, or space of icosahedra. With these facts, the proof consists of checking that the complement of $SO(3,\mathbb{C})/\mathrm{A}_5$ has no real points.

The compactification is achieved by adjoining a union of lower-dimensional orbits of $\mathrm{SO}(3,\mathbb{C})$ and in particular each such point is fixed by a one-parameter subgroup. If X is real this means it is preserved by the Lie algebra action of a vector $u \in U$. But then it is generated by f_u, which is annihilated by u, and weight vectors p, $\bar{p}$ where $u \cdot p = \lambda p$, $\lambda \neq 0$. But then

$$\omega_u(p, \bar{p}) = (u \cdot p, \bar{p}) = \lambda(p, \bar{p}) \neq 0$$

and ω_u does not vanish on X. We deduce that the real subspaces on which ω_u vanishes are all generated by icosahedra. □

6. Vector bundles

From the previous section, we see that to find the icosahedra whose vertices lie in the zero set of a harmonic cubic p, we must determine subspaces $X \subset H$ on which the ω_u vanish. Given $p \in H$, let $W \subset H$ be its 6-dimensional orthogonal complement. We again work first over $\mathbb{C}$ and then specialize to $\mathbb{R}$.

Proposition 6.1. *The form ω_u restricted to W is degenerate if and only if $p(u) = 0$.*

PROOF. We showed in Proposition 5.1 that f_u spans the degeneracy subspace, so ω_u restricted to W is degenerate if and only if $f_u \in W$, i.e., $(p, f_u) = p(u) = 0$. □

The skew form ω_u is linear in $u \in \mathbb{C}^3$ and thus defines a map of bundles on the projective space $\mathrm{P}(U)$:

$$\omega\colon W(-2) \to W^*(-1).$$

The determinant of this is homogeneous of degree 6, but because ω_u is skew, it is the square of a cubic polynomial — the Pfaffian ($\omega_u^3 \in \bigwedge^6 W^*(3)$). Thus Proposition 6.1 tells us that the cubic p is given by the equation of the Pfaffian of ω_u on W.

Remark 6.1. This derivation of the Pfaffian can be applied to *any* irreducible representation space H of $\mathrm{SO}(3)$, for, up to a scalar multiple, there is a unique vector annihilated by an $\mathrm{SO}(2) \subset \mathrm{SO}(3)$. Given $v \in H$ the Pfaffian defines a homogeneous polynomial of degree $l = (\dim H - 1)/2$, and so it provides a canonical identification of the projective space of H with the projective space of spherical harmonics of degree l .

Describing a plane curve as a determinant is a classical problem and this, together with the related Pfaffian problem, is discussed from a modern point of view in Beauville's paper [2]. For $u \in C$, the map $\omega_u\colon W(-2) \to W^*(-1)$ is singular and has a nonzero cokernel. When the curve C defined by $p(u) = 0$ is smooth, this defines a rank 2 holomorphic vector bundle E over C. It has the properties (see [2]) $\det E \cong K_C$ and $H^0(C, E) = 0$.

Conversely, given any rank 2 bundle E on a nonsingular plane curve $C \subset \mathrm{P}^2$ of degree d with $\det E \cong K_C$ and $H^0(C, E) = 0$, then C can be obtained as a Pfaffian, i.e., E admits a natural resolution

$$0 \to \mathcal{O}_{\mathrm{P}^2}(\mathbb{C}^{2d}(-2)) \xrightarrow{A} \mathcal{O}_{\mathrm{P}^2}(\mathbb{C}^{2d}(-1)) \to \mathcal{O}_C(E) \to 0$$

where A is skew-symmetric and linear in u and C is defined by the vanishing of the Pfaffian of A.

In our case, $d = 3$, and K_C is trivial so we have a rank 2 vector bundle E with $\det E$ trivial and $H^0(C, E) = 0$. As Atiyah showed in [1], when C is smooth, such a bundle is either of the form

$$E = L \oplus L^*$$

for a nontrivial degree zero line bundle L, or is a nontrivial extension

$$0 \to L \to E \to L \to 0$$

of a nontrivial line bundle L with $L^2 \cong \mathcal{O}$.

Proposition 6.2. *For a smooth cubic curve C, there is a one-to-one correspondence between 3-dimensional subspaces $X \subset W$ on which ω_u vanishes for all u and degree 0 holomorphic line bundles $L \subset E^*$.*

Proof. Suppose that ω_u vanishes on $X \subset W$. The skew form ω_u defines a homomorphism

$$\omega\colon W \to W^*(1)$$

whose kernel is $E^*(-1)$. Restricting to X, we have

$$\omega|_X\colon X \to X^o(1)$$

where $X^o \in W^*$ is the annihilator of X. The cubic p is defined by $\det \omega|_X = 0$, and so as u varies over C, the kernel of $\omega|_X$ describes a line bundle $L(-1) \subset X$ where $\deg L = 0$. Since this is also in the kernel of $\omega\colon W \to W^*(1)$ it follows that $L(-1) \subset E^*(-1)$.

Conversely, suppose $L(-1) \subset E^*(-1) \subset W$. From the exact sequence of bundles on C

$$0 \to E^*(-2) \to W(-1) \to W^* \to E(1) \to 0$$

we have $W^* \cong H^0(C, E(1))$. Since $L^*(1)$ has degree 3, $H^0(C, L^*(1)) \subset H^0(C, E(1))$ defines a three-dimensional subspace $X^o \subset W^*$ and $X \subset W$. Moreover, since a line bundle of degree three embeds a smooth elliptic curve in the plane, we have $L(-1) \subset X$.

The restriction of ω_u to X is a section w of $\bigwedge^2 X^*(1)$. Identifying $\bigwedge^2 X^* \cong X \otimes \bigwedge^3 X^*$, we see that this defines a section w^* of $X(1)$ which at each point lies in the degeneracy subspace of the skew form w on X. But $L(-1) \subset E^*(-1)$ and so $L(-1)$ is the degeneracy subspace of ω_u on X. Thus if w^* is nonzero it defines a section of L. But in the Pfaffian construction E, and hence L, has no holomorphic sections, so we have a contradiction unless $w^* = 0$, i.e., ω_u vanishes on X. □

Now when $E \cong L \oplus L^*$ and L^2 is nontrivial, L and L^* are the only degree zero subbundles. For the nontrivial extension $0 \to L \to E \to L \to 0$, L is the only subbundle. The bundle $E = L \oplus L$ for L^2 trivial clearly has infinitely many subbundles isomorphic to L. Hence, using Propositions 5.3 and 6.2 we see that for a nonsingular cubic Atiyah's theorem offers the alternatives:

- If $E \cong L \oplus L^*$ there are two isotropic subspaces $X_1, X_2 \subset W$. If both X_1 and X_2 are real, p vanishes on the vertices of two icosahedra. If $X_1 = \overline{X}_2$ there are no real icosahedra.
- When E is a nontrivial extension with L nontrivial and $L^2 \cong \mathcal{O}$, then there is a unique icosahedron.
- When $E = L \oplus L$ there are infinitely many icosahedra.

The discussion above is for smooth curves, but the situation for nodal cubics or reducible ones with transverse intersections is similar [3]. Cuspidal cubics present far more problems, but fortunately we do not have to deal with these. In fact, as noted in [7] and proved in [4], singularities of nodal sets are quite simple. If $\Delta f = \lambda f$ and f together with its derivative vanish at a point, then the function is locally approximated by a solution to $\Delta f = 0$, i.e. the real part of cz^n. So the nodal set at a singularity locally consists of n smooth curves meeting at the same angle π/n: for example the three great circles given by $x_1x_2x_3 = 0$ meet at right angles.

Another regularity issue is that for a value of u at which ω_u restricted to W is degenerate, it has rank 4 and no lower. This is because $(u \cdot f, q) = 0$ for all $q \in W$ if and only if $u \cdot f = \lambda p$. But if $u \cdot f_i = \lambda_i p$ for $i = 1, 2$ then $u \cdot (\lambda_2 f_1 - \lambda_1 f_2) = 0$ so $\lambda_2 f_1 - \lambda_1 f_2 = \mu f_u$. Hence there is only a two-dimensional space of polynomials f with $(u \cdot f, q) = 0$ for all $q \in W$. In particular this means that even when the curve C is singular the degeneracy subspace still gives a well-defined vector bundle.

On a nodal curve, one can understand vector bundles by passing to the normalization — a copy of P^1 with two distinguished points which map to the node. If the pull-back of E is trivial, then the identification of the fibres at the two points is given by a matrix $A \in \mathrm{SL}(2, \mathbb{C})$ and if $H^0(C, E) = 0$, A must not have 1 as an eigenvalue. The conjugacy classes are then

$$\begin{pmatrix} a & 0 \\ 0 & a^{-1} \end{pmatrix} \; a \neq \pm 1, \quad \begin{pmatrix} -1 & b \\ 0 & -1 \end{pmatrix} \; b \neq 0, \quad \begin{pmatrix} -1 & 0 \\ 0 & -1 \end{pmatrix}$$

corresponding exactly to Atiyah's classification. If the bundle is nontrivial on P^1, since $H^0(C,E)=0$, it can only be isomorphic to $\mathcal{O}(1)\oplus\mathcal{O}(-1)$. In this case there is no degree zero subbundle, so from Proposition 6.2, there is no icosahedron, real or complex, whose axes lie on C.

We could proceed in this manner to the reducible cubics but these can be dealt with more concretely as we shall see in the next section.

7. Counting icosahedra

Given a harmonic cubic, we want to determine how many icosahedra have all their vertices on its nodal set. For a nonsingular or nodal cubic the previous section gives us the information that this number is zero, one, two or infinitely many. There is an elementary way of going the other way round — associating a harmonic cubic to a pair of icosahedra:

Proposition 7.1. *Given two regular icosahedra with vertices on the unit sphere, then*

- *if they have no vertices in common, there exists a one-dimensional space of harmonic cubics which vanish at all vertices,*
- *if they have one vertex a in common then there is a two-dimensional space of such cubics, each of the form*

$$(u,x)\big((a,x)^2-\tfrac{1}{5}(x,x)(a,a)\big)$$

where $(u,a)=0$.

Proof. The two icosahedra define two 4-dimensional subspaces V_1 and V_2 in the 7-dimensional space H, which must therefore intersect nontrivially.

Suppose that $\dim V_1\cap V_2>1$, then we have a pencil of plane cubic curves passing through the points $[a_1],\ldots,[a_6]$, $[b_1],\ldots,[b_6]$ defined by the axes of the two icosahedra. If two axes coincide then so do the icosahedra, so at most one axis can be common. This means the curves intersect in at least eleven distinct points. But by Bézout's theorem, there can be a maximum of nine unless there is a common component.

Suppose that component is a line. No three of the $[a_i]$ are collinear (or the $[b_i]$) so there is a maximum of four (e.g., $[a_1],[a_2],[b_1],[b_2]$) on a line, leaving at least seven. The pencil of conics which remains has a maximum of four intersections, so that must have a common line. Again that can have at most four new vertices, leaving three for a pencil of lines, which is a contradiction.

If the common component is a conic, then it contains at most five axes from either icosahedron (as we saw in the proof of Proposition 3.1.) The remaining axis must lie in the pencil of lines. If $[a]$ is that axis, then the other five axes make the same angle with a and so satisfy the equation

$$(a,x)^2-\tfrac{1}{5}(x,x)(a,a)=0.$$

This is the fixed conic. The remaining line has equation $(u,x)=0$ and since it passes through $[a]$, $(u,a)=0$. □

We see here that a pair of icosahedra without a common axis determines, up to a scalar multiple, a unique harmonic cubic. We show next:

Proposition 7.2. *Suppose p vanishes on the vertices of more than two icosahedra. Then p is of the form*

$$(u,x)\big((a,x)^2 - \tfrac{1}{5}(x,x)(a,a)\big)$$

with $(u,a)=0$. From Proposition 7.1 it vanishes for infinitely many icosahedra, all with vertex a.

PROOF. Let X_1, X_2 be isotropic subspaces of W for ω_u and assume that p is not of the form in Proposition 7.1. Then from that Proposition, $\dim V_1 \cap V_2 = 1$ and hence $X_1 \cap X_2 = 0$. A third subspace X_3 intersects each of these trivially and hence is the graph of an invertible linear transformation $S : X_1 \to X_2$. Thus for $w_1, w_2 \in X_1$ we have

$$0 = \omega_u(w_1 + Sw_1, w_2 + Sw_2) = \omega_u(w_1, Sw_2) + \omega_u(Sw_1, w_2)$$

since $\omega_u(w_1, w_2) = 0 = \omega_u(Sw_1, Sw_2)$. This means the graph of tS for any real number t is also isotropic, so we have a one-parameter family of icosahedra.

Since icosahedra are all in the same SO(3) orbit, differentiating at $t=0$ we see that $Sw = u \cdot w$ for some u. Moreover, since S is invertible, u is not a multiple of the vertices a_i of the icosahedron defined by X_1. But this means that each element of $X_2 = u \cdot X_1$ is orthogonal to X_1 because

$$(u \cdot w_1, w_2) = \omega_u(w_1, w_2) = 0.$$

In particular if b is a vertex of the icosahedron for X_2, $f_b \in X_2$ is orthogonal to the $f_{a_i} \in X_1$, or equivalently $f_b(a_i) = 0$. However, as remarked in the introduction, no f_u can vanish on the vertices of an icosahedron, so we have a contradiction. □

It remains to discuss the other reducible cubics which pass through the axes of the icosahedron. First consider the case of a line through two axes and a nonsingular conic through the remaining four. In the standard model if we take the line to be $x_3 = 0$, then the conic is of the form

$$a(\phi^4 x_1^2 + x_2^2 - \phi^2 x_3^2) + b x_1 x_2 = 0 \tag{7.1}$$

where $x_1^2 + x_2^2 + x_3^2 = 1 + \phi^2$ and for nonsingularity $a \neq 0, b^2 \neq 4a^2\phi^4$.

Now rotate the icosahedron an angle θ about the x_3-axis. Four vertices still lie in $x_3 = 0$, and we want to see if the remaining eight lie on the conic. By symmetry it is enough to consider

$$(\phi \sin\theta, \phi\cos\theta, 1) \quad (\cos\theta, -\sin\theta, \phi).$$

Substituting in (7.1) this gives either $\sin\theta = 0$—the original icosahedron, or

$$\tan\theta = \frac{b}{a(1-\phi^4)}.$$

There are thus two icosahedra (and no more by Proposition 7.2).

Now consider three lines passing through the six points. There are 15 possibilities which break up into two A_5 orbits. The first are the five cubics $q_1, \dots, q_5$ we started with, and we observed in the introduction that they vanish on the vertices of two icosahedra (this is also the case $a = 0$ in (7.1)). The other 10 are of the form $q_\alpha - q_\beta$ where the cubic curve consists of three lines meeting at a point. On the sphere these are three equally spaced lines of longitude. Since no three axes in $P(U)^r$ are collinear, we must have 4 vertices in each great circle and the North–South axis passes through the midpoint of a face of the icosahedron. But this determines the distances of the vertices from the poles, so the icosahedron is unique.

In discussing these special cases, we have shown that on the line in $\mathrm{P}(V)$ defined by the family (7.1) there are three points $a = 0, b = \pm 2a\phi$ where there is a unique icosahedron: the three degenerate cubics consisting of L_{12} together with $\{L_{34}, L_{56}\}, \{L_{35}, L_{46}\}$ or $\{L_{36}, L_{45}\}$. We shall see this more generally next.

8. The Clebsch cubic revisited

Proposition 8.1. *A real harmonic cubic $p \in V$ vanishes on the vertices of precisely two distinct icosahedra if and only $[p]$ does not lie on the Clebsch cubic surface.*

PROOF. Let $X \subset H$ be the 3-dimensional subspace defining an icosahedron and p be a cubic which vanishes at its vertices. Consider as in Proposition 6.2 $\omega|_X : X \to X^o(1)$, whose determinant vanishes on the plane cubic $C \subset \mathrm{P}(U)$ defined by p. The kernel $L(-1)$ lies in the kernel $E^*(-1)$ of $\omega: W \to W^*(1)$. From Atiyah's classification, for a smooth or nodal cubic there is a unique icosahedron if and only if L^2 is trivial but L is nontrivial.

As u varies in C consider f_u. Now $\omega_u(f_u, q) = (u \cdot f_u, q) = 0$ so that f_u generates a line bundle in the kernel of $\omega: W \to W^*(1)$. We identify this bundle as follows. Consider the map $f: \mathrm{P}(U) \to \mathrm{P}(H)$ defined by $u \mapsto f_u$. Then f_u spans the pull-back $f^*(\mathcal{O}(-1))$ and since f_u (see (2.1)) is homogeneous of degree 3 in u, $f^*(\mathcal{O}(-1)) \cong \mathcal{O}(-3)$. Restricted to C, this is the line bundle we want.

We thus have an inclusion $\mathcal{O}(-3) \subset E^*(-1)$ over C. Projecting to $E^*(-1)/L(-1)$ gives a section of $L^*(2)$. This vanishes when $f_u \in X$ which from Proposition 5.2 is at the six points $[a_1], \dots, [a_6]$ defined by the axes of the icosahedron. Thus, on C we have the relation of divisor classes

$$L \sim \mathcal{O}(2) - \sum_1^6 [a_i]. \tag{8.1}$$

Now take a general line in $P(V)$: this defines a pencil of cubics in $P(U)$ passing through $[a_1], \dots, [a_6]$ and three extra points $[a_7]$, $[a_8]$, $[a_9]$. Blow them up to obtain an elleptic surface Y: the anticanonical bundle K_Y^* has two sections which define a projection $\pi: Y \to \mathrm{P}^1$ with elliptic fibres, which consist of the pencil of cubics through the nine points. Let E_i, $1 \leq i \leq 9$ be the exceptional curves on Y, then

$$K_Y \sim -3H + \sum_1^9 E_i \tag{8.2}$$

where H is the pull-back of the hyperplane divisor on $\mathrm{P}(U)$. Let J be the line bundle on Y whose divisor class is

$$J \sim 2H - \sum_1^6 E_i$$

then from (8.1) J restricts to L on each elliptic fibre of π.

Remark 8.1. This formula identifies L, which has come to us from the consideration of vector bundles, with a more basic geometric object. The elliptic curve $p(x) = 0$ lies in two projective planes: $\mathrm{P}(U)$ and (via its proper transform) a plane section of $S \subset \mathrm{P}(V)$. The two different hyperplane bundles differ by L.

We shall apply Grothendieck-Riemann-Roch to J^2 on Y to determine the number of cubics in the pencil for which L^2 is trivial.

Now $\pi_*[K_Y] = 0$ since it is dual to a fibre, and for the exceptional curves $\pi_*[E_i] = 1$. Hence $\pi_*[H] = 3$ and $\pi_*[J] = 0$. Moreover we have intersection numbers $[J]^2 = -2, [K_Y][J] = 0$. Hence

$$\pi_*(\mathrm{td}(Y)\mathrm{ch}(J^2)) = -3[\mathrm{P}^1].$$

Hence $\mathrm{ch}(\pi_!(J^2)) = -3[\mathrm{P}^1]$ and so L^2 is trivial for three cubics in the pencil.

It follows that a generic line in $\mathrm{P}(V)$ meets the locus of cubics for which L^2 is trivial in three points, so that locus is a cubic surface. But it is invariant under the icosahedral group and hence must be the Clebsch cubic S.

All real cubics in $\mathrm{P}(V)$ pass through the axes of the standard icosahedron, moreover from Proposition 7.2 those that pass through infinitely many form six of the 27 lines on S. We also treated the reducible cases in the previous section, so any $[p]$ in the complement of S must pass through the axes of two distinct icosahedra and no more. □

9. The general case

So far we have mainly focused on the four-dimensional space V of harmonic cubics which vanish on the vertices of a fixed icosahedron, and identified those which vanish on a second. We now ask the main question of the seven-dimensional space H of all harmonic cubics: which ones vanish on the vertices of *two* icosahedra? In what follows we are only considering real points, so we omit the superscript r to denote this.

Clearly if p vanishes for icosahedra I_1, I_2, we can transform I_1 to the standard icosahedron I by the action of $\mathrm{SO}(3)$, so that the polynomials we are seeking are the transforms of the complement of S under the action map

$$F\colon \mathrm{SO}(3) \times \mathrm{P}(V) \to \mathrm{P}(H).$$

The points of $\mathrm{P}(H)$ for which there is either one or infinitely many icosahedra are given by the orbit of the cubic surface S, which is an $SO(3)$-invariant hypersurface in $\mathrm{P}(H)$. Now the ring of $SO(3)$-invariants of the 7-dimensional representation H is generated in degrees $2, 4, 6, 10$, and 15, with the square of the one of degree 15 being a polynomial in the other four generators, which are themselves algebraically independent. (I thank Robert Bryant for this information). There is no invariant of degree three but in the Appendix we calculate a sextic invariant J on H which, when restricted to V, becomes σ_3^2. Consequently for real harmonic cubics, the orbit of $\mathrm{P}(V)$ in $\mathrm{P}(H)$ under the action of $\mathrm{SO}(3)$ is contained in the subset $J(f) \geq 0$.

Proposition 9.1. *If $J(f) > 0$ then $[f]$ is in the orbit of $\mathrm{P}(V) \setminus S$ under $\mathrm{SO}(3)$.*

PROOF. From Proposition 7.1, the orbit of $\mathrm{P}(V) \setminus S$ under the $\mathrm{SO}(3)$ action is the set of unordered pairs of icosahedra having no axis in common. Similarly $\mathrm{P}(V) \setminus S$ consists of icosahedra having no axis in common with the standard one I. The ordering of the two icosahedra is the basis for what follows.

Define an equivalence relation on $\mathrm{SO}(3) \times \mathrm{P}(V)$ by $(g,p) \sim (h,q)$ if $[gp] = [hq]$ and $gI = hI$. Let M be the space of equivalence classes. Clearly $F(g,p) = [gp]$ factors through M. Since $h^{-1}g \in \mathrm{A}_5$ the equivalence class of (g,p) is finite and in one-to-one correspondence with the subgroup of A_5 fixing $[p]$. We may see these fixed points directly by looking at the permutation representation and in all cases the fixed point set in $\mathrm{P}(V)$ is a line, and so of codimension 2. So M is a six-manifold.

If $[p] \notin S$ there are two icosahedra and so given $[gp]$, two equivalence classes. This represents F as a double covering, and the involution interchanging the sheets extends to M with fixed point set the five-dimensional image of $\mathrm{SO}(3) \times S$. Thus F defines a map from an (unoriented) six-manifold $\overline{M} = M/\mathbb{Z}_2$ with boundary to the six-dimensional $\mathrm{P}(H)$ where the boundary maps to the hypersurface Y defined by $J(f) = 0$. The image of F lies in the region $\mathrm{P}(H)^+$ where $J \geq 0$. Since F is one-to-one on the interior, it is of degree one and hence the map

$$F_*\colon H_6(\overline{M}, \partial\overline{M}; \mathbb{Z}_2) \to H_6(\mathrm{P}(H)^+, Y; \mathbb{Z}_2)$$

is an isomorphism. In particular it follows that F is surjective, and so $J(f) \geq 0$ means that $f = gp$ for $[p] \in \mathrm{P}(V) \setminus S$. Hence f vanishes on the vertices of exactly two icosahedra. □

We thus obtain the theorem:

Theorem 9.1. *Let f be a degree three spherical harmonic.*

- *If $J(f) > 0$ (resp. $J(f) < 0$) then the nodal set of f contains the vertices of two (resp. zero) regular icosahedra.*
- *When $J(f) = 0$ and f is of the form $f(x) = (u,x)((a,x)^2 - (x,x)(a,a)/5)$ with $(u,a) = 0$ then any regular icosahedron with a as a vertex lies on the nodal set, otherwise there is a unique such icosahedron.*

Appendix A. Invariants

We calculate explicitly here the $\mathrm{SO}(3)$-invariant polynomial J on H which restricts to σ_3^2 on V.

Let $f(x)$ be a spherical harmonic of degree 3 and consider the symmetric form S on U defined by

$$S(u,v) = \int_{S^2} f^2(x)(u,x)(v,x)\,\mathrm{d}S$$

or

$$S_{ij} = \int_{S^2} f^2(x) x_i x_j \,\mathrm{d}S$$

The $\mathrm{SO}(3)$-invariants $\operatorname{tr} S, \operatorname{tr} S^2, \operatorname{tr} S^3$ define invariants of degree $2,4,6$ of f, and we shall relate these to the A_5-invariants when we restrict to our distinguished four-dimensional space of harmonic cubics of the form

$$f(x) = \sum_1^5 y_\alpha q_\alpha(x)$$

where $\sum_\alpha y_\alpha = 0$ and $\sum_\alpha q_\alpha = 0$.

First consider $f = q_\alpha$. In this case S is invariant under conjugation by the stabilizer $\mathrm{A}_4 \subset \mathrm{A}_5$ of α. But A_4, the tetrahedral group, fixes no axes in $\mathbb{R}^3$, so that

$$\int_{S^2} q_\alpha^2 x_i x_j \,\mathrm{d}S = \lambda \delta_{ij} \tag{A.1}$$

Taking the trace

$$3\lambda = \int_{S^2} q_\alpha^2 \,\mathrm{d}S.$$

To evaluate this and similar integrals, note that each q_α is of the form $(a_1, x) \times (a_2, x)(a_3, x)$, and by the invariant theory of the orthogonal group

$$\int_{S^2} (a_1, x)(a_2, x) \cdots (a_{2n}, x)\, \mathrm{d}S = c_n \sum (a_{i_1}, a_{i_2}) \cdots (a_{i_{2n-1}}, a_{i_{2n}})$$

for a universal constant c_n where the sum is over partitions $\{i_1, i_2\} \ldots \{i_{2n-1}, i_{2n}\}$ of $\{1, 2, 3, \ldots, 2n\}$ into two-element subsets. In fact, taking all the $a_i = (0, 0, 1)$ we evaluate $c_n = 2^{n+2}\pi n!/(2n+1)!$

If $q_\alpha(x) = x_1 x_2 x_3 = (e_1, x)(e_2, x)(e_3, x)$ then by orthogonality there is only one partition of $\{e_1, e_2, e_3, e_1, e_2, e_3\}$ which gives a nonzero result. Thus

(A.2) $$\lambda = \tfrac{1}{3} c_3.$$

Now consider

$$\int_{S^2} q_\alpha q_\beta x_i x_j\, \mathrm{d}S.$$

This is invariant by the stabilizer S_3 of the unordered pair $\{\alpha, \beta\}$.

The (dihedral) action of S_3 has one invariant axis, which we represent by a unit vector $v^{\alpha\beta}$ and then

(A.3) $$\int_{S^2} q_\alpha q_\beta x_i x_j\, \mathrm{d}S = \mu \delta_{ij} + \nu v_i^{\alpha\beta} v_j^{\alpha\beta}$$

Lemma A.1. $\mu = 0$, $\nu = -3\lambda/4$.

Proof. Taking the trace of (A.3) gives

$$3\mu + \nu = \int_{S^2} q_\alpha q_\beta\, \mathrm{d}S$$

and using $\sum_\alpha q_\alpha = 0$ we obtain

$$\int_{S^2} q_1 q_5\, \mathrm{d}S = -\int_{S^2} q_1^2 + q_1 q_2 + q_1 q_3 + q_1 q_4\, ddS$$

and so

(A.4) $$3\lambda + 12\mu + 4\nu = 0.$$

The action of A_5 on the 10 unordered pairs $\{\alpha, \beta\}$ is, in the icosahedron, the action on opposite pairs of the 20 faces, and $v^{\alpha\beta}$ is the axis joining their centres. In Figure 1 such an axis points towards the reader, and passes through the centre of the triangle ABC. In Fig. 2 it defines an Eckard point.

The three shaded planes in the diagram form the zero locus of q_α. The face ABC has three adjacent faces and each shaded plane provides one of its edges. The remaining three edges define the second triple of planes which is the zero locus of q_β.

Analytically suppose that $q_\alpha(x) = x_1 x_2 x_3$ and $v = v^{\alpha\beta} = (1, 1, 1)/\sqrt{3}$. The vertices A, B, C are then $(0, \phi, 1)$, $(\phi, 1, 0)$, $(1, 0, \phi)$. Then q_α and q_β are interchanged by a rotation by π in S_3 which takes v to $-v$ and the vertex $(0, \phi, 1)$ to its opposite $-(0, \phi, 1)$. This is the rotation

(A.5) $$g = \frac{1}{2}\begin{pmatrix} -\phi & \phi - 1 & -1 \\ \phi - 1 & -1 & -\phi \\ -1 & -\phi & \phi - 1 \end{pmatrix}$$

Now from (A.3) we see that

$$S(v^{\alpha\beta},v^{\alpha\beta})=\int_{S^2}q_\alpha q_\beta(v^{\alpha\beta},x)^2\,\mathrm{d}S=\mu+\nu$$

and we can evaluate this by partitioning $\{e_1,e_2,e_3,ge_1,ge_2,ge_3,v,v\}$ into pairs. If v is paired with v then since $(v,v)=1$ we get as the sum of these contributions

$$c_3^{-1}\int_{S^2}q_\alpha q_\beta\,\mathrm{d}S=c_3^{-1}(3\mu+\nu)=-\frac{1}{4} \tag{A.6}$$

from (A.4) and (A.2).

If both v terms are paired with e_i terms or ge_j terms then by orthogonality one of the remaining inner products is zero. Thus the only nonzero contributions to the integral have factors $(v,e_i)(v,ge_j)$ which (since $gv=-v$) is $-(v,e_i)^2=-\frac{1}{3}$. This gets multiplied by terms $(e_k,ge_l)(e_m,ge_n)$. These are products of entries in (A.5) and the sum of these 18 terms is readily evaluated to be 3. Because of the repeated terms v,v there are two partitions which give the same contribution, and so we find these terms producing a contribution -2. Adding in (A.6) we find

$$\mu+\nu=-\frac{9}{4}c_4.$$

But $3\mu+\nu=-3\lambda/4=-c_3/4$ and $c_3/c_4=9$, so $3\mu+\nu=\mu+\nu$ and therefore $\mu=0$ and $\nu=-3\lambda/4$. □

From the lemma we may as well normalize S and take $\lambda=4,\mu=0,\nu=-3$. Consider a general $f=\sum_\alpha y_\alpha q_\alpha$. Then

$$\begin{aligned}S_{ij}&=\sum_{\alpha,\beta}\int_{S^2}y_\alpha y_\beta q_\alpha q_\beta x_i x_j\,\mathrm{d}S\\&=\lambda\sum_1^5 y_\alpha^2\delta_{ij}+\sum_{\alpha\neq\beta}y_\alpha y_\beta(\mu\delta_{ij}+\nu v_i^{\alpha\beta}v_j^{\alpha\beta})\\&=-8\sigma_2\delta_{ij}-3\sum_{\alpha\neq\beta}y_\alpha y_\beta v_i^{\alpha\beta}v_j^{\alpha\beta}\end{aligned}$$

using the elementary symmetric polynomials σ_i in y_α and the condition $\sigma_1=0$. In particular $\operatorname{tr}S=-30\sigma_2$. Define the symmetric matrix M by

$$M_{ij}=\sum_{\alpha\neq\beta}y_\alpha y_\beta v_i^{\alpha\beta}v_j^{\alpha\beta}$$

then we have

$$\begin{aligned}\operatorname{tr}M^2&=\sum_{\alpha\neq\beta}y_\alpha y_\beta\sum_{\gamma\neq\delta}y_\gamma y_\delta(v^{\alpha\beta},v^{\gamma\delta})^2\\\operatorname{tr}M^3&=\sum_{\alpha\neq\beta}y_\alpha y_\beta\sum_{\gamma\neq\delta}y_\gamma y_\delta\sum_{\mu\neq\nu}y_\mu y_\nu(v^{\mu\nu},v^{\alpha\beta})(v^{\alpha\beta},v^{\gamma\delta})(v^{\gamma\delta},v^{\mu\nu})\end{aligned}$$

and for this we need to know the inner product terms involving the $v^{\alpha\beta}$.

As we noted, the action of A_5 on unordered pairs $\{\alpha,\beta\}$ is the action on opposite pairs of faces. There are 15 pairs of these $\{\{\alpha,\beta\},\{\gamma,\delta\}\}$ that have no number in common, and these correspond to *adjacent* faces — or, considering the common edges, to opposite pairs of the 30 edges. The other 30 pairs $\{\{\alpha\beta\},\{\beta\gamma\}\}$ correspond to pairs with no common edge. In Figure 1 $D=(0,\phi,-1)$, so a unit vector passing through the centre of triangle ADB is $(\phi, 2\phi, 0)/\sqrt{3}(\phi+1)$. Since ABC, ADB are adjacent this gives $(v^{\alpha\beta}, v^{\gamma\delta})^2 = 5/9$. The vertex E is $(1, 0, -\phi)$ and BDE is not adjacent to ABC or its opposite. The vector through the centre of triangle BDE is $(1, 1, -1)/\sqrt{3}$ so here we obtain $(v^{\alpha\beta}, v^{\beta\gamma})^2 = 1/9$.

We can now evaluate $\operatorname{tr} M^2$ in terms of elementary symmetric functions σ_i. Using $\sigma_1 = 0$ we obtain

$$\operatorname{tr} M^2 = 4\sigma_2^2 + \frac{160}{9}\sigma_4.$$

To evaluate $\operatorname{tr} M^3$ we need to consider triples of two-element subsets of $\{1, \dots, 5\}$. There are 5 orbits represented by

$$\{1,2\}\{1,2\}\{1,2\} \quad \{1,2\}\{1,2\}\{1,3\} \quad \{1,2\}\{1,2\}\{3,4\} \quad \{1,2\}\{1,3\}\{1,4\}$$

and $\{1,2\}\{1,3\}\{4,5\}$ where the last one contributes zero since it occurs with a coefficient given by the symmetric polynomial $\sigma_1\sigma_5$ and in our case $\sigma_1 = 0$. The orbit of $\{1,2\}\{1,3\}\{1,4\}$ introduces a new expression $(v^{12}, v^{13})(v^{13}, v^{14})(v^{14}, v^{12})$ which we must calculate. To do this, note that $\{1,2\}\{1,3\}\{1,4\}$ is stabilized by the rotation g of order 3 which fixes 1 and 5 and permutes 2, 3 and 4. Thus

$$(v^{12}, v^{13})(v^{13}, v^{14})(v^{14}, v^{12}) = (v^{12}, gv^{12})(gv^{12}, g^2v^{12})(g^2v^{12}, v^{12}) = (v^{12}, gv^{12})^3.$$

If $\{1,5\}$ defines the face ABC, then the rotation is $g(x, y, z) = (y, z, x)$ so that taking $v^{12} = (1, 1, -1)/\sqrt{3}$ to pass through the centre of BDE, we obtain $(v^{12}, gv^{12})^3 = -1/27$.

We now calculate

$$\operatorname{tr} M = 2\sigma_2, \quad \operatorname{tr} M^2 = 4\left(\sigma_2^2 + \frac{40}{9}\sigma_4\right), \quad \operatorname{tr} M^3 = 8\left(\sigma_2^3 - \frac{20}{9}\sigma_2\sigma_4 + 2\sigma_3^2\right).$$

Denoting by τ_i the symmetric functions in the eigenvalues of M, we obtain

$$16\sigma_3^2 = 3\tau_3 - 4\tau_1\tau_2$$

and this gives explicitly the invariant J.

References

1. M. F. Atiyah, *Vector bundles over an elliptic curve*, Proc. London Math. Soc. (3) **7** (1957), 414–452.
2. A. Beauville, *Determinantal hypersurfaces*, Michigan Math. J. **48** (2000), 39–64.
3. I. Burban and Y. Drozd, *Coherent sheaves on rational curves with simple double points and transversal intersections*, Duke Math. J. **121** (2004), no. 2, 189–229.
4. S.-Y. Cheng, *Eigenfunctions and nodal sets*, Comment. Math. Helv. **51** (1976), no. 1, 43–55.
5. J. Fauvel, R. Flood, and R. Wilson (eds.), *Oxford figures: 800 years of the mathematical sciences*, Oxford Univ. Press, Oxford, 2000.
6. F. Hirzebruch, *The Hilbert modular group for the field $Q(\sqrt{5})$, and the cubic diagonal surface of Clebsch and Klein*, Russian Math. Surveys **31** (1976), no. 5, 96–110.
7. J. Leydold, *On the number of nodal domains of spherical harmonics*, Topology **35** (1996), no. 2, 301–321.
8. S. Mukai, *Fano 3-folds*, Complex Projective Geometry (Trieste/Bergen, 1989) (G. Ellingsrud, C. Peskine, G. Sacchiero, and S. A. Strømme, eds.), London Math. Soc. Lecture Note Ser., vol. 179, Cambridge Univ. Press, Cambridge, 1992, pp. 255–263.

Mathematical Institute, Oxford University, 24-29 St. Giles, Oxford OX1 3LB, UK
E-mail address: hitchin@maths.ox.ac.uk

Centre de Recherches Mathématiques
CRM Proceedings and Lecture Notes
Volume **47**, 2009

Alternating Group and Multivariate Exponential Functions

Anatoliy Klimyk and Jiří Patera

ABSTRACT. We define and study multivariate exponential functions, symmetric with respect to the alternating group A_n, which is a subgroup of the permutation (symmetric) group S_n. These functions are connected with multivariate exponential functions, determined as the determinants of matrices whose entries are exponential functions of one variable. Our functions are eigenfunctions of the Laplace operator. By means of alternating multivariate exponential functions three types of Fourier transforms are constructed: expansions into corresponding Fourier series, integral Fourier transforms, and multivariate finite Fourier transforms. Alternating multivariate exponential functions are used as a kernel in all these Fourier transforms. Eigenfunctions of the integral Fourier transforms are obtained.

1. Introduction

Mathematical and theoretical physics regularly deal with functions on the Euclidean space $\mathbb{R}^n$ that are symmetric or antisymmetric with respect to the permutation (symmetric) group S_n. For example, such functions describe collections of identical particles. Symmetric and antisymmetric solutions appear in the theory of integrable systems.

The symmetric group S_n contains the alternating group A_n. It is an invariant subgroup of S_n of index 2 (that is, the group S_n/A_n has 2 elements). The alternating group A_n consists of transformations w of the Euclidean space $\mathbb{R}^n$ with $\det w = 1$ and, therefore, is a subgroup of the rotation group $SO(n)$ (note that S_n does not belong to $SO(n)$). The group A_n is simple. For studying multivariate exponential functions, the group A_n is more fundamental than the symmetric group, since multivariate exponential functions symmetric or antisymmetric with respect to S_n can be constructed by means of the multivariate exponential functions symmetric with respect to A_n.

2000 *Mathematics Subject Classification.* Primary 42B05; Secondary 20B30, 33E99, 42B10.

The research of the first author was partially supported by Grant 14.01/016 of the State Foundation of Fundamental Research of Ukraine.

Work partially supported by the National Science and Engineering Research Council of Canada, by the MIND Institute and by MITACS..

This is the final form of the paper.

The aim of this paper is to describe and study multivariate exponential functions symmetrized by the alternating group and the corresponding Fourier transforms. We call these functions *alternating multivariate exponential functions* and denote them by $E_\lambda(x)$, $\lambda = (\lambda_1, \lambda_2, \dots, \lambda_n)$, $x = (x_1, x_2, \dots, x_n) \in \mathbb{R}^n$. Such a function is a sum of terms in the determinant of an $n \times n$ matrix (whose entries are usual exponential functions of one variable) which enter in the expression for the determinant with sign $+$. We call this sum a semideterminant.

Alternating multivariate exponential functions are connected with symmetric and antisymmetric multivariate exponential functions studied in [8]. Symmetric and antisymmetric multivariate exponential functions can be considered as a generalization of cosine and sine functions of one variable, respectively, whereas alternating multivariate exponential functions are a generalization of the usual exponential function of one variable. The connection of symmetric and antisymmetric multivariate exponential functions with alternating multivariate exponential functions is the same as that of the cosine and sine functions with the exponential function of one variable (see Section 2 below).

As in the case of the exponential functions of one variable, we may consider three types of alternating multivariate exponential functions:

(a) functions $E_m(x)$ with $m = (m_1, m_2, \dots, m_n)$, $m_i \in \mathbb{Z}$, which determine Fourier series expansions in alternating multivariate exponential functions;
(b) functions $E_\lambda(x)$ with $\lambda = (\lambda_1, \lambda_2, \dots, \lambda_n)$, $\lambda_i \in \mathbb{R}$, which determine integral multivariate Fourier transforms;
(c) functions $E_\lambda(x)$, where $x = (x_1, x_2, \dots, x_n)$ take a finite set of values; they determine multivariate finite Fourier transforms.

Functions (b) are symmetric with respect to elements of the alternating group A_n. Since exponential functions $\mathrm{e}^{2\pi i m x}$, $m \in \mathbb{Z}$, of one variable are invariant with respect to shifts $x \to x + k$, $k \in \mathbb{Z}$, then symmetries of functions (a) are described by a wider group, which is called the *affine alternating group* $\mathrm{A}_n^{\mathrm{aff}}$. This group is a product of the group A_n and the group T_n, consisting of shifts in the space $\mathbb{R}^n$ by vectors $r = (r_1, r_2, \dots, r_n)$, $r_j \in \mathbb{Z}$. A fundamental domain $F(\mathrm{A}_n^{\mathrm{aff}})$ of the group $\mathrm{A}_n^{\mathrm{aff}}$ is a certain bounded subset of $\mathbb{R}^n$.

Functions on the fundamental domain $F(\mathrm{A}_n^{\mathrm{aff}})$ can be expanded into series in the functions (a). These expansions are an analogue of the usual Fourier series for functions of one variable. Functions (b) determine Fourier integral transforms on the fundamental domain $F(\mathrm{A}_n)$ of the alternating group A_n. This domain consists of points $x \in \mathbb{R}^n$ such that $x_1, x_2 > x_3 > \cdots > x_n$, where $x_1, x_2 > x_3$ means that $x_1 > x_3$ and $x_2 > x_3$.

Functions (c) are used to determine finite (on a finite set) Fourier transforms. These Fourier transforms are given on grids consisting of points in the fundamental domain $F(\mathrm{A}_n^{\mathrm{aff}})$.

Alternating multivariate exponential functions are closely related to symmetric and antisymmetric exponential functions of [8]. The symmetric and antisymmetric exponential functions are connected with symmetric and antisymmetric orbit functions defined in [11, 12] and studied in detail in [7, 9]. Discrete orbit function transforms, corresponding to Coxeter – Dynkin diagrams of low order, were studied in detail, and it was shown that they are very useful for applications [1–6, 13–15].

The discrete Fourier transforms, determined by (anti)symmetric multivariate exponential functions, studied in [8], and by alternating multivariate exponential functions have a number of practically useful properties. In particular, continuous extension of the discrete transforms smoothly interpolate digital data in any dimension. Examples show that relative to the amount of available data, these transforms provide much smoother interpolation than the conventional Fourier transforms.

Symmetric and antisymmetric multivariate exponential functions, studied in [8], satisfy certain boundary conditions (antisymmetric exponential functions vanish on the boundary of the corresponding fundamental domain and the derivative of the symmetric exponential functions with respect to the normal to the boundary of the fundamental domain vanishes on the boundary). This means that smooth functions, which are expanded in these functions, have to satisfy these conditions, that is, not each smooth function can be expanded in (anti)symmetric exponential functions. Alternating multivariate exponential functions satisfy no boundary conditions and any smooth function can be expanded in these exponential functions.

Alternating multivariate exponential functions are also related to the so-called E-orbit functions exposed in [6, 11, 12]. The exposition of the theory of orbit functions strongly depends on the theory of Weyl groups, properties of root systems, etc. In this paper we avoid this dependence. We use only the alternating group and its affinization.

The best way to define alternating multivariate exponential functions is to use the semideterminant of a finite matrix. The semideterminant of a matrix is closely related to the determinant and the antideterminant of the same matrix. It is well-known that the determinant $\det(a_{ij})_{i,j=1}^n$ of an $n \times n$ matrix $(a_{ij})_{i,j=1}^n$ is defined as

$$\begin{aligned} \det(a_{ij})_{i,j=1}^n &= \sum_{w \in \mathrm{S}_n} (\det w) a_{1,w(1)} a_{2,w(2)} \cdots a_{n,w(n)} \\ &= \sum_{w \in \mathrm{S}_n} (\det w) a_{w(1),1} a_{w(2),2} \cdots a_{w(n),n}, \end{aligned} \tag{1}$$

where S_n is the symmetric group of n symbols $1, \dots, n$, the set $\big(w(1), \dots, w(n)\big)$ denotes the set $w(1, \dots, n)$, and $\det w$ denotes the determinant of the transform w, that is, $\det w = 1$ if w is an even permutation and $\det w = -1$ otherwise. The antideterminant $\det^+$ of a matrix $(a_{ij})_{i,j=1}^n$ is the sum of all terms, entering in the expression for the corresponding determinant taken with sign $+$,

$$\det{}^+(a_{ij})_{i,j=1}^n = \sum_{w \in S_n} a_{1,w(1)} a_{2,w(2)} \cdots a_{n,w(n)} = \sum_{w \in \mathrm{S}_n} a_{w(1),1} a_{w(2),2} \cdots a_{w(n),n}.$$

For the semideterminant sdet of a matrix $(a_{ij})_{i,j=1}^n$ we have

$$\mathrm{sdet}(a_{ij})_{i,j=1}^n = \frac{1}{2}\big(\det(a_{ij})_{i,j=1}^n + \det{}^+(a_{ij})_{i,j=1}^n\big).$$

Clearly,

$$\begin{aligned} \mathrm{sdet}(a_{ij})_{i,j=1}^n &= \sum_{w \in A_n} a_{1,w(1)} a_{2,w(2)} \cdots a_{n,w(n)} \\ &= \sum_{w \in \mathrm{A}_n} a_{w(1),1} a_{w(2),2} \cdots a_{w(n),n}. \end{aligned} \tag{2}$$

In the text, we use formulas of the type

$$N \geq a_1, a_2 \geq a_3 \geq a_4 \geq \cdots \geq a_n.$$

Here $N \geq a_1, a_2 \geq a_3$ means that $N \geq a_1 \geq a_3$ and $N \geq a_2 \geq a_3$. Formulas of the type $a_1, a_2 > a_3 > \cdots > a_n$ are of the same sense.

2. Alternating multivariate exponential functions

An alternating multivariate exponential function $E_\lambda(x)$ of $x = (x_1, x_2, \dots, x_n)$ is defined as the function

$$(3)\qquad E_\lambda(x) \equiv E_{(\lambda_1,\lambda_2,\dots,\lambda_n)}(x) = \mathrm{sdet}\big(\mathrm{e}^{2\pi\mathrm{i}\lambda_i x_j}\big)_{i,j=1}^n$$

$$= \mathrm{sdet}\begin{pmatrix} \mathrm{e}^{2\pi\mathrm{i}\lambda_1 x_1} & \mathrm{e}^{2\pi\mathrm{i}\lambda_1 x_2} & \dots & \mathrm{e}^{2\pi\mathrm{i}\lambda_1 x_n} \\ \mathrm{e}^{2\pi\mathrm{i}\lambda_2 x_1} & \mathrm{e}^{2\pi\mathrm{i}\lambda_2 x_2} & \dots & \mathrm{e}^{2\pi\mathrm{i}\lambda_2 x_n} \\ \vdots & \vdots & \ddots & \vdots \\ \mathrm{e}^{2\pi\mathrm{i}\lambda_n x_1} & \mathrm{e}^{2\pi\mathrm{i}\lambda_n x_2} & \dots & \mathrm{e}^{2\pi\mathrm{i}\lambda_n x_n} \end{pmatrix}$$

$$\equiv \sum_{w\in A_n} \mathrm{e}^{2\pi\mathrm{i}\lambda_1 x_{w(1)}}\mathrm{e}^{2\pi\mathrm{i}\lambda_2 x_{w(2)}}\cdots\mathrm{e}^{2\pi\mathrm{i}\lambda_n x_{w(n)}} = \sum_{w\in A_n} \mathrm{e}^{2\pi\mathrm{i}\langle\lambda, wx\rangle},$$

where $\lambda = (\lambda_1, \lambda_2, \dots, \lambda_n)$ is a set of real numbers, which determines the function $E_\lambda(x)$, and $\langle\lambda, x\rangle$ denotes the scalar product in the n-dimensional Euclidean space E_n, $\langle\lambda, x\rangle = \sum_{i=1}^n \lambda_i x_i$. When $\lambda_1, \lambda_2, \dots, \lambda_n$ are integers, we denote this set of numbers as $m \equiv (m_1, m_2, \dots, m_n)$,

$$(4)\qquad E_{m_1,m_2,\dots,m_n}(x) = \mathrm{sdet}\big(\mathrm{e}^{2\pi\mathrm{i}m_i x_j}\big)_{i,j=1}^n.$$

From expression (3), it follows that the alternating exponential functions $E_\lambda(x)$ satisfy the relation

$$(5)\qquad E_\lambda(x_1 + a, x_2 + a, \dots, x_n + a) = \mathrm{e}^{2\pi\mathrm{i}(\lambda_1+\lambda_2+\cdots+\lambda_n)a}E_\lambda(x).$$

It is therefore sufficient to consider the function $E_\lambda(x)$ on the hyperplane

$$x_1 + x_2 + \cdots + x_n = b,$$

where b is a fixed number (we denote this hyperplane by $\mathcal{H}_b$). A transition from one hyperplane $\mathcal{H}_b$ to another $\mathcal{H}_c$ is fulfilled by multiplication by a usual exponential function $\mathrm{e}^{2\pi\mathrm{i}|\lambda|(c-b)}$, where $|\lambda| = \lambda_1 + \lambda_2 + \cdots + \lambda_n$,

$$E_\lambda(x)|_{x\in\mathcal{H}_b} = \mathrm{e}^{2\pi\mathrm{i}(\lambda_1+\lambda_2+\cdots+\lambda_n)(c-b)}E_\lambda(x)|_{x\in\mathcal{H}_c}.$$

It is useful to consider the functions $E_\lambda(x)$ on the hyperplane $\mathcal{H}_0$. For $x \in \mathcal{H}_0$ we have the relation

$$(6)\qquad E_{\lambda_1+\nu,\lambda_2+\nu,\dots,\lambda_n+\nu}(x) = E_\lambda(x).$$

We conclude from (2) that the expression of semideterminant sdet does not change when applying a permutation from A_n to rows or to columns. This means that for any permutation $w \in A_n$ we have

$$(7)\qquad E_{w\lambda}(x) = E_\lambda(x), \quad E_\lambda(wx) = E_\lambda(x).$$

Therefore, it is sufficient to consider only alternating exponential functions $E_\lambda(x)$ with $\lambda = (\lambda_1, \lambda_2, \dots, \lambda_n)$ such that

$$\lambda_1, \lambda_2 \geq \lambda_3 \geq \cdots \geq \lambda_n.$$

Such λ are called *semidominant*. The set of all semidominant λ is denoted by D_+^e. Below, when considering alternating exponential functions $E_\lambda(x)$, we assume that $\lambda \in D_+^e$.

Alternating exponential functions are related to symmetric and antisymmetric exponential functions $E^+_\lambda(x)$ and $E^-_\lambda(x)$, which are studied in [8]. They are determined by the formulas

$$E^+_\lambda(x) = \det{}^+\big(\mathrm{e}^{2\pi\mathrm{i}\lambda_i x_j}\big)_{i,j=1}^n, \quad E^-_\lambda(x) = \det\big(\mathrm{e}^{2\pi\mathrm{i}\lambda_i x_j}\big)_{i,j=1}^n,$$

where λ and x are such as in (3). This relation will be considered in Section 4. Here we consider the case $n = 2$.

The case $n = 2$. We examine the alternating multivariate exponential functions $E_\lambda(x)$ for $n = 2$ and $\lambda_2 = -\lambda_1 \equiv \lambda$. Then

$$E_{(\lambda,-\lambda)}(x_1, x_2) = \mathrm{e}^{2\pi\mathrm{i}\lambda(x_1-x_2)}, \quad E_{(-\lambda,\lambda)}(x_1, x_2) = \mathrm{e}^{-2\pi\mathrm{i}\lambda(x_1-x_2)}.$$

For antisymmetric and symmetric multivariate exponential functions with these n and λ_2 we have

$$E^-_{(\lambda,-\lambda)}(x_1, x_2) = \mathrm{e}^{2\pi\mathrm{i}\lambda(x_1-x_2)} - \mathrm{e}^{-2\pi\mathrm{i}\lambda(x_1-x_2)} = 2\mathrm{i}\sin 2\pi\lambda(x_1 - x_2),$$

$$E^+_{(\lambda,-\lambda)}(x_1, x_2) = \mathrm{e}^{2\pi\mathrm{i}\lambda(x_1-x_2)} + \mathrm{e}^{-2\pi\mathrm{i}\lambda(x_1-x_2)} = 2\cos 2\pi\lambda(x_1 - x_2).$$

Thus,

$$E^-_{(\lambda,-\lambda)}(x_1, x_2) = E_{(\lambda,-\lambda)}(x_1, x_2) - E_{(-\lambda,\lambda)}(x_1, x_2), \tag{8}$$

$$E^+_{(\lambda,-\lambda)}(x_1, x_2) = E_{(\lambda,-\lambda)}(x_1, x_2) + E_{(-\lambda,\lambda)}(x_1, x_2). \tag{9}$$

Therefore, for $n = 2$ and $\lambda_2 = -\lambda_1$ the alternating multivariate exponential function is the usual exponential function of one variable, whereas antisymmetric and symmetric multivariate exponential functions are sine and cosine functions of one variable. As we shall see below, the relations (8) and (9) can be generalized for any $n \in \mathbb{Z}_+$.

3. Affine alternating group and fundamental domains

The functions $E_\lambda(x)$ are symmetric with respect to the alternating group A_n, that is, $E_\lambda(wx) = E_\lambda(x)$, $w \in \mathrm{A}_n$. The functions $E_m(x)$ with integral $m = (m_1, m_2, \ldots, m_n)$ admit additional symmetries related to the periodicity of the exponential functions $\mathrm{e}^{2\pi\mathrm{i}ry}$, $r \in \mathbb{Z}$, $y \in \mathbb{R}$. These symmetries are described by the discrete group of shifts in the space $\mathbb{R}^n$ by vectors

$$r_1\mathbf{e}_1 + r_2\mathbf{e}_2 + \cdots + r_n\mathbf{e}_n, \quad r_i \in \mathbb{Z},$$

where $\mathbf{e}_1, \mathbf{e}_2, \ldots, \mathbf{e}_n$ are the unit vectors along the corresponding coordinate axes. We denote this group by T_n.

Permutations of A_n and shifts of T_n generate a group which is denoted as $\mathrm{A}_n^{\mathrm{aff}}$ and referred to as the *affine alternating group*. The group $\mathrm{A}_n^{\mathrm{aff}}$ is a semidirect product of its subgroups A_n and T_n,

$$\mathrm{A}_n^{\mathrm{aff}} = \mathrm{A}_n \circledS \mathrm{T}_n,$$

where T_n is an invariant subgroup, that is, $wtw^{-1} \in \mathrm{T}_n$ for $w \in \mathrm{A}_n$ and $t \in \mathrm{T}_n$.

An open connected simply connected set $F \subset \mathbb{R}^n$ is called a *fundamental domain* for the group $\mathrm{A}_n^{\mathrm{aff}}$ (for the group A_n) if it does not contain equivalent points (that is, points x and x' such that $x' = wx$, where w belongs to $\mathrm{A}_n^{\mathrm{aff}}$ or A_n, respectively) and if its closure contains at least one point from each $\mathrm{A}_n^{\mathrm{aff}}$-orbit (from each A_n-orbit). Recall that an $\mathrm{A}_n^{\mathrm{aff}}$-orbit of a point $x \in \mathbb{R}^n$ is the set of points wx, $w \in \mathrm{A}_n^{\mathrm{aff}}$.

It is evident that the set D^e_{++} of all points $x=(x_1,x_2,\dots,x_n)$ such that

$$x_1, x_2 > x_2 > \cdots > x_n,$$

is a fundamental domain for the group A_n (we denote it as $F(\mathrm{A}_n)$). The set of points $x=(x_1,x_2,\dots,x_n)\in D^e_{++}$ such that

$$1 > x_1, x_2 > x_2 > \cdots > x_n > 0$$

is a fundamental domain for the affine group $\mathrm{A}_n^{\rm aff}$ (we denote it as $F(\mathrm{A}_n^{\rm aff})$).

As previously seen, the functions $E_\lambda(x)$ are symmetric with respect to the alternating group A_n. This means that it is sufficient to consider the functions $E_\lambda(x)$ only on the closure of the fundamental domain $F(\mathrm{A}_n)$. Values of $E_\lambda(x)$ on other points are obtained by using the symmetry.

The symmetry of functions $E_m(x)$, $m=(m_1,m_2,\dots,m_n)$, $m_i\in\mathbb{Z}$, with respect to the affine alternating group $\mathrm{A}_n^{\rm aff}$,

$$E_m(wx+r)=E_m(x),\quad w\in\mathrm{A}_n,\ r\in\mathrm{T}_n, \tag{10}$$

means that we may consider $E_m(x)$ only on the closure of the fundamental domain $F(\mathrm{A}_n^{\rm aff})$, that is, on the set of points x such that $1\ge x_1, x_2\ge x_2\ge\cdots\ge x_n\ge 0$. Values of $E_m(x)$ on other points are obtained by using the relation (10).

4. Relation to symmetric and antisymmetric exponential functions

The alternating multivariate exponential functions $E_\lambda(x)$ are related to symmetric and antisymmetric multivariate exponential functions $E^+_\lambda(x)$ and $E^-_\lambda(x)$ defined in [8] and determined as

$$E^-_\lambda(x)=\det\left(e^{2\pi i\lambda_i x_j}\right)^n_{i,j=1},\quad E^+_\lambda(x)=\det{}^+\left(e^{2\pi i\lambda_i x_j}\right)^n_{i,j=1},$$

where $\det^+\left(e^{2\pi i\lambda_i x_j}\right)^n_{i,j=1}$ is the antideterminant of the matrix $\left(e^{2\pi i\lambda_i x_j}\right)^n_{i,j=1}$.

It follows from the definitions of alternating and symmetric and antisymmetric multivariate exponential functions that for λ such that $\lambda_1>\lambda_2>\lambda_3>\cdots>\lambda_n$ we have

$$E^-_\lambda(x)=E_\lambda(x)-E_{r_{12}\lambda}(x), \tag{11}$$

$$E^+_\lambda(x)=E_\lambda(x)+E_{r_{12}\lambda}(x), \tag{12}$$

where r_{12} means the permutation of λ_1 and λ_2. It follows from here that

$$E_\lambda(x)=\tfrac12\left(E^+_\lambda(x)+E^-_\lambda(x)\right), \tag{13}$$

$$E_{r_{12}\lambda}(x)=\tfrac12\left(E^+_\lambda(x)-E^-_\lambda(x)\right). \tag{14}$$

It is directly derived from (11)–(14) that

$$\left(E^+_\lambda(x)\right)^2-\left(E^-_\lambda(x)\right)^2=4E_\lambda(x)E_{r_{12}\lambda}(x),$$

$$\left(E^+_\lambda(x)\right)^2+\left(E^-_\lambda(x)\right)^2=2\left(E_\lambda(x)\right)^2+2\left(E_{r_{12}\lambda}(x)\right)^2.$$

If there are two coinciding numbers in the set $\lambda_1,\lambda_2,\dots,\lambda_n$, we get

$$E_\lambda(x)=E^+_\lambda(x). \tag{15}$$

5. Properties

The symmetry of alternating multivariate exponential functions $E_\lambda(x)$ with respect to the alternating group A_n is a main property of these functions. However, they possess many other interesting properties.

Continuity. The functions $E_\lambda(x)$ are finite sums of multivariate exponential functions. Therefore, they are continuous functions of $x_1, x_2, \ldots, x_n$ and have continuous derivatives of all orders in $\mathbb{R}^n$.

Complex conjugation. Let there be two coinciding numbers in the set $\lambda = (\lambda_1, \lambda_2, \ldots, \lambda_n)$. Since in this case $E_\lambda(x) = E^+_\lambda(x)$, then due to the behavior of $E^+_\lambda(x)$ under complex conjugation (see [8, Section 4]) we obtain

$$E_{(\lambda_1,\lambda_2,\ldots,\lambda_n)}(x) = \overline{E_{-(\lambda_n,\lambda_{n-1},\ldots,\lambda_1)}(x)}. \tag{16}$$

Let $\lambda = (\lambda_1, \lambda_2, \ldots, \lambda_n)$ be such that $\lambda_1 > \lambda_2 > \cdots > \lambda_n$. Then due to relations (17) and (18) in [8] and formulas (13) and (14) we derive that

$$E_{(\lambda_1,\lambda_2,\ldots,\lambda_n)}(x) = \overline{E_{-(\lambda_n,\lambda_{n-1},\ldots,\lambda_1)}(x)}, \tag{17}$$

$$E_{r_{12}(\lambda_1,\lambda_2,\ldots,\lambda_n)}(x) = \overline{E_{-r_{12}(\lambda_n,\lambda_{n-1},\ldots,\lambda_1)}(x)} \tag{18}$$

for $n = 4k, 4k+1$ and

$$E_{(\lambda_1,\lambda_2,\ldots,\lambda_n)}(x) = \overline{E_{-r_{12}(\lambda_n,\lambda_{n-1},\ldots,\lambda_1)}(x)}, \tag{19}$$

$$E_{r_{12}(\lambda_1,\lambda_2,\ldots,\lambda_n)}(x) = \overline{E_{-(\lambda_n,\lambda_{n-1},\ldots,\lambda_1)}(x)} \tag{20}$$

for $n = 4k-1, 4k-2$, where k is a positive integer.

Scaling symmetry. For $c \in \mathbb{R}$, let $c\lambda = (c\lambda_1, c\lambda_2, \ldots, c\lambda_n)$. Then

$$E_{c\lambda}(x) = \sum_{w \in \mathrm{A}_n} \mathrm{e}^{2\pi \mathrm{i}\langle cw\lambda, x\rangle} = \sum_{w \in \mathrm{A}_n} \mathrm{e}^{2\pi \mathrm{i}\langle w\lambda, cx\rangle} = E_\lambda(cx).$$

The equality $E_{c\lambda}(x) = E_\lambda(cx)$ expresses the *scaling symmetry of exponential functions* $E_\lambda(x)$.

Duality. Due to invariance of the scalar product $\langle\cdot,\cdot\rangle$ with respect to the alternating group A_n, $\langle w\mu, wy\rangle = \langle \mu, y\rangle$, we have

$$E_\lambda(x) = \sum_{w \in \mathrm{A}_n} \mathrm{e}^{2\pi \mathrm{i}\langle \lambda, w^{-1}x\rangle} = \sum_{w \in \mathrm{A}_n} \mathrm{e}^{2\pi \mathrm{i}\langle x, w\lambda\rangle} = E_x(\lambda).$$

This relation expresses the *duality* of alternating exponential functions.

Orthogonality on the fundamental domain $F(\mathrm{A}_n^{\mathrm{aff}})$. Alternating exponential functions $E_m(x)$ with $m = (m_1, m_2, \ldots, m_n) \in D^e_{++}$, $m_j \in \mathbb{Z}$, are orthogonal on $F(\mathrm{A}_n^{\mathrm{aff}})$ with respect to the Euclidean measure,

$$|F(\mathrm{A}_n^{\mathrm{aff}})|^{-1} \int_{\overline{F(\mathrm{A}_n^{\mathrm{aff}})}} E_m(x)\overline{E_{m'}(x)}\,\mathrm{d}x = |\mathrm{A}_n|\delta_{mm'},$$

where $|\mathrm{A}_n|$ means the number of elements in the set A_n, $\overline{F(\mathrm{A}_n^{\mathrm{aff}})}$ is the closure of $F(\mathrm{A}_n^{\mathrm{aff}})$, and $|F(\mathrm{A}_n^{\mathrm{aff}})|$ is the area of the fundamental domain $F(\mathrm{A}_n^{\mathrm{aff}})$. This relation follows from the equality

$$\int_{\mathsf{T}} E_m(x)\overline{E_{m'}(x)}\,\mathrm{d}x = |\mathrm{A}_n|\delta_{mm'} \tag{21}$$

(where T is the torus in E_n consisting of points $x = (x_1, x_2, \ldots, x_n)$, $0 \le x_i < 1$), which is a consequence of the orthogonality of the exponential functions $\mathrm{e}^{2\pi \mathrm{i}\langle \mu, x\rangle}$ (entering into the definition of $E_m(x)$) for different sets μ.

Assuming that an area of T is equal to 1, $|\mathsf{T}|=1$, we have $|F(\mathrm{A}_n^{\mathrm{aff}})| = |\mathrm{A}_n|^{-1}$, and formula (21) takes the form

$$\int_{F(\mathrm{A}_n^{\mathrm{aff}})} E_m(x)\overline{E_{m'}(x)}\,\mathrm{d}x = \delta_{mm'}. \tag{22}$$

There can be coinciding summands in expression (3) for alternating exponential functions. For this reason, for functions $E_m(x)$ with $m=(m_1,m_2,\dots,m_n)\in D_+^e$, $m_j\in\mathbb{Z}$, the relation (22) is replaced by

$$\int_{F(\mathrm{A}_n^{\mathrm{aff}})} E_m(x)\overline{E_{m'}(x)}\,\mathrm{d}x = |G_m|\delta_{mm'}. \tag{23}$$

where $|G_m|$ is the number of elements in the subgroup G_m of A_n consisting of elements $w\in\mathrm{A}_n$ such that $wm=m$, i.e., it is the order of the stabilizer G_m of m in A_n.

Solutions of the Laplace equation. The Laplace operator on the Euclidean space $\mathbb{R}^n$ in the Cartesian coordinates $x=(x_1,x_2,\dots,x_n)$ takes the form

$$\Delta = \frac{\partial^2}{\partial x_1^2} + \frac{\partial^2}{\partial x_2^2} + \cdots + \frac{\partial^2}{\partial x_n^2}.$$

Taking any summand in the expression for the alternating multivariate exponential function $E_\lambda(x)$, we get

$$\Delta \mathrm{e}^{2\pi\mathrm{i}((w(\lambda))_1x_1+\cdots+(w(\lambda))_nx_n)} = -4\pi^2\langle\lambda,\lambda\rangle \mathrm{e}^{2\pi\mathrm{i}((w(\lambda))_1x_1+\cdots+(w(\lambda))_nx_n)},$$

where $\lambda=(\lambda_1,\lambda_2,\dots,\lambda_n)$ determines $E_\lambda(x)$. Since this action of Δ does not depend on a summand from the expression for the alternating exponential function, we have

$$\Delta E_\lambda(x) = -4\pi^2\langle\lambda,\lambda\rangle E_\lambda(x), \tag{24}$$

The formula (24) can be generalized in the following way. Let $\sigma_k(y_1,y_2,\dots,y_n)$ be the kth elementary symmetric polynomial of degree k, that is,

$$\sigma_k(y_1,y_2,\dots,y_n) = \sum_{1\le k_1<k_2<\cdots<k_n\le n} y_{k_1}y_{k_2}\cdots y_{k_n}.$$

Then for $k=1,2,\dots,n$ we have

$$\sigma_k\left(\frac{\partial^2}{\partial x_1^2},\frac{\partial^2}{\partial x_2^2},\dots,\frac{\partial^2}{\partial x_n^2}\right)E_\lambda(x) = (-4\pi^2)^k\sigma_k(\lambda_1^2,\lambda_2^2,\dots,\lambda_n^2)E_\lambda(x). \tag{25}$$

Note that n differential equations (25) are algebraically independent.

Symmetric and antisymmetric multivariate exponential functions also satisfy these equations. They satisfy the certain boundary conditions (antisymmetric exponential functions vanish on the boundary of the corresponding fundamental domain and the derivative of the symmetric exponential functions with respect to the normal to the boundary of the fundamental domain vanishes on the boundary). Alternating exponential functions do not satisfy these conditions.

6. Expansions in alternating exponential functions on $F(A_n^{\rm aff})$

Alternating exponential functions determine symmetric (with respect to A_n) multivariate Fourier transforms that generalize the usual Fourier transform. There are three types of such transforms:

(a) Fourier transforms related to the functions $E_m(x)$ with $m = (m_1, \dots, m_n)$, $m_j \in \mathbb{Z}$ (Fourier series);
(b) Fourier transforms related to $E_\lambda(x)$ with $\lambda \in D_+^{\rm e}$;
(c) Multivariate finite Fourier transforms.

In this section, we consider expansions in alternating exponential functions on the fundamental domain $F(A_n^{\rm aff})$. These expansions are constructed in the same way as in the case of (anti)symmetric exponential functions in [8].

Let $f(x)$ be a symmetric (with respect to the group $A_n^{\rm aff}$) continuous function on the n-dimensional Euclidean space E_n which has continuous derivatives. We may consider this function on the torus T which can be identified with a closure of the union of the sets $wF(A_n^{\rm aff})$, $w \in A_n$. The function $f(x)$, as a function on T, can be expanded in exponential functions $e^{2\pi i m_1 x_1} e^{2\pi i m_2 x_2} \dots e^{2\pi i m_n x_n}$, $m_i \in \mathbb{Z}$. We have

$$f(x) = \sum_{m_i \in \mathbb{Z}} c_m e^{2\pi i m_1 x_1} e^{2\pi i m_2 x_2} \dots e^{2\pi i m_n x_n}, \tag{26}$$

where $m = (m_1, m_2, \dots, m_n)$. It follows from the symmetry $f(wx) = f(x)$, $w \in A_n$, that

$$\begin{aligned} f(wx) &= \sum_{m_i \in \mathbb{Z}} c_m e^{2\pi i m_1 x_{w(1)}} \dots e^{2\pi i m_n x_{w(n)}} = \sum_{m_i \in \mathbb{Z}} c_m e^{2\pi i m_{w^{-1}(1)} x_1} \dots e^{2\pi i m_{w^{-1}(n)} x_n} \\ &= \sum_{m_i \in \mathbb{Z}} c_{wm} e^{2\pi i m_1 x_1} \dots e^{2\pi i m_n x_n} = f(x) = \sum_{m_i \in \mathbb{Z}} c_m e^{2\pi i m_1 x_1} \dots e^{2\pi i m_n x_n}. \end{aligned}$$

Therefore, the coefficients c_m satisfy the conditions $c_{wm} = c_m$, $w \in A_n$. Collecting exponential functions in (26) at the same c_{wm}, $w \in A_n$, we obtain the expansion

$$f(x) = \sum_{m \in P_+^{\rm e}} c_m E_m(x), \tag{27}$$

where $P_+^{\rm e} = D_+^{\rm e} \cap \mathbb{Z}^n$. Thus, *any symmetric (with respect to A_n) continuous function f on T which has continuous derivatives (that is, any continuous function on $D_+^{\rm e}$ with continuous derivatives) can be expanded in alternating exponential functions $E_m(x)$, $m \in P_+^{\rm e}$.*

By the orthogonality relation (23), the coefficients c_m in the expansion (27) are determined by the formula

$$c_m = |G_m|^{-1} \int_{F(A_n^{\rm aff})} f(x) \overline{E_m(x)}\, dx, \tag{28}$$

where, as before, $|G_m|$ is the number of elements in the subgroup G_m of A_n consisting of $w \in A_n$ such that $wm = m$. Moreover, the Plancherel formula holds:

$$\sum_{m \in P_+^{\rm e}} |c_m|^2 = |G_m|^{-1} \int_{F(A_n^{\rm aff})} |f(x)|^2\, dx. \tag{29}$$

Formula (28) is the symmetrized (with respect to the group A_n) Fourier transform of the function $f(x)$. Formula (27) gives an inverse transform. Formulas (27) and (28) give the *multivariate Fourier transforms* corresponding to the alternating exponential functions $E_m(x)$, $m \in P_+^{\rm e}$.

Let $\mathcal{L}^2(\overline{F(\mathrm{A}_n^{\rm aff})})$ denote the Hilbert space of functions on the domain $\overline{F(\mathrm{A}_n^{\rm aff})}$ with the scalar product

$$\langle f_1, f_2 \rangle = \int_{\overline{F(\mathrm{A}_n^{\rm aff})}} f_1(x)\overline{f_2(x)}\,\mathrm{d}x.$$

The formulas (27)–(29) show that *the set of alternating multivariate exponential functions* $E_m(x)$, $m \in P_+^{\rm e}$, *forms an orthogonal basis of* $\mathcal{L}^2(\overline{F(\mathrm{A}_n^{\rm aff})})$.

7. Fourier transforms on the fundamental domain $F(\mathrm{A}_n)$

The expansions (27) of functions on the fundamental domain $F(\mathrm{A}_n^{\rm aff})$ are expansions in the exponential functions $E_m(x)$ with integral $m = (m_1, m_2, \ldots, m_n)$. The exponential functions $E_\lambda(x)$ with λ lying in the fundamental domain $F(\mathrm{A}_n)$ (and not obligatory integral) are not invariant with respect to the corresponding affine alternating group $\mathrm{A}_n^{\rm aff}$. They are invariant only with respect to the alternating group A_n. A closure of the fundamental domain of A_n coincides with the set $D_+^{\rm e}$ consisting of points x such that $x_1, x_2 \geq x_3 \geq \cdots \geq x_n$. The functions $E_\lambda(x)$, $\lambda \in D_+^{\rm e}$, determine Fourier transforms on $D_+^{\rm e}$.

We begin with the usual Fourier transforms on $\mathbb{R}^n$:

$$\tilde{f}(\lambda) = \int_{\mathbb{R}^n} f(x) \mathrm{e}^{2\pi \mathrm{i}\langle \lambda, x\rangle}\,\mathrm{d}x, \tag{30}$$

$$f(x) = \int_{\mathbb{R}^n} \tilde{f}(\lambda) \mathrm{e}^{-2\pi \mathrm{i}\langle \lambda, x\rangle}\,\mathrm{d}\lambda. \tag{31}$$

Let the function $f(x)$ be invariant with respect to the alternating group A_n, $f(wx) = f(x)$, $w \in \mathrm{A}_n$. It is easy to verify that $\tilde{f}(\lambda)$ is also invariant with respect to A_n. By replacing λ by $w\lambda$, $w \in \mathrm{A}_n$, in (30) and summing up both sides over $w \in \mathrm{A}_n$, we obtain, instead of (30), the equality

$$\tilde{f}(\lambda) = \int_{D_+^{\rm e}} f(x) E_\lambda(x)\,\mathrm{d}x, \quad \lambda \in D_+^{\rm e}, \tag{32}$$

where we have taken into account that $f(x)$ is invariant with respect to A_n.

Similarly, starting from (31), we obtain the inverse formula:

$$f(x) = \int_{D_+^{\rm e}} \tilde{f}(\lambda)\overline{E_\lambda(x)}\,\mathrm{d}\lambda. \tag{33}$$

For the transforms (32) and (33), the Plancherel formula

$$\int_{D_+^{\rm e}} |f(x)|^2\,\mathrm{d}x = \int_{D_+^{\rm e}} |\tilde{f}(\lambda)|^2\,\mathrm{d}\lambda$$

holds. Formulas (32) and (33) determine the symmetric (with respect to A_n) multivariate Fourier transforms on the domain $\overline{F(\mathrm{A}_n)}$, which are referred to as the *alternating Fourier transforms*.

Let $\mathcal{L}^2(D_+^{\rm e})$ denote the Hilbert space of functions on the domain $D_+^{\rm e}$ with the scalar product

$$\langle f_1, f_2 \rangle = \int_{D_+^{\rm e}} f_1(x)\overline{f_2(x)}\,\mathrm{d}x$$

The formulas (32) and (33) show that *the set of alternating multivariate exponential functions* $E_\lambda(x)$, $\lambda \in D_+^e$, *form a full system of functions of* $\mathcal{L}^2(D_+^e)$.

8. Eigenfunctions of the alternating Fourier transform

Let $H_n(x)$, $n = 0, 1, 2, \ldots$, be the well-known Hermite polynomials of one variable. They satisfy the relation

$$\int_{-\infty}^{\infty} \mathrm{e}^{2\pi \mathrm{i} px} \mathrm{e}^{-\pi p^2} H_m(\sqrt{2\pi} p)\, \mathrm{d}p = \mathrm{i}^{-m} \mathrm{e}^{-\pi x^2} H_m(\sqrt{2\pi} x) \tag{34}$$

(see, for example, [16, Section 12.2.4]).

We generate polynomials of many variables

$$H_{\mathbf{m}}(\mathbf{x}) \equiv H_{m_1, m_2, \ldots, m_n}(x_1, x_2, \ldots, x_n) := H_{m_1}(x_1) H_{m_2}(x_2) \cdots H_{m_n}(x_n). \tag{35}$$

The functions

$$\mathrm{e}^{-|\mathbf{x}|^2/2} H_{\mathbf{m}}(\mathbf{x}), \quad m_i = 0, 1, 2, \ldots, \ i = 1, 2, \ldots, n, \tag{36}$$

where $|\mathbf{x}|$ is a length of the vector $\mathbf{x}$, form an orthogonal basis of the Hilbert space $\mathcal{L}^2(\mathbb{R}^n)$ with the scalar product $\langle f_1, f_2 \rangle := \int_{\mathbb{R}^n} f_1(\mathbf{x}) \overline{f_2(\mathbf{x})}\, \mathrm{d}\mathbf{x}$, where $\mathrm{d}\mathbf{x} = \mathrm{d}x_1\, \mathrm{d}x_2 \cdots \mathrm{d}x_n$.

We symmetrize the functions

$$\mathcal{H}_{\mathbf{m}}(\mathbf{x}) := \mathrm{e}^{-\pi |\mathbf{x}|^2} H_{\mathbf{m}}(\sqrt{2\pi}\mathbf{x})$$

(obtained from (36) by replacing $\mathbf{x}$ by $\sqrt{2\pi}\mathbf{x}$) by means of alternating multivariate exponential functions:

$$\int_{\mathbb{R}^n} E_\lambda(\mathbf{x}) \mathrm{e}^{-\pi|\mathbf{x}|^2} H_{\mathbf{m}}(\sqrt{2\pi}\mathbf{x})\, \mathrm{d}\mathbf{x} = \mathrm{i}^{-|\mathbf{m}|} \mathrm{e}^{-\pi|\lambda|^2} \mathbf{H}_{\mathbf{m}}(\sqrt{2\pi}\lambda), \tag{37}$$

It is easy to see that the polynomials $\mathbf{H}_{\mathbf{m}}$ are indeed symmetric with respect to the group A_n,

$$\mathbf{H}_{\mathbf{m}}(w\lambda) = \mathbf{H}_{\mathbf{m}}(\lambda), \quad w \in \mathrm{A}_n.$$

Therefore, we may consider $\mathbf{H}_{\mathbf{m}}(\lambda)$ for values of $\lambda = (\lambda_1, \lambda_2, \ldots, \lambda_n)$ such that $\lambda_1, \lambda_2 \geq \lambda_3 \geq \cdots \geq \lambda_n$. The polynomials $\mathbf{H}_{\mathbf{m}}$ are of the form

$$\mathbf{H}_{\mathbf{m}}(\lambda) = \operatorname{sdet}\big(H_{m_i}(\lambda_j)\big)_{i,j=1}^n, \tag{38}$$

that is, it is sufficient to consider the polynomials $\mathbf{H}_{\mathbf{m}}(\lambda)$ for integer n-tuples $\mathbf{m}$, such that $m_1, m_2 \geq m_3 \geq \cdots \geq m_n$.

Let us apply the alternating Fourier transform (32) (we denote it as $\mathfrak{F}$) to the functions (38). Taking into account formula (37) we derive

$$\begin{aligned}\mathfrak{F}\big(\mathrm{e}^{-\pi|\mathbf{x}|^2} \mathbf{H}_{\mathbf{m}}(\sqrt{2\pi}\mathbf{x})\big) &= \frac{1}{|\mathrm{A}_n|} \int_{\mathbb{R}^n} E_\lambda(\mathbf{x}) \mathrm{e}^{-\pi|\mathbf{x}|^2} \mathbf{H}_{\mathbf{m}}(\sqrt{2\pi}\mathbf{x})\, \mathrm{d}\mathbf{x} \\ &= \mathrm{i}^{-|\mathbf{m}|} \mathrm{e}^{-\pi|\lambda|^2} \mathbf{H}_{\mathbf{m}}(\sqrt{2\pi}\lambda),\end{aligned}$$

that is, the *functions* $\mathrm{e}^{-\pi|\mathbf{x}|^2} \mathbf{H}_{\mathbf{m}}(\sqrt{2\pi}\mathbf{x})$ *are eigenfunctions of the alternating Fourier transform* $\mathfrak{F}$. Since these functions for $m_i = 0, 1, 2, \ldots$, $i = 1, 2, \ldots, n$, such that $m_1, m_2 \geq m_3 \geq \cdots \geq m_n$, form an orthogonal basis of the Hilbert space $\mathcal{L}^2(D_+^e)$, they constitute a complete set of eigenfunctions of this transform. Thus, this transform has only four eigenvalues $\mathrm{i}, -\mathrm{i}, 1, -1$ in $\mathcal{L}^2(D_+^e)$. This means that we have $\mathfrak{F}^4 = 1$.

9. Finite alternating Fourier transforms

Along with the integral Fourier transform in one variable, there exists a discrete Fourier transform in one variable. It is given by the kernel

$$e_{mn} := N^{-1/2} \exp(2\pi i mn/N), \quad m, n = 1, 2, \cdots, N, \tag{39}$$

where N is a fix positive integer. The matrix $(e_{mn})_{m,n=1}^N$ is unitary, that is,

$$\sum_k e_{mk}\overline{e_{nk}} = \delta_{mn}, \quad \sum_k e_{km}\overline{e_{kn}} = \delta_{mn}. \tag{40}$$

Let $f(n)$ be a function of $n \in \{1, 2, \ldots, N\}$. Then the function

$$\tilde{f}(m) = \sum_{n=1}^N f(n) e_{mn} \equiv N^{-1/2} \sum_{n=1}^N f(n) \exp(2\pi i mn/N) \tag{41}$$

is a finite Fourier transform of $f(n)$. The inverse transform is given by

$$f(n) = N^{-1/2} \sum_{m=1}^N \tilde{f}(m) \exp(-2\pi i mn/N). \tag{42}$$

The Plancherel formula $\sum_{m=1}^N |\tilde{f}(m)|^2 = \sum_{n=1}^N |f(n)|^2$ holds for transforms (41) and (42). This means that the finite Fourier transform conserves the norm introduced in the space of functions on $\{1, 2, \ldots, N\}$.

In order to derive the finite alternating multivariate Fourier transform, we use the discrete exponential function (39) in the form

$$e_m(s) := N^{-1/2} \exp(2\pi i ms), \quad s \in F_N \equiv \left\{ \frac{1}{N}, \frac{2}{N}, \ldots, \frac{N-1}{N}, 1 \right\}, \quad m \in \mathbb{Z}^{\geq 0}, \tag{43}$$

and generate a multivariate discrete exponential function by taking a product of n copies of these functions,

$$\begin{aligned} e_{\mathbf{m}}(\mathbf{s}) &:= e_{m_1}(s_1) e_{m_2}(s_2) \cdots e_{m_n}(s_n) \\ &= N^{-n/2} \exp(2\pi i m_1 s_1) \exp(2\pi i m_2 s_2) \cdots \exp(2\pi i m_n s_n) \end{aligned} \tag{44}$$

where $\mathbf{s} = (s_1, s_2, \ldots, s_n) \in F_N^n$ and $\mathbf{m} = (m_1, m_2, \ldots, m_n) \in (\mathbb{Z}^{\geq 0})^n$. We introduce a scalar product in the space of linear combinations of the functions (44) by the formula

$$\langle e_{\mathbf{m}}(\mathbf{s}), e_{\mathbf{m}'}(\mathbf{s}) \rangle \equiv \prod_{i=1}^n \langle e_{m_i}(s_i), e_{m_i'}(s_i) \rangle := \prod_{i=1}^n \sum_{s_i \in F_N} e_{m_i}(s_i) \overline{e_{m_i'}(s_i)} = \delta_{\mathbf{m}\mathbf{m}'} \tag{45}$$

where $m_i, m_i' \in \{1, 2, \ldots, N\}$. Here we used the relations (40).

We now take multivariate functions (44) for integers m_i such that

$$N \geq m_1, m_2 \geq m_3 \geq \cdots \geq m_n \geq 1$$

and symmetrize them with respect to the alternating group A_n. We obtain a finite version of the alternating exponential functions (3),

$$\widetilde{E}_{\mathbf{m}}(\mathbf{s}) := |\mathrm{A}_n|^{-1/2} \operatorname{sdet}^+ (e_{m_i}(s_j))_{i,j=1}^n = |\mathrm{A}_n|^{-1/2} N^{-n/2} E_{\mathbf{m}}(\mathbf{s}), \tag{46}$$

where the discrete functions $e_m(s)$ are given by (43).

The n-tuples $\mathbf{s}$ in (46) run over $F_N^n \equiv F_N \times \cdots \times F_N$ (n times). We denote by $\breve{F}_N^n$ the subset of F_N^n consisting of $\mathbf{s} = (s_1, s_2, \ldots, s_n) \in F_N^n$ such that

$$s_1, s_2 \geq s_2 \geq \cdots \geq s_n.$$

The set $\breve{F}_N^n$ is a finite subset of the closure of the fundamental domain $F(\mathrm{A}_n^{\mathrm{aff}})$.

Acting by permutations $w \in \mathrm{A}_n$ upon $\breve{F}_N^n$ we obtain the whole set F_N^n, where each point, having some coinciding coordinates m_i, is repeated several times. Namely, a point $\mathbf{s}$ is contained $|G_{\mathbf{s}}|$ times in $\{w\breve{F}_N^n; w \in \mathrm{A}_n\}$, where $G_{\mathbf{s}}$ is the subgroup of A_n consisting of elements $w \in \mathrm{A}_n$ such that $w\mathbf{s} = \mathbf{s}$.

By $\breve{D}_N^n$ we denote the set of integer n-tuples $\mathbf{m} = (m_1, m_2, \dots, m_n)$ such that

$$N \geq m_1, m_2 \geq m_3 \geq \cdots \geq m_n \geq 1.$$

Proposition. *For $\mathbf{m}, \mathbf{m}' \in \breve{D}_N^n$ the discrete functions (46) satisfy the orthogonality relation*

$$\langle \widetilde{E}_{\mathbf{m}}(\mathbf{s}), \widetilde{E}_{\mathbf{m}'}(\mathbf{s}) \rangle = |\mathrm{A}_n| \sum_{\mathbf{s} \in \breve{F}_N^n} |G_{\mathbf{s}}|^{-1} \widetilde{E}_{\mathbf{m}}(\mathbf{s}) \overline{\widetilde{E}_{\mathbf{m}'}(\mathbf{s})} = |G_{\mathbf{m}}| \delta_{\mathbf{m}\mathbf{m}'}. \tag{47}$$

PROOF. Due to the definition of the scalar product we get

$$\begin{aligned} \langle \widetilde{E}_{\mathbf{m}}(\mathbf{s}), \widetilde{E}_{\mathbf{m}'}(\mathbf{s}) \rangle &= \sum_{\mathbf{s} \in F_N^n} \widetilde{E}_{\mathbf{m}}(\mathbf{s}) \overline{\widetilde{E}_{\mathbf{m}'}(\mathbf{s})} \\ &= |\mathrm{A}_n|^{-1} |G_{\mathbf{m}}| \sum_{w \in \mathrm{A}_n} \prod_{i=1}^{n} \sum_{s_i \in F_N} e_{m_{w(i)}}(s_i) \overline{e_{m'_{w(i)}}(s_i)} = |G_{\mathbf{m}}| \delta_{\mathbf{m}\mathbf{m}'}, \end{aligned}$$

where we have taken into account that some summands in the expression for sdet in (46) can coincide.

Since functions $\widetilde{E}_{\mathbf{m}}(\mathbf{s})$ are symmetric with respect to A_n, then

$$\sum_{\mathbf{s} \in F_N^n} \widetilde{E}_{\mathbf{m}}(\mathbf{s}) \overline{\widetilde{E}_{\mathbf{m}'}(\mathbf{s})} = |\mathrm{A}_n| \sum_{\mathbf{s} \in \breve{F}_N^n} |G_{\mathbf{s}}|^{-1} \widetilde{E}_{\mathbf{m}}(\mathbf{s}) \overline{\widetilde{E}_{\mathbf{m}'}(\mathbf{s})},$$

where we have taken into account that under the action by A_n upon $\breve{F}_N^n$ a point $\mathbf{s}$ appears $|G_{\mathbf{s}}|$ times in F_N^n. This proves the proposition. □

Let f be a function on $\breve{F}_N^n$ (or a symmetric function on F_N^n). Then it can be expanded in functions (46) as

$$f(\mathbf{s}) = \sum_{\mathbf{m} \in \breve{D}_N^n} a_{\mathbf{m}} \widetilde{E}_{\mathbf{m}}(\mathbf{s}). \tag{48}$$

The coefficients $a_{\mathbf{m}}$ are determined by the formula

$$a_{\mathbf{m}} = |\mathrm{A}_n||G_{\mathbf{m}}|^{-1} \sum_{\mathbf{s} \in \breve{F}_N^n} |G_{\mathbf{s}}|^{-1} f(\mathbf{s}) \overline{\widetilde{E}_{\mathbf{m}}(\mathbf{s})}. \tag{49}$$

The expansions (48) and (49) follow from the facts that numbers of elements in $\breve{D}_N^n$ and in $\breve{F}_N^n$ are the same and from the orthogonality relation (47) (see also [10]). We refer to expansions (48) and (49) as the *alternating multivariate finite Fourier transforms*.

Acknowledgments. One of us (A. K.) is grateful for the hospitality repeatedly extended to him at the Centre de recherches mathématiques, Université de Montréal.

References

1. A. Atoyan and J. Patera, *Properties of continuous Fourier extension of the discrete cosine transform and its multidimensional generalization*, J. Math. Phys. **45** (2004), no. 6, 2468–2491.
2. ———, *Continuous extension of the discrete cosine transform, and its applications to data processing*, Group Theory and Numerical Analysis (P. Winternitz, D. Gomez-Ullate, A. Iserles, D. Levi, P. J. Olver, R. Quispel, and P. Tempesta, eds.), CRM Proc. Lecture Notes, vol. 39, Amer. Math. Soc., Providence, RI, 2005, pp. 1–15.
3. A. Atoyan, J. Patera, V. Sahakian, and A. Akhperjanian, *Fourier transform method for imaging atmospheric Cherenkov telescopes*, Astroparticle Physics **23** (2005), 79–95.
4. M. Germain, J. Patera, and Y. Allard, *Cosine transform generalized to Lie groups* $SU(2) \times SU(2)$, $O(5)$, *and* $SU(2) \times SU(2) \times SU(2)$: *application to digital image processing*, Computational Imaging IV (San Jose, CA, 2006) (C. A. Bouman, E. L. Miller, and I. Pollak, eds.), Proc. SPIE, vol. 6065, 2006. 606519.
5. M. Germain, J. Patera, and A. Zaratsyan, *Multiresolution analysis of digital images using the continuous extension of discrete group transform*, Computational Imaging IV (San Jose, CA, 2006). 606503.
6. I. Kashuba and J. Patera, *Discrete and continuous exponential transforms of simple Lie groups of rank* 2, J. Phys. A **40** (2007), no. 18, 4751–4774.
7. A. U. Klimyk and J. Patera, *Orbit functions*, SIGMA Symmetry Integrability Geom. Methods Appl. **2** (2006), Paper 006.
8. ———, *(Anti)symmetric multivariate exponential functions and corresponding Fourier transforms*, J. Phys. A **40** (2007), no. 34, 10473–10489.
9. ———, *Antisymmetric orbit functions*, SIGMA Symmetry Integrability Geom. Methods Appl. **3** (2007), Paper 023.
10. R. V. Moody and J. Patera, *Orthogonality within the families of C-, S-, and E-functions of any compact semisimple Lie group*, SIGMA Symmetry Integrability Geom. Methods Appl. **2** (2006), Paper 076.
11. J. Patera, *Orbit functions of compact semisimple Lie groups as special functions*, Symmetry in Nonlinear Mathematical Physics (Kiev, 2003), Pr. Inst. Mat. Nats. Akad. Nauk Ukr. Mat. Zastos., vol. 50, Natsīonal. Akad. Nauk Ukraïni, Īnst. Mat., Kiev, 2004, pp. 1152–1160.
12. ———, *Compact simple Lie groups and their C-, S-, and E-transforms*, SIGMA Symmetry Integrability Geom. Methods Appl. **1** (2005), paper 025.
13. J. Patera and A. Zaratsyan, *Discrete and continuous cosine transform generalized to Lie groups* $SU(2) \times SU(2)$ *and* $O(5)$, J. Math. Phys. **46** (2005), no. 5, 053514.
14. ———, *Discrete and continuous cosine transform generalized to Lie groups* $SU(3)$ *and* $G(2)$, J. Math. Phys. **46** (2005), no. 11, 113506.
15. J. Patera, A. Zaratsyan, and H.-M. Zhu, *New class of interpolation methods based on discretized Lie group transform*, Image Processing: Algorithms and Systems V (San Jose, CA, 2006). 6064A-06.
16. N. Ja. Vilenkin and A. U. Klimyk, *Representations of Lie groups and special functions*, Vol. 2, Math. Appl. (Soviet Ser.), vol. 74, Kluwer, Dordrecht, 1993.

Bogolyubov Institute for Theoretical Physics, Kiev 03680, Ukraine
E-mail address: `aklimyk@bitp.kiev.ua`

Centre de recherches mathématiques, Université de Montréal, C.P. 6128, succ. Centre-ville, Montréal, H3C 3J7, QC Canada
E-mail address: `patera@crm.umontreal.ca`

Centre de Recherches Mathématiques
CRM Proceedings and Lecture Notes
Volume **47**, 2009

Moonshine Elements in Elliptic Cohomology

Jack Morava

ABSTRACT. This is a historical talk about the recent confluence of two lines of research in equivariant elliptic cohomology, one concerned with *connected* Lie groups, the other with the *finite* case. These themes come together in (what seems to me remarkable) work of N. Ganter, relating replicability of McKay-Thompson series to the theory of exponential cohomology operations.

Introduction

Moonshine, the Monster, conformal field theory, and elliptic cohomology have all been linked since birth. This survey foregrounds the last of these topics; its theme is the emergence, under the influence of Hopkins and Miller and their coworkers, of a coordinate-free perspective on the moduli stacks of bundles over elliptic curves.

I could never have approached this subject but for the untiring interest of John McKay, who has often seemed to me an emissary from some advanced Galactic civilization, sent here to speed up our evolution. I also want to acknowledge many helpful conversations about this material with Matt Ando, Andy Baker, Jorge Devoto, and Nora Ganter.

1. 'Classical' elliptic cohomology

1.1. The orbifold

$$\overline{\mathcal{M}}_{\mathrm{ell}} = (\mathfrak{h} \cup \mathbb{Q}) / \mathrm{PSl}_2(\mathbb{Z})$$

($\mathfrak{h}$ is the complex upper half-plane) has one point with isotropy $\mathbb{Z}/2$, one with isotropy $\mathbb{Z}/3$, and a third (the cusp) with (infinite) isotropy $\mathbb{Z}$; it's a good model over $\mathbb{C}$ for the stack $\overline{\mathcal{M}}_{1,1}(\mathbb{Z})$ of stable genus zero curves with one marked point, which is what the real experts work with. These moduli objects are *not* affine; there is a canonical line bundle ω over $\overline{\mathcal{M}}_{\mathrm{ell}}$, defined by the cotangent line $T^*_{E,0}$ at the origin of the elliptic curve $E \ni 0$. Modular forms are sections of powers of this bundle; they provide a graded substitute for an affine coordinate ring.

2000 *Mathematics Subject Classification.* 55N34, 55N91, 20D08.
Key words and phrases. equivariant elliptic cohomology, Monster, Moonshine.
The author was supported in part by the NSF.
This is the final form of the paper.

I want to discuss a cohomology theory [23, 30] taking values in the abelian category of quasicoherent sheaves over $\mathcal{M}_{\text{ell}}$, such that for example

$$\mathbf{Ell}(S^{2n}) = \omega^{\otimes n}.$$

Regarding ω_E as dual to the Lie algebra of E leads to a refinement of this picture which incorporates odd-dimensional spheres naturally, but that's one of the many topics I'll leave aside in this talk.

There is an associated cohomology theory Ell^* of more classical type, taking values in modules over a certain graded ring. Away from small primes (two and three) its coefficient ring is a graded algebra

$$\text{Ell}^*(\text{pt}) \cong \bigoplus_{n \geq 0} \Gamma(\mathcal{M}_{\text{ell}}, \omega^{\otimes n}) \quad (\otimes \mathbb{Z}[\tfrac{1}{6}])$$

of classical modular forms, but in general this isomorphism must be replaced by a spectral sequence

$$R^*\Gamma(\mathcal{M}_{\text{ell}}, \omega^{*/2}) \implies \text{tmf}^*(\text{pt})$$

with the derived functor of global sections as its E_2 term. Away from six these derived functors vanish; but in general the spectral sequence is complicated, with *lots* of two- and three-torsion, which may look strange to arithmetic geometers but turns out to be quite familiar to algebraic topologists. At the prime two, for example, the Hurewicz homomorphism from the stable homotopy ring of spheres to the resulting cohomology theory is injective up to dimension fifty or so [25].

There are other differences, which leads homotopy theorists to distinguish the abutment of this spectral sequence, which they call the ring of *topological* modular forms, from its classical analog. The discriminant Δ, for example, is a modular form but not a topological one: it supports a nontrivial (torsion-valued) differential in the spectral sequence. However, both 24Δ and Δ^{24} survive, and *do* represent elements of $\text{tmf}^*(\text{pt})$. These facts seem to be related to some congruences of Borcherds which are beyond my depth [10; 23, Theorem 5.10]; I mention them only to indicate the arithmetic power of the theory of topological modular forms.

There are at present two approaches to constructing topological modular forms, both based a systematic rigidifications of Landweber's exact functor theorem. The original approach of Hopkins and Miller builds a homotopy limit of spectra whose connective cover is tmf, while Lurie's more recent approach constructs this object as the global sections of a sheaf of spectra over a certain kind of derived moduli stack. The subject is very much under construction (see Section 2.4 below), and both approaches require deep new ideas which change the way we think of both homotopy theory and algebraic geometry; but in this talk I will take these ideas to some extent for granted, concentrating instead on their applications to Moonshine and the Monster.

1.2. In practice these constructions have many variants, associated to related moduli problems: away from two, for example, curves with $\Gamma_0(2)$-level structures correspond to Jacobi/Igusa quadrics

$$Y^2 = 1 - 2\delta X^2 + \epsilon X^4,$$

which was the example which really crystallized this whole subject [22, 29]. Away from two, the associated modular forms have Fourier coefficients in $\mathbb{Z}[\frac{1}{2}]$ and poles only at the cusp $\tau \to i\infty, q = e^{2\pi i\tau} \to 0$, where Tate's elliptic curve over $\mathbb{Z}[\![q]\!]$

lives. Over the localization $\mathbb{Z}((q))$ its formal group law has multiplicative type, and restriction to a (formal, *affine*) neighborhood of $i\infty$ defines a natural transformation

$$\mathbf{Ell} \to K^{\mathrm{Tate}}$$

of cohomology theories, which on coefficients becomes the q-expansion map

$$\mathrm{Ell}^*(\mathrm{pt}) \sim \{\text{modular forms}\} \to \mathbb{Z}((q)).$$

The target is a variant of classical K-theory, defined by 'extension of scalars' from $\mathbb{Z}$ to $\mathbb{Z}((q))$, suitably oriented: this means that the Chern class of a complex line bundle L in K^{Tate} is essentially the Weierstrass σ-function [4, 35].

It is possible to think of K-theory as the truncation of Ell defined by taking the leading term of a modular form [5, 26, 29]. Some of the more interesting calculations in K-theory involve Bernoulli numbers; in Ell their analogs involve the Eisenstein series having those numbers as constant term [7]. The natural operations in K-theory are the Adams operations; their analogs in Ell are Hecke operators [1, 6], as we shall see.

1.3. Physicists (Witten in particular) became interested in these matters because Ell^* is the best approximation by a cohomology theory

$$\begin{array}{ccc} K^*_{\mathbb{T}}(LM) & \longrightarrow & \mathrm{Ell}^*(M) \\ & \searrow & \Big\downarrow {\scriptstyle q\text{-exp'n}} \\ & & K^{\mathrm{Tate}}(M) \end{array}$$

to the $\mathbb{T}$-equivariant K-theory of a free loopspace: that construction does not preserve cofibrations, so the group on the left is very complicated. The completion

$$u \mapsto 1 - q \colon K_{\mathbb{T}}(\mathrm{pt}) = \mathbb{Z}[u, u^{-1}] \to \mathbb{Z}[\![q]\!] \to K^{\mathrm{Tate}}(\mathrm{pt})$$

restricts to the neighborhood of the fixed point set and so simplifies things greatly [27]; on the other hand, the resulting theory has a coefficient ring much larger than Ell^*.

Witten studied the index of an analog of the Dirac operator on LM, and showed that when the loopspace possesses the proper analog of a spin structure, the resulting invariant is a modular form. This precipitated an enormous amount of research; in particular, such a spin structure is now understood as the natural form of orientation [4] for elliptic cohomology. Witten further conjectured the rigidity of his invariant, under actions on the manifold of a *connected* Lie group G. This is now understood in topological terms [3] as a property of the Thom isomorphism in a suitable form of equivariant elliptic cohomology.

2. The equivariant picture

2.1. Sheaf-valued cohomology theories defined over non-affine objects are unfamiliar in algebraic topology, but working with them is usually routine, perhaps involving higher derived functors of global sections but not too much more. Extending this framework to the equivariant context, however, seems to lead us into a genuinely new world: it poses deep *conceptual* questions. Ganter's recent progress on these questions is the real focus of this talk.

It seems very reasonable to expect that for a compact Lie group G,

- there is an equivariant version $X \mapsto \mathbf{Ell}_G(X)$ of elliptic cohomology taking values in quasicoherent sheaves over some moduli stack

$$\mathcal{M}_{\rm ell}(G) = \{G\text{-bundles over } \check{E} \mid E \in \mathcal{M}_{\rm ell}\} \to \mathcal{M}_{\rm ell}$$

of *principal G-bundles* over (the dual of) the 'universal' elliptic curve.

(The appearance here of the dual Abelian variety is technical, having to do with naturality. When $G = A$ is finite abelian, A-bundles over $\check{E}$ are closely related to extensions of $\check{E}$ by A, and thus (by some Cartier duality or Cartan exchange isomorphism) to homomorphisms from the Pontrjagin dual $\hat{A}$ to E; this is related to another approach [20] to equivariance.)

2.2. For example, when $G = \mathbb{T}$ is the circle group, bundles over $\check{E}$ are classified by elements of the Picard group

$$\mathrm{Pic}(\check{E}) \cong E.$$

A $\mathbb{T}$-space X thus defines a sheaf $\mathbf{Ell}_{\mathbb{T}}(X)$ over the universal curve $\mathbf{E} = \mathcal{M}_{\rm ell}(\mathbb{T})$.

For *connected* groups this program was pioneered by Grojnowski and Ginzburg–Kapranov–Vasserot, followed by Ando and others. For example, the two-sphere $\mathbb{C} \cup \infty$ with $z \in \mathbb{T}$ acting as multiplication by z^n has (reduced) equivariant elliptic cohomology defined by the sheaf

$$\widetilde{\mathbf{Ell}}_{\mathbb{T}}(S^2(n)) = \mathcal{O}_{\mathbf{E}}\Big(-\sum E[n]\Big)$$

[41, §3.11]; where $\sum E[n]$ represents the divisor of n-torsion points on $\mathbf{E}$ (equal as such to $n^2 \cdot [0]$), and $\mathcal{O}_{\mathbf{E}}$ is the structure sheaf of the universal curve.

The grading of a classical G-equivariant cohomology theory E extends naturally from $\mathbb{Z}$ to a Grothendieck group of E-orientable linear representations of G, but sheaf-valued cohomology theories over non-affine objects are naturally graded by the Picard group of line bundles over the base object. When G is connected and (for example) simply-connected, Grojnowkski [21, §3.3] and Ando [2, §§9.20, 10.11] use earlier work of Looijenga to construct a line bundle $\mathcal{L}$ over $\mathcal{M}_{\rm ell}(G)$ such that the $\mathbb{Z}((q))$-module of cusp expansions of sections of $\mathcal{L}^{\otimes k}$ can be identified with the corresponding module generated by graded characters of level k representations of the loop group LG.

2.3. In this talk, however, I want to focus on recent developments in the case of G *finite*. In spite of deep early work of Devoto, this seems to have received less attention than for G connected: perhaps because of Witten's rigidity conjecture, and perhaps because of its intrinsic subtlety.

We can think of G-bundles over $\check{E}$ as classified by elements of the (zero-dimensional!) orbifold

$$\mathrm{Pairs}_G = \mathrm{Hom}(\mathbb{Z}^2, G)/G^{\rm conj} \cong \mathrm{Hom}(\pi_1(\check{E}), G)/G^{\rm conj} \cong H^1(\check{E}, G)$$

of conjugacy classes of commuting pairs of elements in G [19]. There is an action of $\mathrm{Sl}_2(\mathbb{Z})$ on such pairs, by

$$\begin{bmatrix} a & b \\ c & d \end{bmatrix} [h, g] = [h^a g^b, h^c g^d],$$

and we can take (see [16, §2.1] for a more precise statement)

$$\mathcal{M}_{\rm ell}(G) \sim \mathrm{Pairs}_G \times_{\mathrm{Sl}_2(\mathbb{Z})} \mathfrak{h} \to \mathrm{pt} \times_{\mathrm{PSl}_2(\mathbb{Z})} \mathfrak{h} \sim \mathcal{M}_{\rm ell}$$

as a model for the stack of G-bundles. Experienced Moonshiners will recognize the space on the left as a natural habitat for Norton's work.

(Note that this map is *not* the identity when $G = \{1\}$ is the trivial group: the object on the left is then an *un*reduced orbifold, with nontrivial isotropy coming from the central element -1 in $\mathrm{Sl}_2(\mathbb{Z})$ acting as the involution $[g,h] \mapsto [g^{-1}, h^{-1}]$ on the pair $[1,1]$. This is part of an action of the multiplicative monoid $\mathbb{Z}^\times$ on pairs by 'Adams operations' $[g,h] \mapsto [g^n, h^n]$ (perhaps related to the fixing groups involved in generalized Moonshine (cf. Section 3.1 below).)

Conjugacy classes in the centralizers $C_G(g) = \{h \in G \mid gh = hg\}$ form a kind of coordinate atlas on conjugacy classes of pairs, by the map

$$h \mapsto [h,g] \colon C_G(g)^{\mathrm{conj}} \to \mathrm{Pairs}_G$$

which extends to a parametrization

$$C_G(g) \times_{\mathbb{Z}} (\text{neighborhood of } i\infty) \to \mathrm{Pairs}_G \times_{\mathrm{Sl}_2(\mathbb{Z})} \mathfrak{h} = \mathcal{M}_{\mathrm{ell}}(G)$$

of the neighborhoods of the cusps in $\mathcal{M}_{\mathrm{ell}}(G)$; here $\mathbb{Z}$ acts as the subgroup

$$\begin{bmatrix} 1 & * \\ 0 & 1 \end{bmatrix}$$

of translations on pairs, sending $[h,g]$ to $[hg^n, g]$. Restriction to such neighborhoods of infinity in $\mathcal{M}_{\mathrm{ell}}(G)$ defines an equivariant lifting

$$\mathbf{Ell}_G(\mathrm{pt}) \to K_G^{\mathrm{Dev}}(\mathrm{pt})$$

of the q-expansion map, K^{Dev} being Ganter's reformulation [15, §3.1] of Devoto's equivariant version of K^{Tate}, which accomodates some subtle Galois-theoretic properties of the coefficients of McKay – Thompson series, both classical and generalized [12]. I'll omit some details here, and describe that construction as the collection, indexed by conjugacy classes of elements $g \in G$, of functors of the form

$$X \mapsto K^*_{C_G(g)}(X^g)((q^{1/|g|}))|_0,$$

where X^g is the fixed point set of g, $|g|$ is the order of that element, and the symbol $|_0$ denotes the degree zero component of a certain auxiliary $|g|$-grading defined as follows:

There are orthogonal projections

$$P_k = |g|^{-1} \sum_{1 \leq n \leq |g|} \mathrm{e}^{-2\pi i n k/|g|} g^n, \; |g| - 1 \geq k \geq 0,$$

in the complex group ring of the centralizer, which split any $C_G(g)$-equivariant bundle over X^g into a sum of eigenspaces in which g acts as multiplication by $\exp(2\pi i k/|g|)$; this makes $K^*_{C_G(g)}(X^g)$ into a bigraded algebra (with one grading cyclic of order two, the other cyclic of order $|g|$). On the other hand $\mathbb{Z}((q^{1/|g|}))$ is also naturally $|g|$-graded, and the group above carries a tensor product grading. The boundary maps of a cofibration preserve this extra structure, so taking its degree zero component is again a cohomology theory.

At the 'classical' cusp (with $g = 1$, i.e., associated to pairs of the form $[h,1]$), this construction simplifies to $K_G(X)((q))$.

2.4. Relations of this form between Ell and K-theory are part of a more general picture: we can think of classical K_G as taking values in sheaves over the moduli space of principal G-bundles over the circle, which by reasoning like that sketched above is just the space

$$H^1(S^1, G) \cong \mathrm{Hom}(\pi_1(S^1), G) \cong G/G^{\mathrm{conj}} \cong \mathrm{Spec}(R(G) \otimes \mathbb{C})$$

of conjugacy classes in G. In fact Quillen showed, back in the 70s, that the classical cohomology of G can be naturally regarded as a sheaf over the category whose objects are the abelian subgroups of G, with morphisms $A \to A'$ being the homomorphisms induced by conjugation by elements of G. There is a natural 'chromatic' filtration on the objects of this category (by their rank), with cyclic subgroups at the top. The elliptic picture sees the first *two* layers of this filtration; the general case is the concern of the generalized character theory of Hopkins, Kuhn, and Ravenel [24, 36]

The Hopkins – Miller theorem [19, 39] constructs topological modular forms in terms of a sheaf of E_∞-algebra spectra over the moduli stack of elliptic curves. Work of Lurie [30, §4] formulates this very naturally in terms of a general representability theorem in enriched (or derived) algebraic geometry; in particular, he constructs elliptic cohomology as a sheaf of spectra over a derived moduli stack of suitably oriented elliptic curves over E_∞-algebras. It seems likely that similar techniques (cf. also [12]) can be used to define an equivariant version of this theory as a sheaf of spectra over a derived moduli stack, now of G-torsors over Lurie's generalized oriented elliptic curves; but a precise definition of some such object has yet to appear.

3. Ganter's formulation of replicability

3.1. One concise (but historically misleading) way to tell the Moonshine story is to say that the graded character of the representation defined by the Frenkel – Lepowsky – Meurman vertex operator algebra defines a (McKay – Thompson) map J from conjugacy classes in the Monster to modular functions [9, 11]; on $1 \in \mathbb{M}$, for example, this construction yields the value $j(q) - 744$. In the interpretation presented here, Norton's generalized Moonshine sees J as the restriction to the classical cusp of a section of ω^0 over $\mathcal{M}_{\mathrm{ell}}(\mathbb{M})$.

It is known, essentially by case-by-case verification, that the invariance group of a generalized Moonshine function is a genus zero subgroup of $\mathrm{PGl}_2(\mathbb{Z})$, commensurable with the modular group. This most mysterious property of McKay – Thompson series is believed (known, in the classical case) to be equivalent to a condition called *replicability* [11, 37]. The rest of this note is an introduction to recent work of Ganter, who shows that at the classical cusp, replicability can be expressed very naturally in terms of cohomology operations on a sheaf-valued theory.

3.2. This involves at least three separate issues, the first being the classification of line bundles over the G-bundle stack. Based on earlier work on Chern – Simons theory, Ganter [16, §2.3] defines a homomorphism

$$H^4(BG, \mathbb{Z}) \to \mathrm{Pic}\big(\mathcal{M}_{\mathrm{ell}}(G)\big)$$

which associates to a degree four cohomology class in G, an element of the Picard group of line bundles over the moduli space of G-bundles. When G is connected and simple, this cohomology group is infinite cyclic, and its elements can be naturally identified with the *levels* which occur in the theory of positive-energy representations of loop groups [2, 13]. On the other hand this cohomology group is finite when G is; for example, there is reason to suspect [33, §11] that the fourth integral cohomology group of the Monster contains a nontrivial element of order 48.

She then constructs generalizations [16, §6.3]

$$T_k \colon \Gamma_{\mathcal{M}}(\mathcal{L}^{\alpha}) \to \Gamma_{\mathcal{M}}(\mathcal{L}^{k\alpha})$$

of Hecke operations on the line bundles $\mathcal{L}^{\alpha}$ corresponding to $\alpha \in H^4(BG, \mathbb{Z})$, which in the case $\alpha = 0$ restrict at the cusp to the classical form

$$T_k(f(\tau)) = \frac{1}{k} \sum_{\substack{ad=k \\ d>b\geq 0}} \psi^a(f)\Big(\frac{a\tau + b}{d}\Big);$$

where $\psi^a(f)$ denotes the result of applying an Adams operation to the coefficients of the power series f.

3.3. The third topic concerns power operations in generalized cohomology theories. In K-theory the formal sum

$$V \mapsto \Lambda_t(V) = \sum_{k\geq 0} \Lambda^k(V)t^k \colon K(X) \to (1 + tK(X)[\![t]\!])^{\times}$$

of exterior powers of vector bundles defines a homomorphism, because the total exterior power satisfies the identity

$$\Lambda_t(V \oplus W) = \Lambda_t(V) \otimes \Lambda_t(W).$$

The total symmetric power $S_t(V) = \sum_{k\geq 0} S^k(V)t^k$ is similarly exponential, in particular because these operations are related by a formal identity

$$\Lambda_{-t}(V) = S_t(V)^{-1}$$

(for example, in some ring of symmetric functions [31]). In that context, Newton's relations lead to Adams' identity

$$S_t(V) = \exp\Big(\sum_{k\geq 1} \psi^k(V)\frac{t^k}{k}\Big).$$

Exponential operations are extremely important in algebraic topology, going back to work of Atiyah and Steenrod: in modern terms, an orbifold $[X/G]$ has an nth orbifold symmetric power $[X^n/(G \wr \Sigma_n)]$, and a good (multiplicative, equivariant) cohomology theory E^* will admit 'external' power operations

$$x \mapsto x^{[\otimes n]} \colon E^*_G(X) \to E^*_{G\wr\Sigma_n}(X^n)$$

which, restricted to the diagonal, yield 'internal' power operations: those of Steenrod and Adams, as well as Baker's Hecke operations and Ando's higher generalizations (involving sums over isogenies of formal groups).

Ganter defined exterior and symmetric power operations for Lubin – Tate theories in her thesis [14, §§6.11, 7.15], and constructed Hecke operations satisfying the analog [14, §9.2] of the formula

$$\mathbf{S}_t(x) = \exp\Big(\sum_{k\geq 1} T_k(x)\frac{t^k}{k}\Big)$$

of Newton and Adams. (At about the same time C. Rezk [40, §1.12], working in a related but different context, constructed a kind of universal *logarithmic* operation behaving much like an inverse to these symmetric powers.) Using the generalized Hecke operations mentioned above, Ganter extended these constructions to elliptic cohomology and established the

3.4. Theorem ([16, §6.4]). *Replicability of the classical Moonshine function J is equivalent to the equation*

$$t\big(J(t) - J(q)\big) = \mathbf{\Lambda}_{-t}\big(J(q)\big) \in K_{\mathbb{M}}^{\mathrm{Dev}}(\mathrm{pt})[\![t]\!].$$

PROOF. The argument uses the theory of Faber polynomials, which associates to a function

$$f(q) = q^{-1} + \sum_{k\geq 1} a_k q^k \in \mathbb{C}((q)),$$

the sequence of polynomials $P_{n,f}(X) = X^n - na_1 X^{n-1} + \cdots - na_n$ characterized by the property

$$q^{-n} - P_{n,f}(f(q)) \in q\mathbb{C}[\![q]\!].$$

A generalization [34] of Newton's relations then imply the identity

$$\log\big[q\big(f(q) - f(p)\big)\big] = -\sum_{n\geq 1} P_{n,f}\big(f(p)\big)\frac{q^n}{n};$$

the assertion of the theorem is thus equivalent to the original form

$$P_{k,J}(J(q)) = kT_k\big(J(\tau)\big)$$

of the replicability condition. □

3.5. I will close with some remarks, perhaps too vague to be very useful: The first concerns the beautifully simple consequence

$$q\mathbf{\Lambda}_{-t}\big(J(q)\big) = -t\mathbf{\Lambda}_{-q}\big(J(t)\big)$$

of Ganter's formula, reminiscent in many ways of a product formula of Borcherds; which, however, at second sight becomes increasingly mysterious. It implies that the generalized exterior power of $J(q)$ is modular in the purely formal auxiliary variable t. The two sides of the equation are cusp expansions around very different points: the left-hand side of the equation lies in $K_{\mathbb{M}}(\mathrm{pt})((q))[\![t]\!]$, while the right-hand side lies in $K_{\mathbb{M}}(\mathrm{pt})((t))[\![q]\!]$, and their equality implies that both lie in $K_{\mathbb{M}}(\mathrm{pt})[\![q,t]\!]$. A coordinate-free interpretation of this, or some similar relation (cf. [15, just after §5.15]), might be very enlightening.

The second is that this replication formula, rewritten as

$$J(t) = J(q) + t^{-1}\mathbf{\Lambda}_{-t}\big(J(q)\big),$$

looks like some kind of evolution equation; but finding an infinitesimal version seems to be difficult, because of the presence of poles. Related integrable systems will be the subject of Devoto's talk.

The final remark concerns naturality. Once replicability is formulated in terms of cohomology operations, we can study its behavior under restriction; for example any cyclic subgroup of $\mathbb{M}$ pulls J back to a replicable element of $K_{\mathbb{Z}/n\mathbb{Z}}^{\mathrm{Dev}}(\mathrm{pt})$. (Of course the resulting element must be the restriction of a Moonshine class for the centralizer of $\mathbb{Z}/n\mathbb{Z}$. The $n = 3$ case, for example, is related to the Thompson group, and is studied in [22]. But even the pullback to a cyclic group is interesting.)

There is evidence [42] for the existence of analogs of J for other sporadic simple groups, and one might even speculate about symmetric groups. In that context the replicability equation calls to mind an equation

$$T(z) = z \cdot \exp\left(\sum_{k\geq 1} \frac{T(z^k)}{k}\right)$$

satisfied by the generating function for unlabelled trees (originating with Cayley, but in this form due to Pólya). Current thinking [8] interprets equations of this sort as generalized *Dyson – Schwinger* equations, taking values in some combinatorially-defined Hopf algebra (e.g., of trees). It is easier to draw a picture to explain where such equations come from, than to spell it out in words.

One might speculate that some such equation, in some ring of symmetric functions, might lie behind the integrable systems investigated in current work of Devoto and McKay.

Finally, though it seems to have no overt connection to Moonshine, I can't bring myself to end this talk without mentioning A. Ogg's observation [38] that the primes p dividing the order of $\mathbb{M}$ are precisely those for which every supersingular elliptic curve over a finite field of characteristic p has j-invariant in $\mathbb{F}_p$. This seems to be a property neither of $\mathbb{M}$ nor of $\mathcal{M}_{\text{ell}}$ but rather of $\mathcal{M}_{\text{ell}}(\mathbb{M})$; it deserves further attention.

References

1. M. Ando, *Isogenies of formal group laws and power operations in the cohomology theories* E_n, Duke Math. J. **79** (1995), no. 2, 423 – 485.
2. ———, *Power operations in elliptic cohomology and representations of loop groups*, Trans. Amer. Math. Soc. **352** (2000), no. 12, 5619 – 5666.
3. ———, *The sigma orientation for analytic circle-equivariant elliptic cohomology*, Geom. Topol. **7** (2003), 91–153.
4. M. Ando, M. Hopkins, and N. Strickland, *Elliptic spectra, the Witten genus and the theorem of the cube*, Invent. Math. **146** (2001), no. 3, 595 – 687.
5. A. Baker, *Operations and cooperations in elliptic cohomology.* I: *Generalized modular forms and the cooperation algebra*, New York J. Math. **1** (1994/95), 39–74.
6. ———, *Hecke algebras acting on elliptic cohomology*, Homotopy Theory via Algebraic Geometry and Group Representations (Evanston, IL, 1997) (M. Mahowald and S. Priddy, eds.), Contemp. Math., vol. 220, Amer. Math. Soc, Providence, RI, 1998, pp. 17 – 26.
7. ———, *Hecke operations and the Adams* E_2*-term based on elliptic cohomology*, Canad. Math. Bull. **42** (1999), no. 2, 129–138.
8. J. P. Bell, S. N. Burris, and K. A. Yeats, *Counting rooted trees: the universal law* $t(n) \sim C\rho^{-n}n^{-3/2}$, Electron. J. Combin. **13** (2006), no. 1, R63.
9. R. E. Borcherds, *Vertex algebras, Kac – Moody algebras, and the Monster*, Proc. Nat. Acad. Sci. U.S.A. **83** (1986), no. 10, 3068–3071.
10. ———, *Monstrous moonshine and monstrous Lie superalgebras*, Invent. Math. **109** (1992), no. 2, 405 – 444.
11. C. J. Cummins and T. Gannon, *Modular equations and the genus zero property of moonshine functions*, Invent. Math. **129** (1997), no. 3, 413–443.
12. J. A. Devoto, *Equivariant elliptic homology and finite groups*, Michigan Math. J. **43** (1996), no. 1, 3–32.
13. D. Freed, M. Hopkins, and C. Teleman, *Twisted* K*-theory and loop group representations*, available at `arXiv:math/0312155`.
14. N. Ganter, *Orbifold genera, product formulas and power operations*, Adv. Math. **205** (2006), no. 1, 84 – 133, available at `arXiv:math/0407021`.
15. ———, *Stringy power operations in Tate* K*-theory*, available at `arXiv:math/0701565`.

16. ———, *Hecke operators in equivariant elliptic cohomology and generalized moonshine*, available at arXiv:0706.2898.
17. D. Gepner, *Homotopy topoi and equivariant elliptic cohomology*, dissertation, Univ. of Illinois at Urbana-Champaign, 2005.
18. V. Ginzburg, M. Kapranov, and É. Vasserot, *Elliptic algebras and equivariant elliptic cohomology*, available at arXiv:q-alg/9505012.
19. P. Goerss and M. Hopkins, *Moduli problems for structured ring spectra*, in progress.
20. J. P. C. Greenlees, *Equivariant formal group laws and complex oriented cohomology theories*, Homology Homotopy Appl. **3** (2001), no. 2, 225–263.
21. I. Grojnowski, *Delocalized equivariant elliptic cohomology*, Elliptic Cohomology (Cambridge, 2002) (H. R. Miller and D. C. Ravenel, eds.), London Math. Soc. Lecture Note Ser., vol. 342, Cambridge Univ. Press, Cambridge, 2007, pp. 114–121.
22. Hirzebruch F., T. Berger, and R. Jung, *Manifolds and modular forms*, Aspects Math., vol. E20, Vieweg, Braunschweig, 1992.
23. M. Hopkins, *Algebraic topology and modular forms*, Proceedings of the International Congress of Mathematicians, Vol. I (Beijing, 2002) (T. Li, ed.), Higher Ed. Press, Beijing, 2002, pp. 291–317, available at arXiv:math/0212397.
24. M. Hopkins, N. Kuhn, and D. Ravenel, *Generalized group characters and complex oriented cohomology theories*, J. Amer. Math. Soc. **13** (2000), no. 3, 553–594.
25. M. Hopkins and M. Mahowald, *From elliptic curves to homotopy theory*, available at http://hopf.math.purdue.edu//Hopkins-Mahowald/eo2homotopy.pdf.
26. N. M. Katz, *Higher congruences between modular forms*, Ann. of Math. (2) **101** (1975), 332–367.
27. N. Kitchloo and J. Morava, *Thom prospectra for loopgroup representations*, Elliptic Cohomology (Cambridge, 2002) (H. R. Miller and D. C. Ravenel, eds.), London Math. Soc. Lecture Note Ser., vol. 342, Cambridge Univ. Press, Cambridge, 2007, pp. 214–238, available at arXiv:math/040454.
28. P. S. Landweber, *Elliptic cohomology and modular forms*, Elliptic Curves and Modular Forms in Algebraic Topology (Princeton, NJ, 1986) (P. S. Landweber, ed.), Lecture Notes in Math., vol. 1326, Springer, Berlin, 1988, pp. 55–68.
29. G. Laures, *The topological q-expansion principle*, Topology **38** (1999), no. 2, 387–425.
30. J. Lurie, *A survey of elliptic cohomology*, available at www-math.mit.edu/~lurie/papers/survey.pdf.
31. I. G. MacDonald, *Symmetric functions and Hall polynomials*, 2nd ed., Oxford Math. Monogr., Oxford Univ. Press, New York, 1995.
32. G. Mason, *Finite groups and modular functions*, with an appendix by S. Norton, The Arcata Conference on Representations of Finite Groups (Arcata, CA, 1986), Proc. Sympos. Pure Math., vol. 47, Part 1, Amer. Math. Soc, Providence, RI, 1987, pp. 181–210.
33. ———, *Orbifold conformal field theory and cohomology of the Monster*, available at http://www.newton.cam.ac.uk/programmes/NST/Mason.pdf.
34. J. McKay, *The essentials of Monstrous Moonshine*, Groups and Combinatorics — in Memory of Michio Suzuki (E. Bannai, H. Suzuki, H. Yamaki, and T. Yoshida, eds.), Adv. Stud. Pure Math., vol. 32, Math. Soc. Japan, Tokyo, 2001, pp. 347–353.
35. H. Miller, *The elliptic character and the Witten genus*, Algebraic Topology (Evanston, IL, 1988) (M. Mahowald and S. Priddy, eds.), Contemp. Math., vol. 96, Amer. Math. Soc, Providence, RI, 1989, pp. 281–289.
36. J. Morava, *HKR characters and higher twisted sectors*, Gromov–Witten Theory of Spin Curves and Orbifolds (San Francisco, CA, 2003) (T. J. Jarvis, T. Kimura, and A. Vaintrob, eds.), Contemp. Math., vol. 403, Amer. Math. Soc., Providence, RI, 2006, pp. 143–152, available at arXiv:math/0208235.
37. S. P. Norton, *More on moonshine*, Computational Group Theory (Durham, 1982) (M. D. Atkinson, ed.), Academic Press, London, 1984, pp. 185–193.
38. A. P. Ogg, *Modular functions*, The Santa Cruz Conference on Finite Groups (Santa Cruz, CA, 1979) (B. Cooperstein and G. Mason, eds.), Proc. Sympos. Pure Math., vol. 37, Amer. Math. Soc, Providence, RI, 1980, pp. 521–532.
39. C. Rezk, *Notes on the Hopkins–Miller theorem*, Homotopy Theory via Algebraic Geometry and Group Representations (Evanston, IL, 1997) (M. Mahowald and S. Priddy, eds.), Contemp. Math., vol. 220, Amer. Math. Soc, Providence, RI, 1998, pp. 313–366.

40. ———, *The units of a ring spectrum and a logarithmic cohomology operation*, J. Amer. Math. Soc. **19** (2006), no. 4, 969–1014, available at `arXiv:math/0407022`.
41. I. Rosu, *Equivariant elliptic cohomology and rigidity*, Amer. J. Math. **123** (2001), no. 4, 647–677, available at `arXiv:math/9912089`.
42. C. B. Thomas, *Elliptic cohomology*, Univ. Ser. Math., Kluwer Academic/Plenum, New York, 1999.

Department of Mathematics, The Johns Hopkins University, Baltimore, MD 21218, USA

E-mail address: `jack@math.jhu.edu`

Centre de Recherches Mathématiques
CRM Proceedings and Lecture Notes
Volume **47**, 2009

The Generalized Artin Conjecture and Arithmetic Orbifolds

M. Ram Murty and Kathleen L. Petersen

In honor of John McKay

ABSTRACT. Let K be a number field with positive unit rank, and let $\mathcal{O}_K$ denote the ring of integers of K. A generalization of Artin's primitive root conjecture is that that $\mathcal{O}_K^\times$ is a primitive root set for infinitely many prime ideals. We prove this with additional conjugacy conditions in the case when K is Galois with unit rank greater than three. This was previously known under the assumption of the Generalized Riemann Hypothesis. From our result, we deduce a topological corollary about the structure of quotients of $\mathrm{PSL}_2(\mathcal{O}_K)$.

1. Introduction

It is a classical result that the modular group, $\mathrm{PSL}_2(\mathbb{Z})$, does not have the congruence subgroup property (CSP). That is, that there are finite index subgroups which do not contain the kernel of the entrywise modulo n map for any integer $n \in \mathbb{N}$. Petersson [13] proved that, although there are infinitely many subgroups of the modular group whose quotient by $\mathbb{H}$ has only one cusp, only finitely many of these are congruence subgroups. He showed that the index of these groups are exactly the divisors of $55440 = 2^4 \cdot 3^2 \cdot 5 \cdot 7 \cdot 11$, a divisor of the order of the monster group. This theorem is a consequence of the fact, famously noticed by Galois, that $\mathrm{PSL}_2(\mathbb{F}_p)$ has an index p subgroup only for $p \leq 11$. There appears to be a connection between this and moonshine. Moreover, the groups commensurable with the modular group that appear in moonshine have a cusp transitivity condition which suggests a connection with the one cusped congruence subgroups (see [3]). Petersson's result has been generalized to the Bianchi groups (see [11]) which also fail to have the CSP. In contrast, for number fields K with positive unit rank, such that $\sqrt{-1} \notin K$, there are infinitely many maximal (necessarily congruence) subgroups of $\mathrm{PSL}_2(\mathcal{O}_K)$ whose quotient has a minimal number of cusps. (See Section 2 for details.) This result is conditional, under the assumption of a form of Artin's primitive root conjecture (to be specified below). In this paper we prove

2000 *Mathematics Subject Classification.* 11A07, 11N36, 22E40.
Key words and phrases. Large sieve, hyperbolic spaces, orbifolds, Euclidean algorithm.
The research of the first author is partially supported by an NSERC Research grant.
The second author is a Coleman post-doctoral fellow.
This is the final form of the paper.

this primitive root conjecture for Galois K with unit rank greater than 3, which was previously known only under the assumption of the generalized Riemann hypothesis (GRH) (see [12]).

We now describe the primitive root conjectures. An integer s is called a primitive root modulo the prime p if s has order $p-1$ modulo p. To be primitive modulo $p > 2$ it is necessary that s is not a square or -1. Artin conjectured a specific density for the set of primes for which s is a primitive root. This density is positive if s is not a square or -1. We will refer to the conjecture that given an integer s other than -1 or a square there are infinitely many primes p for which s is a primitive root as Artin's primitive root conjecture. In 1967 this was proved conditionally, under the assumption of the GRH by Hooley [6]. There are many unconditional results related to Artin's primitive root conjecture (See [9] for a good exposition.) but at present there are no integers for which the conjecture is known to be true.

This classical primitive root conjecture has been generalized to number fields as follows. Let K be a number field with ring of integers $\mathcal{O}_K$. Let π be a prime ideal in $\mathcal{O}_K$ and φ_π be the reduction modulo π map from $\mathcal{O}_K$ to $\mathcal{O}_K/\pi$. Given a multiplicative set S in $\mathcal{O}_K$, $\varphi_\pi(S)$ is contained in $(\mathcal{O}_K/\pi)^\times$. We say that S is a *primitive root set modulo* π if $\varphi_\pi(S) = (\mathcal{O}_K/\pi)^\times$. If S is generated by an element s, we say that s is a *primitive root modulo* π. One generalization of the classical conjecture is that if K has positive unit rank, then $\mathcal{O}_K^\times$ is a primitive root set for infinitely many prime ideals. One can add congruence conditions to the above to get the following conjecture. We will use ζ_q to denote a primitive qth root of unity.

Conjecture 1. *Let K be a number field with positive unit rank. Then $\mathcal{O}_K^\times$ is a primitive root set for infinitely many prime ideals. Moreover, given integers $1 \le a < q$ with $(a,q) = 1$, if $\mathbb{Q}(\zeta_q) \cap K = \mathbb{Q}$ then there are infinitely many prime ideals π such that $|N_{K/\mathbb{Q}}(\pi)| \equiv a \pmod q$ and $\mathcal{O}_K^\times$ is a primitive root set modulo π.*

Extending Hooley's conditional proof, this was proven under the assumption of the GRH by Weinberger in 1973 [15]. We provide an unconditional proof of Conjecture 1, in the case where K is Galois with unit rank greater than three. Specifically, we show the following.

Theorem 2. *Let $K/\mathbb{Q}$ be a finite Galois extension of unit rank greater than three. Let a and q be integers such that $1 \le a < q$ and $(a,q) = 1$. Assume that $\mathbb{Q}(\zeta_q) \cap K = \mathbb{Q}$. Then there are infinitely many degree one prime ideals π in $\mathcal{O}_K$ such that $|N_{K/\mathbb{Q}}(\pi)| \equiv a \pmod q$ and $\mathcal{O}_K^\times$ is a primitive root set modulo π.*

In Section 2 we describe a topological application of Theorem 2, in Section 3 we discuss some preliminaries needed for the proof, in Section 4 we prove Theorem 2, and in Section 5 we make some concluding remarks.

2. A topological application

Let K be a number field with r_1 real places, and r_2 complex places, and let $\mathcal{O}_K$ denote the ring of integers of K. Then group $\mathrm{PSL}_2(\mathcal{O}_K)$ embeds discretely in $\mathrm{PSL}_2(\mathbb{R})^{r_1} \times \mathrm{PSL}_2(\mathbb{C})^{r_2}$ via the map

$$\pm\begin{pmatrix} \alpha & \beta \\ \gamma & \delta \end{pmatrix} \mapsto \prod_\sigma \pm \begin{pmatrix} \sigma(\alpha) & \sigma(\beta) \\ \sigma(\gamma) & \sigma(\delta) \end{pmatrix}$$

where the product is taken over all infinite places σ of K. The product $\mathrm{PSL}_2(\mathbb{R})^{r_1} \times \mathrm{PSL}_2(\mathbb{C})^{r_2}$ is isomorphic to a subgroup of the orientation preserving isometries of

$H_{r_1,r_2} = [\mathbb{H}^2]^{r_1} \times [\mathbb{H}^3]^{r_2}$ where $\mathbb{H}^2$ is the hyperbolic plane, and $\mathbb{H}^3$ is hyperbolic 3-space. This follows from the fact that $\mathrm{Isom}^+(\mathbb{H}^2) \cong \mathrm{PSL}_2(\mathbb{R})$ and $\mathrm{Isom}^+(\mathbb{H}^3) \cong \mathrm{PSL}_2(\mathbb{C})$. The quotient $M_K = H_{r_1,r_2}/\mathrm{PSL}_2(\mathcal{O}_K)$ is a finite volume $(2r_1+3r_2)$-dimensional orbifold equipped with a metric inherited from H_{r_1,r_2}. This orbifold has h_K cusps where h_K is the class number of K. If Γ is a finite index subgroup of $\mathrm{PSL}_2(\mathcal{O}_K)$ the quotient $M_\Gamma = H_{r_1,r_2}/\Gamma$ is a finite volume $(2r_1+3r_2)$-dimensional orbifold with finitely many cusps. If M_Γ has n cusps, we say that Γ is *n-cusped.* Any such Γ has at least h_K cusps and M_Γ covers M_K.

The groups $\mathrm{PSL}_2(\mathcal{O}_K)$ and the quotients M_K exhibit vastly different behavior depending on whether or not the unit rank of K is positive. Only when $K = \mathbb{Q}$ or an imaginary quadratic field is the unit rank zero, and these quotients are the prototypes for non-compact arithmetic hyperbolic 2- and 3-orbifolds, as all such orbifolds are commensurable to one of these. (Two orbifolds are commensurable if they share a finite sheeted cover.) A principal congruence subgroup of $\mathrm{PSL}_2(\mathcal{O}_K)$ is the kernel of the entry-wise modulo $\mathcal{J}$ map for some non-zero ideal $\mathcal{J}$ in $\mathcal{O}_K$ and is denoted $\Gamma(\mathcal{J})$. A finite index subgroup of $\mathrm{PSL}_2(\mathcal{O}_K)$ is a *congruence subgroup* if it contains a principal congruence subgroup, and $\mathrm{PSL}_2(\mathcal{O}_K)$ is said to have the *congruence subgroup property (CSP)* if "most" index subgroups are congruence subgroups. Only when the unit rank of K is zero does $\mathrm{PSL}_2(\mathcal{O}_K)$ fail to have the CSP.

In 1971 Petersson [13] proved that in the case of the modular group, $\mathrm{PSL}_2(\mathbb{Z})$, there are only finitely many one-cusped congruence subgroups, although there are infinitely many maximal one-cusped subgroups. The Bianchi groups are the groups $\mathrm{PSL}_2(\mathcal{O}_d)$ where $\mathcal{O}_d$ is the ring of integers of the imaginary quadratic $\mathbb{Q}(\sqrt{-d})$. The quotients $M_{\mathbb{Q}(\sqrt{-d})}$ are hyperbolic 3-orbifolds. The class number of $\mathcal{O}_d$ is one precisely when $d = 1, 2, 3, 7, 11, 19, 43, 67$, or 163, so only for these values of d can $\mathrm{PSL}_2(\mathcal{O}_d)$ have one-cusped subgroups. Of these, if $d \neq 1$ or 3 there are infinitely many maximal one-cusped subgroups, and for $d = 1$ and 3 there are infinitely many one-cusped subgroups. However, for $d = 11, 19, 43, 67$, or 163 there are only finitely many one-cusped congruence subgroups, and for $d = 1, 2, 3$, and 7, there are only finitely many maximal one-cusped subgroups (see [11]).

If $\mathcal{O}_K$ has positive unit rank, the situation is quite different. Let π be a prime ideal in $\mathcal{O}_K$ with odd norm $|N_{K/\mathbb{Q}}(\pi)| = q$, let $\varphi_\pi \colon \mathcal{O}_K \to \mathcal{O}_K/\pi$ denote the reduction modulo π map, and $\Phi_\pi \colon \mathrm{PSL}_2(\mathcal{O}_K) \to \mathrm{PSL}_2(\mathbb{F}_q)$ be the induced map with the identification $\mathcal{O}_K/\pi \cong \mathbb{F}_q$. Since φ_π maps $\mathcal{O}_K^\times$ into $(\mathcal{O}_K/\pi)^\times$ we will also use φ_π to denote this restriction and let $l = [(\mathcal{O}_K/\pi)^\times : \varphi_\pi(\mathcal{O}_K^\times)]$. If q is odd, there is a subgroup of $\mathrm{PSL}_2(\mathbb{F}_q)$ isomorphic to the dihedral group with $q+1$ elements. Let $\Gamma_\pi < \mathrm{PSL}_2(\mathcal{O}_K)$ be the φ_π pull-back of this group. For π large enough, the principal congruence subgroup $\Gamma(\pi)$ has $h_K l(q+1)$ cusps where h_K is the class number of K. From this, one can show that Γ_π has $h_K l$ cusps if $q \equiv 3 \pmod 4$. (See [12] for details.) Therefore, $\mathrm{PSL}_2(\mathcal{O}_K)$ contains infinitely many maximal h_K-cusped (congruence) subgroups if there are infinitely many prime ideals π in $\mathcal{O}_K$ such that $|N_{K/\mathbb{Q}}(\pi)| \equiv 3 \pmod 4$ and $\varphi_\pi(\mathcal{O}_K^\times) = (\mathcal{O}_K/\pi)^\times$. If $\sqrt{-1} \notin K$, a conditional proof of this under the assumption of the GRH, is a corollary of Weinberger's result [15]. Theorem 2 immediately implies the following corollary.

Corollary 3. *Let $K/\mathbb{Q}$ be a finite Galois extension with unit rank greater than three, and such that $\sqrt{-1} \notin K$. There are infinitely many maximal h_K-cusped subgroups of $\mathrm{PSL}_2(\mathcal{O}_K)$, where h_K is the class number of K.*

3. Preliminaries

In this section we collect various results that we will use in the proof of Theorem 2. Let $K/\mathbb{Q}$ be a finite Galois extension with Galois group $G = \mathrm{Gal}(K/\mathbb{Q})$ and let C be a conjugacy class in G. Let a and q be positive integers with $1 \le a < q$ and $(a, q) = 1$. We denote by $\pi_C(x, q, a)$ the number of primes $p \le x$ which are unramified in K such that $p \equiv a \pmod q$ and such that the Artin symbol $(p, K/\mathbb{Q}) = C$.

Classically, in the case where $K = \mathbb{Q}$, Dirichlet density theorems state that

$$\pi(x, q, a) \sim \frac{\pi(x)}{\phi(q)}$$

where $\pi(x, q, a)$ denotes the number of primes $\le x$ with the required congruence condition, $\pi(x)$ denotes the number of primes $\le x$, and ϕ is Euler's function. Estimates of the error term, $\pi(x, q, a) - \pi(x)/\phi(q)$ have proven to be very important in many applications. The Riemann Hypothesis for Dirichlet L-functions implies that it is $O\big(x^{1/2} \log qx\big)$. Bombieri [1] and Vinogradov [14] have proven that this estimate holds on the average. That is, for any $A > 0$ there is a $B > 0$ depending on A so that

$$\sum_{q \le Q} \max_{y \le x} \max_{(a,q)=1} \left| \pi(y, q, a) - \frac{\pi(y)}{\phi(q)} \right| \ll \frac{x}{(\log x)^A}$$

where $Q = x^{1/2}(\log x)^{-B}$. (The notation $f \ll g$ means that $|f/g|$ is bounded.)

In general, the Chebotarev density theorem states that

$$\pi_C(x, q, a) \sim \delta(C, q, a)\phi(q)$$

for some density $\delta(C, q, a) \ge 0$. In fact, if $K \cap \mathbb{Q}(\zeta_q) = \mathbb{Q}$ where ζ_q is a primitive qth root of unity, then $\delta(C, q, a) = |C|/|G|\phi(q)$. Murty and Murty [10] have shown the following average formulation.

Proposition 4. *Let $L/\mathbb{Q}$ be a finite Galois extension with Galois group $G = \mathrm{Gal}(L/\mathbb{Q})$. Let C be a conjugacy class in G and let a and q be positive integers with $1 \le a < q$, $(a, q) = 1$. There exist numbers $\delta(C, q, a) \ge 0$ such that for any $\epsilon > 0$ and $A > 0$, we have*

$$\sum_{q \le Q} \max_{(a,q)=1} \max_{y \le x} |\pi_C(y, q, a) - \delta(C, q, a)\pi(y)| \ll \frac{x}{(\log x)^A}$$

with $Q = x^{1/\eta - \epsilon}$ where $\eta \ge \max(|G|/2, 2)$ and the summation is over q such that $L \cap \mathbb{Q}(\zeta_q) = \mathbb{Q}$.

If A is an abelian subgroup of G with $A \cap C \ne \varnothing$, and $d = [G : A]$. We can improve the above by taking

$$\eta = \begin{cases} d - 2 & \text{if } d \ge 4 \\ 2 & \text{if } d \le 4. \end{cases}$$

Moreover, we may replace d above by d^* where $d^* = \min_H \max_\omega [G{:}H]\omega(1)$ where the minimum is taken over all subgroups H satisfying the following two conditions

- *$H \cap C \ne \varnothing$*
- *For every irreducible character ω of H and any non-trivial Dirichlet character χ, then Artin L-series $L(s, \omega \otimes \chi)$ is entire.*

The maximum above is over all irreducible characters of H. In particular, if Artin's holomorphy conjecture is true for $L/\mathbb{Q}$ then we may take $d^* = \max_\chi \chi(1)$ where the maximum is over all irreducible characters of G.

If $\pi_q(x)$ denotes the number of primes $p \le x$ which split completely in $\mathbb{Q}(\zeta_q, \sqrt[q]{s})$, an estimate of the form

$$\sum_{q \le Q} \left| \pi_q(x) - \frac{\pi(x)}{q(q-1)} \right| \ll \frac{x}{(\log x)^2}$$

with Q about $x^{1/2}$ is sufficient to imply Artin's primitive root conjecture for s (see [8]). At present, such a result seems to be beyond the reach of modern technology.

We will also need the following, whose proof is analogous to that of Lemma 2 in [4].

Proposition 5. *Let K be an algebraic number field. Let M be a monoid in $\mathcal{O}_K$ and π be coprime to the elements of M. If M contains r multiplicatively independent elements, then*

$$\#\{\pi : |\varphi_\pi(M)| \le Y\} \ll Y^{(r+1)/r}.$$

The proof is essentially combinatorial in nature. If $\{b_1, \dots, b_r\}$ is an independent set of generators for M, and if $|\varphi_\pi(M)| \le Y$ then for $|N_{K/\mathbb{Q}}(\pi)|$ large enough, there are r-tuples $b_1^{a_1} \cdots b_r^{a_r} \equiv b_1^{\alpha_1} \cdots b_r^{\alpha_r} \pmod{\pi}$. Hence $N_{K/\mathbb{Q}}(\pi)$ divides the numerator of $b_1^{a_1-\alpha_1} \cdots b_r^{a_r-\alpha_r} - 1$. The fact that $|\varphi_\pi(M)| \le Y$ bounds $|a_1| + \cdots + |a_r|$. Together with the divisibility condition this results in the required bound.

We also require an application of the lower bound sieve. The conditions we need are a slight modification of those in [2]. The proof relies upon the linear sieve in the form given by Iwaniec [7], and the estimates from the Bombieri – Vinogradov theorem, as extended by Murty and Murty [10].

Proposition 6. *Let $K/\mathbb{Q}$ be a finite Galois extension, such that $\mathbb{Q}(\zeta_q) \cap K = \mathbb{Q}$ and let $t_K = \max_m\{m : K \supseteq \mathbb{Q}(\zeta_m)\}$. Denote by L the Galois extension $K(\zeta_q)$. For a such that $(a, q) = 1$, let C be a conjugacy class of the form $(a, 1)$ in $\mathrm{Gal}(L/\mathbb{Q})$. For η as in Proposition 4 the number of primes $p \le x$ with Artin symbol $(p, L/\mathbb{Q}) = C$ such that for all primes l dividing $(p-1)/t_K$, $l > x^{1/2\eta - \epsilon}$ is $\gg x/\log^2 x$.*

4. Proof of Theorem 2

We form the compositum, L, of $\mathbb{Q}(\zeta_q)$ and K, which is Galois over $\mathbb{Q}$ as both K and $\mathbb{Q}(\zeta_q)$ are Galois. Let $G = \mathrm{Gal}(L/\mathbb{Q})$ and $H = \mathrm{Gal}(K/\mathbb{Q})$. Notice that $G \cong \mathbb{Z}/\phi(q)\mathbb{Z} \times H$. Let $C \in G$ be the conjugacy class $(a, 1)$. For η and ϵ as in Proposition 4, let $\mathcal{S}$ be the set of all primes $p \le x$ such that $(p, L/\mathbb{Q}) \in C$ and such that for all primes l which divide $(p-1)/t_K$, $l > x^{1/2\eta - \epsilon}$. The value η is given in Proposition 4 for the number field L. Proposition 6 implies that $\#\mathcal{S} \gg x/\log^2 x$. Notice that the condition that $(p, L/\mathbb{Q}) = C$ ensures that all prime ideals π in K lying over p are degree one and have the required congruence condition.

For p in $\mathcal{S}$, t_K divides $|\varphi_\pi(\mathcal{O}_K^\times)|$ as the roots of unity in $\mathcal{O}_K^\times$ inject into $\mathcal{O}_K/\pi$ for $p = N_{K/\mathbb{Q}}(\pi)$ large enough. Therefore, if l divides $(p-1)/|\varphi_\pi(\mathcal{O}_K^\times)|$ then l divides $(p-1)/t_K$ and is greater than $x^{1/2\eta-\epsilon}$. As a result, the index $(p-1)/|\varphi_\pi(\mathcal{O}_K^\times)|$ is either one or greater than $x^{1/2\eta-\epsilon}$. If the index is not one then $|\varphi_\pi(\mathcal{O}_K^\times)| < x^{1-1/2\eta+\epsilon}$.

By Proposition 5 $\#\{\pi : |\mathcal{O}_K^\times/\pi| \leq x^{1-1/2\eta+\epsilon}\} \ll x^{(1-1/2\eta+\epsilon)(r+1)/r}$ where r is the rank of $\mathcal{O}_K^\times$. It suffices to show that

$$O(x^{(1-1/2\eta+\epsilon)(r+1)/r}) = o\left(\frac{x}{\log^2 x}\right).$$

This occurs when $2\eta < r+1$.

Recall from Proposition 4 that if A is an abelian subgroup of G such that $A \cap C \neq \varnothing$, setting $d = [G{:}A]$

$$\eta = \begin{cases} d-2 & \text{if } d \geq 4 \\ 2 & \text{if } d \leq 4. \end{cases}$$

Taking A isomorphic to $\mathbb{Z}/\phi(q)\mathbb{Z} \times B$ where B is an abelian subgroup of H, $d = [H : B]$. Notice that $A \cap C \neq \varnothing$. If $d \leq 4$, then the inequality is satisfied when $r > 3$. If $d \geq 4$ then the inequality is

$$2|H| < |B|(r+5).$$

Since K is Galois, K is totally real or totally complex. In the former case, $|H|$ is $r+1$. The inequality is $2|H| < |B|(|H|+4)$ and can always be satisfied. In the totally complex case, $|H| = 2(r+1)$ and the inequality is $4|H| < |B|(|H|+8)$ and is satisfied if there is an abelian subgroup B of H with $|B| \geq 4$. This is the case unless $K/\mathbb{Q}$ is an S_3 extension, which has unit rank only 2.

The following two corollaries follow immediately.

Corollary 7. *Let $K/\mathbb{Q}$ be a finite Galois extension of unit rank greater than three. Then there are infinitely many degree one prime ideals π in $\mathcal{O}_K$ such that $\mathcal{O}_K^\times$ is a primitive root set modulo π.*

Corollary 8. *Let $K/\mathbb{Q}$ be a finite Galois extension of unit rank greater than three such that $\sqrt{-1} \notin K$. Fix $a = \pm 1$. Then there are infinitely many degree one prime ideals π in $\mathcal{O}_K$ such that $|N_{K/\mathbb{Q}}(\pi)| \equiv a \pmod 4$ and $\mathcal{O}_K^\times$ is a primitive root set modulo π.*

5. Concluding remarks

There are two possible avenues for further research. The first is the problem when the unit rank is 1, 2 or 3. This problem may be approached by refining the techniques of [5], but this refinement will not be straightforward. These techniques may allow us to treat the case of rank 3. The second problem to address is the case when K is not Galois. To treat this case, one would have to extend the results of [10] to the non-Galois situation. This can be done, in principle, though again there are several technical difficulties that need to be addressed. We hope to investigate both of these problems in future work.

Acknowledgments. We thank the referee for useful comments.

References

1. E. Bombieri, *On the large sieve*, Mathematika **12** (1965), 201–225.
2. D. A. Clark and M. R. Murty, *The Euclidean algorithm for Galois extensions of* $\mathbb{Q}$, J. Reine Angew. Math. **459** (1995), 151–162.
3. J. Conway, J. McKay, and A. Sebbar, *On the discrete groups of Moonshine*, Proc. Amer. Math. Soc. **132** (2004), no. 8, 2233–2240.

4. R. Gupta and M. R. Murty, *A remark on Artin's conjecture*, Invent. Math. **78** (1984), no. 1, 127–130.
5. M. Harper and M. R. Murty, *Euclidean rings of algebraic integers*, Canad. J. Math. **56** (2004), no. 1, 71–76.
6. C. Hooley, *On Artin's conjecture*, J. Reine Angew. Math. **225** (1967), 209–220.
7. H. Iwaniec, *Rosser's sieve*, Acta Arith. **36** (1980), no. 2, 171–202.
8. M. R. Murty, *On Artin's conjecture*, J. Number Theory **16** (1983), no. 2, 147–168.
9. ———, *Artin's conjecture for primitive roots*, Math. Intelligencer **10** (1988), no. 4, 59–67.
10. M. R. Murty and V. K. Murty, *A variant of the Bombieri–Vinogradov theorem*, Number Theory (Montréal, QC, 1985) (H. Kisilevsky and J. Labute, eds.), CMS Conf. Proc., vol. 7, Amer. Math. Soc., Providence, RI, 1987, pp. 243–272.
11. K. L. Petersen, *One-cusped congruence subgroups of Bianchi groups*, Math. Ann. **338** (2007), no. 2, 249–282.
12. ———, *Counting cusps of subgroups of* $\mathrm{PSL}_2(\mathcal{O}_K)$, Proc. Amer. Math. Soc., to appear.
13. H. Petersson, *Über die Konstruktion zykloider Kongruenzgruppen in der rationalen Modulgruppe*, J. Reine Angew. Math. **250** (1971), 182–212.
14. A. I. Vinogradov, *The density hypothesis for Dirichlet L-series*, Izv. Akad. Nauk SSSR Ser. Mat. **29** (1965), 903–934 (Russian).
15. P. J. Weinberger, *On Euclidean rings of algebraic integers*, Analytic Number Theory (St. Louis, MO, 1972) (H. G. Diamond, ed.), Proc. Sympos. Pure Math., vol. 24, Amer. Math. Soc., Providence, RI, 1973, pp. 321–332.

DEPARTMENT OF MATHEMATICS AND STATISTICS, QUEEN'S UNIVERSITY, KINGSTON, ON K7L 3N6, CANADA

E-mail address, M. R. Murty: `murty@mast.queensu.ca`

E-mail address, K. L. Petersen: `petersen@mast.queensu.ca`

Centre de Recherches Mathématiques
CRM Proceedings and Lecture Notes
Volume **47**, 2009

McKay Correspondence

Iku Nakamura

ABSTRACT. We discuss the two-dimensional McKay correspondence from the view point of Hilbert schemes.

Introduction

There is a whole series of apparently unrelated phenomena that are governed by the so-called ADE Dynkin diagram scheme. It is widely believed that, despite the diverse nature of the objects concerned, there must be some hidden reasons for these coincidences. The ADE Dynkin diagrams provide a classification of the following types of objects:

(1) simple singularities (rational double points) of complex surfaces,
(2) finite subgroups of $\mathrm{SL}(2, \mathbb{C})$,
(3) simple Lie groups and simple Lie algebras,
(4) the following three finite simple groups, the derived group $\mathbb{F}_{24}'$ of the Fischer $\mathbb{F}_{24}$, the Baby monster $\mathbb{B}$ and the Monster $\mathbb{M}$ are related with E_6, E_7 and E_8 respectively.

Meanwhile there are three outstanding McKay observations. The first McKay observation made in November, 1978 was concerned with the so-called moonshine, and the second in December, 1978 with the connection between the above items (1) and (2), while the third in February, 1979 with the above item (4). It is the second McKay observation that we discuss in this article, which we refer to as the McKay correspondence. The purpose of this article is to discuss the McKay correspondence in detail, partially based on [6].

For a given finite subgroup G of $\mathrm{SL}(2, \mathbb{C})$, the McKay correspondence is incorporated into a quiver, called the McKay quiver, in the coinvariant algebra. The McKay quiver gives the Dynkin diagram of the exceptional set of the corresponding simple singularity ADE. Moreover, the natural strata of the exceptional set are understood via the subquivers of the McKay quiver, which are easily described by directed Dynkin diagrams (See Figures 7 – 16). The McKay correspondence is also

---

2000 *Mathematics Subject Classification.* Primary 14-02, 14B05, 14J17; 17B10, 17B67, 17B68, 17C20; Secondary 20C05, 20C15.

The present work is partially supported by the Grant-in-aid (No. 16204001) for Scientific Research, JSPS.

This is the final form of the paper.

understood as a natural bijective correspondence in the irreducible decomposition of a certain coherent sheaf over the G-orbit Hilbert scheme. Any extended Dynkin diagram of ADE is realized by a quiver in the symmetric algebra extending the McKay quiver.

The present article is organized as follows. In Section 1 we recall the simple singularities, the McKay correspondence and related notions. In Section 2 we recall the notion of G-orbit Hilbert schemes. For a finite subgroup G of $\mathrm{SL}(2,\mathbb{C})$, the G-orbit Hilbert scheme is a minimal resolution of the quotient $\mathbb{A}^2/G$ by [6]. In Section 3 we recall the main theorem of [6] and state a new theorem which sharpens the main theorem. In Section 4 we prove the McKay correspondence for D_5 in full detail, partially based on [6]. We introduce the (extended) McKay quiver, and the subquivers of it so that we will have a transparent overview of the strata of the exceptional set of the G-orbit Hilbert scheme. In Section 5 we briefly discuss E_6 along the same line. In Section 6 we prove our new theorem mentioned above.

1. Simple singularities and McKay correspondence

1.1. Simple singularities. We first recall the definition of simple singularities. A germ of a two-dimensional isolated hypersurface singularity is called a *simple singularity* if it is isomorphic to one of the following germs at the origin

$$\begin{aligned}
&\mathrm{A}_n\colon x^{n+1}+y^2+z^2=0 && \text{for } n\geq 1,,\\
&\mathrm{D}_n\colon x^{n-1}+xy^2+z^2=0 && \text{for } n\geq 4,\\
&\mathrm{E}_6\colon x^4+y^3+z^2=0,\\
&\mathrm{E}_7\colon x^3y+y^3+z^2=0,\\
&\mathrm{E}_8\colon x^5+y^3+z^2=0.
\end{aligned}$$

It is also a quotient of the germ $(\mathbb{C}^2,0)$ by a finite subgroup of $\mathrm{SL}(2,\mathbb{C})$. Moreover it has a minimal resolution of singularities with exceptional set consisting of smooth rational curves of self-intersection -2 intersecting transversally. See Figure 1 for the Dynkin diagram involved.

1.2. Finite subgroups of $\mathrm{SL}(2,\mathbb{C})$. Up to conjugacy, any finite subgroup of $\mathrm{SL}(2,\mathbb{C})$ is one of the subgroups listed in Table 1; see [7]. The triple (d_1,d_2,d_3) specifies the degrees of the generators of the G-invariant polynomial ring. The integer h is the Coxeter number of the Lie algebra of the type involved (see Table 1), which we also call the Coxeter number of the simple singularity $(\mathbb{A}^2/G,0)$.

TABLE 1. Finite subgroups of $\mathrm{SL}(2,\mathbb{C})$

Type	G	name	$\lvert G\rvert$	h	(d_1,d_2,d_3)
A_n	$\mathbb{Z}_{n+1}$	cyclic	$n+1$	$n+1$	$(2,n+1,n+1)$
D_n	$\mathbb{D}_{n-2}$	binary dihedral	$4(n-2)$	$2n-2$	$(4,2n-4,2n-2)$
E_6	$\mathbb{T}$	binary tetrahedral	24	12	$(6,8,12)$
E_7	$\mathbb{O}$	binary octahedral	48	18	$(8,12,18)$
E_8	$\mathbb{I}$	binary icosahedral	120	30	$(12,20,30)$

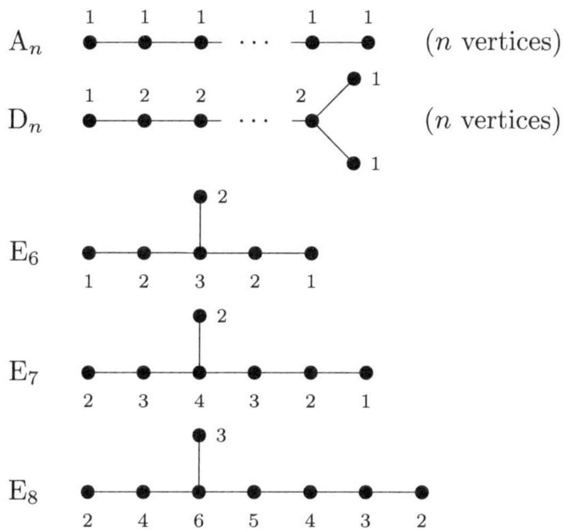

FIGURE 1. The Dynkin diagrams ADE

1.3. Dynkin diagrams. Let $(S, 0)$ be a germ of a simple singularity, $\pi\colon X \to S$ its minimal resolution, $E := \pi^{-1}(0)_{\mathrm{red}}$ and E_i for $1 \le i \le r$ the irreducible component of E. It is known that $E_i \simeq \mathbb{P}^1$ with self-intersection (-2). Let $\operatorname{Irr} E$ be the set $\{E_i : 1 \le i \le r\}$ and $H^2 := H^2(X, \mathbb{Z})$. We see that $H_2 = \bigoplus_{1\le i\le r} \mathbb{Z}[E_i]$. Then H_2 has a negative definite intersection pairing $(\ ,\ )_{\mathrm{SING}}\colon H_2 \times H_2 \to \mathbb{Z}$. Since $(E_iE_j)_{\mathrm{SING}} = 0$ or 1 for $i \neq j$, the pairing $(\ ,\ )_{\mathrm{SING}}$ can be expressed by a finite graph with simple edges. We rephrase this as follows: we associate a vertex $v(E')$ to any irreducible component E' of E, and join two vertices $v(E')$ and $v(E'')$ if and only if $(E'E'')_{\mathrm{SING}} = 1$. Thus we have a finite graph with simple edges. We call this graph the *dual graph* of E, and denote it by $\Gamma_{\mathrm{SING}}(S)$ or $\Gamma(\operatorname{Irr} E)$.

There exists a unique divisor E_{fund}, called the *fundamental cycle* of X, which is the minimal nonzero effective divisor such that $E_{\mathrm{fund}}E_i \le 0$ for all i. Let $E_{\mathrm{fund}} := \sum_{i=1}^r m_i^{\mathrm{SING}} E_i$ and $E_0 := -E_{\mathrm{fund}}$. For the simple singularities we have $E_0E_i = 0$ or 1 for any $E_i \in \operatorname{Irr} E$ (except for the case A_1, when $E_0E_1 = 2$). Thus we can draw a new graph $\widetilde{\Gamma}_{\mathrm{SING}}$ by adding the vertex $v(E_0)$ to $\Gamma_{\mathrm{SING}}(S)$. By abuse of notation we denote $\operatorname{Irr} E \cup \{E_0\}$ by $\operatorname{Irr}_* E$. Also for a given finite subgroup G of $\mathrm{SL}(2, \mathbb{C})$, we have a quotient singularity $(\mathbb{A}^2/G, 0)$, which is one of simple singularities so that we have a Dynkin diagram as a dual graph $\Gamma_{\mathrm{SING}}(\mathbb{A}^2/G, 0)$ of the exceptional set. Also we denote $\Gamma_{\mathrm{SING}}(\mathbb{A}^2/G, 0)$ by $\Gamma_{\mathrm{SING}}(G)$ and similarly $\widetilde{\Gamma}_{\mathrm{SING}}(\mathbb{A}^2/G, 0)$ by $\widetilde{\Gamma}_{\mathrm{SING}}(G)$.

In the D_5 case, we have $E = E_1 + E_2 + E_3 + E_4 + E_5$ with $E_i^2 = -2$ and

$$-E_0 = E_{\mathrm{fund}} = E_1 + 2E_2 + 2E_3 + E_4 + E_5.$$

Then $E_0E_2 = E_1E_2 = E_2E_3 = E_3E_4 = E_3E_5 = 1$, and all other $E_iE_j = 0$. Hence $(m_1^{\mathrm{SING}}, \ldots, m_5^{\mathrm{SING}}) = (1, 2, 2, 1, 1)$, as indicated in Figure 2.

1.4. McKay correspondence. Any simple singularity is a quotient singularity by a finite subgroup G of $\mathrm{SL}(2, \mathbb{C})$, and so has a corresponding Dynkin diagram of exceptional set. McKay [8] showed how one can recover the same graph purely

FIGURE 2. The Dynkin diagrams D_5 and $\widetilde{\mathrm{D}}_5$

in terms of the representation theory of G, without passing through the geometry of the germ $(\mathbb{A}^2/G, 0)$.

To be more precise, let G be a finite subgroup of $\mathrm{SL}(2,\mathbb{C})$. Clearly, G has a two-dimensional representation, which maps G injectively into $\mathrm{SL}(2,\mathbb{C})$; we call this the *natural representation* ρ_{nat}. Let $\mathrm{Irr}_* G$, respectively $\mathrm{Irr}\, G$, be the set of all equivalence classes of irreducible representations, respectively nontrivial ones. (Caution: note that this goes against the familiar notation of group theory.) Thus by definition, $\mathrm{Irr}_* G = \mathrm{Irr}\, G \cup \{\rho_0\}$, where ρ_0 is the one-dimensional trivial representation. Any representation of G over $\mathbb{C}$ is completely reducible, that is, is a direct sum of irreducible representations up to equivalence. Therefore for any $\rho \in \mathrm{Irr}_* G$, we have

$$\rho \otimes \rho_{\mathrm{nat}} = \sum_{\rho' \in \mathrm{Irr}_* G} a_{\rho,\rho'} \rho',$$

where $a_{\rho,\rho'}$ are certain nonnegative integers. In our situation, we see that $a_{\rho,\rho'} = 0$ or 1 (except for the case A_1, when $a_{\rho,\rho'} = 2$).

Let us look at the example D_5, the case of the binary dihedral group $G := \mathbb{D}_3$ of order 12. The group G is generated by σ and τ:

$$\sigma = \begin{pmatrix} \varepsilon & 0 \\ 0 & \varepsilon^{-1} \end{pmatrix}, \quad \tau = \begin{pmatrix} 0 & 1 \\ -1 & 0 \end{pmatrix} \quad \text{where } \varepsilon = \mathrm{e}^{2\pi \mathrm{i}/6}.$$

We note that $\mathrm{Tr}(\sigma) = 1$, $\mathrm{Tr}(\tau) = 0$, hence in this case, the natural representation is ρ_2 in Table 2.

1.5. Definition. The graph $\widetilde{\Gamma}_{\mathrm{GROUP}}(G)$ is defined to be the graph consisting of vertices $v(\rho)$ for $\rho \in \mathrm{Irr}_* G$, and simple edges connecting any pair of vertices $v(\rho)$ and $v(\rho')$ with $a_{\rho,\rho'} = 1$. We denote by $\Gamma_{\mathrm{GROUP}}(G)$ the full subgraph of $\widetilde{\Gamma}_{\mathrm{GROUP}}(G)$ consisting of the vertices $v(\rho)$ for $\rho \in \mathrm{Irr}\, G$ and all the edges between them.

For example, let us look at the D_5 case. Let $\chi_j := \mathrm{Tr}(\rho_j)$ be the character of ρ_j. Then from Table 2 we see that

$$\chi_2(g)\chi_{\mathrm{nat}}(g) = \chi_2(g)\chi_2(g) = \chi_0(g) + \chi_1(g) + \chi_3(g), \quad \text{for } g = 1, \sigma \text{ or } \tau.$$

TABLE 2. Character table of $\mathbb{D}_3$ (of type D_5)

ρ	$\mathrm{Tr}\,\rho$	1	σ	τ
ρ_0	χ_0	1	1	1
ρ_1	χ_1	1	1	-1
ρ_2	χ_2	2	1	0
ρ_3	χ_3	2	-1	0
ρ_4	χ_4	1	-1	i
ρ_5	χ_5	1	-1	$-$i

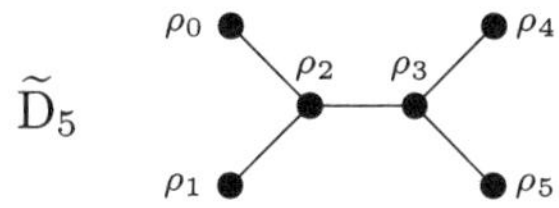

FIGURE 3. $\widetilde{\Gamma}_{\mathrm{GROUP}}(\mathbb{D}_3)$

Hence $\chi_2\chi_{\mathrm{nat}} = \chi_0 + \chi_1 + \chi_3$. General representation theory says that an irreducible representation of G is uniquely determined up to equivalence by its character. Therefore $\rho_2 \otimes \rho_{\mathrm{nat}} = \rho_0 + \rho_1 + \rho_3$. Hence $a_{\rho_2,\rho_j} = 1$ for $j = 0, 1, 3$ and $a_{\rho_2,\rho_j} = 0$ for $j = 2, 4, 5$. Similarly, we see that

$$\chi_0\chi_{\mathrm{nat}} = \chi_2, \quad \chi_1\chi_{\mathrm{nat}} = \chi_2, \quad \chi_3\chi_{\mathrm{nat}} = \chi_2 + \chi_4 + \chi_5,$$
$$\chi_4\chi_{\mathrm{nat}} = \chi_3 \quad \text{and} \quad \chi_5\chi_{\mathrm{nat}} = \chi_3.$$

In this way we obtain a graph — the extended Dynkin diagram $\widetilde{\mathrm{D}}_5$ of Figure 3. Thus we see that there are two completely different ways to obtain the same extended Dynkin diagram $\widetilde{\mathrm{D}}_5$ as $\widetilde{\Gamma}_{\mathrm{SING}}(\mathbb{A}^2/G, 0)$ and $\widetilde{\Gamma}_{\mathrm{GROUP}}(G)$, while D_5 as $\Gamma_{\mathrm{SING}}(\mathbb{A}^2/G, 0)$ and $\Gamma_{\mathrm{GROUP}}(G)$.

The same is true in the other cases. Namely the two graphs $\Gamma_{\mathrm{SING}}(\mathbb{A}^2/G, 0)$ and $\Gamma_{\mathrm{GROUP}}(G)$ turn out to be one of the Dynkin diagrams ADE and coincide with each other, while both $\widetilde{\Gamma}_{\mathrm{SING}}(\mathbb{A}^2/G, 0)$ and $\widetilde{\Gamma}_{\mathrm{GROUP}}(G)$ are the corresponding extended Dynkin diagram (See Figure 4). It is also interesting to note that the degrees of the characters $\deg \rho_j = \chi_j(1)$ are equal to the multiplicities of the fundamental cycle we computed in Section 1.3.

This is the second observation of McKay that we are going to discuss in this article.

2. The G-orbit Hilbert schemes

2.1. Hilbert schemes. The Hilbert scheme of a given projective (or quasiprojective) scheme X is the scheme parametrizing all the subschemes of X. More precisely, let X be a projective scheme embedded in a projective space $\mathbb{P}^N$, and L the restriction of $\mathcal{O}_{\mathbb{P}^N}(1)$ to X. The *Hilbert scheme* is the scheme representing the functor

$$\mathrm{Hilb}_X : S \mapsto \{\text{flat families } Y \text{ of subschemes of } X \text{ over } S\}.$$

The Euler–Poincaré characteristic $P(m) := \sum_{q\in\mathbb{Z}}(-1)^q h^q(Y_s, L_{Y_s}^{\otimes m})$ (called the *Hilbert polynomial*) of the sheaf $L_{Y_s}^{\otimes m}$ is constant on each connected component of S. Therefore the Hilbert scheme decomposes as the disjoint union of open subsets labeled by Hilbert polynomials.

The point set Hilb_X^P classifies the subschemes of X with Hilbert polynomial equal to P. Let $U \subset X$ be an open subscheme. Then Hilb_U is an open subscheme of Hilb_X consisting of the subschemes of X whose supports are contained in U. This means that Hilb_U^P is empty or an open subscheme of Hilb_X^P for a fixed Hilbert polynomial P.

2.2. The Hilbert scheme of n points. In what follows we denote Hilb_X^P by $\mathrm{Hilb}^P(X)$. Write $\mathrm{S}^n(\mathbb{A}^2)$ for the nth symmetric product of the affine plane $\mathbb{A}^2$. This is by definition the quotient of the products of n copies of $\mathbb{A}^2$ by the natural permutation action of the symmetric group S_n on n letters. It is the set

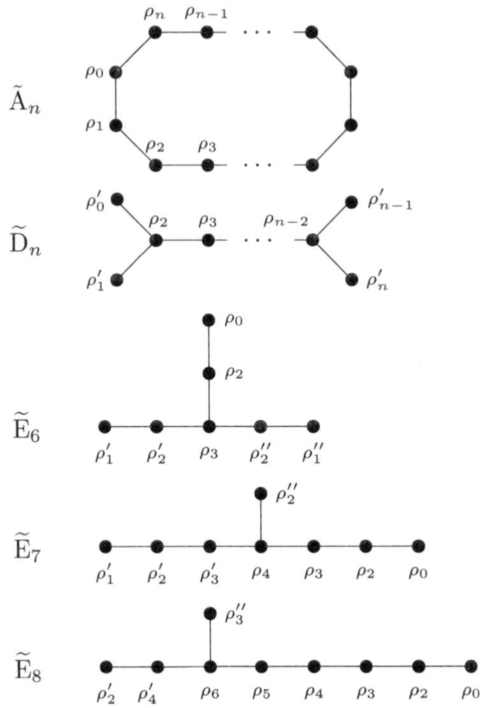

FIGURE 4. The extended Dynkin diagrams and representations

of formal sums of n points, in other words, the set of unordered n-tuples of points. Let $P(m) = n$ for any $m \in \mathbb{Z}$, namely let $P(m)$ be a constant polynomial. We call $\mathrm{Hilb}^n(\mathbb{A}^2) := \mathrm{Hilb}^P(\mathbb{A}^2)$ the *Hilbert scheme of n points in* $\mathbb{A}^2$. It is a quasiprojective scheme of dimension $2n$. Any $Z \in \mathrm{Hilb}^n(\mathbb{A}^2)$ is a zero-dimensional subscheme with $h^0(Z, \mathcal{O}_Z) = \dim(\mathcal{O}_Z) = n$. Suppose that Z is reduced. Then Z is the union of n distinct points. Since being reduced is an open and generic condition, $\mathrm{Hilb}^n(\mathbb{A}^2)$ contains a Zariski open subset consisting of formal sums of n distinct points.

We have a natural morphism π from $\mathrm{Hilb}^n(\mathbb{A}^2)$ onto $\mathrm{S}^n(\mathbb{A}^2)$ defined by

$$\pi\colon Z \mapsto \sum_{p \in \mathrm{Supp}(Z)} \dim(\mathcal{O}_{Z,p})p$$

which is called the *Hilbert – Chow morphism*. One of the most remarkable features of $\mathrm{Hilb}^n(\mathbb{A}^2)$ is the following result.

2.3. Theorem ([3]). *$\mathrm{Hilb}^n(\mathbb{A}^2)$ is a smooth quasi-projective scheme, and the Hilbert – Chow morphism $\pi\colon \mathrm{Hilb}^n(\mathbb{A}^2) \to \mathrm{S}^n(\mathbb{A}^2)$ is a resolution of singularities of the symmetric product.*

We note that smoothness of $\mathrm{Hilb}^n(\mathbb{A}^2)$ is peculiar to $\dim \mathbb{A}^2 = 2$.

2.4. The *G*-orbit Hilbert scheme. For any finite subgroup G of $\mathrm{SL}(2, \mathbb{C})$ of order n, we consider the Hilbert – Chow morphism π from $\mathrm{Hilb}^n(\mathbb{A}^2)$ onto $\mathrm{S}^n(\mathbb{A}^2)$. Since the morphism $\pi\colon \mathrm{Hilb}^n(\mathbb{A}^2) \to \mathrm{S}^n(\mathbb{A}^2)$ is G-equivariant, we have a natural morphism between G-fixed point loci. We note that the G-fixed point set of $\mathrm{S}^n(\mathbb{A}^2)$ is nothing but $\mathbb{A}^2/G$ because $n = |G|$. The G-fixed point set $\mathrm{Hilb}^n(\mathbb{A}^2)^G$

in $\mathrm{Hilb}^n(\mathbb{A}^2)$ is always nonsingular, but could a priori be disconnected. There is however a unique irreducible component of $\mathrm{Hilb}^n(\mathbb{A}^2)^G$ dominating $\mathrm{S}^n(\mathbb{A}^2)^G$, which we denote by $\mathrm{Hilb}^G(\mathbb{A}^2)$ and call it the *G-orbit Hilbert scheme.* It is a G-invariant subscheme of $\mathrm{Hilb}^n(\mathbb{A}^2)$ that parametrizes all smoothable G-invariant subschemes of length $|G|$.

Now we recall the following theorem proved in [6].

2.5. Theorem. *Let $G \subset \mathrm{SL}(2,\mathbb{C})$ be a finite subgroup of order n. Then there is a unique irreducible component $\mathrm{Hilb}^G(\mathbb{A}^2)$ of $\mathrm{Hilb}^n(\mathbb{A}^2)^G$ dominating $\mathbb{A}^2/G$, which is a minimal resolution of $\mathbb{A}^2/G$. In particular, the dualizing sheaf of $\mathrm{Hilb}^G(\mathbb{A}^2)$ is trivial.*

PROOF. Any point of $\mathrm{S}^n(\mathbb{A}^2)^G\backslash\{0\}$ is a G-orbit of a point $0 \neq \mathfrak{p} \in \mathbb{A}^2$, which is a reduced zero-dimensional subscheme invariant under G. It follows that $\mathrm{Hilb}^G(\mathbb{A}^2)$ is isomorphic to $\mathrm{S}^n(\mathbb{A}^2)^G$ ($\simeq \mathbb{A}^2/G$) over $\mathrm{S}^n(\mathbb{A}^2)^G\backslash\{0\}$ under the Hilbert – Chow morphism. Hence $\mathrm{Hilb}^G(\mathbb{A}^2)$ is birationally equivalent to $\mathbb{A}^2/G$), so that it is a resolution of $\mathbb{A}^2/G$. Moreover by [4, Proposition 2.6], $\mathrm{Hilb}^G(\mathbb{A}^2)$ inherits a canonical holomorphic symplectic structure from $\mathrm{Hilb}(\mathbb{A}^2)$. Since $\dim \mathrm{Hilb}^G(\mathbb{A}^2) = \dim \mathbb{A}^2/G = 2$, this implies that the dualizing sheaf of $\mathrm{Hilb}^G(\mathbb{A}^2)$ is trivial. This proves the theorem. □

We denote the natural morphism from $\mathrm{Hilb}^G(\mathbb{A}^2)$ onto $\mathbb{A}^2/G$ by the same letter π. There are two corollaries to Theorem 2.5, useful for explicit computations. We only quote these from [6]. See [6] for proofs.

2.6. Corollary. *Let G be a finite subgroup of $\mathrm{SL}(2,\mathbb{C})$, and I an ideal of $\mathcal{O}_{\mathbb{A}^2}$ with $I \in \mathrm{Hilb}^G(\mathbb{A}^2)$ (to be exact, the subscheme defined by I belonging to $\mathrm{Hilb}^G(\mathbb{A}^2)$). Then as G-modules $\mathcal{O}_{\mathbb{A}^2}/I \simeq \mathbb{C}[G] \simeq \sum_\rho(\deg\rho)\rho$, the group algebra (the regular representation).*

2.7. Corollary. *Let I be an ideal of $\mathcal{O}_{\mathbb{A}^2}$ with $I \in \mathrm{Hilb}^G(\mathbb{A}^2)$. Any G-invariant function vanishing at the origin is contained in I.*

3. Theorems

3.1. A link from $\mathrm{Hilb}^G(\mathbb{A}^2)$ to McKay correspondencc. For any finite subgroup G of $\mathrm{SL}(2,\mathbb{C})$ of order n, the *G-orbit Hilbert scheme* $\mathrm{Hilb}^G(\mathbb{A}^2)$ of $\mathrm{Hilb}^n(\mathbb{A}^2)$ is a minimal resolution of a simple singularity $S_G := \mathbb{A}^2/G$. Let $\pi\colon \mathrm{Hilb}^G(\mathbb{A}^2) \to S_G$ be the natural morphism.

Let $S := \mathbb{C}[x,y]$. Let $\mathfrak{m}$ be the ideal of S generated by x and y, $\mathfrak{n}$ the ideal of S generated by all the G-invariant polynomials vanishing at the origin. A point $\mathfrak{p}$ of $\mathrm{Hilb}^G(\mathbb{A}^2)$ is a G-invariant zero-dimensional subscheme Z of $\mathbb{A}^2$. Since $\mathbb{A}^2$ is affine, we can associate to it the G-invariant ideal $I := I_Z$ of $S = \mathbb{C}[x,y]$ defining Z with $S/I \simeq \mathbb{C}[G]$, the regular representation of G. Thus often we write $I \in \mathrm{Hilb}^G(\mathbb{A}^2)$ instead of $Z \in \mathrm{Hilb}^G(\mathbb{A}^2)$ or $\mathfrak{p} \in \mathrm{Hilb}^G(\mathbb{A}^2)$. We also have the exact sequence

$$0 \to I \to S \to \Gamma(Z, \mathcal{O}_Z) \to 0.$$

Any point of the exceptional set E of π is a G-invariant zero-dimensional subscheme Z of $\mathbb{A}^2$ with support the origin. In other words, the ideal I of S defining Z is an infinite-dimensional G-module contained in $\mathfrak{m}$. By deriving a finite-dimensional G-module from it naturally, we will give the McKay correspondence as described in Section 1. We shall introduce the McKay quiver (in the coinvariant algebra) and subquivers so as to understand the natural stratification of $\mathrm{Hilb}^G(\mathbb{A}^2)$. Then the corresponding extended Dynkin diagram emerges naturally (See Figure 6) by extending the McKay quiver in the polynomial ring. We also have a natural irreducible decomposition of a certain universal coherent sheaf on $\mathrm{Hilb}^G(\mathbb{A}^2)$. This somewhat sharpens the McKay correspondence (Theorem 3.9).

3.2. A stratification of $\mathrm{Hilb}^G(\mathbb{A}^2)$ by $\mathrm{Irr}\, G$. Let G be a finite subgroup of $\mathrm{SL}(2,\mathbb{C})$. In what follows we assume that G is not cyclic because A_n-case is much easier. As in Section 1.4, we write $\mathrm{Irr}\, G$ for the set of all the equivalence classes of nontrivial irreducible G-modules, and $\mathrm{Irr}_*\, G$ for the union of $\mathrm{Irr}\, G$ and the trivial one-dimensional G-module.

Let E the exceptional set of the minimal resolution, $\mathrm{Irr}\, E$ the set of irreducible components of E.

Any $I \in X$ contained in E is a G-invariant ideal of S which contains $\mathfrak{n}$ by Corollary 2.7. For any ρ, ρ', and $\rho'' \in \mathrm{Irr}\, G$, we define

$$\begin{aligned} V(I) &:= I/(\mathfrak{m}I + \mathfrak{n}), \\ E(\rho) &:= \{I \in \mathrm{Hilb}^G(\mathbb{A}^2); V(I) \supset \rho\}, \\ P(\rho,\rho') &:= \{I \in \mathrm{Hilb}^G(\mathbb{A}^2); V(I) \supset \rho \oplus \rho'\}, \\ Q(\rho,\rho',\rho'') &:= \{I \in \mathrm{Hilb}^G(\mathbb{A}^2); V(I) \supset \rho \oplus \rho' \oplus \rho''\}. \end{aligned}$$

Any nonempty stratum of $\mathrm{Hilb}^G(\mathbb{A}^2)$ will be described in a simple manner by using Dynkin diagrams with directed edges, called McKay subquivers. See Section 4.8, Figures 7 – 12.

3.3. Definition. Two irreducible G-modules ρ and ρ' are said to be *adjacent* if $\rho \otimes \rho_{\mathrm{nat}} \supset \rho'$ or vice versa.

The *Dynkin diagram* $\Gamma(\mathrm{Irr}\, G)$ of $\mathrm{Irr}\, G$ is the graph whose vertices are $\mathrm{Irr}\, G$, with ρ and ρ' joined by a simple edge if and only if ρ and ρ' are adjacent.

Then the following theorem was proved in [6].

3.4. Theorem. *Let G be a finite subgroup of* $\mathrm{SL}(2,\mathbb{C})$. *Then*

(1) *the map $\rho \mapsto E(\rho)$ is a bijective correspondence between* $\mathrm{Irr}\, G$ *and* $\mathrm{Irr}\, E$;
(2) *$E(\rho)$ is a smooth rational curve with $E(\rho)^2 = -2$ for any $\rho \in \mathrm{Irr}\, G$*;
(3) *$P(\rho,\rho') \neq \varnothing$ if and only if ρ and ρ' are adjacent. In this case $P(\rho,\rho')$ is a single (reduced) point, at which $E(\rho)$ and $E(\rho')$ intersect transversally*;
(4) *$P(\rho,\rho) = Q(\rho,\rho',\rho'') = \varnothing$ for any $\rho,\rho',\rho'' \in \mathrm{Irr}\, G$.*

By Theorem 3.4(3), $\Gamma(\mathrm{Irr}\, G)$ is the same as the dual graph $\Gamma(\mathrm{Irr}\, E)$ of E, in other words, the Dynkin diagram of the singularity $(\mathbb{A}^2/G, 0)$.

In what follows we assume that G is not cyclic. We define nonnegative integers $d(\rho)$ for any $\rho \in \mathrm{Irr}\, G$ as follows. Since G is not cyclic, $\Gamma(\mathrm{Irr}\, G)$ is star-shaped with a unique center. For any $\rho \in \mathrm{Irr}\, G$, we define $d(\rho)$ to be the distance from the vertex ρ to the center, where $d(\rho) = 0$ for the center ρ. It is obvious that $d(\rho) = d(\rho') \pm 1$ if ρ and $\rho' \in \mathrm{Irr}\, G$ are adjacent.

Let $S = \mathbb{C}[x,y]$. For any positive integer k let S_k be the subspace of homogeneous polynomials of degree k in S. We say that a G-submodule W of $\mathfrak{m}/\mathfrak{n}$ is *homogeneous* of degree k if it is generated over $\mathbb{C}$ by homogeneous polynomials of degree k. The G-module $\mathfrak{m}/\mathfrak{n}$ splits as a direct sum of irreducible homogeneous

G-modules. If W is a direct sum of homogeneous G-submodules, then we denote the homogeneous part of W of degree k by $S_k(W)$. For any G-module W in some $S_k(\mathfrak{m}/\mathfrak{n})$, we write $S_j \cdot W$ for the G-submodule of $S_{k+j}(\mathfrak{m}/\mathfrak{n})$ generated over $\mathbb{C}$ by the products of $S_j(\mathfrak{m}/\mathfrak{n})$ and W. We denote by $W[\rho]$ the ρ factor of W, that is, the sum of all the copies of ρ in W; and similarly, we denote by $[W{:}\rho]$ the multiplicity of $\rho \in \operatorname{Irr} G$ in a G-module W.

3.5. Definition. The quotient algebra $S/\mathfrak{n}$ is called the coinvariant algebra of G, denoted by $\operatorname{Coinv}(G)$ (or denoted by Coinv(the type of S_G) with the notation of Table 1). Let h be the Coxeter number of the simple singularity S_G. Then we define the very positive part $S^\dagger := S^\dagger(\mathfrak{m}/\mathfrak{n})$ of $S/\mathfrak{n}$ to be

$$S^\dagger := \sum_{\substack{k>h/2+d(\rho)\\ \rho\in\operatorname{Irr} G}} S_k(\mathfrak{m}/\mathfrak{n})[\rho].$$

We also define the McKay quiver of G by

$$S_{\mathrm{McKay}}(\mathfrak{m}/\mathfrak{n}) = \sum_{\rho\in\operatorname{Irr} G} S_{h/2\pm d(\rho)}(\mathfrak{m}/\mathfrak{n})[\rho] + S^\dagger/S^\dagger,$$

which we also denote bu $S_{\mathrm{McKay}}(G)$. Let $V_k(\rho) = S_k(\mathfrak{m}/\mathfrak{n})[\rho]$ and $V(\rho) = S_{\mathrm{McKay}}(\mathfrak{m}/\mathfrak{n})[\rho]$.

The following Lemma 3.6 and Lemma 3.7 describe the structure of the McKay quiver $S_{\mathrm{McKay}}(\mathfrak{m}/\mathfrak{n})$. See [6] for the proofs.

3.6. Lemma. *Let φ_i be three generators of G-invariants, $d_i = \deg\varphi_i$ and h the Coxeter number of the simple singularity S_G. Then $d_1+d_2 = d_3+2$ and $d_3 = h$, and moreover $S_k(\mathfrak{m}/\mathfrak{n}) = 0$ for $k \geq d_3$.*

3.7. Lemma. *Assume that G is not cyclic. Let h be the Coxeter number of the simple singularity S_G. Then as G-modules, we have*

(1) $\mathfrak{m}/\mathfrak{n} = \sum_{\rho\in\operatorname{Irr} G} 2(\deg\rho)\rho$;
(2) $S_{\mathrm{McKay}}(\mathfrak{m}/\mathfrak{n}) \simeq \sum_{\rho\in\operatorname{Irr} G} 2\rho$;
(3) $V_{h/2-d(\rho)}(\rho) \simeq V_{h/2+d(\rho)}(\rho) = \rho$ *if* $d(\rho)\geq 1$ *and* $V_{h/2}(\rho) = \rho^{\oplus 2}$ *if* $d(\rho) = 0$;
(4) $S_{h/2-k}(\mathfrak{m}/\mathfrak{n}) \simeq S_{h/2+k}(\mathfrak{m}/\mathfrak{n})$ *for any* k.

Using this notation we see by Lemma 3.7(2) that $V(\rho) = V_{h/2-d(\rho)}(\rho) + V_{h/2+d(\rho)}(\rho) \simeq \rho^{\oplus 2}$ and that the subset $E(\rho)$ is $\mathbb{P}(V(\rho))$, the set of all nontrivial G-submodules of $\rho^{\oplus 2}$. This is isomorphic to the projective line, or a smooth rational curve by Schur's lemma. This proves Theorem 3.4, (2).

3.8. The ideals $\mathfrak{n}_X$ and $\pi^*\overline{\mathfrak{m}}$. Since $(\mathbb{A}^2/G)\times_{(\mathbb{A}^2/G)} X \simeq X$, X is a closed subscheme of $(\mathbb{A}^2/G)\times X$, which is defined by an ideal I_X of $O_{\mathbb{A}^2/G}\otimes O_X$. Let $\mathfrak{n}_X$ be the ideal of $O_{\mathbb{A}^2}\otimes O_X$ generated by I_X. Let Z_{univ} be the universal subscheme of $\mathbb{A}^2\times X$, and I_{univ} the ideal sheaf of $O_{\mathbb{A}^2\times X}$ defining Z_{univ}. Since we have a commutative diagram

$$\begin{array}{ccc} Z_{\mathrm{univ}} & \xrightarrow{\mathrm{pr}_2} & X \\ {\scriptstyle \mathrm{pr}_1}\downarrow & & \downarrow{\scriptstyle \pi} \\ \mathbb{A}^2 & \xrightarrow{\phi} & \mathbb{A}^2/G, \end{array}$$

the morphism $\phi\times\mathrm{id}_X : \mathbb{A}^2\times X \to \mathbb{A}^2/G\times X$ maps Z_{univ} into $(\mathbb{A}^2/G)\times_{(\mathbb{A}^2/G)} X \simeq X$. This implies that $\mathfrak{n}_X = I_X O_{\mathbb{A}^2} \otimes O_X \subset I_{\mathrm{univ}}$. Now we define

$$\mathcal{V} := V(I_{\mathrm{univ}}) = I_{\mathrm{univ}}/\mathfrak{m} I_{\mathrm{univ}} + \mathfrak{n}_X.$$

Let $\overline{\mathfrak{m}}$ be the maximal ideal of $O_{\mathbb{A}^2/G}$ of the unique singular point. We note $\mathfrak{n} = \Gamma(\mathbb{A}^2, \phi^*\overline{\mathfrak{m}})$. We also note $\pi^*(\overline{\mathfrak{m}}) \cap I_{\mathrm{univ}} = \{0\}$. In fact, suppose $\pi^*(F) \in I_{\mathrm{univ}}$. Then $\pi^*(F) = 0$ on Z_{univ}. Since Z_{univ} is surjective over $\mathbb{A}^2/G$, $F = 0$. It follows $\pi^*(\overline{\mathfrak{m}}) \cap I_{\mathrm{univ}} = \{0\}$.

Since $\pi^*\overline{\mathfrak{m}} := \overline{\mathfrak{m}} O_X$ is the defining ideal of E_{fund} by [1], we see that $\mathcal{V}$ is a finite $O_{E_{\mathrm{fund}}}$-module. See section 6 for the proof. However we prove a little stronger

3.9. Theorem. *The $O_{\mathbb{A}^2\times_{\mathbb{C}} X}$-module $\mathcal{V}$ is a finite O_E-module; as an $O_{\mathbb{A}^2\times_{\mathbb{C}} X}$-module with G-action, we have an isomorphism*

$$\mathcal{V} \simeq \bigoplus_{\rho\in\mathrm{Irr}\,G} \rho \otimes_{\mathbb{C}} O_{E(\rho)}(-1).$$

In Section 6 we will prove Theorem 3.9. In order to prepare for the proof of Theorem 3.9, we recall in Section 4 the proof of Theorem 3.4 and related constructions in the case of D_5 in full detail. In the cases D_n $(n \neq 5)$ and E_6 we give only a sketch of proofs of Theorem 3.9. Though we give almost no proof for E_7 because we need more notation, we remark that we can prove it in the same manner. It remains to check E_8.

4. The simple singularity D_5

In this section we explain the case of D_5 in full detail.

4.1. The binary dihedral group $\mathbb{D}_3$. The simple singularity D_5 is the quotient singularity of $\mathbb{A}^2$ by the binary dihedral group $G := \mathbb{D}_3$ of order 12, which is generated by σ and τ:

$$\sigma = \begin{pmatrix} \varepsilon & 0 \\ 0 & \varepsilon^{-1} \end{pmatrix}, \quad \tau = \begin{pmatrix} 0 & 1 \\ -1 & 0 \end{pmatrix},$$

where $\varepsilon := e^{2\pi i/6}$. We have $\sigma^6 = \tau^4 = 1$, $\sigma^3 = \tau^2$ and $\tau\sigma\tau^{-1} = \sigma^{-1}$. The ring of G-invariants in $\mathbb{C}[x,y]$ is generated by three elements $A_6 := x^6 + y^6$, $A_8 := xy(x^6 - y^6)$ and $A_4 := x^2y^2$. The quotient $\mathbb{A}^2/G$ is isomorphic to the hypersurface $4A_4^4 + A_8^2 - A_4A_6^2 = 0$.

4.2. The McKay quiver, tables and figures. We consider the case of D_5. By an elementary computation we have the irreducible decomposition of the coinvariant algebra $\mathrm{Coinv}(\mathrm{D}_5)$ in Table 3. As Tables 3 – 5 indicate, $\mathrm{Coinv}(\mathrm{D}_5)$ consists of dual pairs. The McKay submodule S_{McKay} of $\mathrm{Coinv}(\mathrm{D}_5)$, that is, the irreducible factors in the parentheses in Tables 3 – 5 consist of dual pairs, exactly one pair for each equivalence class $\rho \in \mathrm{Irr}\,G$ except ρ_3, while $V_4(\rho_3) = 2\rho_3$ is self-dual.

The coinvariant algebra $\mathrm{Coinv}(\mathrm{D}_5)$ admits a quiver structure induced from multiplication of the symmetric algebra. This induces a quiver structure on S_{McKay}, which we call the McKay quiver of D_5. It yields naturally a corresponding Dynkin diagram of the minimal resolution $X := \mathrm{Hilb}^G(\mathbb{A}^2)$, as we see later. In other words, the McKay quiver of D_5 gives the Dynkin diagram D_5 of the exceptional set E of X by replacing each pair of arrows by an edge. To describe the strata of E precisely we also introduce the subquivers $\mathrm{Quiv}(\rho)$ and $\mathrm{Quiv}(\rho, \rho')$, which are Dynkin diagrams

TABLE 3. Irreducible decompositions of S and $\mathrm{Coinv}(\mathrm{D}_5)$

k	S_k	$S_k(S/\mathfrak{n})$
0	ρ_0	ρ_0
1	ρ_2	ρ_2
2	$\rho_1+\rho_3$	$(\rho_1)+\rho_3$
3	$\rho_2+\rho_4+\rho_5$	$(\rho_2+\rho_4+\rho_5)$
4	$[\rho_0]+2\rho_3$	$(2\rho_3)$
5	$2\rho_2+\rho_4+\rho_5$	$(\rho_2+\rho_4+\rho_5)$
6	$[\rho_0]+2\rho_1+2\rho_3$	$(\rho_1)+\rho_3$
7	$3\rho_2+\rho_4+\rho_5$	ρ_2

TABLE 4. $\mathrm{Coinv}(\mathrm{D}_5)$

k	$V_k(\rho)\subset S_k(S/\mathfrak{n})$	equiv. class
0	$\{1\}_{\rho_0}$	ρ_0
1	$\{x,y\}_{\rho_2}$	ρ_2
2	$\{xy\}_{\rho_1}\oplus\{x^2,y^2\}_{\rho_3}$	$(\rho_1)+\rho_3$
3	$\{x^2y,-xy^2\}_{\rho_2}$ $\oplus\{x^3+\mathrm{i}y^3\}_{\rho_4}\oplus\{x^3-\mathrm{i}y^3\}_{\rho_5}$	$(\rho_2+\rho_4+\rho_5)$
4	$\{y^4,x^4\}_{\rho_3}\oplus\{x^3y,-xy^3\}_{\rho_3}$	$(2\rho_3)$
5	$\{y^5,-x^5\}_{\rho_2}$ $\oplus\{xy(x^3-\mathrm{i}y^3)\}_{\rho_4}\oplus\{xy(x^3+\mathrm{i}y^3)\}_{\rho_5}$	$(\rho_2+\rho_4+\rho_5)$
6	$\{x^6-y^6\}_{\rho_1}\oplus\{xy^5,-x^5y\}_{\rho_3}$	$(\rho_1)+\rho_3$
7	$\{xy^6,x^6y\}_{\rho_2}$	ρ_2

TABLE 5. Dual pairs of $\mathrm{Coinv}(\mathrm{D}_5)$

degree	1	2	3	4
equiv. class	ρ_2	$(\rho_1)+\rho_3$	$(\rho_2+\rho_4+\rho_5)$	$(2\rho_3)$
degree	7	6	5	
equiv. class	ρ_2	$(\rho_1)+\rho_3$	$(\rho_2+\rho_4+\rho_5)$	

with certain directed edges. See Figures 7 – 16. These enable us to specify which of the G-invariants A_4, A_6 and A_8 generates the ideal $\pi^*(\mathfrak{n})$ along each $E(\rho)$.

Moreover in order to obtain the extended Dynkin diagram $\widetilde{D}_5$ it seems natural to add to Figure 5 the G-submodules $V_4(\rho_0):=S_4[\rho_0]$ and $V_6(\rho_0):=S_6[\rho_0]$ as in Figure 6, where $V_4(\rho_0)$ and $V_6(\rho_0)$, the parts of Table 3 in the bracket, are generated by the generators $A_4=x^2y^2$ and $A_6=x^6+y^6$ of G-invariants respectively.

We note that the arrows between ρ_2 to ρ_0 are exceptional in the sense that both the arrows between them are directed from ρ_2 to ρ_0, while in any other cases, say for ρ_1 and ρ_2, both directions are taken by arrows between them. As its consequence $\deg V_4(\rho_0)+\deg V_6(\rho_0)=10>8=2+6=3+5=4+4$, the sum of degrees for the other pairs or the Coxeter number of D_5.

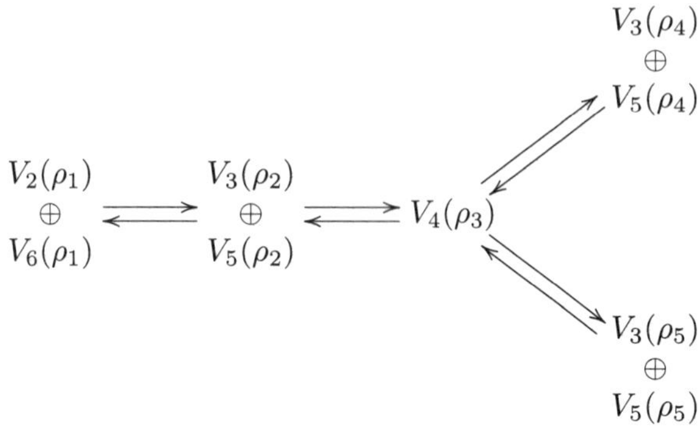

FIGURE 5. The McKay quiver of D_5

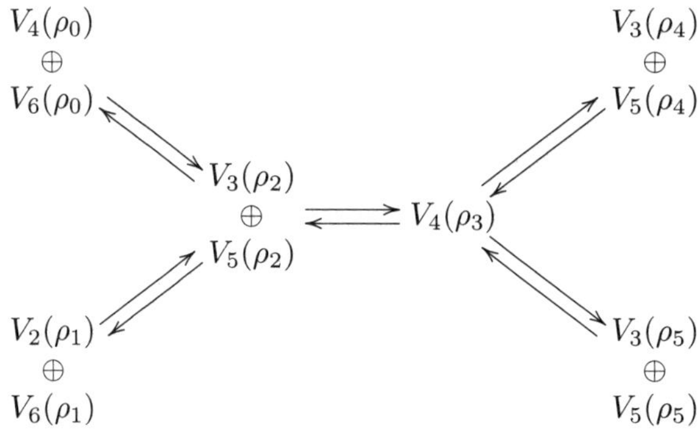

FIGURE 6. The extended McKay quiver of D_5

4.3. The subset $E(\rho_1)$.

4.3.1. We first classify $I \in E(\rho_1)$. Now we recall

$$V_2(\rho_1) = \{xy\}, \qquad V_6(\rho_1) = \{x^6 - y^6\},$$
$$V_3(\rho_2) = \{x^2y, xy^2\}, \qquad V_5(\rho_2) = \{x^5, y^5\}.$$

For any nonzero G-submodule W of $V_2(\rho_1) \oplus V_6(\rho_1)$ with $W \neq V_6(\rho_1')$, let $I(W) := SW + \mathfrak{n}$ where $S = \mathbb{C}[x, y]$. The G-module W is generated by $xy + t(x^6 - y^6)$ for some $t \in \mathbb{C}$. Then since $S_8 \subset \mathfrak{n}$ by Lemma 3.6, we see

$$S_2W + \mathfrak{n} = S_2 \cdot \left(xy + t(x^6 - y^6)\right) + \mathfrak{n} = S_2 \cdot xy + \mathfrak{n},$$
$$S_kW + \mathfrak{n} = S_kV_2(\rho_1) + \mathfrak{n} \quad \text{for any } k \geq 2.$$

By Table 4, we see in $S/\mathfrak{n}$

$$S_2V_2(\rho_1) = \{x^3y, x^2y^2, xy^3\} = \{x^3y, -xy^3\} = V_4(\rho_3),$$
$$S_3V_2(\rho_1) = \{x^4y, xy^4\} = V_5(\rho_4) \oplus V_5(\rho_5),$$
$$S_4V_2(\rho_1) = V_6(\rho_3), \;\; S_5V_2(\rho_1) = V_7(\rho_2).$$

It follows that

$$S_1W + S_5W + \mathfrak{n} = S_1V_2(\rho_1) + S_5V_2(\rho_1) + \mathfrak{n}.$$

Hence we see

$$\begin{aligned} I(W)/\mathfrak{n} &= W + \sum_{k=1}^{5} S_k W + \mathfrak{n}/\mathfrak{n} = W + \sum_{k=1}^{5} S_k V_2(\rho_1) + \mathfrak{n}/\mathfrak{n} \\ &\simeq W + \rho_2 + \rho_3 + (\rho_4 + \rho_5) + \rho_3 + \rho_2 \simeq \sum_{\rho \in \operatorname{Irr} G} \deg(\rho)\rho. \end{aligned}$$

Thus we have $S/I(W) \simeq S/\mathfrak{n} \ominus I(W)/\mathfrak{n} \simeq \sum_{\rho\in \operatorname{Irr}_* G} \deg(\rho)\rho$. Hence $I(W)$ belongs to $\operatorname{Hilb}^{|G|}(\mathbb{A}^2)^{G\text{-inv}}$.

4.3.2. Next we prove that $I\big(V_2(\rho_1)\big)$ belongs to $\operatorname{Hilb}^G(\mathbb{A}^2)$. For this purpose we consider local deformations of $I\big(V_2(\rho_1)\big)$ in $\operatorname{Hilb}^{|G|}(\mathbb{A}^2)^{G\text{-inv}}$. Let $I = I\big(V_2(\rho_1)\big)$. Then $I = xyS + A_6 S$ because $\mathfrak{n}$ is generated by $A_4 = x^2y^2$, $A_6 = x^6 + y^6$ and $A_8 = xy(x^6 - y^6)$. General deformation theory says that G-equivariant local deformations of I are captured by the ρ_0-part of $\operatorname{Hom}_S(I/I^2, S/I)$. We see

$$(S/I)[\rho_1] = \{xy, x^6 - y^6\} = V_2(\rho_1) + V_6(\rho_1), \quad (S/I)[\rho_0] = \{1\} = \mathbb{C}.$$

Since I/I^2 is generated by the elements xy and $A_6 = x^6 + y^6$, the G-module $\operatorname{Hom}_S(I/I^2, S/I)[\rho_0]$ is spanned by the elements ψ_1 and ψ_2:

$$\psi_1(xy) = x^6 - y^6, \quad \psi_1(A_6) = 0, \quad \psi_2(xy) = 0, \quad \psi_2(A_6) = 1.$$

In fact, let $\phi \in \operatorname{Hom}_S(I/I^2, S/I)[\rho_0]$. Let $a = xy$ be a generator of $V_2(\rho_1)$. Then we may assume $\phi(a) = s(x^6 - y^6)$. Then we have $(x^6 - y^6)\phi(a) = \phi(A_8) + s_0\phi(A_6^2 - 4A_4^3) = \phi(A_8)$ because $A_jA_k \in I^2$, while $(x^6 - y^6)\phi(a) = s(A_6^2 - 4A_4^3) = 0$ in S/I because $A_k \in I$. Hence $\phi(A_8) = 0$. Moreover, $\phi(A_4) = \phi(xya) = sA_8 = 0$. Since $\operatorname{Hom}_S(I/I^2, S/I)[\rho_0]$ is two-dimensional and I is generated by a and A_6, it follows that letting $\phi(A_6) = t$, then $\phi = s\psi_1 + t\psi_2$, whence ψ_1 and ψ_2 span $\operatorname{Hom}_S(I/I^2, S/I)[\rho_0]$.

Thus the tangent space $T_{[I]}\big(\operatorname{Hilb}^{|G|}(\mathbb{A}^2)\big)^{G\text{-inv}}$ at the point $[I]$ is spanned by $\partial/\partial s$ and $\partial/\partial t$. In other words, s and t are local (regular) parameters of $\operatorname{Hilb}^G(\mathbb{A}^2)$ at I. This implies the following. Let $\mathbb{C}[\![s,t]\!]$ be the formal power series ring of two variables s and t, and $R = \mathbb{C}[\![s,t]\!][x,y]$. Now we define $\mathcal{I}$ to be the ideal of R generated by the elements

$$xy + t(x^6 - y^6), \quad A_6 + s, \quad A_4 + tA_8, \quad A_8 + s^2t + 4t^4A_8^3.$$

Let $\phi_1(s,t)$ be a power series with initial term $-s^2t$ satisfying the equation

$$\phi_1 + s^2t + 4t^4\phi_1^3 = 0,$$

which comes from $A_8^2 = A_4A_6^2 - 4A_4^4$. The power series ϕ is uniquely determined by this property. Then $\mathcal{I}$ is also generated by the elements

$$xy + t(x^6 - y^6), \quad A_6 + s, \quad A_4 + t\phi_1(s,t), \quad A_8 - \phi_1(s,t).$$

We note that

$$S/I = \bigoplus_{k=1}^{5} \{x^k, y^k\} \oplus \{x^6 - y^6\}.$$

By the upper semi-continuity $R/\mathcal{I}$ is generated over $\mathbb{C}[\![s,t]\!]$ by S/I (regarded as a G-submodule of S). Then it is almost clear that $R/\mathcal{I}$ is a free $\mathbb{C}[\![s,t]\!]$-module with the same basis S/I because of the forms of the four generators of $\mathcal{I}$. Hence $R/\mathcal{I}$ gives a $\mathbb{C}[\![s,t]\!]$-flat family of zero-dimensional subschemes deforming $Z := \operatorname{Spec}(S/I)$. It is obvious that the support of the subscheme over $\mathbb{C}(\!(s,t)\!)$ is away from the origin (if necessary by restricting the deformation to a small disc of the (s,t) space near the origin in the complex topology). In other words, Z is deformed into a G-invariant reduced subscheme of $\mathbb{A}^2$. Therefore Z, hence I, belongs to $\mathrm{Hilb}^G(\mathbb{A}^2)$.

4.3.3. Since I belongs to $\mathrm{Hilb}^G(\mathbb{A}^2)$, any deformation I' of I over a connected base belongs to $\mathrm{Hilb}^G(\mathbb{A}^2)$ because $\mathrm{Hilb}^G(\mathbb{A}^2)$ is connected. In particular, $I(W)$ defined above and all I' we are going to construct in this (sub)section belong to $\mathrm{Hilb}^G(\mathbb{A}^2)$.

4.3.4. Next we consider the case where $W = V_6(\rho_1)$. Let $W_\infty = V_6(\rho_1)$ and $W_t = \big(xy + t(x^6 - y^6)\big)S + \mathfrak{n}$. Then $\lim_{t\to\infty} W_t = W_\infty$ in $\mathbb{P}^1 = \mathbb{P}\big(V(\rho_1)\big) = \mathbb{P}\big(V_2(\rho_1) \oplus V_6(\rho_1)\big)$. Now we compute $\lim_{t\to\infty} I(W_t)$ in $\mathrm{Hilb}^G(\mathbb{A}^2)$. For any t, hence for $t = \infty$ we have

$$I(W_t)/\mathfrak{n} = W_t + \sum_{k=1}^{5} S_k W_t + \mathfrak{n}/\mathfrak{n} = W_t + \sum_{k=1}^{5} S_k V_2(\rho_1) + \mathfrak{n}/\mathfrak{n},$$

$$I(W_\infty)/\mathfrak{n} = V_6(\rho_1) + \sum_{k=1}^{5} S_k V_2(\rho_1) + \mathfrak{n}/\mathfrak{n} \simeq \sum_{\rho\in\operatorname{Irr} G} \deg(\rho)\rho.$$

Since $I(W_\infty)$ is a limit of $I(W_t)$, to be more precise, we have a flat family over $\mathbb{P}^1$ deforming $I(W_t)$ into $I(W_\infty)$, the limit $I(W_\infty)$ belongs to $\mathrm{Hilb}^G(\mathbb{A}^2)$.

We note that

$$V\big(I(W_\infty)\big) = V_6(\rho_1) + S_1 V_2(\rho_1) = V_6(\rho_1) + S_3(\rho_2).$$

Hence $I(W_\infty)$ belongs to the subset $P(\rho_1, \rho_2)$ of $\mathrm{Hilb}^G(\mathbb{A}^2)$.

4.3.5. Now we shall prove the converse. Let $I \in \mathrm{Hilb}^{|G|}(\mathbb{A}^2)^{G\text{-inv}}$. Suppose $S/I \simeq \mathbb{C}[G]$ and $\rho_1 \subset V(I)$. Then $\mathfrak{n} \subset I$ by (the same reason as) Corollary 2.7. This implies that a nonzero G-submodule W of $V_2(\rho_1)\oplus V_6(\rho_1)$ is contained in $V(I)$, and in I as part of generators of I. If $W \neq V_6(\rho_1)$, then I contains $I(W)$ defined above. Since $I(W)/\mathfrak{n} \simeq I/\mathfrak{n}$, we have $I = I(W)$. Suppose $W = W_\infty = V_6(\rho_1)$, or equivalently $V_6(\rho_1) \subset V(I)$. Since W is part of generators of I, $V_5(\rho_2)$ is not contained in I. There are only $V_3(\rho_2)$ and $V_5(\rho_2)$ in $S(\mathfrak{m}/\mathfrak{n})[\rho_2]$. Hence $V(I) \supset \{x^2y + ty^5, xy^2 - tx^5\}$ for some t, whence $t\{x^6, -y^6\} + \mathfrak{n} \subset S_1 V(I) + \mathfrak{n} \subset I$ because $x^2y^2 \in \mathfrak{n}$. If $t \neq 0$, this implies that $\{x^6, -y^6\} = V_6(\rho_1)$ is not part of generators of I, which contradicts the assumption $W = V_6(\rho_1)$. It follows that $t = 0$, and $V_3(\rho_2)$ is contained in I. Let $I(W) = I(W_\infty) = I\big(V_6(\rho_1)\big) := V_6(\rho_1)S + V_3(\rho_2)S + \mathfrak{n}$. Then $I(W_\infty) \subset I$. As we saw above, $I(W_\infty) \in \mathrm{Hilb}^G(\mathbb{A}^2)$ and $I(W_\infty)/\mathfrak{n} \simeq \mathbb{C}[G]$. Hence $I(W_\infty)/\mathfrak{n} \simeq I/\mathfrak{n}$, whence $I = I(W_\infty)$.

Thus we have proved that if $S/I \simeq \mathbb{C}[G]$ and $V(I) \supset \rho_1$, then $I = I(W)$ for some nonzero G-submodule $W \subset V_2(\rho_1) \oplus V_6(\rho_1)$.

4.3.6. To summarize the above, we define for a nonzero G-submodule W of $V_2(\rho_1) + V_6(\rho_1)$,

$$I_1(W) = \begin{cases} V_6(\rho_1)S + V_3(\rho_2)S + \mathfrak{n} & \text{if } W = V_6(\rho_1), \\ SW + \mathfrak{n} & \text{otherwise,} \end{cases}$$

$$= W + \sum_{k=1}^{5} S_k V_2(\rho_1) + \mathfrak{n}.$$

The above classification of I shows that

$$E(\rho_1) = \{I_1(W); \rho_1 \simeq W \subset V_2(\rho_1) \oplus V_6(\rho_1)\},$$
$$P(\rho_1, \rho_2) = \{I_1(V_6(\rho_1))\} = \{V_6(\rho_1)S + V_3(\rho_2)S + \mathfrak{n}\}.$$

We note that $E(\rho_1)$ is identified with the set of all nonzero G-submodules of $V_2(\rho_1) \oplus V_6(\rho_1)$, which is (at least set-theoretically) isomorphic to $\mathbb{P}^1$ by Schur's lemma. Moreover

$$\lim_{W \to V_6(\rho_1)} I_1(W) = I_1(V_6(\rho_1)) = V_6(\rho_1)S + V_3(\rho_2)S + \mathfrak{n}.$$

The family $I_1(W)$, $W \in \mathbb{P}(V_2(\rho_1) \oplus V_6(\rho_1)) \simeq \mathbb{P}^1$ is flat over $\mathbb{P}^1$ because $\mathcal{I}$ with $\mathcal{I}_0 = I_1(V_6(\rho_1))$ is $\mathbb{C}[\![s,t]\!]$-flat (See Section 4.4). Hence we have an injective morphism $\phi : \mathbb{P}^1 \to \mathrm{Hilb}^G(\mathbb{A}^2)$ with injective homomorphism of tangent spaces, whence $E(\rho_1)$, the image of ϕ, is a smooth rational curve. Since $\mathrm{Hilb}^G(\mathbb{A}^2)$ is a smooth surface with trivial dualizing sheaf by Theorem 2.5, this proves that the self-intersection of $E(\rho_1)$ is -2.

4.4. The subset $E(\rho_2)$.

4.4.1. First we consider the tangent space $T_{[I]}(\mathrm{Hilb}^G(\mathbb{A}^2))$ at $I := I_1(V_6(\rho_1))$. Recall $V(I) = V_6(\rho_1) + V_3(\rho_2)$. Since $(S/I)[\rho_1] = V_2(\rho_1)$ and $(S/I)[\rho_2] = V_1(\rho_2) \oplus V_5(\rho_2)$, the tangent space $T_{[I]}(\mathrm{Hilb}^G(\mathbb{A}^2))$ at I is spanned as before by the elements ψ_1 and ψ_2 :

$$\psi_1(x^6 - y^6) = xy, \quad \psi_1(V_3(\rho_2)) = 0,$$
$$\psi_2(V_6(\rho_1)) = 0, \quad \psi_2(x^2y) = y^5, \quad \psi_2(-xy^2) = -x^5$$

Because if $\psi_2(x^2y) = ax + by^5$ and $\psi_2(-xy^2) = ay - bx^5$ for nonzero a, then deformations by ϕ_2 of I contain S, a contradiction. Hence $a = 0$.

We define $\mathcal{I}$ to be the ideal of $\mathbb{C}[\![s,t]\!][x,y]$ generated by the elements

$$B := x^6 - y^6 + sxy,$$
$$C_1 := x^2y + ty^5 - \frac{st}{2}x, \quad C_2 := -xy^2 - tx^5 - \frac{st}{2}y,$$
$$2A_4 + tA_6, A_6 - \phi_2, \quad A_8 + sA_4,$$

where $\phi_2 \in \mathbb{C}[\![s,t]\!]$ is the unique power series with initial term $-\frac{1}{2}s^2t$ satisfying the equation

$$\phi_2 + \frac{s^2t}{2} + \frac{t^3}{2}\phi_2^2 = 0.$$

We note $2A_4 + tA_6 = yC_1 - xC_2$, $A_8 + sA_4 = xyB$ and $A_8 - \frac{st}{2}A_6 = x^5C_1 + y^5C_2$. The ideal $\mathcal{I}$ is therefore generated by the elements

$$(1) \qquad B, \quad C_1, \quad C_2, \quad A_4 + \frac{t}{2}\phi_2, \quad A_6 - \phi_2, \quad A_8 - \frac{st}{2}\phi_2.$$

One can show that $\mathcal{I}$ defines a $\mathbb{C}[\![s,t]\!]$-flat family of subschemes, which is a local universal deformation of $\mathrm{Spec}(S/I)$ or simply I.

4.4.2. One can read from this that $s = 0$ yields a new deformation of subschemes. This leads us to the following definitions.

For W, any nonzero G-submodule of $V(\rho_2) := V_3(\rho_2) \oplus V_5(\rho_2)$, we define

$$I_2(W) = \begin{cases} V_6(\rho_1)S + V_3(\rho_2)S + \mathfrak{n} & \text{if } W = V_3(\rho_2), \\ V_5(\rho_2)S + S_1V_3(\rho_2)S + \mathfrak{n} & \text{if } W = V_5(\rho_2), \\ SW + \mathfrak{n} & \text{otherwise.} \end{cases}$$

Suppose W to be any nonzero G-submodule of $V(\rho_2)$ such that $W \neq V_3(\rho_2)$ and $W \neq V_5(\rho_2)$. Then we see $S_3W + S_4W + \mathfrak{n} = V_6(\rho_3) + V_7(\rho_2) + \mathfrak{n}$ as $S_8 \subset \mathfrak{n}$. Since $W \neq V_3(\rho_2)$ and $W \neq V_5(\rho_2)$, we infer

$$\begin{aligned} SW + \mathfrak{n} &= W + S_1W + S_2W + V_6(\rho_3) + V_7(\rho_2) + \mathfrak{n} \\ &= W + S_1V_3(\rho_2) + V_5(\rho_4) + V_5(\rho_5) + V_6(\rho_1) + V_6(\rho_3) + V_7(\rho_2) + \mathfrak{n}, \end{aligned}$$

whence $I_2(W)/\mathfrak{n} \simeq \mathbb{C}[G]$. This implies that $I_2(W) \in \mathrm{Hilb}^G(\mathbb{A}^2)$. In the same manner as above one can prove that $I_2(W)$ belongs to $\mathrm{Hilb}^G(\mathbb{A}^2)$ for $W = V_3(\rho_2)$ or $W = V_5(\rho_2)$. We also see that

$$\begin{aligned} I_2(W) &= W + S_1W + S_2W + V_6(\rho_3) + V_7(\rho_2) + \mathfrak{n} \\ &= W + S_1V_3(\rho_2) + V_5(\rho_4) + V_5(\rho_5) + V_6(\rho_1) + V_6(\rho_3) + V_7(\rho_2) + \mathfrak{n} \end{aligned}$$

for any nonzero G-submodule W of $V(\rho_2)$.

4.4.3. Moreover in the same manner as before one can check that if I satisfies $S/I \simeq \mathbb{C}[G]$ and $V(I)$ contains ρ_2, then $I = I_2(W)$ for some nonzero G-submodule W of $V_3(\rho_2) \oplus V_5(\rho_2)$.

This proves

$$\begin{aligned} E(\rho_2) &= \{I_2(W); \rho_2 \simeq W \subset V_3(\rho_2) \oplus V_5(\rho_2)\} \simeq \mathbb{P}^1, \\ P(\rho_1, \rho_2) &= \{I_2(V_3(\rho_2))\} = \{V_6(\rho_1)S + V_3(\rho_2)S + \mathfrak{n}\}, \\ P(\rho_2, \rho_3) &= \{I_2(V_5(\rho_2))\} = \{V_5(\rho_2)S + S_1V_3(\rho_2)S + \mathfrak{n}\}. \end{aligned}$$

The subset $E(\rho_2)$ is proved as before to be a smooth rational curve with self-intersection -2. We also see that

$$\begin{aligned} I_2(V_3(\rho_2)) &= \lim_{W \to V_3(\rho_2)} I_2(W) = V_3(\rho_2)S + V_6(\rho_1)S + \mathfrak{n}, \\ I_2(V_5(\rho_2)) &= \lim_{W \to V_5(\rho_2)} I_2(W) = V_5(\rho_2)S + S_1V_3(\rho_2)S + \mathfrak{n}. \end{aligned}$$

4.4.4. Now we focus on the intersection point $P(\rho_1, \rho_2)$ of two rational curves $E(\rho_1)$ and $E(\rho_2)$. The computation of limits shows that

$$\begin{aligned} I_1(V_6(\rho_1)) &= \lim_{W \to V_6(\rho_1)} I_1(W) = \{V_6(\rho_1) + S_1V_2(\rho_1)\}S + \mathfrak{n} \\ &= \{V_6(\rho_1) + V_3(\rho_2)\}S + \mathfrak{n}, \\ I_2(V_3(\rho_2)) &= \lim_{W \to V_3(\rho_2)} I_2(W) = \{V_3(\rho_2)S + S_1V_5(\rho_2)\}S + \mathfrak{n} \\ &= \{V_3(\rho_2) + V_6(\rho_1)\}S + \mathfrak{n}. \end{aligned}$$

The reason why $I_1(V_6(\rho_1)) = I_2(V_3(\rho_2))$ holds true is just the fact

$$S_1V_2(\rho_1) = V_3(\rho_2), S_1V_5(\rho_2) = V_6(\rho_1) \mod \mathfrak{n},$$

where $V_2(\rho_1)$ is the dual partner of $V_6(\rho_1)$, while $V_3(\rho_2)$ is the dual partner of $V_5(\rho_2)$. Since $S_1 \simeq \rho_{\mathrm{nat}}$, the natural representation of G, this is part of the McKay rule of irreducible decompositions by tensoring with ρ_{nat}, relevant to ρ_1 and ρ_2: $\rho_1\rho_{\mathrm{nat}} = \rho_2$, $\rho_2\rho_{\mathrm{nat}} = \rho_1 + \cdots$.

4.5. The subset $E(\rho_3)$. The subset $E(\rho_3)$ is computed in the same manner as before. Let W be a nonzero irreducible G-submodule of $V_4(\rho_3)$. Then we define in the same manner as before

$$\begin{aligned} I_3(W) &= \begin{cases} \big(S_1V_3(\rho_2)[\rho_3] + V_5(\rho_2)\big)S + \mathfrak{n} & \text{if } W = S_1V_3(\rho_2)[\rho_3],\\ \big(S_1V_3(\rho_4) + V_5(\rho_4)\big)S + \mathfrak{n} & \text{if } W = S_1V_3(\rho_4),\\ \big(S_1V_3(\rho_5) + V_5(\rho_5)\big)S + \mathfrak{n} & \text{if } W = S_1V_3(\rho_5),\\ SW + \mathfrak{n} & \text{otherwise,} \end{cases}\\ &= W + \sum_{k=5}^{7} S_k(\mathfrak{m}/\mathfrak{n}) + \mathfrak{n} \quad \text{for any } W. \end{aligned}$$

Then one can prove in the same manner as before

$$\begin{aligned} E(\rho_3) &= \{I_3(W); \rho_3 \simeq W \subset V_4(\rho_3)\},\\ P(\rho_2,\rho_3) &= \big\{I_3\big(S_1V_3(\rho_2)\big)\big\} = \{S_1V_3(\rho_2)S + V_5(\rho_2)S + \mathfrak{n}\},\\ P(\rho_3,\rho_4) &= \big\{I_3\big(S_1V_3(\rho_4)\big)\big\} = \{S_1V_3(\rho_4)S + V_5(\rho_4)S + \mathfrak{n}\},\\ P(\rho_3,\rho_5) &= \big\{I_3\big(S_1V_3(\rho_5)\big)\big\} = \{S_1V_3(\rho_5)S + V_5(\rho_5)S + \mathfrak{n}\}. \end{aligned}$$

We have also similar formulas of limits

$$\begin{aligned} P(\rho_2,\rho_3) &= I_3(S_1V_3(\rho_2)[\rho_3]) = S_1V_3(\rho_2) + \sum_{k=5}^{7} S_k + \mathfrak{n}\\ &= I_2\big(V_5(\rho_2)\big) = V_5(\rho_2)S + S_1V_3(\rho_2)S + \mathfrak{n},\\ P(\rho_3,\rho_4) &= I_3\big(S_1V_3(\rho_4)\big) = \lim_{W\to S_1V_3(\rho_4)} I_3(W),\\ P(\rho_3,\rho_5) &= I_3\big(S_1V_3(\rho_5)\big) = \lim_{W\to S_1V_3(\rho_5)} I_3(W). \end{aligned}$$

The first formula is true because $SS_1W + \mathfrak{n} = \sum_{k=5}^{7} S_k + \mathfrak{n}$ for any general irreducible G-submodule W of $V_4(\rho_3)$. Hence, the reason why $I_2\big(V_5(\rho_2)\big) = I_3\big(S_1V_3(\rho_2)\big)$ holds true is just the fact

$$\begin{aligned} S_1V_3(\rho_2) &= \{x^3y, -xy^3\} + \{x^2y^2\} = \{x^3y, -xy^3\} \simeq \rho_3 \bmod \mathfrak{n},\\ S_1\{x^4, y^4\}_{\rho_3} &= V_5(\rho_2) + V_5(\rho_4) + V_5(\rho_5) \simeq \rho_2 + \rho_4 + \rho_5, \end{aligned}$$

where $V_3(\rho_2)$ is the dual partner of $V_5(\rho_2)$, while $\{x^3y, -xy^3\}_{\rho_3}$ is the dual partner of $\{x^4, y^4\}_{\rho_3}$ with respect to the natural pairing (See [6, Lemma 11.5]). This reminds us of the McKay rule for representations explained in Section 1.4.

4.6. The subsets $E(\rho_4)$ and $E(\rho_5)$. Since ρ_4 and ρ_5 are one-dimensional, we can argue in the same manner as ρ_1. Then we define

$$\begin{aligned} I_4(W) &= \begin{cases} SW + S_1V_3(\rho_4)S + \mathfrak{n} & \text{if } W = V_5(\rho_4),\\ SW + \mathfrak{n} & \text{otherwise,} \end{cases}\\ &= W + S_1V_3(\rho_4) + V_5(\rho_2) + V_5(\rho_5) + S_6 + S_7 + \mathfrak{n} \end{aligned}$$

for any nonzero G-submodule W of $V(\rho_4) := V_3(\rho_4) + V_5(\rho_4)$, and

$$I_5(W) = \begin{cases} SW + S_1V_3(\rho_5)S + \mathfrak{n} & \text{if } W = V_5(\rho_5), \\ SW + \mathfrak{n} & \text{otherwise,} \end{cases}$$
$$= W + S_1V_3(\rho_5) + V_5(\rho_2) + V_5(\rho_4) + S_6 + S_7 + \mathfrak{n}$$

for any nonzero G-submodule W of $V(\rho_5) := V_3(\rho_5) + V_5(\rho_5)$. In the same manner as before we see

$$E(\rho_4) = \{I_4(W); \rho_4 \simeq W \subset V(\rho_4)\} \simeq \mathbb{P}^1,$$
$$E(\rho_5) = \{I_5(W); \rho_5 \simeq W \subset V(\rho_5)\} \simeq \mathbb{P}^1,$$
$$P(\rho_3, \rho_4) = \{I_4\big(V_5(\rho_4)\big)\}, \quad P(\rho_3, \rho_5) = \{I_5\big(V_5(\rho_5)\big)\}.$$

By Section 4.5

$$P(\rho_3, \rho_4) = I_4\big(V_5(\rho_4)\big) = I_3\big(S_1V_3(\rho_4)\big),$$
$$P(\rho_3, \rho_5) = I_5\big(V_5(\rho_5)\big) = I_3\big(S_1V_3(\rho_5)\big).$$

These formulas come from the relations

$$S_1V_3(\rho_4) \simeq \rho_3, \quad S_1V_3(\rho_5) \simeq \rho_3 \bmod \mathfrak{n},$$
$$S_1\{x^4, y^4\} = V_5(\rho_2) + V_5(\rho_4) + V_5(\rho_5),$$

which reminds us of the McKay rule for representations in Section 1.4.

4.7. Versal deformations.

4.7.1. Let $I = I_2\big(V_5(\rho_2)\big) = I_3\big(S_1V_3(\rho_2)\big) = V_5(\rho_2)S + S_1V_3(\rho_2)S + \mathfrak{n}$. Let $R = \mathbb{C}[\![s,t]\!][x,y]$, and $\lambda = s^2t/(1+t^2)$. We define a deformation $\mathcal{I}_3$ of I as the ideal of R generated by the elements

$$y^5 + sx^2y + \lambda x, \quad -x^5 - sxy^2 + \lambda y, \quad x^3y + ty^4 + stx^2, \quad -xy^3 + tx^4 + sty^2,$$
$$A_6 + 2sA_4, \quad A_8 - 2\lambda A_4, \quad A_8 + stA_6 + 2tA_4^2, \quad 2A_4^2 - tA_8, \quad A_4 - t\lambda.$$

It turns out that $\mathcal{I}_3$ is generated by the elements

$$y^5 + sx^2y + \lambda x, \quad -x^5 - sxy^2 + \lambda y, \quad x^3y + ty^4 + stx^2, \quad -xy^3 + tx^4 + sty^2,$$
$$A_6 + 2st\lambda, \quad A_8 - 2t\lambda^2, \quad A_4 - t\lambda,$$

which gives a $\mathbb{C}[\![s,t]\!]$-flat deformation of the subscheme defined by I.

4.7.2. Next let $I = I_3\big(S_1V_3(\rho_4)\big) = I_4\big(V_5(\rho_4)\big) = V_5(\rho_4)S + S_1V_3(\rho_4) + \mathfrak{n}$. Then we define a $\mathbb{C}[\![s,t]\!]$-(co)flat deformation $\mathcal{I}_4$ of I as the ideal of R generated by the elements

$$y^4 - \mathrm{i}x^3y + sy^4 + \mathrm{i}stx^2, \quad x^4 + \mathrm{i}xy^3 + sx^4 + \mathrm{i}sty^2, \quad x^4y - \mathrm{i}xy^4 + t(x^3 + \mathrm{i}y^3),$$
$$(1+s)A_8 + 2\mathrm{i}A_4^2, \quad (1+s)A_6 + 2\mathrm{i}stA_4,$$
$$A_8 - 2\mathrm{i}A_4^2 + tA_6, \quad (2+2s)A_4^2 - \mathrm{i}A_8 + \mathrm{i}stA_6.$$

It turns out that the ideal $\mathcal{I}_4$ is generated by the elements

$$y^4 - \mathrm{i}x^3y + sy^4 + \mathrm{i}stx^2, \quad x^4 + \mathrm{i}xy^3 + sx^4 + \mathrm{i}sty^2, \quad x^4y - \mathrm{i}xy^4 + t(x^3 + \mathrm{i}y^3),$$
$$A_4 + \frac{st^2}{2+s}, \quad A_6 - \frac{2\mathrm{i}s^2t^3}{(1+s)(2+s)}, \quad A_8 + \frac{2\mathrm{i}s^2t^4}{(1+s)(2+s)^2},$$

which gives a $\mathbb{C}[\![s,t]\!]$-flat deformation of the subscheme defined by I.

4.7.3. Let $I = I_3(S_1V_3(\rho_5)) = I_4(V_5(\rho_5)) = V_5(\rho_5)S + S_1V_3(\rho_5) + \mathfrak{n}$. We can construct a versal deformation $\mathcal{I}_5$ of I in the same manner as ρ_4. In fact, we define $\mathcal{I}_5$ by replacing i in the definition of $\mathcal{I}_4$ by $-$i. Then $\mathcal{I}_5$ gives a $\mathbb{C}[\![s,t]\!]$-flat family of deformations of the subscheme defined by I.

4.8. The subquivers of the McKay quiver.

4.8.1. Now we give a simple algorithm for describing $E(\rho)$ for $\rho \in \operatorname{Irr} G$. First we note that the very positive part $S^\dagger$ is contained in any $I \in E$, the exceptional divisor. It remains to describe precisely $I + S^\dagger/S^\dagger \subset S_{\mathrm{McKay}}(\mathfrak{m}/\mathfrak{n})$. This is done by using the subquivers of the McKay quiver as follows.

Now we take ρ_1 as an example. The curve $E(\rho_1)$ consists of all $I_1(W)$, $W \in \mathbb{P}(V_2(\rho_1) \oplus V_6(\rho_1))$, where

$$\begin{aligned} I_1(W) &= W + \sum_{k=1}^{5} S_k V_2(\rho_1) \\ &= W + V_3(\rho_2) + S_1V_3(\rho_2) + V_5(\rho_4) + V_5(\rho_5) + S^\dagger. \end{aligned}$$

We define $\mathrm{Quiv}(\rho_1)$, a subquiver of $S_{\mathrm{McKay}}(\mathfrak{m}/\mathfrak{n})$ associated with ρ_1, to be the sum of all G-submodules V of $S_{\mathrm{McKay}}(\mathfrak{m}/\mathfrak{n})$ in Figure 7 equipped with arrows (= quiver structure) of $S_{\mathrm{McKay}}(\mathfrak{m}/\mathfrak{n})$.

We denote it simply as in Figure 8.

It is easy to recover the original $\mathrm{Quiv}(\rho_1)$ from Figure 8 because between the subspaces of $\mathrm{Quiv}(\rho_1)$ corresponding to the vertices there are no nonzero arrows of $S_{\mathrm{McKay}}(\mathfrak{m}/\mathfrak{n})$ with reverse directions. For instance, in $\mathrm{Quiv}(\rho_1)$ the arrow from $V_2(\rho_1) \oplus V_6(\rho_1)$ to $V_3(\rho_2)$ is precisely the disjoint union of the arrow from $V_2(\rho_1)$ to $V_3(\rho_2)$ and the reverse arrow from $\{0\}$ ($\subset V_5(\rho_2)$) to $V_6(\rho_1)$, or equivalently, just the arrow from $V_2(\rho_1)$ to $V_3(\rho_2)$.

The G-submodule $(I_1(W)/\mathfrak{n} + S^\dagger) \cap S_{\mathrm{McKay}}(\mathfrak{m}/\mathfrak{n})$ is the sum of W and all G-submodules of $\mathrm{Quiv}(\rho_1)$ inequivalent to ρ_1. Therefore any $I_1(W) \in E(\rho_1)$ is given

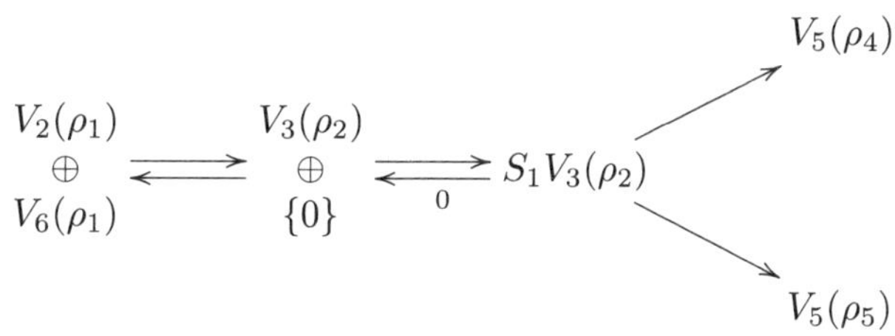

FIGURE 7. The original $\mathrm{Quiv}(\rho_1)$

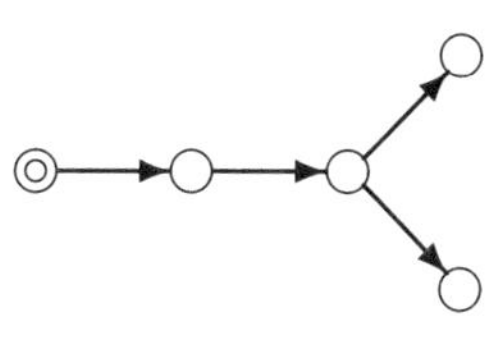

FIGURE 8. $\mathrm{Quiv}(\rho_1)$

by

$$I_1(W) = W + \sum_{\substack{V \subset \mathrm{Quiv}(\rho_1) \\ V \not\simeq \rho_1}} V + S^\dagger \quad \text{for } W \in \mathbb{P}(V_2(\rho_1) \oplus V_6(\rho_1)),$$

where the summation ranges over all irreducible G-submodules V of $\mathrm{Quiv}(\rho_1)$ inequivalent to ρ_1.

4.8.2. This is generalized in the other cases in the obvious manner. The subquivers $\mathrm{Quiv}(\rho_i)$ $(i = 2, 3, 4, 5)$ are given in Figure 9.

Let $\rho = \rho_i$. Any $I(W) \in E(\rho)$ is given by

$$I(W) = W + \sum_{\substack{V \subset \mathrm{Quiv}(\rho) \\ V \not\simeq \rho}} V + S^\dagger \quad \text{for} \quad W \in \mathbb{P}(V_{h-d(\rho)}(\rho) + V_{h+d(\rho)}(\rho)),$$

where the summation ranges over all irreducible G-submodules V of $\mathrm{Quiv}(\rho)$ inequivalent to ρ. The first diagram in Figure 9 is denoted $\mathrm{Quiv}(\rho_2)$, which means the subquiver diagram in Figure 10. We note that the quiver S_{McKay} restricted to $S_1V_3(\rho_2)$ automatically reduces to zero to the direction of $V_5(\rho_2)$.

With these diagrams, the point $P(\rho_1, \rho_2)$ is understood as the first diagram of Figure 11 which can be embedded into both of $\mathrm{Quiv}(\rho_1)$ and $\mathrm{Quiv}(\rho_2)$, to which it may be natural to assign the second graph in Figure 11. Thus the four intersection points $P(\rho, \rho')$ are described by the subquivers in Figure 12.

4.9. The extended McKay quiver revisited. The subquiver $\mathrm{Quiv}(\rho_1)$ extends itself in the extended McKay quiver by adding an arrow from $V_3(\rho_2)$ to $V_4(\rho_0)$ as Figure 13 indicates.

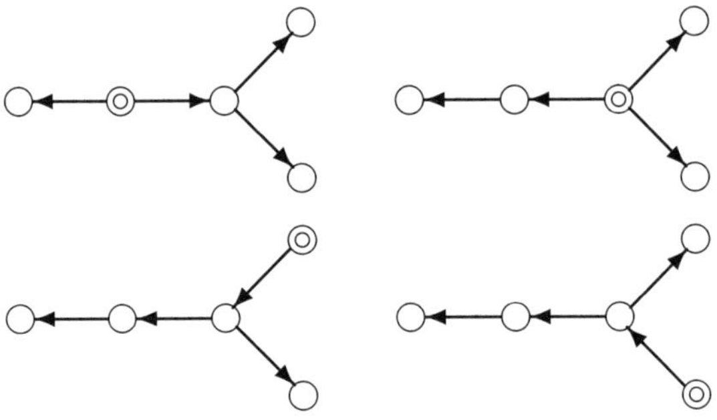

FIGURE 9. $\mathrm{Quiv}(\rho)$ for D_5

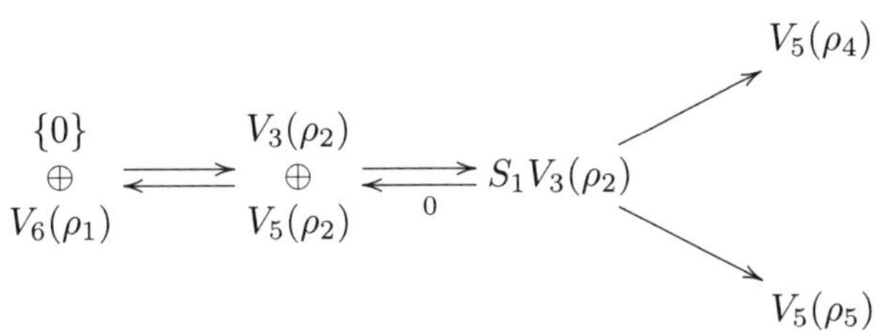

FIGURE 10. The original $\mathrm{Quiv}(\rho_2)$

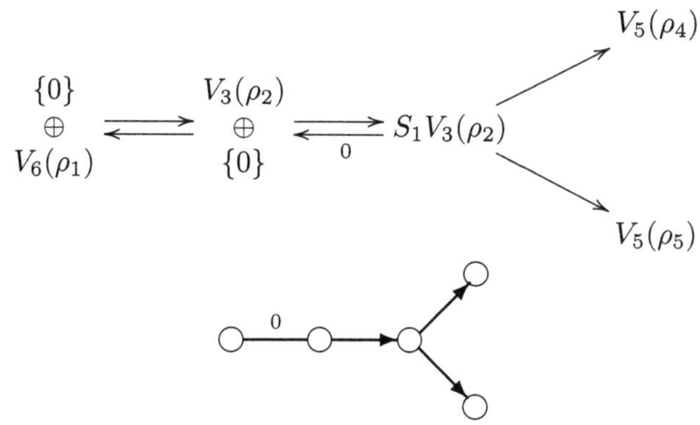

FIGURE 11. Quiv(ρ_1, ρ_2)

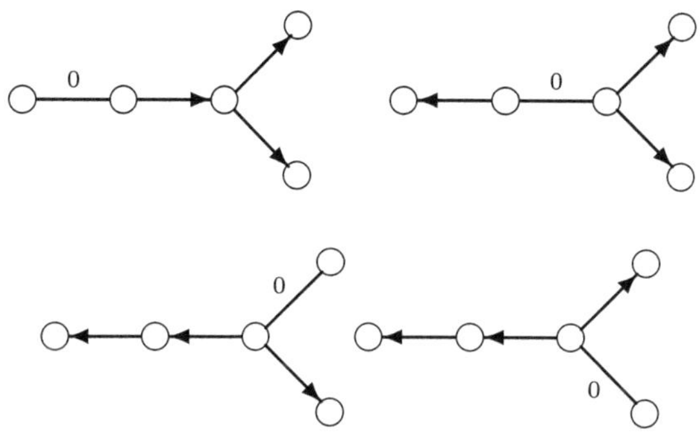

FIGURE 12. Quiv(ρ, ρ') for D_5

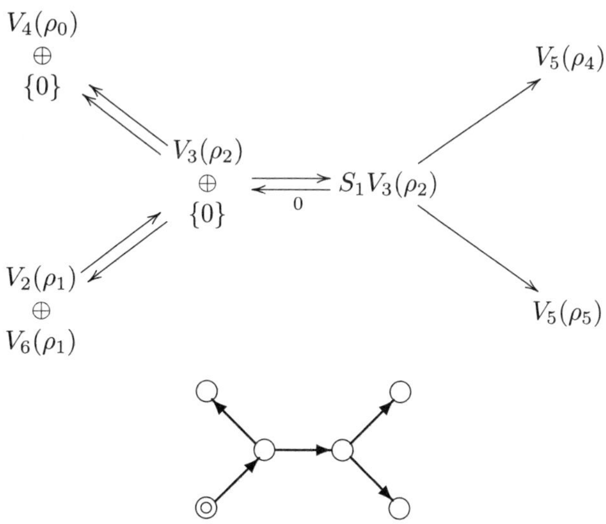

FIGURE 13. The extended Quiv(ρ_1)

This implies the following. By Table 4, $V_3(\rho_2) = S_1V_2(\rho_1)$ and $V_4(\rho_0) \subset S_1V_3(\rho_2)$. The subquiver $\mathrm{Quiv}(\rho_1)$ does not generate $V_6(\rho_0)$ in $I_1(W)$, which explains why $V_6(\rho_0)$ is always included as a generator of $I_1(W)$. This is true over a Zariski open subset of $\mathrm{Hilb}^G(\mathbb{A}^2)$ containing $\mathrm{Spec}\,\mathbb{C}[s]$ $(\subset E(\rho_1))$ in Section 4.3.

Similarly, the subquiver $\mathrm{Quiv}(\rho_2)$ extends itself by adding an arrow from $V_3(\rho_2) \oplus V_5(\rho_2)$ to $V_4(\rho_0) \oplus V_6(\rho_0)$ as well. This means that the image of the arrow spans a one-dimensional subspace of $V_4(\rho_0) \oplus V_6(\rho_0)$, whose one-dimensional complement is necessary as a part of the generators of $I_2(W)$.

In the other cases, the subquiver $\mathrm{Quiv}(\rho_k)$ $(k = 3, 4, 5)$ extends itself by adding an arrow from $V_5(\rho_2)$ to $V_6(\rho_0)$ as well. This shows that $V_4(\rho_0)$ is always necessary as a part of the generators of $I_k(W)$.

5. The simple singularity $\mathbf{E}_6$

5.1. The binary tetrahedral group $\mathbb{T}$. The simple singularity E_6 is the quotient singularity of $\mathbb{A}^2$ by the binary tetrahedral group $G := \mathbb{T}$, which is the subgroup of $\mathrm{SL}(2, \mathbb{C})$ of order 24 generated by $\mathbb{D}_2 = \langle \sigma, \tau \rangle$ and μ:

$$\sigma = \begin{pmatrix} \mathrm{i}, & 0 \\ 0, & -\mathrm{i} \end{pmatrix}, \quad \tau = \begin{pmatrix} 0, & 1 \\ -1, & 0 \end{pmatrix}, \quad \mu = \frac{1}{\sqrt{2}} \begin{pmatrix} \varepsilon^7, & \varepsilon^7 \\ \varepsilon^5, & \varepsilon \end{pmatrix},$$

where $\varepsilon = e^{2\pi \mathrm{i}/8}$ [11, p. 74]. The group G acts on $\mathbb{A}^2$ from the right by $(x, y) \mapsto (x, y)g$ for $g \in G$ and $\mathbb{D}_2$ is a normal subgroup of G with the following exact sequence:

$$1 \to \mathbb{D}_2 \to G \to \mathbb{Z}/3\mathbb{Z} \to 1.$$

See Table 6 for the character table of G [10] where $h = 12$ and $\omega = (-1 + \sqrt{3}\mathrm{i})/2$.

5.2. Symmetric tensors modulo $\mathfrak{n}$. Let S_m be the space of homogeneous polynomials in x and y of degree m. The G-modules S_m and $S_m(S/\mathfrak{n})$ via ρ_{nat} decompose into irreducible G-submodules. We define a G-submodule of $\mathfrak{m}/\mathfrak{n}$ by $V_i(\rho_j) := S_i(\mathfrak{m}/\mathfrak{n})[\rho_j]$ the sum of all copies of ρ in $S_i(\mathfrak{m}/\mathfrak{n})$, which we always choose as a homogeneous G-submodule of S_i. For a G-module W we define $W[\rho]$ to be the sum of all the copies of ρ in W. It is known by [7, p. 51] that there are G-invariant polynomials A_6, A_8, A_6^2 and A_{12} respectively of degrees 6, 8, 12 and 12. We may assume that $A_6 = T$, $A_8 = W$ and $A_{12} = U := \varphi^3 + \psi^3$. Let $V_6(\rho_0) = S_6[\rho_0] = \{A_6\}$ and $V_8(\rho_0) = S_8[\rho_0] = \{A_8\}$, where $U^2 = 4W^3 - 27T^4$. See [6, Section 14.3] for the notation. We caution that A_{12} in [6] is different from the present one.

TABLE 6. Character table of $\mathbb{T}$

ρ	1	2	3	4	5	6	7	d	$(h/2 \pm d)$
	1	-1	τ	μ	μ^2	μ^4	μ^5		
$(\sharp)$	1	1	6	4	4	4	4		
ρ_0	1	1	1	1	1	1	1	(2)	–
ρ_2	2	-2	0	1	-1	-1	1	1	$(5,7)$
ρ_3	3	3	-1	0	0	0	0	0	$(6,6)$
ρ_2'	2	-2	0	ω^2	$-\omega$	$-\omega^2$	ω	1	$(5,7)$
ρ_1'	1	1	1	ω^2	ω	ω^2	ω	2	$(4,8)$
ρ_2''	2	-2	0	ω	$-\omega^2$	$-\omega$	ω^2	1	$(5,7)$
ρ_1''	1	1	1	ω	ω^2	ω	ω^2	2	$(4,8)$

TABLE 7. Irreducible decompositions of S and Coinv(E_6)

k	S_k	V_k
0	ρ_0	0
1	ρ_2	ρ_2
2	ρ_3	ρ_3
3	$\rho_2' + \rho_2''$	$\rho_2' + \rho_2''$
4	$\rho_1' + \rho_1'' + \rho_3$	$(\rho_1' + \rho_1'') + \rho_3$
5	$\rho_2 + \rho_2' + \rho_2''$	$(\rho_2 + \rho_2' + \rho_2'')$
6	$[\rho_0] + 2\rho_3$	$(2\rho_3)$
7	$2\rho_2 + \rho_2' + \rho_2''$	$(\rho_2 + \rho_2' + \rho_2'')$
8	$[\rho_0] + \rho_1' + \rho_1'' + 2\rho_3$	$(\rho_1' + \rho_1'') + \rho_3$
9	$\rho_2 + 2\rho_2' + 2\rho_2''$	$\rho_2' + \rho_2''$
10	$\rho_1' + \rho_1'' + 3\rho_3$	ρ_3
11	$2\rho_2 + 2\rho_2' + 2\rho_2''$	ρ_2
12	$2\rho_0 + \rho_1' + \rho_1'' + 3\rho_3$	0

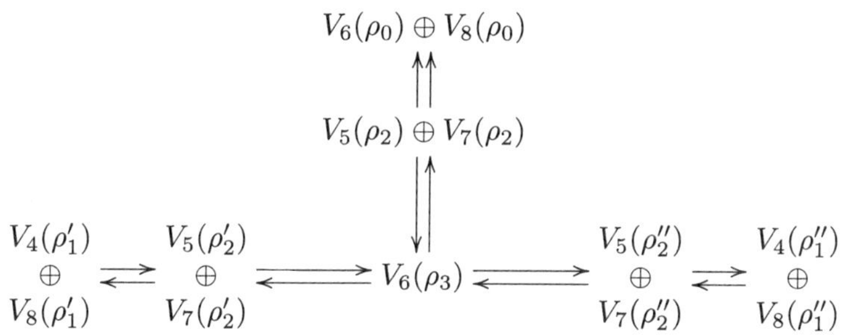

FIGURE 14. The extended McKay quiver of E_6

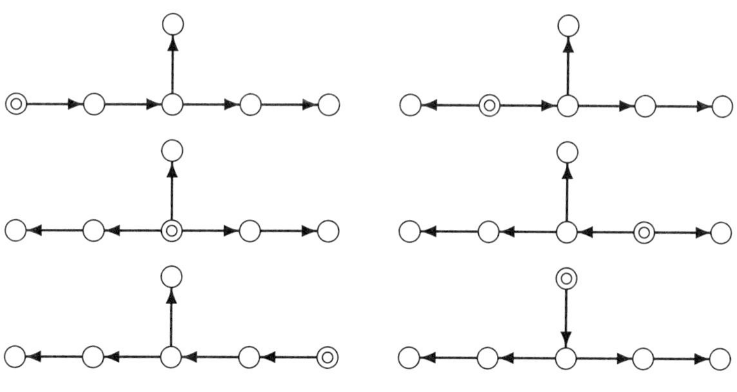

FIGURE 15. Quiv(ρ) for E_6

The decomposition of S_k and $S_k(\mathfrak{m}/\mathfrak{n})$ for small values of k are given in Table 7. The factors of $S_k(\mathfrak{m}/\mathfrak{n})$ in the parentheses are those in S_{McKay}. We see by Table 7 that $V_{6\pm d(\rho)}(\rho) \simeq \rho^{\oplus 2}$ if $d(\rho) = 0$, or ρ if $d(\rho) \geq 1$. We also see that $S_{6-k}(\mathfrak{m}/\mathfrak{n}) \simeq S_{6+k}(\mathfrak{m}/\mathfrak{n})$ for any k. Thus Lemma 3.7 for E_6 follows from Table 7 immediately.

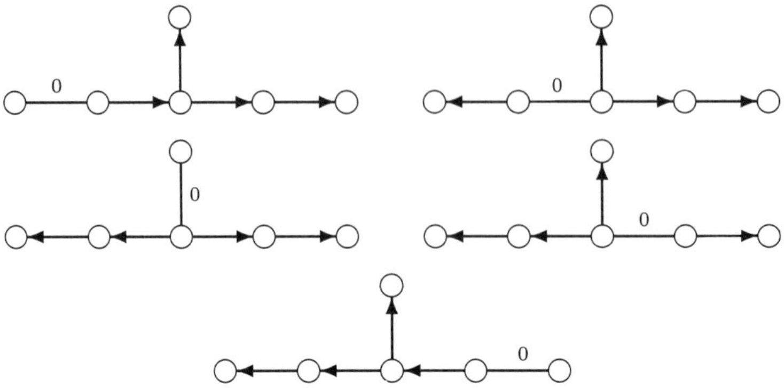

FIGURE 16. Quiv(ρ, ρ') for E_6

The (extended) McKay quiver and subquivers of E_6 are given in Figures 14-16. See also Figure 4.

6. Proof of Theorem 3.9

Now we prove Theorem 3.9 mainly for D_5.

6.1. The sheaf $\mathcal{V}$. Let Z_{univ} be the universal subscheme of $\mathbb{A}^2$ of G-orbits parametrized by $\mathrm{Hilb}^G(\mathbb{A}^2)$, and $X = \mathrm{Hilb}^G(\mathbb{A}^2)$. The natural morphism $\pi : X \to \mathbb{A}^2/G$ is known by Theorem 2.5 to be the minimal resolution. We define I_{univ} to be the ideal sheaf of $O_{\mathbb{A}^2\times X}$ defining Z_{univ} in $\mathbb{A}^2 \times X$. Then we have an exact sequence $0 \to I_{\mathrm{univ}} \to O_{\mathbb{A}^2\times X} \to O_{Z_{\mathrm{univ}}} \to 0$.

We define I_X to be the defining ideal of X as a subscheme of $X \simeq (\mathbb{A}^2/G)\times_{(\mathbb{A}^2/G)} X$ of $(\mathbb{A}^2/G)\times X$, $\mathfrak{n}_X := I_X O_{\mathbb{A}^2\times X}$ and

$$V(I_{\mathrm{univ}}) := I_{\mathrm{univ}}/\mathfrak{m}I_{\mathrm{univ}} + \mathfrak{n}_X,$$

which we denote by $\mathcal{V}$. The sheaf $\mathcal{V}$ is a finite $O_{\mathbb{A}^2}\otimes O_X$-module supported by $\{0\}\times E$ because $\mathfrak{m}O_{\mathbb{A}^2} = O_{\mathbb{A}^2}$ outside the origin and $\{0\}\times X \cap \mathrm{Supp}(Z_{\mathrm{univ}})_{\mathrm{red}} = \{0\}\times E$. It is clear that $\mathfrak{m}\mathcal{V} = \mathfrak{n}_X\mathcal{V} = 0$ from the definition of $\mathcal{V}$.

Let $\overline{\mathfrak{m}}$ be the maximal ideal of the unique singular point of $\mathbb{A}^2/G$. By the definition of $\mathfrak{n}_X$, $\phi^*F = \pi^*F \mod \mathfrak{n}_X$ for any $F \in \overline{\mathfrak{m}}$.

We prove next $\pi^*(\overline{\mathfrak{m}})\mathcal{V} = 0$. Let $H \in I_{\mathrm{univ}}$ and $a \in \pi^*\overline{\mathfrak{m}}$. Then there are some $F_k \in O_X$ and $A_k \in \overline{\mathfrak{m}}$ such that $a = \sum_k F_k\pi^*(A_k)$. Since $\mathfrak{n}_X \subset I_{\mathrm{univ}}$ (See Section 3.8), we have $aH \equiv \sum_k F_k\phi^*(A_k)H$ in $I_{\mathrm{univ}}/\mathfrak{n}_X$. However $\sum_k F_k\phi^*(A_k)H \in \mathfrak{m}I_{\mathrm{univ}}$, which proves $aH = 0$ in $\mathcal{V}$. It follows $\pi^*(\overline{\mathfrak{m}})\mathcal{V} = 0$. Since $\pi^*\overline{\mathfrak{m}}$ is the ideal of O_X defining the fundamental cycle E_{fund} of E, this implies that $\mathcal{V}$ is a finite $O_{E_{\mathrm{fund}}}$-module.

Since $E(\rho)$ is a subscheme of E_{fund}, $\mathcal{V}\otimes O_{E(\rho)}$ is a finite $O_{E(\rho)}$-module and we have a natural homomorphism

$$\mathcal{V} \to \sum_{\rho\in\mathrm{Irr}\,G} \mathcal{V}\otimes O_{E(\rho)}. \tag{2}$$

We prove this is an isomorphism in Sections 6.2 and 6.3.

6.2. Freeness outside Sing(E). First we prove that (2) is an isomorphism at a nonsingular point of E. Let $S = \mathbb{C}[x, y]$.

Let $I \in E(\rho) \setminus \operatorname{Sing}(E)$. Then the ideal I is generated by a nonzero irreducible G-submodule W of $V(\rho) = V_{h-d(\rho)}(\rho) + V_{h+d(\rho)}(\rho)$ and $\mathfrak{n}$ by section 4 or by [6, Theorem 10.7].

First we consider the case D_5.

6.2.1. Let $\rho = \rho_1$. Then $W \neq V_6(\rho_1)$. As $\operatorname{Hilb}^G(\mathbb{A}^2)$ is nonsingular and two-dimensional, the tangent space $T_{[I]}(\operatorname{Hilb}^G(\mathbb{A}^2))$ is exactly two-dimensional. By Section 4.3

$$\operatorname{Hom}_S(I/I^2, S/I)[\rho_0] = \operatorname{Hom}_{\mathbb{C}}(W, V_6(\rho_1)) \oplus \operatorname{Hom}_{\mathbb{C}}(V_6(\rho_0), V_0(\rho_0)).$$

We note that I is generated by W and A_6 by Section 4.9. Since $T_{[I]}(E(\rho_1)) = \operatorname{Hom}_{\mathbb{C}}(W, V_6(\rho_1))$, the parameter t of $\operatorname{Hom}_{\mathbb{C}}(V_6(\rho_0), V_0(\rho_0))$ gives a defining equation of $E(\rho_1)$.

By Section 4.3 the ideal I_{univ} of Z_{univ} is over $E(\rho_1) \setminus \operatorname{Sing}(E)$ generated by $xy + s(x^6 - y^6)$ and $A_6 + t$. Since $A_6 + t \in \mathfrak{n}_X$, the quotient $\mathcal{V}$ is $S \otimes \mathbb{C}[s,t]/t$-free of rank one, hence O_E-free of rank one.

6.2.2. Let us consider next the case $\rho = \rho_4$. In this case, $I = I_4(W)$, $W \neq V_5(\rho_4)$. Then I is generated by W and A_4 by Section 4.9. Let $a = x^3 + \mathrm{i}y^3 + s_0(x^4y - \mathrm{i}xy^4) \in W$ and take $\phi \in \operatorname{Hom}_S(I/I^2, S/I)[\rho_0]$. We may assume $\phi(a) = s(x^4y - \mathrm{i}xy^4) \in V_5(\rho_4)$. Then we see $A_8 + 2\mathrm{i}A_4^2 + s_0A_4A_6 = (x^4y + \mathrm{i}xy^4)a$. Since $A_jA_k \in I^2$, we have $\phi(A_8) = (x^4y + \mathrm{i}xy^4)\phi(a) = sA_4A_6 = 0$. Similarly, $\phi(A_6 + s_0A_8 - 2\mathrm{i}s_0A_4^2) = (x^3 - \mathrm{i}y^3)\phi(a) = s(A_8 - 2\mathrm{i}A_4^2) = 0$. Hence we have $\phi(A_6) = \phi(A_8) = 0$. It follows that letting $\phi(A_4) = t$, then s and t are local (regular) parameters of $\operatorname{Hilb}^G(\mathbb{A}^2)$ at I. Thus we see

$$\operatorname{Hom}_S(I/I^2, S/I)[\rho_0] = \operatorname{Hom}_{\mathbb{C}}(W, V_3(\rho_4)) \oplus \operatorname{Hom}_{\mathbb{C}}(V_4(\rho_0), V_0(\rho_0)).$$

Thus we have generators $A := A_4 + t$ and $B := x^3 + \mathrm{i}y^3 + sxy(x^3 - \mathrm{i}y^3)$ and of I_{univ}, whose derivation span $T_{[I]}(\operatorname{Hilb}^G(\mathbb{A}^2))$. Moreover $A_6 + sA_8 - 2\mathrm{i}sA_4^2 = (x^3 - \mathrm{i}y^3)A \in I_{\text{univ}}$, and $A_8 + 2\mathrm{i}A_4^2 + sA_4A_6 = (x^3 + \mathrm{i}y^3)xyA \in I_{\text{univ}}$. Thus I_{univ} is generated by A and B, whence over $E(\rho_4) \setminus \operatorname{Sing}(E) \simeq \operatorname{Spec}\mathbb{C}[s,t]/(t)$,

$$I_{\text{univ}} = (x^3 + \mathrm{i}y^3 + sxy(x^3 - \mathrm{i}y^3), A_4 + t).$$

Since $\pi^*(\overline{\mathfrak{m}})$ is the defining ideal of E_{fund}, we have $\pi^*(\overline{\mathfrak{m}}) = (t)$. Therefore $\mathcal{V}$ is $S \otimes \mathbb{C}[s,t]/t$-free of rank one.

6.2.3. The above arguments are easily generalized to any one-dimensional irreducible representation of D_n, E_6 and E_7 by using [6, Sections 13–15]. We note that there is no one-dimensional irreducible representation for E_8.

6.2.4. Next we consider ρ_2 in the D_5-case. Let $I = I_2(W)$, $W \neq V_3(\rho_2), V_5(\rho_2)$. Then we note that $I_6(\rho_1) \subset SW + \mathfrak{n}$ and that $V_4(\rho_0) + V_6(\rho_0)/W$ is a part of generators of I/I^2 by the property of $\operatorname{Quiv}(\rho_2)$ mentioned in Section 4.9. Thus I/I^2 is generated by $b_1 = x^2y + s_0y^5$, $b_2 = -xy^2 - s_0x^5$ and A_k. The condition $W \neq V_3(\rho_2), V_5(\rho_2)$ is just $s_0 \neq 0, \infty$.

We compute now $T_{[I]}(\operatorname{Hilb}^G(\mathbb{A}^2)) = \operatorname{Hom}_S(I/I^2, S/I)[\rho_0]$, relying on these facts. It is clear that $(S/I)[\rho_2] \simeq V_5(\rho_2) \oplus V_1(\rho_2)$. We define ψ_1 and ψ_2 to be the elements of $\operatorname{Hom}_S(I/I^2, S/I)[\rho_0]$

$$\psi_1(b_1) = y^5, \quad \psi_1(b_2) = -x^5, \quad \psi_2(b_1) = -x, \quad \psi_2(b_2) = -y.$$

We prove that ψ_1 and ψ_2 span $\mathrm{Hom}_S(I/I^2, S/I)[\rho_0]$. So we take an element ϕ of $\mathrm{Hom}_S(I/I^2, S/I)[\rho_0]$ and let $\phi(b_1) = sy^5 - tx$, $\phi(b_2) = -sx^5 - ty$.

We shall prove $\phi(A_k) = 0$ $(k = 4, 6, 8)$. We note $\phi(A_k) \in \mathbb{C} = (S/I)[\rho_0]$, a constant. First $\phi(A_8) = x^5\phi(b_1) + y^5\phi(b_2) = -tA_6 = 0$. Secondly, $\phi(2A_4 + s_0A_6) = y\phi(b_1) - x\phi(b_2) = sA_6 = 0$. Thirdly, we see

$$xy\phi(A_6) = x^5\phi(b_1) - y^5\phi(b_2) - 2s_0xy\phi(A_4^2) = -t(x^6 - y^6) = 0$$

because $(x^6 - y^6) \in V_6(\rho_1) \subset SW + \mathfrak{n} = I_2(W) = I$. Meanwhile, $xy \in V_2(\rho_1)$ which is nonzero in S/I. Hence $\phi(A_6) = 0$. Hence we have $\phi(A_k) = 0$ for all A_k. It follows that $\phi = s\psi_1 + t\psi_2$.

Since I/I^2 is generated by b_1, b_2 and A_k, this proves

$$\mathrm{Hom}_S(I/I^2, S/I)[\rho_0] = \mathrm{Hom}_{\mathbb{C}}\big(W, V_5(\rho_2) \oplus V_1(\rho_2)\big).$$

Since $T_{[I]}\big(E(\rho_2)\big) = \mathrm{Hom}_{\mathbb{C}}\big(W, V_5(\rho_2)\big)$, the parameter t of $\mathrm{Hom}_{\mathbb{C}}\big(W, V_1(\rho_2)\big)$ gives the equation of $E(\rho_2)$ locally along $E(\rho_2) \setminus \mathrm{Sing}(E)$. Now we prove that the ideal I_{univ} is generated by the elements

$$B_1 := x^2y + sy^5 - tx, \quad B_2 := -xy^2 - sx^5 - ty, \quad A := A_4 - \eta,$$

where $\eta := \pi^*A_4$ is a power series of t with initial term t^2 satisfying $s^2\eta^2 - \eta + t^2 = 0$. In fact, Noetherian property shows that I_{univ} is generated by B_1, B_2 and some elements with initial terms being A_k. We see $2A_4 + sA_6 = yB_1 - xB_2$, $sA_8 + 2tA_4 = -xy(yB_1 + xB_2)$ and $A_8 - tA_6 = x^5B_1 + y^5B_2$. Hence all of these belong to I_{univ}. Since $\pi^*(\overline{\mathfrak{m}}) \cap I_{\mathrm{univ}} = \{0\}$, we have $\pi^*(2A_4 + sA_6) = \pi^*(A_8 - tA_6) = 0$, and therefore the ideal $\pi^*\overline{\mathfrak{m}} = (\pi^*A_4, \pi^*A_6, \pi^*A_8)$ of O_X is generated by π^*A_4 because s is invertible over $E(\rho_2) \setminus \mathrm{Sing}(E)$. We note $E(\rho_2) \setminus \mathrm{Sing}(E) \simeq \mathrm{Spec}\,\mathbb{C}[s, s^{-1}]$. It is easy to infer from $A_8^2 = A_4A_6^2 - 4A_4^4$ that η satisfies $s^2\eta^2 - \eta + t^2 = 0$.

Since $-t^2 = s^2\eta^2 - \eta$ and $A_4 - \eta \in \mathfrak{n}_X$, we have in $\mathcal{V}$

$$tB_1 = tx^2y + sty^5 + (s^2A_4^2 - A_4)x = -xyB_1 - sy^4B_2 = 0,$$
$$tB_2 = xyB_2 + sx^4B_1 = 0.$$

Hence $\mathcal{V}$ is an O_E-module with B_1 and B_2 generators. By Section 4.4 that $V(I)$ is of rank two for any $I \in E(\rho_2) \setminus \mathrm{Sing}(E)$, hence for each $s \in \mathbb{C}$, $s \neq 0$. This implies that $\mathcal{V} \otimes S[s, s^{-1}, t]/(t)$ is $S[s, s^{-1}]$-free of rank two.

6.2.5. As the final case of D_5, we consider ρ_3. Let $I = I_3(W)$, $W \neq S_1V_3(\rho_k)$ $(k = 2, 4, 5)$. Then we see

$$\mathrm{Hom}_S(I/I^2, S/I)[\rho_0] = \mathrm{Hom}_{\mathbb{C}}(W, V_4(\rho_3)/W) \oplus \mathrm{Hom}_{\mathbb{C}}\big(W, V_2(\rho_3)\big).$$

The proof in this case is however rather tricky. Let $b_1 = y^4 + \mathrm{i}s_0x^3y$ and $b_2 = x^4 - \mathrm{i}s_0xy^3$. By the condition $W \neq S_1V_3(\rho_k)$, we have $s_0 \neq \pm 1, \infty$. Let $\phi \in \mathrm{Hom}_S(I/I^2, S/I)[\rho_0]$. Since $(S/I)[\rho_3] = \{x^3y, -xy^3\} \oplus V_2(\rho_3)$, we may assume $\phi(b_1) = \mathrm{i}sx^3y + tx^2$ and $\phi(b_2) = -\mathrm{i}sxy^3 + ty^2$. First we note that $\phi(b_i) \in (S_4 + S_2)(\mathfrak{m}/I)$, whence $\phi(S_2b_i) \in (S_6 + S_4)(\mathfrak{m}/I) = S_4(\mathfrak{m}/I)$. Hence we see $\phi(y^2b_1) = \mathrm{i}sx^3y^3 + tA_4 = tA_4 = 0$, $\phi(x^2b_2) = -\mathrm{i}sx^3y^3 + tA_4 = tA_4 = 0$. Similarly $\phi(x^2b_1) = \mathrm{i}sx^5y + tx^4 = tx^4$, $\phi(xyb_1) = tx^3y$, $\phi(xyb_2) = txy^3$ and $\phi(y^2b_2) = ty^4$.

Then $\phi(xy^3b_1 - x^3yb_2) = -\phi(A_8) + 2\mathrm{i}s_0\phi(A_4^2) = -\phi(A_8)$, while $\phi(xy^3b_1 - x^3yb_2) = xy\phi(y^2b_1) - xy\phi(x^2b_2) = 0$. Hence $\phi(A_8) = 0$. Similarly $\phi(A_6) = \phi(y^2b_1 + x^2b_2) = 2tA_4 = 0$. We also see that $(1 - s_0^2)\phi(x^2y^4) = \phi(x^2b_1 - \mathrm{i}s_0xyb_2) = tx^4 - \mathrm{i}s_0txy^3 = tb_2 \in W \subset I$. Hence $(1 - s_0^2)\phi(x^2y^4) = 0$ in S/I, whence $\phi(x^2y^4) = 0$ because $1 - s_0^2 \neq 0$. Similarly we see $\phi(x^4y^2) = 0$. It follows that $\{x^2, y^2\}\phi(A_4) = 0$. Hence

$\phi(A_4) = 0$ because $0 \neq \{x^2, y^2\} \subset (S/I)[\rho_2]$. This proves $\phi(A_k) = 0$ for any A_k. Hence $\mathrm{Hom}_S(I/I^2, S/I) = \mathrm{Hom}_{\mathbb{C}}(W, V_4(\rho_3)/W) \oplus \mathrm{Hom}_{\mathbb{C}}\big(W, V_2(\rho_3)\big)$.

We know by the subquiver $\mathrm{Quiv}(\rho_3)$ that I is generated by W and A_4. We have B_1, B_2 and A as generators of I_{univ} as follows :

$$B_1 = y^4 + \mathrm{i}sx^3y + tx^2, \quad B_2 = x^4 - \mathrm{i}sxy^3 + ty^2, \quad A = A_4 - \pi^*(A_4),$$

where $\pi^*(A_4) = t^2/(1-s^2)$. We note $A_6 + 2tA_4$ and $A_8 - 2\mathrm{i}sA_4^2 \in I_{\mathrm{univ}}$. Since t is the parameter of $\mathrm{Hom}_{\mathbb{C}}\big(W, V_2(\rho_3)\big)$, $E(\rho_3)$ is defined by $t = 0$.

Now we prove in $S[s,t]$

$$tB_1 = y^2B_2 + \mathrm{i}sxyB_1 - (1-s^2)x^2A,$$
$$tB_2 = x^2B_1 - \mathrm{i}sxyB_2 - (1-s^2)y^2A.$$

Hence $tB_1 = tB_2 = 0$. This shows that $\mathcal{V}$ is $\mathbb{C}[s, 1/(1-s^2)]$-free of rank two, where $E(\rho_3) \setminus \mathrm{Sing}(E) = \mathrm{Spec}\,\mathbb{C}[s, 1/(1-s^2)]$.

This completes the proof of freeness of $\mathcal{V}$ over $E \setminus \mathrm{Sing}(E)$ in the D_5-case.

6.2.6. It is clear that one can generalize the above arguments to D_n for the other n. To settle the E_6 and E_7-cases, we need to discuss three-dimensional or four-dimensional representations ρ. Since the discussion below on the three-dimensional representation ρ_3 of E_6 shows the general features of the arguments for the proof sufficiently, we take up only ρ_3 of E_6 in order to avoid the overwhelming notation for E_7.

6.2.7. We consider the E_6-case. Let G be the binary tetrahedral group $\mathbb{T}$, and ρ_3 the unique three-dimensional representation of G. Let A_6, A_8 and A_{12} be the homogeneous generators of the ring of G-invariants:

$$A_6 = T = p_1p_2p_3, \quad A_8 = W = \varphi\psi, \quad A_{12} = U = \varphi^3 + \psi^3,$$

where $U^2 = 4W^3 - 27T^4$. See [6, Section 14.3] for the notation. We note $\varphi^3 - \psi^3 = 3(2\omega+1)T^2$ and that both φ^3 and ψ^3 are G-invariants.

Then any point $I = I_3(W) \in E(\rho_3) \setminus \mathrm{Sing}(E)$ is given by an irreducible G-submodule of $V_6(\rho_3)$ with $W \neq S_1V_5(\rho_2), S_1V_5(\rho_2'), S_1V_5(\rho_2'')$ under the notation of Section 5. Then we see

$$\mathrm{Hom}_S(I/I^2, S/I) = \mathrm{Hom}_{\mathbb{C}}(W, V_6(\rho_3)/W) \oplus \mathrm{Hom}_{\mathbb{C}}\big(W, V_4(\rho_3)\big).$$

Moreover a versal deformation I_{univ} of I is generated by six elements

$$B_1, \quad B_2, \quad B_3, \quad A_6 - \pi^*(A_6), \quad A_8 - \pi^*(A_8), \quad A_{12} - \pi^*(A_{12}),$$

where with the notation of [6, Section 14.3],

$$B_1 = p_1(\varphi + \omega\psi) + sp_1\varphi + tp_2p_3 + up_1,$$
$$B_2 = \omega p_2(\varphi + \omega^2\psi) + s\omega p_2\varphi - tp_3p_1 + up_2,$$
$$B_3 = \omega^2p_3(\varphi + \psi) + s\omega^2p_3\varphi + tp_1p_2 + up_3,$$

where s, t and u are parameters. We note that

$$V_6(\rho_3) = \{p_1\varphi, \omega p_2\varphi, \omega^2p_3\varphi\} \oplus \{p_1\psi, \omega^2p_2\psi, \omega p_3\psi\},$$
$$V_4(\rho_3) = \{p_2p_3, -p_3p_1, p_1p_2\}, \quad V_2(\rho_3) = \{p_1, p_2, p_3\}.$$

Then we see $p_1B_1 - \omega^2p_2B_2 + \omega p_3B_3 = \omega(1-\omega)(\psi^2 - \omega u\varphi)$. Hence $\psi^2 - \omega u\varphi \in I_{\mathrm{univ}}$, whence $\psi^3 - \omega uW \in I_{\mathrm{univ}}$. Similarly we see $p_1B_1 - p_2B_2 + p_3B_3 = (\omega-1)sW - 3tT$, which belongs to I_{univ}. We also have $(1-\omega^2)\{(1+s)\varphi^2 - \omega u\psi\} = p_1B_1 -$

$\omega p_2B_2+\omega^2p_3B_3\in I_{\mathrm{univ}}$. Hence $(1+s)\varphi^3-\omega uW\in I_{\mathrm{univ}}$. From $\pi^*\overline{\mathfrak{m}}\cap I_{\mathrm{univ}}=\{0\}$ it follows that

$$\pi^*\psi^3=\omega u\pi^*W,\quad (1+s)\pi^*\varphi^3=\omega u\pi^*W,$$
$$(\omega-1)s\pi^*W=3t\pi^*T,\quad \pi^*\varphi^3-\pi^*\psi^3=3(2\omega+1)\pi^*T^2.$$

It follows that

$$\pi^*W=\frac{\omega^2u^2}{1+s},\quad \pi^*\varphi^3=\frac{u^3}{(1+s)^2},\quad \pi^*\psi^3=\frac{u^3}{1+s},$$
$$3\pi^*T=(1-\omega^2)\frac{u^2s}{(1+s)t},\quad (1-\omega)su=t^2.$$

Though the relation $(1-\omega)su=t^2$ looks singular at $P(\rho_2,\rho_3)$, the point $s=t=0$ of $\mathrm{Hilb}^G(\mathbb{A}^2)$, it is not singular at all because s and t/s are regular parameters at $P(\rho_2,\rho_3)$. See Section 6.3.3.

The condition $W\neq S_1V_5(\rho_2),S_1V_5(\rho_2'),S_1V_5(\rho_2'')$ implies $s(1+s)\neq 0$, $s\neq\infty$. The parameters s, t and u are related by $(1-\omega)su=t^2$. Hence $\pi^*\overline{\mathfrak{m}}=(\pi^*T)=(u^2/t)=(t^3)$ along $E(\rho_3)\setminus\mathrm{Sing}(E)$, whence $E_{\mathrm{fund}}=3E(\rho_3)$ there.

We see mod $\mathfrak{n}_X$

$$\begin{aligned}
tB_1&=tp_1(\varphi+\omega\psi)+stp_1\varphi+t^2p_2p_3+tup_1\\
&=tp_1(\varphi+\omega\psi)+stp_1\varphi+t^2p_2p_3-3\omega(1+s)p_1T\\
&=-\omega^2(s-\omega)p_3B_2+(s-\omega^2)p_2B_3,\\
tB_2&=\omega(s-\omega)p_1B_3-(s-\omega^2)p_3B_1,\\
tB_3&=-\omega(s-\omega)p_2B_1+(s-\omega^2)p_1B_2.
\end{aligned}$$

This proves $tB_i=0$ in $\mathcal{V}$. Hence $\mathcal{V}$ is $S\big[s,1/\big(s(1+s)\big)\big]$-free of rank three. This completes the proof of freeness of $\mathcal{V}$ over $E(\rho_3)\setminus\mathrm{Sing}(E)$ for E_6.

6.2.8. We explain very briefly the most complicated case of E_7, that is, the ρ_4-case. The finite group G involved is the binary octahedral group $\mathbb{O}$, and the invariant ring of G is generated by homogeneous polynomials of degree 8, 12, and 18, where we note that 18 is also the Coxeter number of E_7.

Any point $I=I_4(W)\in E(\rho_4)\setminus\mathrm{Sing}(E)$ is given by an irreducible G-submodule of $V_9(\rho_4)$ with $W\neq S_1V_8(\rho_2''),S_1V_8(\rho_3),S_1V_8(\rho_3')$ under the notation of Section 5. Then we see

$$\mathrm{Hom}_S(I/I^2,S/I)=\mathrm{Hom}_{\mathbb{C}}(W,V_9(\rho_4)/W)\oplus\mathrm{Hom}_{\mathbb{C}}\big(W,V_7(\rho_4)\big).$$

The versal deformation I_{univ} is generated over $E(\rho_4)\setminus\mathrm{Sing}(E)$ by five elements $B_1,B_2,B_3,B_4,A:=A_8-\pi^*(A_8)$, where $A_8=W$ is the same as W of E_6 and the elements B_i are of the form

$$B_i=B_{i1}+sB_{i2}+tB_{i3}+uB_{i4}+vB_{i5}$$

such that

$$V_9(\rho_4)=\{B_{i1},B_{i2};i=1,2,3,4\},$$
$$V_{13-2k}(\rho_4)=\{B_{ik};i=1,2,3,4\}\qquad (k=3,4,5).$$

See [6, Table 13] for ρ_4-factors of $S_{\mathrm{McKay}}(\mathfrak{m}/\mathfrak{n})$. Moreover over $E(\rho_4)\setminus\mathrm{Sing}(E)$, u (resp. v) is a unit multiple of t^2 (or resp. t^3), while $\pi^*(A_8)$ is a unit multiple of t^4, and $\pi^*\overline{\mathfrak{m}}=(A_8)=(t^4)$. Then we can prove $tB_i=0$ in $\mathcal{V}$ in the same manner as before. In fact, the proof goes roughly as follows. The term tB_1 is the sum of B_{ik}, whose last term tvB_{i5} is a multiple of t^4. Hence tvB_{i5} can be replaced mod $\mathfrak{n}_X$ by a unit multiple of A_8B_{i5}. Then we see that $tB_1 - tvB_{i5} +$(the unit multiple of A_8B_{i5}) is a sum of B_j over $\mathfrak{m}$. In other words, tB_1 is a sum of B_j and A over $\mathfrak{m}$. Since ρ_4 is irreducible, this implies that tB_i is also a sum of B_j and A over $\mathfrak{m}$ for any i. Thus we can prove that $\mathcal{V}$ is $O_{E(\rho_4)}$-free over $E(\rho_4) \setminus \mathrm{Sing}(E)$.

We can write down precisely the versal deformations of $I \in E(\rho) \setminus \mathrm{Sing}(E)$ similarly for any ρ of E_7. By this, we can prove freeness of $\mathcal{V}$ along $E \setminus \mathrm{Sing}(E)$. We omit the details of E_7-case because we need more notation.

6.3. Isomorphism at $I(\rho, \rho')$. In this section we prove that (2) is an isomorphism at any singular point $I := I(\rho, \rho')$ of E.

6.3.1. First we consider the pair $\rho = \rho_1$ and $\rho' = \rho_2$ in the D_5-case. Then I_{univ} is given in (1). It is clear that $\mathcal{V}$ is generated by those elements whose specializations at $s = t = 0$ are just the generators of I belonging to $V_6(\rho_1) + V_3(\rho_2)$. Let $B = x^6 - y^6 + sxy$, $C_1 = x^2y + ty^5 - \frac{st}{2}x$ and $C_2 = -xy^2 - tx^5 - \frac{st}{2}y$. The ideal $\pi^*(\overline{\mathfrak{m}})$ is generated by ϕ_2, hence by s^2t. We first see

$$tB = -(yC_1 + xC_2) = 0 \quad \text{in } \mathcal{V}.$$

Next we prove $sC_1 = 0$ in $\mathcal{V}$. We compute mod $\mathfrak{m}I_{\mathrm{univ}} + \mathfrak{n}_X$:

$$\begin{aligned}
sC_1 &= sx^2y + sty^5 - \frac{s^2t}{2}x \\
&= xB - x(x^6 - y^6) + sty^5 - \frac{s^2t}{2}x \\
&= -xA_6 + 2xy^6 + sty^5 - \frac{s^2t}{2}x \qquad (\because xB \in \mathfrak{m}I_{\mathrm{univ}}) \\
&= -x\left(A_6 + \frac{s^2t}{2} + \frac{t^3}{2}A_6^2\right) + 2xy^6 + sty^5 + \frac{t^3}{2}xA_6^2 \\
&= 2xy^6 + sty^5 + 2txA_4^2 = -2y^4C_2 = 0,
\end{aligned}$$

and similarly $sC_2 = 0$. This proves

$$\mathcal{V} = O_{E(\rho_1)}B + O_{E(\rho_2)}C_1 + O_{E(\rho_2)}C_2.$$

Since $\mathcal{V}$ is $O_{E(\rho)}$-free of rank $\deg \rho$ over $E(\rho) \setminus \mathrm{Sing}(E)$, the upper-semicontinuity shows that $O_{E(\rho_1)}B$ is $O_{E(\rho_1)}$-free of rank $\deg \rho_1$ $(= 1)$ at $s = t = 0$, while $O_{E(\rho_2)}C_1 + O_{E(\rho_2)}C_2$ is $O_{E(\rho_2)}$-free of rank $\deg \rho_2$ $(= 2)$ at $s = t = 0$. This proves that (2) is an isomorphism at $I(\rho_1, \rho_2) : s = t = 0$.

6.3.2. If $\rho = \rho_2$ and $\rho' = \rho_3$ in the D_5-case, then I_{univ} is generated by those elements whose specializations at $s = t = 0$ belong to $V_5(\rho_1) + S_1V_3(\rho_2)$.

Let $R = \mathbb{C}[\![s,t]\!][x,y]$. By Section 4.7, the versal deformation I_{univ} of I is given by $\mathcal{I}_3$, which is, as the ideal of R, generated by the elements

$$\begin{gathered}
B_1 := y^5 + sx^2y + \lambda x, B_2 := -x^5 - sxy^2 + \lambda y, \\
C_1 := x^3y + ty^4 + stx^2, C_2 := -xy^3 + tx^4 + sty^2, \\
A_6 + 2sA_4, \quad A_8 - 2\lambda A_4, \quad A_4 - t\lambda,
\end{gathered}$$

where $\lambda = s^2t/(1+t^2)$. Let $A := A_4 - t\lambda$. We will check $sC_1 = sC_2 = 0$ and $tB_1 = tB_2 = 0$ in $\mathcal{V}$. In fact, in $S[\![s,t]\!]$ we have

$$tB_1 = yC_1 - xA,\ tB_2 = -xC_2 - yA,$$
$$sC_1 = (1+t^2)xB_1 - txyC_1 + y^2C_2,$$
$$sC_2 = (1+t^2)yB_2 - txyC_2 + x^2C_1.$$

It follows that

$$\mathcal{V} = O_{E(\rho_1)}B + O_{E(\rho_2)}C_1 + O_{E(\rho_2)}C_2.$$

Since $\mathcal{V}$ is $O_{E(\rho)}$-free of rank $\deg\rho$ over $E(\rho)\setminus\mathrm{Sing}(E)$, the upper-semicontinuity shows that $O_{E(\rho_2)}B_1 + O_{E(\rho_2)}B_2$ is $O_{E(\rho_2)}$-free of rank $\deg\rho_2$ at $s=t=0$, while $O_{E(\rho_3)}C_1 + O_{E(\rho_3)}C_2$ is $O_{E(\rho_3)}$-free of rank $\deg\rho_3$ at $s=t=0$. This proves that (2) is an isomorphism at $I(\rho_2,\rho_3) : s=t=0$.

6.3.3. In this section we consider one of the most complicated case of E_6 where $\rho=\rho_2$ and $\rho'=\rho_3$ with $\deg(\rho_2)=2$ and $\deg(\rho_3)=3$. For the notation see Figure 4. Let $I := P(\rho_2,\rho_3)$ and $W = \big(S_1V_5(\rho_2)\big)[\rho_3]$. Then $I := I(\rho_2,\rho_3) = W + \sum_{k=7}^{11} S_k + \mathfrak{n}$ and $V(I) = W \oplus V_7(\rho_2)$. Moreover

$$\mathrm{Hom}_S(I/I^2, S/I) = \mathrm{Hom}_{\mathbb{C}}\big(W, V_6(\rho_3/W)\big) \oplus \mathrm{Hom}_{\mathbb{C}}\big(V_7(\rho_2), V_5(\rho_2)\big).$$

The versal deformation I_{univ} of $I(\rho_2,\rho_3)$ is generated by

$$B_1,\quad B_2,\quad B_3,\quad C_1,\quad C_2,\quad A,$$

where B_i $(i=1,2,3)$ is the same as in Section 6.2.8, and

$$C_1 := -s_2\varphi + v\gamma_1 + wx,\quad C_2 := s_1\varphi + v\gamma_2 + wy,$$
$$A := T - \pi^*(T) = T - \frac{1}{2\omega+1}sw = T - \frac{1}{1+s}s^2v^3.$$

The parameters s and v are regular parameters of $\mathrm{Hilb}^G(\mathbb{A}^2)$ at I. The other parameters are related as follows:

$$t = -(1-\omega)sv,\quad u = (1-\omega)sv^2,\quad w = -\frac{2\omega+1}{1+s}sv^3,$$

whence $u = -tv$, $uv = -\omega^2(1+s)w$ and $(2\omega+1)\pi^*(T) = sw$.

Then we see in $S[s,v]$,

$$vB_1 = -\omega^2(1+s)(xC_1 - yC_2) + \frac{1}{1-\omega}(\omega p_2B_3 - p_3B_2),$$
$$vB_2 = -\omega^2(1+s)(xC_1 + yC_2) - \frac{1}{1-\omega}(\omega p_3B_1 - p_1B_3),$$
$$vB_3 = -\omega^2(1+s)(yC_1 + xC_2) + \frac{1}{1-\omega}(\omega p_1B_2 - p_2B_1),$$
$$sC_1 = \frac{1}{1-\omega}(yB_1 + yB_2 - xB_3) - (2\omega+1)xA,$$
$$sC_2 = \frac{1}{1-\omega}(xB_1 - xB_2 + yB_3) - (2\omega+1)yA.$$

This proves $\mathcal{V} = \mathcal{V}\otimes O_{E(\rho_2)} + \mathcal{V}\otimes O_{E(\rho_3)}$.

6.3.4. In general, let $I \in P(\rho, \rho')$ be a singular point of E. Then as we saw above, we have an isomorphism

$$\mathcal{V}_{[I]} = \mathcal{V}_{[I]} \otimes O_{E(\rho)} \oplus \mathcal{V}_{[I]} \otimes O_{E(\rho')}$$

locally at I. This proves the global isomorphism over E

$$\mathcal{V} \simeq \bigoplus_{\rho \in \mathrm{Irr}(G)} \mathcal{V} \otimes O_{E(\rho)}.$$

6.4. The sheaf $\mathcal{V} \otimes O_{E(\rho)}$. It remains to prove $\mathcal{V} \otimes O_{E(\rho)} \simeq \rho \otimes O_{E(\rho)}(-1)$.

6.4.1. Let V be a two-dimensional $\mathbb{C}$-vector space, and $\mathbb{P}(V)$ the projective line of one-dimensional subspaces of V. There is a universal family of one-dimensional subspaces of V parametrized by $\mathbb{P}(V)$, which we denote W_{univ}. This is a line bundle (an invertible sheaf) on $\mathbb{P}(V)$. There is an exact sequence of $O_{\mathbb{P}(V)}$-modules:

$$0 \to W_{\mathrm{univ}} \to V \otimes O_{\mathbb{P}(V)} \to V \otimes O_{\mathbb{P}(V)}/W_{\mathrm{univ}} \to 0.$$

This implies that $W_{\mathrm{univ}} \simeq O_{\mathbb{P}(V)}(-1)$ because the bundle is twisted linearly.

Let $V(\rho) := S_{\mathrm{McKay}}(\mathfrak{m}/\mathfrak{n})[\rho] = V_{h-d(\rho)} + V_{h+d(\rho)} \simeq \rho^{\oplus 2}$, and $\mathbb{P}(V(\rho))$ the projective line of nonzero irreducible G-submodules of $V(\rho)$. Then $V(\rho) \simeq \rho \otimes V$ and $\mathbb{P}(V(\rho)) \simeq \mathbb{P}(V)$. It is obvious that $\rho \otimes W_{\mathrm{univ}}$ yields a universal family $W_{\mathrm{univ}}(\rho)$ of nonzero G-submodules of $V(\rho)$ parametrized by $\mathbb{P}(V(\rho))$ ($\simeq \mathbb{P}(V)$). We see $W_{\mathrm{univ}}(\rho) \simeq \rho \otimes O_{\mathbb{P}(V(\rho))}(-1)$.

6.4.2. Let $I \in E(\rho)$. Then by identifying $E(\rho)$ with $\mathbb{P}(V)$, on any Zariski open subset $U := \operatorname{Spec} A$ of $E(\rho)$ we have

$$I_{\mathrm{univ}} \otimes A = W_{\mathrm{univ}}(\rho) + \sum_{V \subset \mathrm{Quiv}(\rho), v \not\simeq \rho} V \otimes A + S^{\dagger} \otimes A,$$

and by Section 6.3

$$\mathcal{V} \otimes A = W_{\mathrm{univ}}(\rho) \otimes A = W_{\mathrm{univ}}(\rho).$$

In other words, $\mathcal{V} \otimes O_{E(\rho)} \simeq W_{\mathrm{univ}}(\rho) \simeq \rho \otimes O_{E(\rho)}(-1)$.

This completes the proof of Theorem 3.9. □

References

1. M. Artin, *On isolated rational singularities of surfaces*, Amer. J. Math. **88** (1966), 129–136.
2. N. Bourbaki, *Éléments de mathématique*: *Groupes et algèbres de Lie. Chapitres* 4, 5 *et* 6, Masson, Paris, 1981.
3. J. Fogarty, *Algebraic families on an algebraic surface*, Amer. J. Math. **90** (1968), 511–521.
4. A. Fujiki, *On primitively symplectic compact Kähler V-manifolds of dimension four*, Classification of Algebraic and Analytic Manifolds (Katata, 1982) (K. Ueno, ed.), Progr. Math., vol. 39, Birkhäuser, Boston, MA, 1983, pp. 71–250.
5. G. Gonzalez-Springberg and J.-L. Verdier, *Construction géométrique de la correspondence de McKay*, Ann. Sci. École Norm. Sup. (4) **16** (1983), no. 3, 409–449.
6. Y. Ito and I. Nakamura, *Hilbert schemes and simple singularities*, New Trends in Algebraic Geometry (Warwick, 1996) (Hulek. K., F. Catanese, C. Peters, and M. Reid, eds.), London Math. Soc. Lecture Note Ser., vol. 264, Cambridge Univ. Press, Cambridge, 1999, pp. 151–233.
7. F. Klein, *Vorlesungen über das Ikosaeder und die Auflösung der Gleichungen vom fünften Grade*, with an introduction and a commentary by P. Slodowy, Birkhäuser, Basel; B.G. Teubner, Stuttgart, 1993.
8. J. McKay, *Graphs, singularities, and finite group*, The Santa Cruz Conference on Finite Groups (Santa Cruz, CA, 1979) (B. Cooperstein and G. Mason, eds.), Proc. Symp. Pure Math., vol. 37, Amer. Math. Soc., Providence, RI, 1980, pp. 183–186.

9. I. Nakamura, *Hilbert schemes of abelian group orbits*, J. Algebraic Geom. **10** (2001), no. 4, 757–779.
10. I. Schur, *Untersuchungen über die Darstellung der endlichen Gruppen durch gebrochene lineare Substitutionen*, J. Reine Angew. Math. **132** (1907), 85–137.
11. P. Slodowy, *Simple singularities and simple algebraic groups*, Lecture Note in Math., vol. 815, Springer, Berlin, 1980.

DEPARTMENT OF MATHEMATICS, HOKKAIDO UNIVERSITY, SAPPORO, 060-0810, JAPAN
E-mail address: nakamura@math.sci.hokudai.ac.jp

Centre de Recherches Mathématiques
CRM Proceedings and Lecture Notes
Volume **47**, 2009

On Ground Fields of Arithmetic Hyperbolic Reflection Groups

Viacheslav V. Nikulin

Dedicated to John McKay

Abstract. Using our methods of 1980, 1981, we define some explicit finite sets of number fields containing ground fields of arithmetic hyperbolic reflection groups and give good bounds of their degrees over $\mathbb{Q}$. For example, the degree of the ground field of any arithmetic hyperbolic reflection group in dimension at least 6 is bounded by 56. These results could be important for further classification.

We also formulate a mirror symmetric conjecture to finiteness of the number of arithmetic hyperbolic reflection groups which was established in full generality recently.

1. Introduction

There are three types of simply-connected complete Riemannian manifolds of constant curvature: spheres, Euclidean spaces and hyperbolic spaces. Discrete reflection groups (generated by reflections in hyperplanes of these spaces) were defined by H. S. M. Coxeter. He classified these groups in spheres and Euclidean spaces [7].

There are two types of discrete reflection groups with fundamental domain of finite volume in hyperbolic spaces: general and arithmetic. In this paper, we consider only arithmetic hyperbolic reflection groups.

In [26], Vinberg (1967) stated and proved a criterion for the arithmeticity of discrete reflection groups in hyperbolic spaces in terms of their fundamental chambers. In particular, he introduced the notion of the ground field of such groups. This is a totally real algebraic number field $\mathbb{K}$ of a finite degree over $\mathbb{Q}$. The arithmetic reflection group W is a subgroup of finite index of the automorphism group $O(S)$ of a hyperbolic quadratic form (a hyperbolic lattice) S over the ring of integers $\mathbb{O}$ of this field. See Section 6 for the exact definition. Hyperbolic lattices S which have a reflection subgroup $W \subset O(S)$ of finite index are called *reflective*. One can

---

2000 *Mathematics Subject Classification.* 20F55, 51F15, 22E40, 11E12, 11F06.

This paper was written with the financial support of EPSRC, United Kingdom (grant no. EP/D061997/1).

This is the final form of the paper.

canonically associate to S the hyperbolic space $\mathcal{L}(S)$ of dimension $\operatorname{rank} S - 1$ such that groups $W \subset O^+(S)$ act in $\mathcal{L}(S)$ and define the arithmetic hyperbolic reflection group W. One can take the maximal reflection subgroup $W(S) \subset O(S)$, the reflection group of the lattice S, which contains W. Thus, S is reflective if and only if $[O(S) : W(S)] < \infty$. Two hyperbolic lattices which differ by multiplication of their forms by $k \in \mathbb{K}$ are called *similar*. Their hyperbolic spaces and automorphism groups are identified.

Thus, *classification of similarity classes of reflective hyperbolic lattices* is the key problem for classification of arithmetic hyperbolic reflection groups. It includes classification of maximal arithmetic hyperbolic reflection groups, and it is important for classification of hyperbolic Lie algebras. The degree $N = [\mathbb{K}:\mathbb{Q}]$ of the ground field $\mathbb{K}$ of S and W, and the dimension $n = \dim \mathcal{L}(S) = \operatorname{rank} S - 1 \geq 2$ are the most important parameters for this classification.

In [13, 14], we proved (1980, 1981) that the number of similarity classes of reflective hyperbolic lattices is finite for any fixed parameters n and N. Moreover, the number of the ground fields of any fixed degree N is also finite. In [14], we proved (1981) that there exists an effective constant N_0 such that $N \leq N_0$ if the dimension $n \geq 10$.

In [27,28], Vinberg proved (1981) that $n < 30$: arithmetic hyperbolic reflection groups don't exist in dimensions $n \geq 30$.

Thus, the numbers of similarity classes of reflective hyperbolic lattices and maximal arithmetic hyperbolic reflection groups are finite in dimensions $n \geq 10$ all together.

About these results, see also reports [15, 29] at International Congresses of Mathematicians.

Almost 25 years, boundedness of the degree $N = [\mathbb{K}:\mathbb{Q}]$ remained unknown in small dimensions $2 \leq n \leq 9$. Only in 2005 it was proved in dimension $n = 2$ by Long, Maclachlan and Reid [12], and in dimension $n = 3$ by Agol [1]. In 2006, we showed [17] that the boundedness for remaining dimensions $4 \leq n \leq 9$ can be easily deduced from boundedness for dimensions $n = 2, 3$, and methods and results of [13, 14] (see Theorem 12 and its proof in Section 4.3).

Thus, now, finiteness of the numbers of reflective hyperbolic lattices and of maximal arithmetic hyperbolic reflection groups are established in full generality: for all dimensions of hyperbolic spaces and for all ground fields together.

Unfortunately, these finiteness results are very far from classification of these finite sets. The purpose of this paper is to prove some explicit results in this direction. Perhaps, the first and the most important problem is to enumerate possible ground fields $\mathbb{K}$ and to bound their degrees $[\mathbb{K}:\mathbb{Q}]$.

First explicit results in this direction were obtained by Vinberg [27,28] in 1981. He had shown: in dimensions $n \geq 30$, the set of the ground fields is empty; in dimensions $n \geq 22$, the ground fields belong to the union of $\mathcal{FL}^4 = \{\mathbb{Q}, \mathbb{Q}(\sqrt{2}), \mathbb{Q}(\sqrt{5})\}$ and $\{\mathbb{Q}(\cos(2\pi/7))\}$; in dimensions $n \geq 14$, the ground fields belong to the set $\mathcal{FT}$ of 13 fields which are ground fields of arithmetic triangle (plane) reflection groups classified by Takeuchi [22, 23] in 1977 (their degrees are bounded by 5). See Section 3.2 about arithmetic triangle groups.

Our results can be viewed as some extensions of these explicit Vinberg's results to smaller dimensions.

In Section 3.3, we introduce finite sets of fields $\mathcal{F}\Gamma_i^{(4)}(14)$, $i = 1, 2, 3, 4, 5$. They are ground fields of some V-arithmetic 3-dimensional fundamental edge chambers of minimality 14 described by their connected hyperbolic Gram graphs $\Gamma_i^{(4)}$, $i = 1, 2, 3, 4, 5$, with four vertices, in Figures 3 – 8. They are some special V-arithmetic edge chambers introduced and used in [14]. Using methods of [14], in Theorem 4 and Section 5 we show that the degrees of the fields from these sets are bounded by reasonable constants 22, 39, 53, 56 and 54 respectively.

Using methods of [13, 14, 28], in Theorem 5 we show that in dimensions $n \geq 10$, the ground field of any arithmetic hyperbolic reflection group belongs to one of the sets $\mathcal{F}L^4$, $\mathcal{F}T$, $\mathcal{F}\Gamma_i^{(4)}(14)$, $i = 1, 2, 3, 4$. In particular, its degree is bounded by 56. Thus, our result in [14] becomes very explicit: the constant N_0 which we mentioned above can be taken to be $N_0 = 56$.

In Section 4.2, we introduce one more set $\mathcal{F}\Gamma_{2,4}(14)$ of fields. It is the set of ground fields of arithmetic quadrangles of minimality 14 (that is ground fields of 2-dimensional arithmetic hyperbolic reflection groups with quadrangle fundamental polygon of minimality 14). According to Borel [5] and Takeuchi [24], the set of ground fields of arithmetic quadrangles is finite, and the degrees of these fields are bounded by 11.

Using methods of [13, 14], in Theorem 9 we show that in dimensions $n \geq 6$, the ground field of any arithmetic hyperbolic reflection group belongs to one of sets $\mathcal{F}L^4$, $\mathcal{F}T$, $\mathcal{F}\Gamma_i^{(4)}(14)$, $i = 1, 2, 3, 4, 5$, or $\mathcal{F}\Gamma_{2,4}(14)$. In particular, its degree is bounded by 56.

Unfortunately, now we don't have so explicit results for smaller dimensions. Following [17], in Theorem 12, we only show that in dimensions $n \geq 4$, the ground field of any arithmetic hyperbolic reflection group belongs to one of sets $\mathcal{F}L^4$, $\mathcal{F}T$, $\mathcal{F}\Gamma_i^{(4)}(14)$, $i = 1, 2, 3, 4$, or it is the ground field of one of 3 or 2-dimensional arithmetic hyperbolic reflection group with a fundamental chamber of minimality 14. Thus, its degree is bounded by the maximum of 56 and of the degrees of the ground fields of 3 and 2-dimensional arithmetic hyperbolic reflection groups with fundamental chambers of minimality 14.

For 3-dimensional arithmetic hyperbolic reflection groups, we know finiteness of the number of ground fields by Agol [1]. By our knowledge, no explicit upper bound of their degrees is known.

Using results by Long, Maclachlan and Reid [12] and Borel [5], Takeuchi [24], in Section 4.5 we show that the degrees of the ground fields of 2-dimensional arithmetic hyperbolic reflection groups are bounded by 44.

Using known at that time finiteness of reflective hyperbolic lattices over $\mathbb{Z}$, in [8–10], we (with Gritsenko) obtained some finiteness results and formulated a general finiteness conjecture for type IV (that is of signature $(2, t)$) integer reflective lattices S. Here a lattice S over $\mathbb{Z}$ of signature $(2, t)$ is called reflective if type IV Hermitian symmetric domain associated to S has an $O^+(S)$-automorphic holomorphic form Φ of positive weight such that all components of its divisor are quadratic divisors orthogonal to roots (giving reflections) of S. Such an automorphic form is called *reflective.* One can consider these finiteness statements about type IV reflective lattices over $\mathbb{Z}$ as "mirror symmetric" to finiteness results about hyperbolic reflective lattices over $\mathbb{Z}$. Since we now know finiteness of hyperbolic reflective lattices in general, in Section 6 we formulate the corresponding mirror symmetric conjecture about type IV reflective lattices in general — over arbitrary totally real algebraic number fields. We expect that it is valid.

Some arithmetic hyperbolic reflection groups and some reflective automorphic forms and corresponding hyperbolic and type IV reflective lattices over $\mathbb{Z}$ are important in Borcherds proof [4] of Moonshine Conjecture by Conway and Norton [6] which had been first discovered by John McKay.

We hope that similar objects over arbitrary number fields will find similar astonishing applications in the future. At least, the results and conjectures of this paper show that they are very exceptional even in this very general setting.

A preliminary version of this paper was published as a preprint [18].

2. A reminder of some basic facts about hyperbolic fundamental polyhedra

Here we remind the reader some basic definitions and results about fundamental chambers (always for discrete reflection groups) in hyperbolic spaces and their Gram matrices. See [13, 14, 26, 30].

We consider Klein model of the hyperbolic space $\mathcal{L}$ associated to a hyperbolic form Φ over the field $\mathbb{R}$ of real numbers with signature $(1, n)$ where $n = \dim \mathcal{L}$. Let $V = \{x \in \Phi \mid x^2 > 0\}$ be the positive cone defined by Φ, and V^+ one of the two halves of the cone. The hyperbolic space $\mathcal{L}$ is the set $V^+/\mathbb{R}^+$ of rays in V^+; we let $[x]$ denote the element of $\mathcal{L}$ determined by the ray $\mathbb{R}^+x$ where $x \in V^+$ and $\mathbb{R}^+$ is the set of all positive real numbers. The hyperbolic distance is given by the formula

$$\operatorname{ch}\bigl(\rho([x],[y])\bigr) = (x\cdot y)/\sqrt{x^2y^2}, \quad [x],[y] \in \mathcal{L}.$$

Then the curvature of $\mathcal{L}$ is equal to -1.

Every half-space $\mathcal{H}^+$ in $\mathcal{L}$ determines and is determined by the orthogonal element $e \in \Phi$ with square $e^2 = -2$:

$$\mathcal{H}^+ = \mathcal{H}_e^+ = \{[x] \in \mathcal{L} \mid x\cdot e \geq 0\}.$$

It is bounded by the hyperplane

$$\mathcal{H}^+ = \mathcal{H}_e^+ = \{[x] \in \mathcal{L} \mid x\cdot e = 0\}$$

orthogonal to e. If two half-spaces $\mathcal{H}_{e_1}^+$, $\mathcal{H}_{e_2}^+$ where $e_1^2 = e_2^2 = -2$ have a common non-empty open subset in $\mathcal{L}$, then $\mathcal{H}_{e_1} \cap \mathcal{H}_{e_2}$ is an angle of the value ϕ where $2\cos\phi = e_1\cdot e_2$ if $-2 < e_1\cdot e_2 \leq 2$, and the distance between hyperplanes $\mathcal{H}_{e_1}$ and $\mathcal{H}_{e_2}$ is equal to ρ where $2\operatorname{ch}\rho = e_1\cdot e_2$ if $e_1\cdot e_2 > 2$.

A convex polyhedron $\mathcal{M}$ in $\mathcal{L}$ is the intersection of a finite number of half-spaces $\mathcal{H}_e^+$, $e \in P(\mathcal{M})$, where $P(\mathcal{M})$ is the set of vectors with square -2 which are orthogonal to codimension one faces of $\mathcal{M}$ and directed outward. The Gram matrix

$$A = (a_{ij}) = (e_i\cdot e_j), \quad e_i, e_j \in P(\mathcal{M}), \tag{1}$$

of $P(\mathcal{M})$ is called the Gram matrix $\Gamma(\mathcal{M}) = \Gamma\bigl(P(\mathcal{M})\bigr)$ of $\mathcal{M}$. It determines $\mathcal{M}$ uniquely up to motions of $\mathcal{L}$. If $\mathcal{M}$ is sufficiently general, then $P(\mathcal{M})$ generates Φ, and the form Φ is equal to

$$\Phi = \sum_{e_i, e_j \in P(\mathcal{M})} a_{ij}X_iY_j \bmod \text{Kernel}. \tag{2}$$

The set $P(\mathcal{M})$ is naturally identified with a subset of Φ, and it determines $\mathcal{M}$.

A convex polyhedron $\mathcal{M}$ is a fundamental chamber of a discrete reflection group W in $\mathcal{L}$ if and only if for all $i \neq j$ we have $a_{ij} \geq 0$, and $a_{ij} = 2\cos \pi/m_{ij}$ where $m_{ij} \geq 2$ is an integer if $a_{ij} < 2$. Symmetric real matrices A satisfying these conditions and having all their diagonal elements equal to -2 are called *fundamental* (then the set $P(\mathcal{M})$ formally corresponds to indices of the matrix A). As usual, further we identify fundamental matrices with fundamental graphs Γ. Their vertices correspond to $P(\mathcal{M})$. Two different vertices $e_i \neq e_j \in P(\mathcal{M})$ are connected by the thin edge of the integer weight $m_{ij} \geq 3$ if $0 < a_{ij} = 2\cos \pi/m_{ij} < 2$, by the thick edge if $a_{ij} = 2$, and by the broken edge of the weight a_{ij} if $a_{ij} > 2$. In particular, the vertices e_i and e_j are disjoint if and only if $e_i \cdot e_j = a_{ij} = 2\cos \pi/2 = 0$. Equivalently, e_i and e_j are perpendicular (or orthogonal). See some examples of such graphs in Figures $1-8$.

For a real $t > 0$, we say that a fundamental matrix $A = (a_{ij})$ (and the corresponding fundamental chamber $\mathcal{M}$) *has minimality* t if $a_{ij} < t$ for all a_{ij}. Here we follow [13, 14]. In our further considerations, minimality $t = 14$ will be especially important.

It is known that fundamental domains of arithmetic hyperbolic groups must have finite volume. Let us assume that it is valid for a fundamental chamber $\mathcal{M}$ of a hyperbolic discrete reflection group. Vinberg had shown in [26], that in order for $\mathcal{M}$ to be a fundamental chamber of an arithmetic reflection group W in $\mathcal{L}$, it is necessary and sufficient that all of the cyclic products

$$b_{i_1,\dots,i_m} = a_{i_1 i_2} \cdot a_{i_2 i_3} \cdots a_{i_{m-1} i_m} \cdot a_{i_m i_1} \tag{3}$$

be algebraic integers, that the field $\widetilde{\mathbb{K}} = \mathbb{Q}(\{a_{ij}\})$ be totally real, and that, for any embedding $\widetilde{\mathbb{K}} \to \mathbb{R}$ not the identity on the *ground field* $\mathbb{K} = \mathbb{Q}(\{b_{i_1,\dots,i_m}\})$ generated by all of the cyclic products (3), the form (2) be negative definite.

A fundamental real matrix $A = (a_{ij})$, $a_{ij} = e_i \cdot e_j$, $e_i, e_j \in P(\mathcal{M})$ (or the corresponding graph) is called *V-arithmetic* if A satisfies these Vinberg's conditions and the corresponding form Φ in (2) is hyperbolic (here we don't require that the corresponding hyperbolic polyhedron $\mathcal{M}$ has finite volume). It is well-known and easy to see (see arguments in Section 5.1) that any subset $P \subset P(\mathcal{M})$ also defines a V-arithmetic matrix $(e_i \cdot e_j)$, $e_i, e_j \in P$, with the same ground field $\mathbb{K}$ if the subset P is hyperbolic, that is the corresponding to P form (2) is hyperbolic.

3. V-arithmetic edge polyhedra

We call a fundamental chamber $\mathcal{M}$ (and the corresponding Gram matrix A or graph) *edge chamber (and matrix, graph)* if all hyperplanes $\mathcal{H}_e$, $e \in P(\mathcal{M})$, contain one of two distinct vertices v_1, v_2 of a 1-dimensional edge v_1v_2 of $\mathcal{M}$. Assume that both vertices v_1 and v_2 are finite (further we consider this case only). We call such edge chambers *finite*. Assume that $\dim \mathcal{L} = n$. Then $P(\mathcal{M})$ consists of $n+1$ elements: e_1, e_2 and $n-1$ elements $P(\mathcal{M}) - \{e_1, e_2\}$. Here $P(\mathcal{M}) - \{e_1, e_2\}$ corresponds to hyperplanes which contain the edge v_1v_2 of $\mathcal{M}$. The e_1 corresponds to the hyperplane which contains v_1 and does not contain v_2. The e_2 corresponds to the hyperplane which contains v_2 and does not contain v_1. Then the set $P(\mathcal{M})$ is hyperbolic (it has hyperbolic Gram matrix), but its subsets $P(\mathcal{M}) - \{e_1\}$ and $P(\mathcal{M}) - \{e_2\}$ are negative definite (they have negative definite Gram matrix) and define Coxeter graphs. Only the element $u = e_1 \cdot e_2$ of the Gram matrix of $\mathcal{M}$ can be greater than 2. Thus, $\mathcal{M}$ has minimality $t > 2$ if and only if $u = e_1 \cdot e_2 < t$.

From considerations above, the Gram graph $\Gamma(P(\mathcal{M}))$ of an edge chamber has only one hyperbolic connected component $P(\mathcal{M})^{\text{hyp}}$ (containing e_1 and e_2) and several negative definite connected components. Gram matrix $\Gamma(P(\mathcal{M})^{\text{hyp}})$ evidently also corresponds to an edge chamber of dimension $\#P(\mathcal{M})^{\text{hyp}} - 1$. If $\mathcal{M}$ is V-arithmetic, the ground field $\mathbb{K}$ of $\mathcal{M}$ is the same as for the hyperbolic connected component $\Gamma(P(\mathcal{M})^{\text{hyp}})$

The following result was obtained in [14].

Theorem 1 ([14, Theorem 2.3.1]). *Given $t > 0$, there exists an effective constant $N(t)$ such that every V-arithmetic edge chamber of minimality t with the ground field $\mathbb{K}$ of the degree greater that $N(t)$ over $\mathbb{Q}$ has the hyperbolic connected component of its Gram graph which has less than 4 vertices.*

Considerations in [14] (and also [13]) also show that the set of possible ground fields $\mathbb{K}$ of V-arithmetic edge chambers of minimality t with hyperbolic connected components having at least 4 vertices is finite. Even the set of Gram graphs $\Gamma(P(\mathcal{M})^{\text{hyp}})$ of minimality t with fixed ≥ 4 number of vertices is finite. Taking this under consideration, here we want to formulate and prove more efficient variant of this theorem. We restrict by minimality $t = 14$ to get an explicit estimate for the constant $N(14)$, but the same finiteness result is valid for any $t > 0$.

To formulate this new variant, we need to introduce some fundamental edge graphs.

3.1. Arithmetic Lannér graphs with $\geq$ 4 vertices. Lannér graphs are Gram graphs of bounded fundamental hyperbolic simplexes. They are characterized as hyperbolic fundamental graphs such that all their proper subgraphs are Coxeter graphs. They were classified by Lannér [11]. In Figure 1, we give all arithmetic Lannér graphs with at least 4 vertices (only one Lannér graph with ≥ 4 vertices is

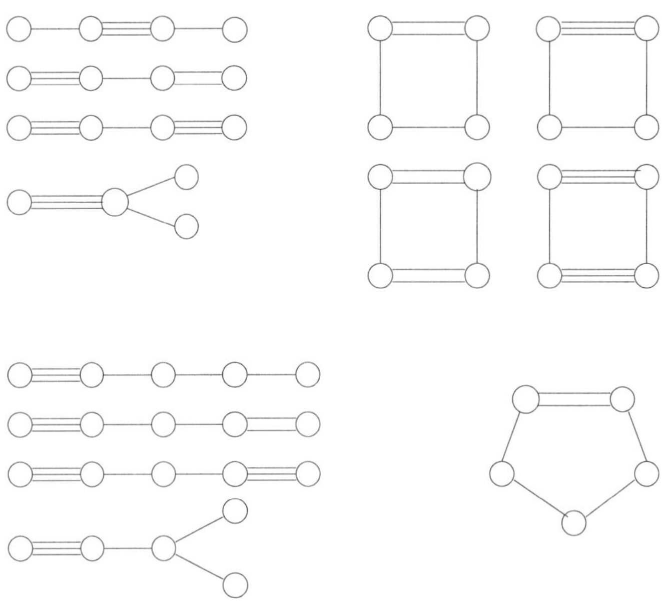

FIGURE 1. All arithmetic Lannér graphs with at least 4 vertices

not arithmetic). As usual, for small k, we replace a thin edge of the weight k by $k-2$-edges. Ground fields of Lannér graphs with ≥ 4 vertices give three fields:

$$\mathcal{F}L^4 = \{\mathbb{Q}, \mathbb{Q}(\sqrt{2}), \mathbb{Q}(\sqrt{5})\}. \tag{4}$$

See [28] for details.

3.2. Arithmetic triangle graphs. Triangle graphs are Gram graphs of bounded fundamental triangles on hyperbolic plane (we don't consider non-bounded triangles). Equivalently, they are Lannér graphs with 3 vertices. They are given in Figure 2 where $2 \leq k, l, m$ and

$$\frac{1}{k} + \frac{1}{l} + \frac{1}{m} < 1.$$

Arithmetic triangles were enumerated by Takeuchi [22]. All bounded arithmetic triangles are given by the following triplets (k, l, m):

$(2,3,7-12)$, $(2,3,14)$, $(2,3,16)$, $(2,3,18)$, $(2,3,24)$, $(2,3,30)$, $(2,4,5-8)$, $(2,4,10)$, $(2,4,12)$, $(2,4,18)$, $(2,5,5)$, $(2,5,6)$, $(2,5,8)$, $(2,5,10)$, $(2,5,20)$, $(2,5,30)$, $(2,6,6)$, $(2,6,8)$, $(2,6,12)$, $(2,7,7)$, $(2,7,14)$, $(2,8,8)$, $(2,8,16)$, $(2,9,18)$, $(2,10,10)$, $(2,12,12)$, $(2,12,24)$, $(2,15,30)$, $(2,18,18)$, $(3,3,4-9)$, $(3,3,12)$, $(3,3,15)$, $(3,4,4)$, $(3,4,6)$, $(3,4,12)$, $(3,5,5)$, $(3,6,6)$, $(3,6,18)$, $(3,8,8)$, $(3,8,24)$, $(3,10,30)$, $(3,12,12)$, $(4,4,4-6)$, $(4,4,9)$, $(4,5,5)$, $(4,6,6)$, $(4,8,8)$, $(4,16,16)$, $(5,5,5)$, $(5,5,10)$, $(5,5,15)$, $(5,10,10)$, $(6,6,6)$, $(6,12,12)$, $(6,24,24)$, $(7,7,7)$, $(8,8,8)$, $(9,9,9)$, $(9,18,18)$, $(12,12,12)$, $(15,15,15)$.

Their ground fields were found by Takeuchi [23]. They give the set of fields

$$\begin{aligned} \mathcal{F}T = \{\mathbb{Q}\} \cup \{\mathbb{Q}(\sqrt{a}) \mid a = 2,3,5,6\} \cup \{\mathbb{Q}(\sqrt{2},\sqrt{3}), \mathbb{Q}(\sqrt{2},\sqrt{5})\} \\ \cup \left\{\mathbb{Q}\left(\cos\frac{2\pi}{b}\right) \,\middle|\, b = 7,9,11,15,16,20\right\}. \end{aligned} \tag{5}$$

3.3. V-arithmetic connected finite edge graphs with 4 vertices for $2 < u < 14$. Using classification of Coxeter graphs, we can draw all possible pictures of connected finite edge graphs $\Gamma^{(4)}$ with 4 vertices and $u = e_1 \cdot e_2 > 2$. They correspond to all 3-dimensional finite fundamental edge polyhedra with connected Gram graph and $u > 2$. They are given in Figure 3 and give five types of graphs $\Gamma = \Gamma_i^{(4)}$, $i = 1,2,3,4,5$. All possible natural parameters $s, k, r, p \geq 2$ for these graphs can be easily enumerated by the condition that $\Gamma - \{e_1\}$, $\Gamma - \{e_2\}$ are Coxeter graphs. They will be given in Section 5 below.

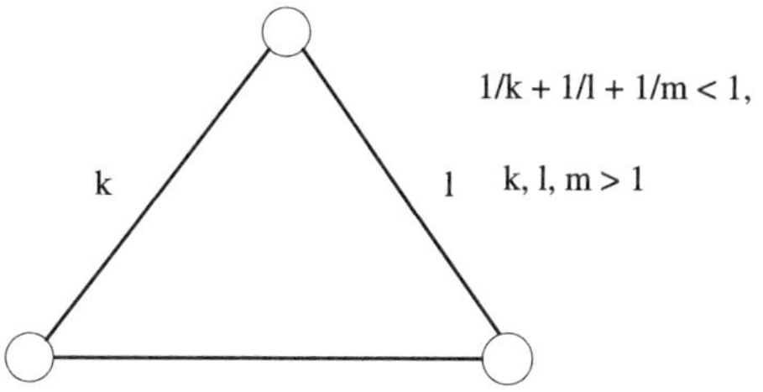

FIGURE 2. Triangle graphs

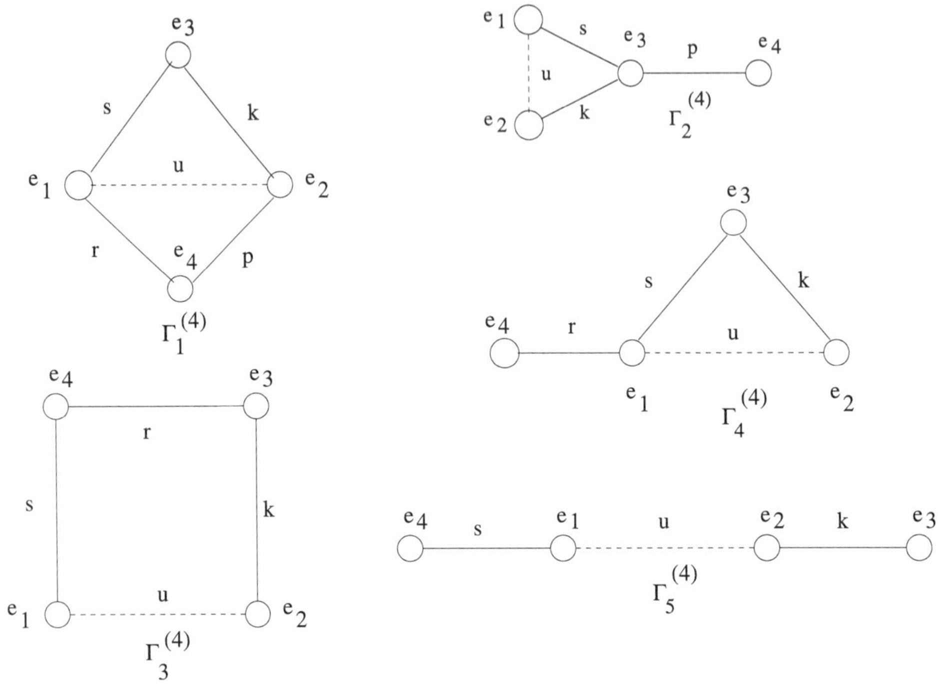

FIGURE 3. Five graphs $\Gamma_i^{(4)}$, $i = 1, 2, 3, 4, 5$.

Definition 2. For $i = 1, 2, 3, 4, 5$ and $t > 0$ we denote by $\Gamma_i^{(4)}(t)$ the set of all V-arithmetic connected finite edge graphs with 4 vertices $\Gamma_i^{(4)}$ of minimality t, that is for $2 < u < t$, and by $\mathcal{F}\Gamma_i^{(4)}(t)$ the set of all their ground fields.

All V-arithmetic graphs $\Gamma_i^{(4)}$ with $2 < u < t$ give particular cases of graphs of V-arithmetic edge polyhedra with hyperbolic connected component having 4 vertices and minimality t. Thus, by Theorem 1, the degrees of the fields from $\mathcal{F}\Gamma_i^{(4)}(t)$ are bounded by an effective constant $N(t)$. It follows that the sets of V-arithmetic graphs $\Gamma_i^{(4)}(t)$ and fields $\mathcal{F}\Gamma_i^{(4)}(t)$ are also finite.

Vice versa, Theorem 2 can be deduced from finiteness of the sets of fields above because of the following easy statement.

Proposition 3. *The ground field of any V-arithmetic edge chamber of minimality $t > 0$ with the hyperbolic connected component of its Gram graph having at least 4 vertices belongs to one of finite sets of fields $\mathcal{F}L^4$, $\mathcal{F}T$ and $\mathcal{F}\Gamma_i^{(4)}(t)$, $1 \leq i \leq 5$, introduced above.*

In particular, Theorem 1 is equivalent to finiteness of the sets of fields $\mathcal{F}\Gamma_i^{(4)}(t)$, $i = 1, 2, 3, 4, 5$.

PROOF. Let $\mathcal{M}$ be a V-arithmetic edge chamber of minimality $t > 0$, and Γ its Gram graph having the hyperbolic connected component Γ^{hyp} with at least 4 vertices. The graph Γ^{hyp} is also V-arithmetic edge graph.

Assume that $u = e_1 \cdot e_2 < 2$. The subset $\{e_1, e_2\}$ gives a negative definite subgraph of Γ^{hyp}. It follows that any subgraph of Γ^{hyp} is also negative definite if it has one or two vertices. Since Γ^{hyp} is hyperbolic, it follows that Γ^{hyp} contains a minimal hyperbolic subgraph L which is Lannér with at least 3 vertices. Since L is hyperbolic, the ground field of Γ is equal to the ground field of L. If L has more than 3 vertices, then the ground field of L is one of fields from $\mathcal{F}L^4$. If L has 3 vertices, then the ground field of L is one of fields from $\mathcal{F}T$.

If $u = e_1 \cdot e_2 = 2$, then the ground field of Γ is equal to $\mathbb{Q}$.

Assume that $2 < u = e_1 \cdot e_2 < t$. Then the subset $\{e_1, e_2\}$ is hyperbolic and connected. Since Γ^{hyp} is connected, contains e_1, e_2 and has at least 4 vertices, there exists a connected subgraph $\Gamma^{(4)}$ of Γ^{hyp} which contains e_1, e_2 and has four vertices. It is hyperbolic since it contains a hyperbolic subset $\{e_1, e_2\}$. Then $\Gamma^{(4)}$ is one of graphs $\Gamma_i^{(4)}$, $i = 1, 2, 3, 4, 5$. Then the ground field of Γ is equal to one of fields of $\Gamma_i^{(4)}(t)$, $1 \le i \le 5$.

This finishes the proof. □

The degrees of the fields from $\mathcal{F}L^4$ are bounded by 2, and the degrees of the fields from $\mathcal{F}T$ are bounded by 5.

For arithmetic hyperbolic reflection groups, minimality $t = 14$ is especially important. Using the same methods as for the proof of Theorem 1 in [14], we can prove the following explicit estimates.

Theorem 4. *The degrees of the fields from $\mathcal{F}\Gamma_1^{(4)}(14)$ are bounded ($\le$) by 22.*

- *The degrees of the fields from $\mathcal{F}\Gamma_2^{(4)}(14)$ are bounded by 39.*
- *The degrees of the fields from $\mathcal{F}\Gamma_3^{(4)}(14)$ are bounded by 53.*
- *The degrees of the fields from $\mathcal{F}\Gamma_4^{(4)}(14)$ are bounded by 56.*
- *The degrees of the fields from $\mathcal{F}\Gamma_5^{(4)}(14)$ are bounded by 54.*
- *Thus, the constant $N(14)$ of Theorem 1 can be taken to be $N(14) = 56$.*

PROOF. The proof requires long considerations and calculations. It will be given later in Section 5. □

In the next section, we shall consider applications of these explicit estimates to arithmetic hyperbolic reflection groups.

4. Applications to the ground fields of arithmetic hyperbolic reflection groups

Let W be an arithmetic hyperbolic reflection group of the dimension $n \ge 2$, and $\mathcal{M}$ its fundamental chamber. If $\mathcal{M}$ is not bounded, then the ground field of W is $\mathbb{Q}$ (see [26]). Since we are interested in possible ground fields of W, further we assume that $\mathcal{M}$ is bounded. Then any edge $r = v_1v_2$ of $\mathcal{M}$ defines a V-arithmetic finite edge chamber $\mathcal{M}(r)$ which is intersection of all half-spaces $\mathcal{H}_\delta^+$, $\delta \in P(\mathcal{M})$, such that the hyperplane $\mathcal{H}_\delta$ contains one of vertices v_1, v_2. The corresponding V-arithmetic edge graph is the Gram graph $\Gamma(r) = \Gamma\big(\mathcal{M}(r)\big)$ of these elements $\delta \in P(\mathcal{M})$.

By [13,14], there exists $e \in P(\mathcal{M})$ which defines a narrow face $\mathcal{M}_e = \mathcal{H}_e \cap \mathcal{M}$ of minimality 14 of $\mathcal{M}$ (the same is valid for any hyperbolic closed convex polyhedron). It means that for the set

$$P(\mathcal{M}, e) = \{\delta \in P(\mathcal{M}) \mid \mathcal{H}_\delta \cap \mathcal{H}_e \neq \varnothing\}$$

of perpendicular vectors to neighbours $\mathcal{M}_\delta$ of $\mathcal{M}_e$, we have

$$\delta_1 \cdot \delta_2 < 14 \quad \forall \delta_1, \delta_2 \in P(\mathcal{M}, e).$$

We consider edges $r = v_1v_2$ in $\mathcal{M}_e$; all of them give V-arithmetic edge chambers $\mathcal{M}(r)$ and V-arithmetic edge graphs of minimality 14.

4.1. The ground fields of arithmetic hyperbolic reflection groups in dimensions $n \geq 10$. Let us assume that the dimension $n \geq 10$. In [14], we showed that the narrow face $\mathcal{M}_e$ contains an edge $r = v_1v_2$ such that the corresponding edge chamber $\mathcal{M}(r)$ and its Gram graph $\Gamma = \Gamma(r)$ (they have minimality 14) have the hyperbolic connected component Γ^{hyp} with at least 4 vertices. By Theorem 1, then the degree of the ground field of W is bounded by the constant $N(14)$. By Proposition 3, we additionally obtain that the ground field of W belongs to one of finite sets of fields $\mathcal{F}L^4$, $\mathcal{F}T$ and $\mathcal{F}\Gamma_i^{(4)}(14)$, $1 \leq i \leq 5$. By Theorem 4, the degree of this field is bounded by 56.

Here we want to give another proof of this result which permits to avoid enumeration of combinatorial types of 3-dimensional polyhedra with small number of vertices and excludes the most difficult and large set of fields $\mathcal{F}\Gamma_5^{(4)}(14)$ from the statement. This proof follows [16]. It additionally uses some important arguments by Vinberg from [29]. In [29], Vinberg showed that for $n \geq 30$ there are no arithmetic hyperbolic reflection groups; for $n \geq 22$ their ground fields belong to the set $\{\mathbb{Q}, \mathbb{Q}(\sqrt{2}), \mathbb{Q}(\sqrt{5}), \mathbb{Q}(\cos 2\pi/7)\}$; for $n \geq 14$ their ground fields belong to $\mathcal{F}T$. Thus, Theorem 5 below can be viewed as an extension of these Vinberg's statements to $n \geq 10$.

Theorem 5. *In dimensions $n \geq 10$, the ground field of any arithmetic hyperbolic reflection group belongs to one of finite sets of fields $\mathcal{F}L^4$, $\mathcal{F}T$, $\mathcal{F}\Gamma_1^{(4)}(14)$, $\mathcal{F}\Gamma_2^{(4)}(14)$, $\mathcal{F}\Gamma_3^{(4)}(14)$ or $\mathcal{F}\Gamma_4^{(4)}(14)$. In particular, its degree is bounded $(\leq)$ by 56.*

PROOF. Let W be an arithmetic hyperbolic reflection group of the dimension $n \geq 2$ and $\mathcal{M}$ its fundamental chamber with the Gram graph $\Gamma(P(\mathcal{M}))$. If $\mathcal{M}$ is not compact, then the ground field is $\mathbb{Q}$. Thus, further we can assume that $\mathcal{M}$ is compact. It is known, [26], that then $\mathcal{M}$ is a simple polyhedron which means that $\mathcal{M}$ is simplicial in its vertices: any vertex is contained in exactly n hyperplanes $\mathcal{H}_\delta$, $\delta \in P(\mathcal{M})$.

Arguing like for the proof of Proposition 3, we obtain

Lemma 6. *Let W be an arithmetic hyperbolic reflection group, $\mathcal{M}$ its fundamental chamber, and their ground field does not belong to $\mathcal{F}L^4$ and $\mathcal{F}T$.*

Then any edge $r = v_1v_2$ of $\mathcal{M}$ defines an edge polyhedron $\mathcal{M}(r)$ such that the corresponding vertices e_1 and e_2 of its Gram graph (that is the hyperplanes $\mathcal{H}_{e_1}$ and $\mathcal{H}_{e_2}$ contain only vertices v_1 and v_2 of the edge respectively) are joined by a broken edge (that is $u = e_1 \cdot e_2 > 2$).

For a vertex v of $\mathcal{M}$ we denote by $C(v)$ the Coxeter graph of v which is Gram graph of all $\delta \in P(\mathcal{M})$ such that the hyperplane $\mathcal{H}_\delta$ contains v. Since $\mathcal{M}$ is compact, $C(v)$ has exactly n vertices.

Lemma 7. *Let W be an arithmetic hyperbolic reflection group, $\mathcal{M}$ its fundamental chamber, and their ground field does not belong to $\mathcal{F}L^4$, $\mathcal{F}T$ and $\mathcal{F}\Gamma_i^{(4)}(14)$, $i = 1, 2, 3, 4$. Let $\mathcal{M}_e$, $e \in P(\mathcal{M})$, be a narrow face of $\mathcal{M}$ of minimality 14.*

Then for any vertex v of $\mathcal{M}$ which is contained in $\mathcal{M}_e$, all connected components of the Coxeter graph $C(v)$ of v have one or two vertices only.

PROOF. Let e_1 be a vertex of $C(v)$ which is different from e. Let r be an edge of $\mathcal{M}$ defined by $C(v)-\{e_1\}$. It means that a hyperplane $\mathcal{H}_\delta$, $\delta \in P(\mathcal{M})$, contains the edge r if and only if $\delta \in C(v)-\{e_1\}$. Since $e \in C(v)-\{e_1\}$, the edge r belongs to $\mathcal{M}_e$. One of terminals of r is v, and $\mathcal{H}_{e_1}$ contains v, and it does not contain r. Let $v_2 \in \mathcal{M}_e$ be another terminal of r, and $e_2 \in P(\mathcal{M})$ gives the hyperplane $\mathcal{H}_{e_2}$ which contains only v_2 and does not contain v. Thus, $\Gamma(r) = C(v) \cup \{e_2\}$ is the Gram graph of the edge polyhedron of $\mathcal{M}$ corresponding to $r = vv_2$. By Lemma 6, e_1, e_2 give the only broken edge of $\Gamma(r)$.

Since $r \subset \mathcal{M}_e$ and $\mathcal{M}_e$ has minimality 14, then $\Gamma(r)$ also has minimality 14.

Assume that the connected component of e_1 in $C(v)$ has more than two vertices. Then there exists a connected subgraph Γ_{e_1} of $C(v)$ which has three vertices and contains e_1. Then $\Gamma_{e_1} \cup \{e_2\}$ is one of graphs $\Gamma_i^{(4)}$, $i = 1, 2, 3, 4, 5$, of minimality 14. It cannot be equal to $\Gamma_5^{(4)}$ because connected components of e_1 in $\Gamma_5^{(4)} - \{e_2\}$ and e_2 in $\Gamma_5^{(4)} - \{e_1\}$ have two vertices. It follows that the ground field of $\mathcal{M}$ belongs to one of sets $\mathcal{F}\Gamma_i^{(4)}(14)$, $i = 1, 2, 3, 4$. We get a contradiction.

Thus, any vertex $e_1 \neq e$ of the graph $C(v)$ has connected component with one or two vertices. Obviously, the same will be valid for e as well.

This finishes the proof. □

The crucial topological argument of the proof of Theorem 5 (additional to Theorem 1 and existence of a narrow face of minimality 14) is as follows. By [14, Theorem 3.2.1], for any $0 \le i \le k$, $2 \le k$ and $2k-1 \le m$, the average number $\alpha_m^{(i,k)}$ of i-dimensional faces in k-dimensional faces of any m-dimensional simple convex polyhedron satisfies the inequality

$$\alpha_m^{(i,k)} < \frac{C_{m-i}^{k-i}\left(C_{[m/2]}^{i} + C_{m-[m/2]}^{i}\right)}{C_{[m/2]}^{k} + C_{m-[m/2]}^{k}}. \tag{6}$$

In particular, for $n \ge 4$, the average number $\alpha_{n-1}^{(0,2)}$ of vertices of 2-dimensional faces of a narrow face $\mathcal{M}_e$ (of dimension $m = n-1$) satisfies

$$\alpha_{n-1}^{(0,2)} < 4 + \begin{cases} 4/(n-2) & \text{if } n \text{ is even,} \\ 4/(n-3) & \text{if } n \text{ is odd.} \end{cases} \tag{7}$$

Now, assume that the ground field does not belong to $\mathcal{F}L$, $\mathcal{F}T$ and $\mathcal{F}\Gamma_i^{(4)}(14)$, $i = 1, 2, 3, 4$. Following Vinberg [28], let us estimate the number A of non-right (that is $\neq \pi/2$) 2-dimensional angles of $\mathcal{M}_e$.

Let v be a vertex of $\mathcal{M}_e$ and $C(v)$ the Coxeter graph of v in $\mathcal{M}$. Any 2-dimensional angle of $\mathcal{M}_e$ with the vertex v is defined by a subset of two distinct vertices $\{\delta_1, \delta_2\} \subset C(v) - \{e\}$. Really, $C(v) - \{\delta_1, \delta_2\}$ gives perpendicular vectors to hyperplanes containing the plane of the angle, and $C(v) - \{\delta_1\}$, $C(v) - \{\delta_2\}$ similarly give edges of the angle. It is easy to see that the angle is not right if and only if δ_1 and δ_2 belong to one connected component of the graph $C(v)$. Thus, the number A_v of non-right 2-dimensional angles of $\mathcal{M}_e$ with the vertex v is equal to the number of subsets of two distinct vertices of $C(v) - \{e\}$ which belong to one connected component of the graph $C(v)$. By Lemma 7, $A_v \le [(n-1)/2]$. Thus,

$$\left[\frac{n-1}{2}\right]\alpha_0 \ge A \tag{8}$$

where α_0 is the number of vertices of $\mathcal{M}_e$.

Each hyperbolic triangle has at least two non-right angles. Each hyperbolic quadrangle has at least one non-right angle. Denoting by α_2^l the number of 2-dimensional faces of $\mathcal{M}_e$ with l vertices, and by α_2 the number of all 2-dimensional faces of $\mathcal{M}_e$, we obtain

$$A \geq \sum_{l\geq 3}(5-l)\alpha_2^l = 5\alpha_2 - \sum_{l\geq 3} l\alpha_2^l = 5\alpha_2 - \alpha_{n-1}^{(0,2)}\alpha_2 = (5-\alpha_{n-1}^{(0,2)})\alpha_2. \tag{9}$$

Since $\mathcal{M}_e$ is simple $(n-1)$-dimensional convex polyhedron, we have

$$\frac{\alpha_0(n-1)(n-2)}{2} = \alpha_2\alpha_{n-1}^{(0,2)}. \tag{10}$$

By (8), (9) and (10), we obtain

$$\alpha_{n-1}^{(0,2)}\left(\frac{(n-1)(n-2)}{2} + \left[\frac{n-1}{2}\right]\right) \geq \frac{5(n-1)(n-2)}{2}.$$

By (7), we get

$$\left(4+\frac{4}{n-2}\right)\left(\frac{(n-1)(n-2)}{2}+\frac{n-2}{2}\right) > \frac{5(n-1)(n-2)}{2}$$

for even n, and

$$\left(4+\frac{4}{n-3}\right)\left(\frac{(n-1)(n-2)}{2}+\frac{n-1}{2}\right) > \frac{5(n-1)(n-2)}{2}$$

for odd n. It follows $n \leq 9$ which contradicts to the assumption $n \geq 10$.

This finishes the proof. □

4.2. The ground fields of arithmetic hyperbolic reflection groups in dimensions $n \geq 6$. We introduce one more set of fields. We consider plane (or Fuchsian) arithmetic hyperbolic reflection groups W with a quadrangle fundamental polygon K of minimality 14. That is $P(K) = \{\delta_1, \delta_2, \delta_3, \delta_4\}$ satisfies

$$\delta_i \cdot \delta_j < 14, \quad \forall \delta_i, \delta_j \in P(K).$$

We call K an *arithmetic quadrangle of minimality* 14 respectively.

Definition 8. We denote by $\Gamma_{2,4}(14)$ the set of Gram graphs $\Gamma\big(P(K)\big)$ of all arithmetic quadrangles K of minimality 14. The set $\mathcal{F}\Gamma_{2,4}(14)$ consists of all their ground fields.

By Borel [5] and Takeuchi [24], for fixed $g \geq 0$ and $t \geq 0$, the number of arithmetic Fuchsian groups with signatures $(g; e_1, e_2, \ldots, e_t)$ is finite. Applying this result to $g = 0$ and $t = 4$, we obtain that sets of arithmetic quadrangles $\Gamma_{2,4}$ and their ground fields $\mathcal{F}\Gamma_{2,4}$ are finite. Then their subsets $\Gamma_{2,4}(14)$ and $\mathcal{F}\Gamma_{2,4}(14)$ are also finite.

Moreover, in [24, pp. 383 – 384] an upper bound n_0 of the degrees of the ground fields of Fuchsian groups with signatures $(g; e_1, e_2, \ldots, e_t)$ is given. It is

$$n_0 = \frac{\big(b + \log_e C(g,t)\big)}{\log_e(a/(2\pi)^{4/3})} \tag{11}$$

where

$$a = 29.099, \quad b = 8.3185, \quad C(g,t) = 2^{2g+t-2}(2g+t-2)^{2/3}$$

(here a and b are due to Odlyzko). It follows that

$$[\mathbb{K}:\mathbb{Q}] \le 11 \quad \text{for } \mathbb{K} \in \mathcal{F}\Gamma_{2,4} \supset \mathcal{F}\Gamma_{2,4}(14). \tag{12}$$

We have the following main result of the paper.

Theorem 9. *In dimensions $n \ge 6$, the ground field of any arithmetic hyperbolic reflection group belongs to one of finite sets of fields $\mathcal{F}L^4$, $\mathcal{F}T$, $\mathcal{F}\Gamma_i^{(4)}(14)$, $1 \le i \le 5$, and $\mathcal{F}\Gamma_{2,4}(14)$. In particular, its degree is bounded by 56.*

PROOF. We use notation of Section 4.1 above. By (7), for $n \ge 6$, a narrow face $\mathcal{M}_e$ has $\alpha_{n-1}^{(0,2)} < 5$. Thus, $\mathcal{M}_e$ has a triangle or quadrangle (2-dimensional) face. We consider both cases.

By Lemma 6, we have

Lemma 10. *Let W be an arithmetic hyperbolic reflection group, $\mathcal{M}$ its fundamental chamber, and their ground field does not belong to $\mathcal{F}L^4$ and $\mathcal{F}T$.*

Then $\mathcal{M}$ has no triangle faces (2-dimensional).

PROOF. Assume $\mathcal{M}$ contains a triangle 2-dimensional face. Then the edge polyhedron of $\mathcal{M}$ corresponding to the edge v_1v_2 of two vertices v_1 and v_2 of this triangle has the corresponding elements e_1 and e_2 (by Lemma 6) such that the hyperplanes $\mathcal{H}_{e_1}$ and $\mathcal{H}_{e_2}$ have a common point which is the third vertex of the triangle. Then $e_1 \cdot e_2 \le 2$. This contradicts Lemma 6.

This finishes the proof. □

Let $\mathcal{M}_e$ has a triangle face. By Lemma 10, then the ground field of $\mathcal{M}$ belongs to $\mathcal{F}L^4$ or $\mathcal{F}T$, as required.

Lemma 11. *Let W be an arithmetic hyperbolic reflection group, $\mathcal{M}$ its fundamental chamber, and their ground field does not belong to $\mathcal{F}L^4$, $\mathcal{F}T$ and $\mathcal{F}\Gamma_i^{(4)}$, $1 \le i \le 5$.*

Let $\mathcal{M}_e$, $e \in P(\mathcal{M})$, be a narrow face of $\mathcal{M}$ of minimality 14. Let $\mathcal{M}_4$ be a quadrangle face of $\mathcal{M}_e$. Let $Q \subset P(\mathcal{M})$ consists of all $n-2$ elements which are perpendicular to the plane of $\mathcal{M}_4$, and $\delta_j \in P(\mathcal{M})$, $j = 1, 2, 3, 4$, are additional 4 elements which are perpendicular to four edges of $\mathcal{M}_4$.

Then all elements δ_j are perpendicular to Q, and $\mathcal{M}_4$ is an arithmetic quadrangle of minimality 14 with $P(\mathcal{M}) = \{\delta_1, \delta_2, \delta_3, \delta_4\}$ and with the same ground field as $\mathcal{M}$. Thus, the ground field of W belongs to $\mathcal{F}\Gamma_{2,4}(14)$.

PROOF. We assume that δ_1, δ_2, δ_3 and δ_4 are perpendicular to four consecutive edges of $\mathcal{M}_4$. The quadrangle $\mathcal{M}_4$ has a non-right angle. We can assume that δ_1 and δ_2 are perpendicular to edges of this angle. Since the angle is not right, δ_1 and δ_2 belong to one connected component of the Coxeter graph of the vertex of the angle. By Lemma 7, then δ_1 and δ_2 give a connected component of the graph. It follows that δ_1 and δ_2 are perpendicular to Q.

Assume that δ_3 is not perpendicular to Q and $\delta_3 \cdot e > 0$ for $e \in Q$. Then δ_1, δ_2, δ_3, e define an edge graph $\Gamma_4^{(4)}$ (if $\delta_2 \cdot \delta_3 > 0$) or $\Gamma_5^{(4)}$ (if $\delta_2 \cdot \delta_3 = 0$) of minimality 14. It follows that the ground field of $\mathcal{M}$ belongs to $\mathcal{F}\Gamma_4^{(4)}(14)$ or $\mathcal{F}\Gamma_5^{(4)}(14)$, and we get a contradiction. Thus, δ_3 is perpendicular to Q. Similarly we can prove that δ_4 is perpendicular to Q.

This finishes the proof of the lemma. □

If $\mathcal{M}_e$ contains a quadrangle, by Lemma 11, the ground field of W belongs to $\mathcal{F}\Gamma_{2,4}(14)$, as required. This finishes the proof of the theorem. $\square$

4.3. The ground fields of arithmetic hyperbolic reflection groups in dimensions $n \geq 4$. Unfortunately, in dimensions $n \geq 4$ we don't know similar results to Theorems 5 and 9. Possibly, the recent preprint by Agol, Belolipetsky, Storm and Whyte [2] contains some similar information. It gives some effective bounds on the degrees and the discriminants of the ground fields of arithmetic hyperbolic reflection groups for $n \geq 4$. Unfortunately, they are not explicit, it seems.

On the other hand, the following result was obtained in [17].

Theorem 12 ([17]). *For $n \geq 4$, the ground field of any n-dimensional arithmetic hyperbolic reflection group is either the ground field of one of $n-1$ or $n-2$-dimensional arithmetic hyperbolic reflection group with a fundamental chamber of minimality* 14, *or a field from $\mathcal{F}L^4$, $\mathcal{F}T$ or $\mathcal{F}\Gamma_i^{(4)}(14)$, $i = 1, 2, 3, 4$.*

In particular, its degree is bounded by the maximum of the degrees of the ground fields of $n-1$ and $n-2$-dimensional arithmetic hyperbolic reflection groups with a fundamental chamber of minimality 14, *and of* 56 (*according to Theorem* 4 *of this paper*).

PROOF. We repeat arguments of [17]. Let $\mathcal{M}$ be a fundamental chamber of W. We can assume that $\mathcal{M}$ is compact and the ground field of $\mathcal{M}$ is not contained in $\mathcal{F}L^4$ and $\mathcal{F}T$. Let $\mathcal{M}_e$, $e \in P(\mathcal{M})$, be a face of $\mathcal{M}$ of minimality 14.

If all hyperplanes $\mathcal{H}_\delta$, $\delta \in P(\mathcal{M}, e) - \{e\}$, are perpendicular to $\mathcal{M}_e$ (equivalently, $\delta \cdot e = 0$), then $\mathcal{M}_e$ is a fundamental chamber of $n-1$-dimensional arithmetic hyperbolic reflection group with the same ground field as $\mathcal{M}$. We have $P(\mathcal{M}_e) = P(\mathcal{M}, e) - \{e\}$, and $\mathcal{M}_e$ has minimality 14.

If this is not the case, there exists $f \in P(\mathcal{M}, e) - \{e\}$ such that $f \cdot e > 0$ (equivalently, f and e are connected by a thin edge in Gram graph of $\mathcal{M}$). Then $\mathcal{M}_{e,f} = \mathcal{M} \cap \mathcal{H}_e \cap \mathcal{H}_f$ is $n-2$-dimensional face of $\mathcal{M}$. Let $P(\mathcal{M}, e, f)$ be the set of all $\delta \in P(\mathcal{M})$ such that the hyperplane $\mathcal{H}_\delta$ intersects the codimension-two subspace $\mathcal{H}_e \cap \mathcal{H}_f$ (then $\mathcal{M} \cap \mathcal{H}_e \cap \mathcal{H}_f \cap \mathcal{H}_\delta$ is a codimension-three face of $\mathcal{M}$ if δ is different from e and f). If $\mathcal{H}_\delta \perp \mathcal{H}_e \cap \mathcal{H}_f$ (equivalently, $\delta \cdot e = \delta \cdot f = 0$) for all $\delta \in P(\mathcal{M}, e, f) - \{e, f\}$, then $\mathcal{M}_{e,f}$ is a fundamental chamber of an arithmetic hyperbolic reflection group of dimension $n-2$ with the same ground field as $\mathcal{M}$. We have $P(\mathcal{M}_{e,f}) = P(\mathcal{M}, e, f) - \{e, f\}$, and $\mathcal{M}_{e,f}$ has minimality 14.

If this is not the case, there is $g \in P(\mathcal{M}, e, f) - \{e, f\}$ such that $\mathcal{H}_g$ is not perpendicular to $\mathcal{H}_e \cap \mathcal{H}_f$. This means that either $g \cdot e > 0$ or $g \cdot f > 0$. Thus, the Gram graph of e, f, g is a connected negative definite (that is connected Coxeter) graph.

We consider an edge r in the face $\mathcal{M}_{e,f} = \mathcal{M} \cap \mathcal{H}_e \cap \mathcal{H}_f$ of $\mathcal{M}$ such that r terminates in the hyperplane $\mathcal{H}_g$. Thus one of vertices of r is contained in $\mathcal{H}_g$ while the other is not (equivalently, r is not contained in $\mathcal{H}_g$). The existence of such an edge is obvious. Let $h \in P(\mathcal{M})$ defines the hyperplane $\mathcal{H}_h$ which contains only the vertex of r which does not belong to $\mathcal{H}_g$. Then g and h are joined by a broken edge in the edge graph $\Gamma(r)$ of r (here we can additionally assume that the ground field does not belong to $\mathcal{F}L^4$ and $\mathcal{F}T$). The four elements $\{e, f, g, h\}$ define a connected hyperbolic subgraph of $\Gamma(r)$ with four vertices. It is one of graphs $\Gamma_i^{(4)}(14)$, $i = 1, 2, 3, 4$. Then the ground field of $\mathcal{M}$ belongs to one of sets $\mathcal{F}\Gamma_i^{(4)}(14)$, $1 \le i \le 4$. This finishes the proof. □

Theorem 12 shows that the ground fields of arithmetic hyperbolic reflection groups which are different from the fields of $\mathcal{F}L^4$, $\mathcal{F}T$ and $\mathcal{F}\Gamma_i^{(4)}(14)$, $i = 1, 2, 3, 4$, come up from 2-dimensional and 3-dimensional arithmetic hyperbolic reflection groups.

4.4. The ground fields of arithmetic hyperbolic reflection groups in dimension $n = 3$. Finiteness of the number of maximal arithmetic hyperbolic reflection groups of dimension $n = 3$ was proved by Agol in [1]. It follows that the number of ground fields of 3-dimensional arithmetic hyperbolic reflection groups is also finite. Unfortunately, an explicit bound of the degrees of these fields is not known (by our knowledge). We remind that by [13], the set of the ground fields of arithmetic hyperbolic reflection groups of a fixed degree is finite and can be effectively found. Thus, an explicit upper bound of the degrees is the crucial problem for finding of the ground fields of 3-dimensional arithmetic hyperbolic reflection groups.

4.5. The ground fields of arithmetic hyperbolic reflection groups in dimension $n = 2$. Finiteness of the number of maximal arithmetic hyperbolic reflection groups of dimension $n = 2$ was proved by Long, Maclachlan and Reid [12]. They proved finiteness of maximal arithmetic Fuchsian groups of genus 0. Their ground fields contain ground fields of all arithmetic hyperbolic reflection groups.

Let Γ be a cocompact maximal arithmetic Fuchsian group of genus 0. Using results by M.-F. Vignéras [25] and Zograf [32], the bound

$$\mathrm{Area}(\mathbb{H}^2/\Gamma) \le \frac{128\pi}{3} \tag{13}$$

of the area of the arithmetic quotient was obtained in [12]. Let Γ has signature $(0; e_1, \ldots, e_t)$ where $e_i \ge 2$. Then the area of the quotient is equal to

$$\mathrm{Area}(\mathbb{H}^2/\Gamma) = 2\pi\left(t - 2 - \sum_{i=1}^{t} 1/e_i\right).$$

Since $e_i \ge 2$, by (13), we obtain $2\pi(t - 2 - t/2) \le 128\pi/3$, and $t \le 46$. By the result of Takeuchi (11), we obtain $n_0 \le 44$.

Thus, *the degrees of the ground fields of arithmetic Fuchsian groups of genus 0 are bounded by* 44. In particular, *the degrees of the ground fields of* 2*-dimensional arithmetic hyperbolic reflection groups are bounded by* 44.

Summarizing above results, we see that an explicit upper bound of the degrees of the ground fields of arithmetic hyperbolic reflection groups remains unknown in dimensions $n = 3, 4, 5$ only. Moreover, the dimension $n = 3$ is crucial for this problem. If one finds this bound for $n = 3$, we will know it for all remaining dimensions $n = 4$ and $n = 5$.

5. The ground fields of V-arithmetic connected finite edge graphs with four vertices, of minimality 14

Here we give explicit upper bounds of the degrees of the fields from the finite sets $\mathcal{F}\Gamma_i^{(4)}(14)$, $1 \le i \le 5$ (see Definition 2), and prove Theorem 4. Moreover, our considerations deliver important information about these sets of fields.

Like for the proof of Theorem 1 from [14], we use the following general results from [14].

Theorem 13 ([14, Theorem 1.2.1]). *Let $\mathbb{F}$ be a totally real algebraic number field, and let each embedding $\sigma : \mathbb{F} \to \mathbb{R}$ corresponds to an interval $[a_\sigma, b_\sigma]$ in $\mathbb{R}$ where*

$$\prod_{\sigma} \frac{b_\sigma - a_\sigma}{4} < 1.$$

In addition, let the natural number m and the intervals $[s_1, t_1], \ldots, [s_m, t_m]$ in $\mathbb{R}$ be fixed. Then there exists a constant $N(s_i, t_i)$ such that, if α is a totally real algebraic integer and if the following inequalities hold for the embeddings $\tau\colon \mathbb{F}(\alpha) \to \mathbb{R}$:

$$\begin{aligned} s_i \le \tau(\alpha) \le t_i \qquad & \text{for } \tau = \tau_1, \ldots, \tau_m,\\ a_{\tau|\mathbb{F}} \le \tau(\alpha) \le b_{\tau|\mathbb{F}} \qquad & \text{for } \tau \ne \tau_1, \ldots, \tau_m, \end{aligned}$$

then

$$[\mathbb{F}(\alpha){:}\mathbb{F}] \le N(s_i, t_i).$$

Theorem 14 ([14, Theorem 1.2.2]). *Under the conditions of Theorem 13, $N(s_i, t_i)$ can be taken to be $N(s_i, t_i) = N$, where N is the least natural number solution of the inequality*

$$NM \ln(1/R) - M \ln(N+1) - \ln B \ge \ln S. \tag{14}$$

Here

$$M = [\mathbb{F}{:}\mathbb{Q}], \qquad B = 2\sqrt{|\operatorname{discr} \mathbb{F}|}; \tag{15}$$

$$R = \sqrt{\prod_{\sigma} \frac{b_\sigma - a_\sigma}{4}}, \quad S = \prod_{i=1}^{m} \frac{2er_i}{b_{\sigma_i} - a_{\sigma_i}} \tag{16}$$

where

$$\sigma_i = \tau_i \mid \mathbb{F}, \quad r_i = \max\{|b_i - a_{\sigma_i}|, |b_{\sigma_i} - a_i|\}. \tag{17}$$

We note that the proof of Theorems 13 and 14 uses a variant of Fekete's Theorem (1923) about existence of non-zero integer polynomials of bounded degree which differ only slightly from zero on appropriate intervals. See [14, Theorem 1.1.1].

5.1. The fields from $\mathcal{F}\Gamma_1^{(4)}(14)$. For $\Gamma_1^{(4)}(14)$ (see Figure 4), we assume that integers $s, k, r, p \ge 3$. Subgraphs $\Gamma_1^{(4)} - \{e_1\}$ and $\Gamma_1^{(4)} - \{e_2\}$ must be Coxeter graphs. It follows that we must consider only the following cases (up to evident symmetries): either $s = k = 3$ and $5 \ge r \ge p \ge 3$, or $s = p = 3$ and $5 \ge r \ge k \ge 4$; the totally real algebraic integer u satisfies the inequality $2 < u < 14$.

The Gram matrix is

$$\begin{pmatrix} -2 & u & 2\cos(\pi/s) & 2\cos(\pi/r)\\ u & -2 & 2\cos(\pi/k) & 2\cos(\pi/p)\\ 2\cos(\pi/s) & 2\cos(\pi/k) & -2 & 0\\ 2\cos(\pi/r) & 2\cos(\pi/p) & 0 & -2 \end{pmatrix}, \tag{18}$$

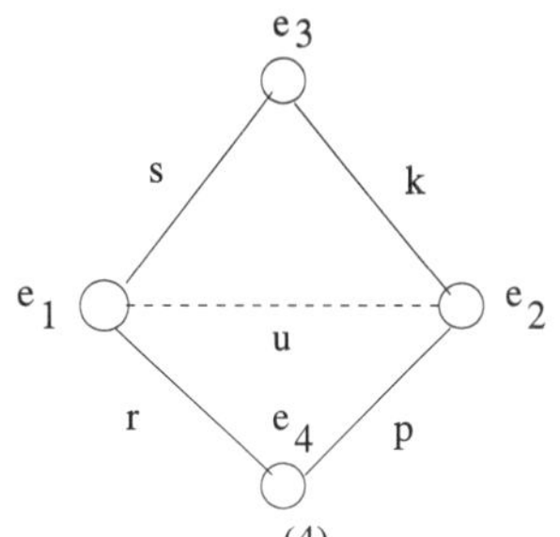

FIGURE 4. The graph $\Gamma_1^{(4)}$

where all entries are algebraic integers, and its determinant $d(u)$ is given by the equality

$$(19)\quad -\frac{d(u)}{4} = \left(u + 2\left(\cos\frac{\pi}{r}\cos\frac{\pi}{p} + \cos\frac{\pi}{k}\cos\frac{\pi}{s}\right)\right)^2 - \left(\cos\frac{2\pi}{k} + \cos\frac{2\pi}{p}\right)\left(\cos\frac{2\pi}{r} + \cos\left(\frac{2\pi}{s}\right)\right).$$

The ground field $\mathbb{K}$ is generated by the cyclic products

$$\mathbb{K} = \left(u^2, u\cos\left(\frac{\pi}{s}\right)\cos\left(\frac{\pi}{k}\right), u\cos\left(\frac{\pi}{p}\right)\cos\left(\frac{\pi}{r}\right), \cos^2\left(\frac{\pi}{s}\right), \cos^2\left(\frac{\pi}{k}\right), \cos^2\left(\frac{\pi}{p}\right), \cos^2\left(\frac{\pi}{r}\right), \cos\left(\frac{\pi}{s}\right)\cos\left(\frac{\pi}{k}\right)\cos\left(\frac{\pi}{p}\right)\cos\left(\frac{\pi}{r}\right)\right).$$

Here and in what follows, we denote $\sigma^{(+)}\colon \mathbb{K} \to \mathbb{R}$ the geometric (the identity) embedding, and $\sigma\colon \mathbb{K} \to \mathbb{R}$ all other embeddings $\sigma \neq \sigma^{(+)}$. We have $\sigma^{(+)}\big(d(u)\big) < 0$ because $\Gamma_1^{(4)}$ is hyperbolic for $\sigma^{(+)}$ since e_1, e_2 define a hyperbolic subgraph because $u = e_1 \cdot e_2 > 2$. And $\sigma\big(d(u)\big) > 0$ since $\Gamma_1^{(4)}$ is negative definite for σ. In particular, $-2 < \sigma(u) < 2$ by Cauchy inequality. Thus, $\Gamma_1^{(4)}$ is V-arithmetic if and only if

$$(20)\quad \sigma\left(u + 2\left(\cos\frac{\pi}{r}\cos\frac{\pi}{p} + \cos\frac{\pi}{k}\cos\frac{\pi}{s}\right)\right)^2 < \sigma\left(\left(\cos\frac{2\pi}{k} + \cos\frac{2\pi}{p}\right)\left(\cos\frac{2\pi}{r} + \cos\frac{2\pi}{s}\right)\right).$$

for each $\sigma \neq \sigma^{(+)}$.

We have $\mathbb{K} = \mathbb{Q}(u^2)$ since $4 < \sigma^{(+)}(u^2) < 14^2$ and $0 < \sigma(u^2) < 4$ if $\sigma \neq \sigma^{(+)}$. We have $[\mathbb{K}(u){:}\mathbb{K}] = 2$ if $u \notin \mathbb{K}$. If $\tau\colon \mathbb{K}(u) \to \mathbb{R}$ gives $\tau \mid \mathbb{K} = \sigma^{(+)}$, then $\tau(u) = \pm u$ (where u is taken for the geometric embedding $\sigma^{(+)}$), and either $2 < \tau(u) < 14$ or $-14 < \tau(u) < -2$. The last inequality is possible, only if u does not belong to $\mathbb{K}$.

If $\tau \mid \mathbb{K} = \sigma \neq \sigma^{(+)}$, then by (20),

$$
\begin{aligned}
- 2\widetilde{\tau}\left(\cos\frac{\pi}{r}\cos\frac{\pi}{p} + \cos\frac{\pi}{k}\cos\frac{\pi}{s}\right)& \\
- \sqrt{\tau\left(\left(\cos\frac{2\pi}{k} + \cos\frac{2\pi}{p}\right)\left(\cos\frac{2\pi}{r} + \cos\frac{2\pi}{s}\right)\right)}& \\
< \tau(u) < -2\widetilde{\tau}\left(\cos\frac{\pi}{r}\cos\frac{\pi}{p} + \cos\frac{\pi}{k}\cos\frac{\pi}{s}\right)& \\
+ \sqrt{\tau\left(\left(\cos\frac{2\pi}{k} + \cos\frac{2\pi}{p}\right)\left(\cos\frac{2\pi}{r} + \cos\frac{2\pi}{s}\right)\right)}&
\end{aligned}
$$

where $\widetilde{\tau}$ extends τ. Thus, $\tau(u)$ belongs to the interval of the length

$$2\sqrt{\tau\left(\left(\cos\frac{2\pi}{k} + \cos\frac{2\pi}{p}\right)\left(\cos\frac{2\pi}{r} + \cos\frac{2\pi}{s}\right)\right)}.$$

We apply Theorems 13 and 14 to

$$\mathbb{F} = \mathbb{Q}\left(\cos^2\left(\frac{\pi}{s}\right), \cos^2\left(\frac{\pi}{k}\right), \cos^2\left(\frac{\pi}{p}\right), \cos^2\left(\frac{\pi}{r}\right)\right)$$

and $\alpha = u$ to bound $[\mathbb{F}(u) : \mathbb{F}]$. Since $\mathbb{K} = \mathbb{F}(u^2)$, then $m = [\mathbb{F}(u):\mathbb{K}] \leq 2$. By considerations above, we have

$$
\begin{gathered}
M = [\mathbb{F}:\mathbb{Q}], \quad B = 2\sqrt{|\operatorname{discr}\mathbb{F}|}, \\
R = \frac{N_{\mathbb{F}/\mathbb{Q}}\big((\cos(2\pi/k) + \cos(2\pi/p))(\cos(2\pi/r) + \cos 2(2\pi/s))\big)^{1/4}}{2^{M/2}}, \\
S = \left(\frac{16 \cdot e}{\sqrt{((\cos(2\pi/k) + \cos(2\pi/p))(\cos(2\pi/r) + \cos(2\pi/s)))}}\right)^m, \\
[\mathbb{K}:\mathbb{Q}] \leq NM/m
\end{gathered}
$$

where N is the least natural solution of the inequality

$$NM\ln(1/R) - M\ln(N+1) - \ln B \geq \ln S.$$

By direct calculations, we obtain the following.

Let $k = s = 3$. Then $m = 1$.

For $r = p = 3$, we obtain $M = 1$, $B = 2$, $R = 1/\sqrt{2}$, $S = 16e$. Then $N = 22$ and $[\mathbb{K}:\mathbb{Q}] \leq 22$.

For $r = 4$, $p = 3$, we obtain $M = 1$, $B = 2$, $R = \frac{1}{2}^{3/4}$, $S = 16e\sqrt{2}$. Then $N = 15$ and $[\mathbb{K}:\mathbb{Q}] \leq 15$.

For $r=5$, $p=3$, we obtain $M=2$, $B=2\sqrt{5}$, $R=\frac{1}{2}^{3/2}$, $S=16e/\sqrt{\frac{1}{2}-\cos(2\pi/5)}$. Then $N = 5$ and $[\mathbb{K}:\mathbb{Q}] \leq 10$.

For $r = p = 4$, we obtain $M = 1$, $B = 2$, $R = \frac{1}{2}$, $S = 32e$. Then $N = 12$ and $[\mathbb{K}:\mathbb{Q}] \leq 12$.

For $r=5$, $p=4$, we obtain $M=2$, $B=2\sqrt{5}$, $R=\frac{1}{4}$, $S=32e/\sqrt{1-2\cos(2\pi/5)}$. Then $N = 4$ and $[\mathbb{K}:\mathbb{Q}] \leq 8$.

For $r = p = 5$, we obtain $M = 2$, $B = 2\sqrt{5}$, $R = \frac{1}{4}$, $S = 16e/\left(\frac{1}{2} - \cos(2\pi/5)\right)$. Then $N = 4$ and $[\mathbb{K}:\mathbb{Q}] \le 8$.

Now let $s = p = 3$. Then $m = 1$ or $m = 2$.

For $r = k = 4$, we obtain $M = 1$, $B = 2$, $R = \frac{1}{2}$, $S = (32e)^m$. Then $N = 13$ and $[\mathbb{K}:\mathbb{Q}] \le 13$ for $m = 1$; $N = 22$ and $[\mathbb{K}:\mathbb{Q}] \le 11$ for $m = 2$.

For $r = 5$, $k = 4$, we obtain $m = 1$, $M = 2$, $B = 2\sqrt{5}$, $R = \frac{1}{4}$, $S = 32e/\sqrt{1 - 2\cos(2\pi/5)}$. Then $N = 4$ and $[\mathbb{K}:\mathbb{Q}] \le 8$.

For $r = k = 5$, we obtain $M = 2$, $B = 2\sqrt{5}$, $R = \frac{1}{4}$, $S = 16e/\left(\frac{1}{2} - \cos(2\pi/5)\right)^m$. Then $N = 4$ and $[\mathbb{K}:\mathbb{Q}] \le 8$ for $m = 1$, and $N = 6$ and $[\mathbb{K}:\mathbb{Q}] \le 6$ for $m = 2$.

Note that the bound for $[\mathbb{K}:\mathbb{Q}]$ is always worse for $m = 1$ than for $m = 2$. It follows from our method. Further, in similar considerations, we can consider $m = 1$ only.

Thus, our upper bound for the degrees of the fields from $\mathcal{F}\Gamma_1^{(4)}(14)$ is 22.

5.2. The fields from $\mathcal{F}\Gamma_2^{(4)}(14)$. For $\Gamma_2^{(4)}(14)$ (see Figure 5), $s, k, p \ge 3$ are natural numbers and $2 < u < 14$ is a totally real algebraic integer. Moreover, we have only the following possibilities: $3 \le s \le k \le 5$, $p = 3$; $s = k = 3$, $p = 4, 5$.

The ground field $\mathbb{K} = \mathbb{Q}(u^2)$ contains cyclic products

$$\cos^2\frac{\pi}{s}, \quad \cos^2\frac{\pi}{k}, \quad \cos^2\frac{\pi}{p}, \quad u^2, \quad u\cos\frac{\pi}{s}\cos\frac{\pi}{k}.$$

This case is similar to $\mathcal{F}\Gamma_1^{(4)}(14)$. The determinant $d(u)$ of the Gram matrix is determined by the equality

$$-\frac{d(u)}{4} = \sin^2\frac{\pi}{p}u^2 + 4\cos\frac{\pi}{s}\cos\frac{\pi}{k}u + 4\left(\cos^2\frac{\pi}{s} + \sin^2\frac{\pi}{k} + \cos^2\frac{\pi}{p} - 1\right).$$

Let

$$D = 16\cos^2\frac{\pi}{s}\cos^2\frac{\pi}{k} + 16\sin^2\frac{\pi}{p}\left(1 - \cos^2\frac{\pi}{s} - \cos^2\frac{\pi}{k} - \cos^2\frac{\pi}{p}\right)$$

be the discriminant of this quadratic polynomial of the variable u. The graph $\Gamma_2^{(4)}(14)$ is V-arithmetic if and only if for $\tau\colon \mathbb{K}(u) \to \mathbb{R}$ which is different from $\sigma^{(+)}$ on $\mathbb{K}$, one has

$$\frac{-4\widetilde{\tau}\big(\cos(\pi/s)\cos(\pi/k)\big) - \sqrt{\tau(D)}}{2\tau\big(\sin^2(\pi/p)\big)} < \tau(u) < \frac{-4\widetilde{\tau}\big(\cos(\pi/s)\cos(\pi/k)\big) + \sqrt{\tau(D)}}{2\tau\big(\sin^2(\pi/p)\big)}$$

where $\widetilde{\tau}$ extends τ.

Thus, $\tau(u)$ belongs to an interval of the length $2\sqrt{\tau\big(D/(4\sin^4(\pi/p))\big)}$.

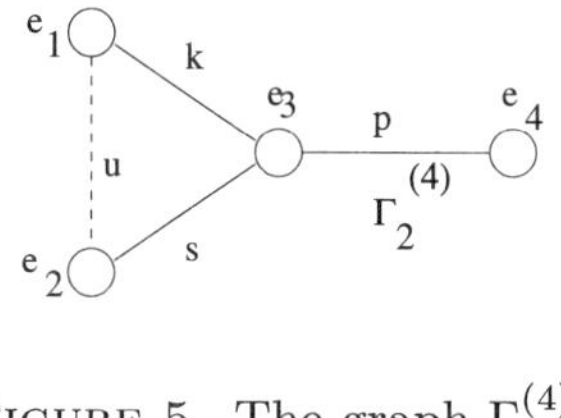

FIGURE 5. The graph $\Gamma_2^{(4)}$

We can apply Theorems 13, 14 to $\mathbb{F} = \mathbb{Q}(\cos^2(\pi/s), \cos^2(\pi/k), \cos^2(\pi/p))$ and $\alpha = u$. Then

$$M = [\mathbb{F}:\mathbb{Q}], \qquad B = 2\sqrt{|\operatorname{discr}\mathbb{F}|},$$
$$R = \frac{N_{\mathbb{F}/\mathbb{Q}}(D)^{1/4}}{N_{\mathbb{F}/\mathbb{Q}}\big(\sin^2(\pi/p)\big)^{1/2}2^M}, \qquad S = \frac{32e\sin^2(\pi/p)}{\sqrt{D}},$$

$[\mathbb{K}:\mathbb{Q}] \le NM$ where N is the smallest natural solution of the inequality

$$NM\ln(1/R) - M\ln(N+1) - \ln B \ge \ln S.$$

We obtain:

if $s = k = p = 3$, then $[\mathbb{K}:\mathbb{Q}] \le 39$; if $s = 3$, $k = 4$, $p = 3$, then $[\mathbb{K}:\mathbb{Q}] \le 21$;
if $s = 3$, $k = 5$, $p = 3$, then $[\mathbb{K}:\mathbb{Q}] \le 14$; if $s = k = 4$, $p = 3$, then $[\mathbb{K}:\mathbb{Q}] \le 14$;
if $s = 4$, $k = 5$, $p = 3$, then $[\mathbb{K}:\mathbb{Q}] \le 10$; if $s = k = 5$, $p = 3$, then $[\mathbb{K}:\mathbb{Q}] \le 10$;
if $s = k = 3$, $p = 4$, then $[\mathbb{K}:\mathbb{Q}] \le 22$; if $s = k = 3$, $p = 5$, then $[\mathbb{K}:\mathbb{Q}] \le 14$.

Thus, our upper bound for the degrees of the fields from $\mathcal{F}\Gamma_2^{(4)}(14)$ is 39.

5.3. The fields from $\mathcal{F}\Gamma_3^{(4)}(14)$. For $\Gamma_3^{(4)}(14)$ (see Figure 6), $s \ge 2$, $k, r \ge 3$ are natural numbers, and $2 < u < 14$ is a totally real algebraic integer. Moreover, we have only the following possibilities: $s = 2$, $k = 3$, $r = 3, 4, 5$; $s = 2$, $k = 4, 5$, $r = 3$; $3 \le s \le k \le 5$, $r = 3$; $s = k = 3$, $r = 4, 5$.

The ground field $\mathbb{K} = \mathbb{Q}(u^2)$ contains cyclic products

$$\cos^2\frac{\pi}{s}, \quad \cos^2\frac{\pi}{k}, \quad \cos^2\frac{\pi}{r}, \quad u^2, \quad u\cos\frac{\pi}{s}\cos\frac{\pi}{k}\cos\frac{\pi}{r}.$$

This case is similar to $\mathcal{F}\Gamma_1^{(4)}(14)$ and $\mathcal{F}\Gamma_2^{(4)}(14)$. The determinant $d(u)$ of the Gram matrix is determined by the equality

$$-\frac{d(u)}{4} = \sin^2\frac{\pi}{r}u^2 + 2\cos\frac{\pi}{s}\cos\frac{\pi}{k}\cos\frac{\pi}{r}u + 4\cos^2\frac{\pi}{r} - 4\sin^2\frac{\pi}{s}\sin^2\frac{\pi}{k}.$$

Let

$$D = 4\cos^2\frac{\pi}{s}\cos^2\frac{\pi}{k}\cos^2\frac{\pi}{r} + 16\sin^2\frac{\pi}{s}\sin^2\frac{\pi}{k}\sin^2\frac{\pi}{r} - 16\sin^2\frac{\pi}{r}\cos^2\frac{\pi}{r}$$

be the discriminant of this quadratic polynomial of the variable u. The graph $\Gamma_3^{(4)}(14)$ is V-arithmetic if and only if for $\tau\colon \mathbb{K}(u) \to \mathbb{R}$ which is different from $\sigma^{(+)}$

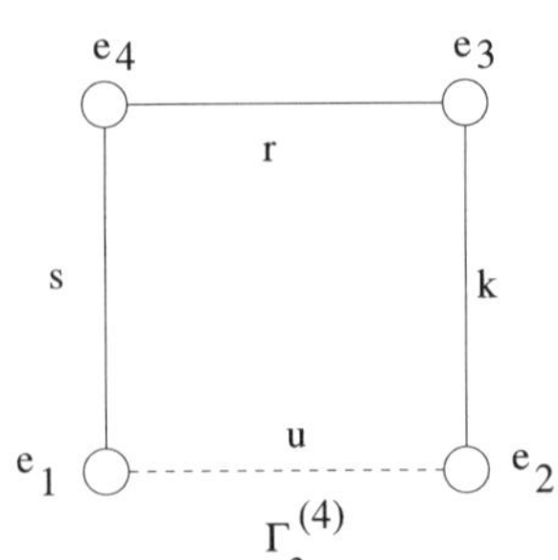

FIGURE 6. The graph $\Gamma_3^{(4)}$.

on $\mathbb{K}$, one has

$$\frac{-2\widetilde{\tau}\big(\cos(\pi/s)\cos(\pi/k)\cos(\pi/r)\big)-\sqrt{\tau(D)}}{2\tau\big(\sin^2(\pi/r)\big)} < \tau(u) < \frac{-2\widetilde{\tau}\big(\cos(\pi/s)\cos(\pi/k)\cos(\pi/r)\big)+\sqrt{\tau(D)}}{2\tau\big(\sin^2(\pi/r)\big)}$$

where $\widetilde{\tau}$ extends τ.

Thus, $\tau(u)$ belongs to an interval of the length $2\sqrt{\tau\big(D/(4\sin^4(\pi/r))\big)}$.

We can apply Theorems 13, 14 to $\mathbb{F}=\mathbb{Q}\big(\cos^2(\pi/s),\cos^2(\pi/k),\cos^2(\pi/r)\big)$ and $\alpha=u$. Then

$$M=[\mathbb{F}:\mathbb{Q}], \qquad B=2\sqrt{|\operatorname{discr}\mathbb{F}|},$$
$$R=\frac{N_{\mathbb{F}/\mathbb{Q}}(D)^{1/4}}{N_{\mathbb{F}/\mathbb{Q}}\big(\sin^2(\pi/r)\big)^{1/2}2^M}, \qquad S=\frac{32e\sin^2(\pi/r)}{\sqrt{D}},$$

$[\mathbb{K}{:}\mathbb{Q}]\le NM$ where N is the smallest natural solution of the inequality

$$NM\ln(1/R)-M\ln(N+1)-\ln B\ge \ln S.$$

We obtain:

if $s=2,k=r=3$, then $[\mathbb{K}{:}\mathbb{Q}]\le 83$; if $s=2,k=3,r=4$, then $[\mathbb{K}{:}\mathbb{Q}]\le 45$;
if $s=2,k=3,r=5$, then $[\mathbb{K}{:}\mathbb{Q}]\le 28$; if $s=2,k=4,r=3$, then $[\mathbb{K}{:}\mathbb{Q}]\le 28$;
if $s=2,k=5,r=3$, then $[\mathbb{K}:\mathbb{Q}]\le 22$; if $s=k=r=3$, then $[\mathbb{K}{:}\mathbb{Q}]\le 37$;
if $s=r=3,k=4$, then $[\mathbb{K}{:}\mathbb{Q}]\le 18$; if $s=r=3,k=5$, then $[\mathbb{K}{:}\mathbb{Q}]\le 10$;
if $s=4,k=4,r=3$, then $[\mathbb{K}{:}\mathbb{Q}]\le 9$; if $s=4,k=5,r=3$, then $[\mathbb{K}{:}\mathbb{Q}]=2$;
if $s=k=5,r=3$, then $[\mathbb{K}{:}\mathbb{Q}]=2$; if $s=k=3,r=4$, then $[\mathbb{K}{:}\mathbb{Q}]\le 17$;
if $s=k=3,r=5$, then $[\mathbb{K}{:}\mathbb{Q}]=2$.

For $s=2$, we can improve these estimates if we consider $\alpha=u^2$. In this case, $0<\tau(u^2)<\tau\big(D/(4\sin^4(\pi/r))\big)$, and we can apply Theorems 13, 14 to the same $\mathbb{F}$ and

$$R=\frac{N_{\mathbb{F}/\mathbb{Q}}(D)^{1/2}}{N_{\mathbb{F}/\mathbb{Q}}\big(\sin^2(\pi/r)\big)4^M}, \qquad S=\frac{2\cdot e\cdot 14^2\cdot 4\cdot\sin^4(\pi/r)}{D}.$$

We obtain:

if $s=2,\ k=r=3$, then $[\mathbb{K}{:}\mathbb{Q}]\le 53$; if $s=2,\ k=3,\ r=4$, then $[\mathbb{K}{:}\mathbb{Q}]\le 31$;
if $s=2,\ k=3,\ r=5$, then $[\mathbb{K}{:}\mathbb{Q}]\le 16$; if $s=2,\ k=4,\ r=3$, then $[\mathbb{K}{:}\mathbb{Q}]\le 19$;
if $s=2,\ k=5,\ r=3$, then $[\mathbb{K}{:}\mathbb{Q}]\le 12$.

Thus, our upper bound for the degrees of the fields from $\mathcal{F}\Gamma_3^{(4)}(14)$ is 53.

5.4. The fields from $\mathcal{F}\Gamma_4^{(4)}(14)$. For $\Gamma_4^{(4)}(14)$ (see Figure 7), $k\ge 2$, $s,r\ge 3$ are natural numbers and $2<u<14$ is a totally real algebraic integer. Moreover, we have the following possibilities only: $s=3$, $r=3,4,5$; $s=4,5$, $r=3$.

The ground field $\mathbb{K}=\mathbb{Q}(u^2)$ contains cyclic products

$$\cos^2\frac{\pi}{s},\quad \cos^2\frac{\pi}{r},\quad \cos^2\frac{\pi}{k},\quad u^2,\quad u\cos\frac{\pi}{s}\cos\frac{\pi}{k}.$$

This case can be considered as a specialization of the graph $\Gamma_1^{(4)}$ when we take $p = 2$. The determinant $d(u)$ of the Gram matrix is determined by the equality

$$\frac{-d(u)}{4} = u^2 + 4\cos\frac{\pi}{s}\cos\frac{\pi}{k}u + 4\cos^2\frac{\pi}{s} - 4\sin^2\frac{\pi}{k}\sin^2\frac{\pi}{r}.$$

Here

$$\tilde{u} = u^2 + 4\cos\frac{\pi}{s}\cos\frac{\pi}{k}u + 4\cos^2\frac{\pi}{s}$$

is a totally positive algebraic integer (since minimum of this quadratic polynomial of u is equal to $4\cos^2(\pi/s)\sin^2(\pi/k)$) which belongs to $\mathbb{K}$, and $\Gamma_4^{(4)}$ is V-arithmetic if and only if

$$0 < \sigma\left(4\cos^2\frac{\pi}{s}\sin^2\frac{\pi}{k}\right) \le \sigma(\tilde{u}) < \sigma\left(4\sin^2\frac{\pi}{r}\sin^2\frac{\pi}{k}\right) < 4$$

for any $\sigma\colon \mathbb{K} \to \mathbb{R}$ which is different from identity $\sigma^{(+)}$. For $\sigma^{(+)}$, we have

$$4 < \sigma^{(+)}(\tilde{u}) < 14^2 + 4\cdot 14 + 4 = 16^2.$$

It follows that $\mathbb{K} = \mathbb{Q}(\tilde{u})$. We can apply Theorems 13 and 14 to $\mathbb{F} = \mathbb{Q}\big(\cos^2(\pi/s), \cos^2(\pi/r), \cos^2(\pi/k)\big)$ and $\alpha = \tilde{u}$ to estimate $[\mathbb{K}{:}\mathbb{Q}]$. We can take $M = [\mathbb{F}{:}\mathbb{Q}]$, $B = 2\sqrt{|\operatorname{discr}\mathbb{F}|}$,

$$R = \sqrt{N_{\mathbb{F}/\mathbb{Q}}\left(\left(\sin^2\frac{\pi}{r} - \cos^2\frac{\pi}{s}\right)\sin^2\frac{\pi}{k}\right)},$$

$$S = \frac{2\cdot 16^2\cdot e}{4\big(\sin^2(\pi/r) - \cos^2(\pi/s)\big)\sin^2(\pi/k)}.$$

Then $[\mathbb{K}{:}\mathbb{Q}] \le MN$ where N is the least natural solution of the inequality

$$NM\ln(1/R) - M\ln(N+1) - \ln B \ge \ln S.$$

It follows that $[\mathbb{K}{:}\mathbb{Q}] \le 31$ for $2 \le k \le 6$ with the maximum bound 31 for $k = 2$ and $s = r = 3$.

When $k \ge 7$, we should additionally use the inequality

$$\frac{16^2 N_{\mathbb{K}/\mathbb{Q}}\big(4\sin^2(\pi/r)\sin^2(\pi/k)\big)}{4\sin^2(\pi/r)\sin^2(\pi/k)} > N_{\mathbb{K}/\mathbb{Q}}(\tilde{u}) \ge 1 \tag{21}$$

which follows from considerations above. This additional argument is very similar to much more difficult case of $\Gamma_5^{(4)}(14)$ which we shall consider below.

Our upper bound for the degrees of the fields from $\mathcal{F}\Gamma_4^{(4)}(14)$ will be 56 (look at the end of the next Section 5.5).

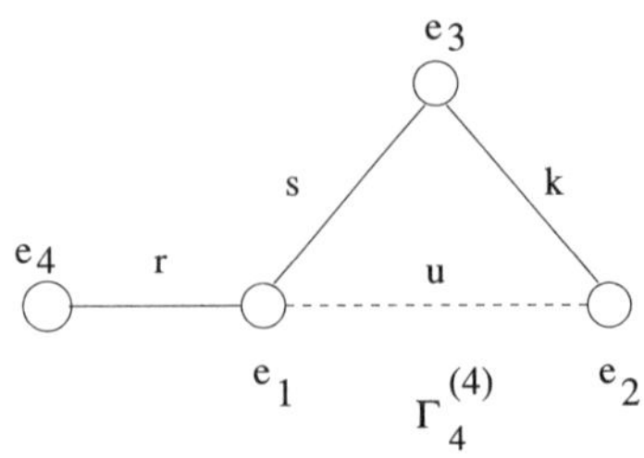

FIGURE 7. The graph $\Gamma_4^{(4)}$

5.5. The fields from $\mathcal{F}\Gamma_5^{(4)}(14)$. For $\Gamma_5^{(4)}(14)$ (see Figure 8), $k \ge s \ge 3$ are natural numbers and $2 < u < 14$ is a totally real algebraic integer.

The ground field $\mathbb{K} = \mathbb{Q}(u^2)$ contains cyclic products

$$\cos^2\frac{\pi}{k}, \quad \cos^2\frac{\pi}{s}, \quad u^2.$$

The determinant $d(u)$ of the Gram matrix is determined by the equality

$$-\frac{d(u)}{4} = u^2 - 4\sin^2\frac{\pi}{k}\sin^2\frac{\pi}{s}.$$

The $\Gamma_5^{(4)}$ is V-arithmetic if and only if

$$\sigma(u^2) < \sigma\left(4\sin^2\frac{\pi}{k}\sin^2\frac{\pi}{s}\right) < 4$$

for any $\sigma\colon \mathbb{K} \to \mathbb{R}$ which is different from identity $\sigma^{(+)}$. For $\sigma^{(+)}$, we have

$$4 < \sigma^{(+)}(u^2) < 14^2.$$

We can apply Theorems 13 and 14 to $\mathbb{F} = \mathbb{Q}\big(\cos^2(\pi/k), \cos^2(\pi/s)\big)$ and $\alpha = u^2$ to estimate $[\mathbb{K}{:}\mathbb{Q}]$. We can take $M = [\mathbb{F}{:}\mathbb{Q}]$, $B = 2\sqrt{|\operatorname{discr}\mathbb{F}|}$,

$$R = \sqrt{N_{\mathbb{F}/\mathbb{Q}}\left(\sin^2\frac{\pi}{k}\sin^2\frac{\pi}{s}\right)}, \quad S = \frac{14^2 \cdot e}{2\big(\sin^2(\pi/k)\sin^2(\pi/s)\big)}.$$

Then $[\mathbb{K}{:}\mathbb{Q}] \le MN_1$ where N_1 is the least natural solution of the inequality

$$N_1 M \ln(1/R) - M\ln(N_1+1) - \ln B \ge \ln S.$$

In this way, for each fixed pair $k \ge s \ge 3$, we can obtain an upper bound of the degrees $[\mathbb{K}{:}\mathbb{Q}]$. Let us call this method as the *Method* A.

By our considerations above, we obtain the inequality

$$\frac{14^2 N_{\mathbb{K}/\mathbb{Q}}\big(4\sin^2(\pi/k)\sin^2(\pi/s)\big)}{4\sin^2(\pi/k)\sin^2(\pi/s)} > N_{\mathbb{K}/\mathbb{Q}}(u^2) \ge 1\,. \tag{22}$$

We use the following elementary facts about cyclotomic fields. Let $\mathbb{F}_l = \mathbb{Q}\big(\cos^2(\pi/l)\big)$ where $l \ge 3$. Then

$$[\mathbb{F}_l{:}\mathbb{Q}] = \varphi(l)/2 \tag{23}$$

where φ is the Euler function. Let $\mathbb{F}_{l,m} = \mathbb{Q}\big(\cos^2(\pi/l), \cos^2(\pi/m)\big)$ where $l, m \ge 3$. Then

$$[\mathbb{F}_{l,m}{:}\mathbb{Q}] = \frac{\varphi([l,m])}{2\rho(l,m)} \tag{24}$$

where $[\ ,\ ]$ is the least common multiple, $\rho(l,m) = 2$ if $(l,m) \mid 2$, and $\rho(l,m) = 1$ otherwise. Here, $(\ ,\ )$ is the greatest common divisor.

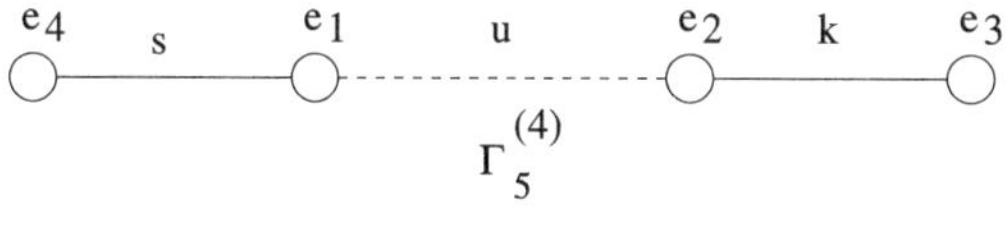

FIGURE 8. The graph $\Gamma_5^{(4)}$

We have for $l \geq 3$

$$(25)\qquad N_{\mathbb{F}_l/\mathbb{Q}}\big(4\sin^2(\pi/l)\big) = \gamma(l) = \begin{cases} p & \text{if } l = p^t > 2 \text{ where } p \text{ is prime,} \\ 1 & \text{otherwise.} \end{cases}$$

We denote $\mathbb{F} = \mathbb{F}_{k,s}$, $N = [\mathbb{K}{:}\mathbb{Q}]$, $n = [k,s]$, $[\mathbb{K}{:}\mathbb{F}_{k,s}] = m$. By (25), we have

$$N_{\mathbb{K}/\mathbb{Q}}\left(4\sin^2\frac{\pi}{k}\sin^2\frac{\pi}{s}\right) = \frac{N_{\mathbb{K}/\mathbb{Q}}\big(4\sin^2(\pi/k)\big)N_{\mathbb{K}/\mathbb{Q}}\big(4\sin^2(\pi/s)\big)}{N_{\mathbb{K}/\mathbb{Q}}(4)}$$
$$= \frac{\gamma(k)^{2N/\varphi(k)}\gamma(s)^{2N/\varphi(s)}}{4^N} = \left(\frac{\gamma(k)^{2/\varphi(k)}\gamma(s)^{2/\varphi(s)}}{4}\right)^N$$

where $\big(\varphi(n)/2\rho(k,s)\big) \mid N$.

Hence, by (22), (23), (24), (25), we obtain

$$(26)\quad N\left(\ln 2 - \frac{\ln\gamma(k)}{\varphi(k)} - \frac{\ln\gamma(s)}{\varphi(s)}\right) < \ln 7 - \ln\sin\frac{\pi}{k} - \ln\sin\frac{\pi}{s},$$
$$\varphi(n)/2\rho(k,s) \mid N.$$

By exact formulae for $\gamma(k)$ and $\varphi(k)$, it is easy to prove that there exists only finite number of *exceptional pairs* (k,s) such that $k \geq s \geq 3$ and

$$(27)\qquad \ln 2 - \frac{\ln\gamma(k)}{\varphi(k)} - \frac{\ln\gamma(s)}{\varphi(s)} \leq 0.$$

They are $s = 3$, $k = 3,4,5,7,8,9,11,13,17,19$; $s = 4$, $k = 4,5$; $s = 5$, $k = 5,7$.

For non-exceptional $k \geq s$, we get

$$(28)\qquad \frac{\ln 7 - \ln\sin(\pi/k) - \ln\sin(\pi/s)}{\ln 2 - \big(\ln\gamma(k)\big)/\varphi(k) - \big(\ln\gamma(s)\big)/\varphi(s)} > \frac{\varphi([k,s])}{2\rho(k,s)},$$

$$(29)\qquad [\mathbb{K}{:}\mathbb{F}_{k,s}] \leq \left[\frac{\ln 7 - \ln\sin(\pi/k) - \ln\sin(\pi/k)}{\ln 2 - \big(\ln\gamma(k)\big)/\varphi(k) - \big(\ln\gamma(s)\big)/\varphi(s)} \Big/ \frac{\varphi([k,s])}{2\rho(k,s)}\right],$$

and

$$(30)\qquad N = [\mathbb{K}{:}\mathbb{Q}]$$
$$\leq \left[\frac{\ln 7 - \ln\sin(\pi/k) - \ln\sin(\pi/s)}{\ln 2 - \big(\ln\gamma(k)\big)/\varphi(k) - \big(\ln\gamma(s)\big)/\varphi(s)} \Big/ \frac{\varphi([k,s])}{2\rho(k,s)}\right] \cdot \frac{\varphi([k,s])}{2\rho(k,s)}.$$

Let us show that there exists only a finite number of non-exceptional pairs $k \geq s \geq 3$ satisfying the inequality (28).

By exact formulae for $\gamma(l)$ and $\varphi(l)$, it is easy to find the minimum of $\ln 2 - \ln\gamma(k)/\varphi(k) - \ln\gamma(s)/\varphi(s)$ for all non-exceptional pairs. It is achieved for $k = 23$, $s = 3$, and it is equal to $\ln 2 - (\ln 23)/22 - (\ln 3)/2 \geq 0.00131857$. Since $x \geq \sin x$ for small x, $\rho(k,s) \leq 2$ and $\varphi([k,s]) \geq \varphi(k)$, we obtain $\ln 7 - 2\ln\pi + 2\ln k \geq 0.00131857 \cdot \varphi(k)/4$. Using the trivial estimate $\varphi(k) \geq \sqrt{k-2}$, we obtain $\ln 7 - 2\ln\pi + 2\ln k \geq 0.0003296425\sqrt{k-2}$. It follows that $s \leq k < 2.1 \cdot 10^{10}$. It follows the finiteness.

Using a better estimate $\varphi(k) \geq Ck/\ln(\ln k)$, one can get a better estimate for k. One can take $C = \varphi(6)\ln(\ln 6)/6 \geq 0.19439$ for $k \geq 6$. See [31]. Then $s \leq k < 10^7$.

It follows, that all non-exceptional pairs $k \geq s$ satisfying (28) can be found (using computer). By (29) and (30), bounds for $[\mathbb{K}{:}\mathbb{F}_{k,s}]$ and for $N = [\mathbb{K}{:}\mathbb{Q}] = [\mathbb{K}{:}\mathbb{F}_{k,s}][\mathbb{F}_{k,s}{:}\mathbb{Q}]$ can be found for each such a pair. This we call the *Method* B.

If (30) gives a poor bound for N because the bound (29) for $[\mathbb{K}{:}\mathbb{F}_{k,s}]$ is poor, we can improve the bound for $[\mathbb{K} : \mathbb{F}_{k,s}]$ using the Method A above. Also we can apply the Method A to all exceptional pairs $k \geq s$.

As a result, we obtain that $[\mathbb{K}{:}\mathbb{Q}] \leq 54$. This is achieved for $k = 109$, $s = 3$. In this case, (30) gives $[\mathbb{K}{:}\mathbb{Q}] \leq 54$, and (29) gives $[\mathbb{K}{:}\mathbb{F}_{k,s}] \leq 1$. Thus, $\mathbb{K} = \mathbb{F}_{109,3}$, and the Method A cannot improve this estimate.

For all other cases when (30) gives a bound $[\mathbb{K}{:}\mathbb{Q}] \leq t$ where $t > 54$, we can improve this bound using the Method A applied to $\mathbb{F} = \mathbb{F}_{k,s}$. For example, for $k = 23$ and $s = 3$, the inequality (30) gives $N = [\mathbb{K}{:}\mathbb{Q}] \leq 3091$, and (29) gives $[\mathbb{K}{:}\mathbb{F}_{23,3}] \leq 281$. Applying the Method A, we obtain $[\mathbb{K}{:}\mathbb{F}_{23,3}] \leq 1$ and $[\mathbb{K}{:}\mathbb{Q}] = 11$. Surprisingly, this strategy works in all cases.

Thus, our upper bound for the degrees of the fields from $\mathcal{F}\Gamma_5^{(4)}(14)$ is 54.

Now, considering the graphs $\Gamma_4^{(4)}(14)$ again, by (21), we similarly get the inequalities

$$\frac{\ln 8 - \ln\sin(\pi/k) - \ln\sin(\pi/r)}{\ln 2 - \big(\ln\gamma(k)\big)/\varphi(k) - \big(\ln\gamma(r)\big)/\varphi(r)} > \frac{\varphi([k,r])}{2\rho(k,r)}, \tag{31}$$

$$[\mathbb{K}{:}\mathbb{F}_{k,r}] \leq \left[\frac{\ln 8 - \ln\sin(\pi/k) - \ln\sin(\pi/r)}{\ln 2 - \big(\ln\gamma(k)\big)/\varphi(k) - \big(\ln\gamma(r)\big)/\varphi(r)} \Big/ \frac{\varphi([k,r])}{2\rho(k,r)}\right], \tag{32}$$

and

$$N = [\mathbb{K}{:}\mathbb{Q}] \leq \left[\frac{\ln 8 - \ln\sin(\pi/k) - \ln\sin(\pi/r)}{\ln 2 - \big(\ln\gamma(k)\big)/\varphi(k) - \big(\ln\gamma(r)\big)/\varphi(r)} \Big/ \frac{\varphi([k,r])}{2\rho(k,r)}\right] \frac{\varphi([k,r])}{2\rho(k,r)} \tag{33}$$

for non-exceptional $k \geq r$ where $r = 3, 4, 5$ and $k \geq 6$ (thus, one should replace s by r and 7 by 8 in our considerations of $\Gamma_5^{(4)}(14)$ above).

Then exactly the same considerations as for $\Gamma_5^{(4)}(14)$ show that $[\mathbb{K}{:}\mathbb{Q}] \leq 56$ for all fields $\mathbb{K}$ from $\Gamma_4^{(4)}(14)$ where 56 is achieved for $k = 113$ and $r = 3$.

This finishes the proof of Theorem 4. □

6. A mirror symmetric finiteness conjecture about reflective automorphic forms on Hermitian symmetric domains of type IV

In [8–10], we (with Gritsenko) obtained some finiteness results and formulated a general conjecture about so called *reflective automorphic forms on symmetric domains of type* IV. They were considered as mirror symmetric statements to finiteness results about arithmetic hyperbolic reflection groups over $\mathbb{Q}$ and corresponding reflective hyperbolic lattices over $\mathbb{Z}$.

Now finiteness results about arithmetic hyperbolic reflection groups are established in full generality. Moreover, results of this paper can be considered as steps to classification in the future. Respectively, it would be interesting to extend and formulate the corresponding finiteness conjecture about reflective automorphic forms on symmetric domains of type IV in full generality too. Let us do it.

Let $\mathbb{K}$ be a totally real algebraic number field and $\mathbb{O}$ its ring of algebraic integers.

We recall that a lattice L over $\mathbb{K}$ is a finitely generated torsion-free $\mathbb{O}$-module L with a symmetric bilinear form defined on L with values in $\mathbb{O}$. Here $\mathbb{K}$ is called the ground field of L, the number $\dim L \otimes_{\mathbb{O}} \mathbb{K}$ is called the rank of L, and the absence of torsion means that $L \subset L \otimes_{\mathbb{O}} \mathbb{K}$. We let $x \cdot y$ denote the value of the bilinear form on L on the pair of elements $x, y \in L$, and we let x^2 denote $x \cdot x$.

A lattice S is said to be *hyperbolic* if the bilinear form $S\otimes_{\mathbb{Q}}\mathbb{R}$ over $\mathbb{R}$ is indefinite for exactly one embedding $\sigma^{(+)}\colon \mathbb{K}\to\mathbb{R}$, and it is hyperbolic under this embedding, i.e., it has signature $(1, t_{(-)})$. For $\operatorname{rank} S\geq 3$ we let $\mathcal{L}(S)$ denote the hyperbolic space of dimension $\operatorname{rank} S-1$ which is canonically associated with the hyperbolic lattice S (and with the form $S\otimes_{\mathbb{Q}}\mathbb{R}$ under the embedding $\sigma^{(+)}$):

$$\mathcal{L}(S)=\{\mathbb{R}^+x\subset S\otimes_{\mathbb{Q}}\mathbb{R}\mid x^2>0\}_0$$

where 0 means that we take a connected component. The automorphism group $O^+(S)$ of the hyperbolic lattice S is discrete and arithmetic in $\mathcal{L}(S)$, and it has a fundamental domain of finite volume.

A lattice T is said to be *of* IV *type* if the bilinear form $T\otimes_{\mathbb{Q}}\mathbb{R}$ over $\mathbb{R}$ is indefinite for exactly one embedding $\sigma^{(+)}\colon\mathbb{K}\to\mathbb{R}$, and it has signature $(2, t_{(-)})$ for this embedding. For $\operatorname{rank} T\geq 3$ we let $\Omega(T)$ denote the Hermitian symmetric domain of type IV (of dimension $\operatorname{rank} T-2$) which is canonically associated with the type IV lattice T (and with the form $T\otimes_{\mathbb{Q}}\mathbb{R}$ under the embedding $\sigma^{(+)}$):

$$\Omega(T)=\{\mathbb{C}\omega\subset T\otimes\mathbb{C}\mid \omega^2=0,\omega\cdot\overline{\omega}>0\}_0.$$

The automorphism group $O^+(T)$ of the type IV lattice T is discrete and arithmetic in $\Omega(T)$, and it has a fundamental domain of finite volume.

For both types of lattices (hyperbolic or of type IV) L we can define reflections as follows. Let $\delta\in L$, where $\sigma^{(+)}(\delta^2)<0$ and $\delta^2\mid 2(L\cdot\delta)$ (such elements are called *roots* of L). Then the formula

$$s_\delta(x)=x-\frac{2(x\cdot\delta)}{\delta^2}\delta,\quad x\in L,$$

defines an involution s_δ of the lattice L which is called reflection relative to the root δ of L. We let $W(L)$ denote the subgroup of $O(L)$ generated by all of the reflections of L. Geometrically, in hyperbolic (respectively IV type) case the reflections s_δ are precisely those automorphisms of L which act as reflections relative to hyperplanes of $\mathcal{L}(L)$ (respectively to quadratic divisors $D_\delta=\{\mathbb{C}\omega\in\Omega(L)\mid\omega\cdot\delta=0\}$ of $\Omega(L)$). They are orthogonal to the roots δ.

A hyperbolic lattice S of the rank no less than three is called *reflective* if $W(S)$ is a subgroup of $O(S)$ of finite index.

By Vinberg's arithmeticity criterion, any arithmetic hyperbolic reflection group W is a subgroup of finite index $W\subset W(S)$ for one of reflective hyperbolic lattices S. The ground field $\mathbb{K}$ of S then coincides with the ground field of W.

It was proved in [13, 14] that for the fixed degree $N=[\mathbb{K}:\mathbb{Q}]$ of the ground field of S and the fixed rank $S\geq 3$ there exists only finitely many reflective hyperbolic lattices S up to similarity (that is up to multiplication of the form of S by elements $k\in\mathbb{K}$). Since it is now established in full generality that the rank S and the degree N are absolutely bounded, it follows that *there exist finitely many similarity classes of reflective hyperbolic lattices* (for all ranks ≥ 3 and for all ground fields together).

Now let T be a lattice of type IV. A holomorphic automorphic form Φ of a positive weight on $\Omega(T)$ which is automorphic relative to $O^+(T)$ is called *reflective automorphic form of the lattice* T if the divisor of Φ is union of quadratic divisors of $\Omega(T)$ which are orthogonal to roots δ of reflections s_δ of T. A lattice T of type IV is called *reflective* if it has at least one reflective automorphic form Φ.

Conjecture 15. *There exist finitely many similarity classes of* IV *type reflective lattices of rank at least* 5 (*for all ranks* ≥ 5 *and all ground fields together*).

For $\mathbb{K} = \mathbb{Q}$ and respectively $\mathbb{O} = \mathbb{Z}$ this conjecture was formulated in [8–10]. Even in this case, it seems, it is not established in full generality.

Some arithmetic hyperbolic reflection groups and some reflective automorphic forms and corresponding hyperbolic and type IV reflective lattices over $\mathbb{Z}$ are important in Borcherds proof [4] of Moonshine Conjecture by Conway and Norton [6] which had been first discovered by John McKay.

We hope that similar objects over arbitrary number fields will find similar astonishing applications in the future. At least, the results and conjectures of this paper show that they are very exceptional even in this very general setting.

After a preliminary version of this paper was published as a preprint [18], further important results related to this paper were obtained in preprints [3, 19, 20].

References

1. I. Agol, *Finiteness of arithmetic Kleinian reflection groups*, International Congress of Mathematicians. Vol. II (Madrid, 2006) (M. Sanz-Solé, J. Soria, J. L. Varona, and J. Verdera, eds.), Eur. Math. Soc., Zürich, 2006, pp. 951–960, available at arXiv:math.GT/0512560.
2. I. Agol, M. Belolipetsky, P. Storm, and K. Whyte, *Finiteness of arithmetic hyperbolic reflection groups*, available at arXiv:math.GT/0612132.
3. M. Belolipetsky, *On fields of definition of arithmetic Kleinian reflection groups*, available at arXiv:0710.5108.
4. R. E. Borcherds, *Monstrous moonshine and monstrous Lie superalgebras*, Invent. Math. **109** (1992), no. 2, 405–444.
5. A. Borel, *Commensurability classes and volumes of hyperbolic 3-manifolds*, Ann. Scuola Norm. Sup. Pisa Cl. Sci. (4) **8** (1981), no. 1, 1–33.
6. J. H. Conway and S. P. Norton, *Monstrous moonshine*, Bull. London Math. Soc. **11** (1979), no. 3, 308–339.
7. H. M. S. Coxeter, *Discrete groups generated by reflections*, Ann. of Math. (2) **35** (1934), no. 3, 588–621.
8. V. A. Gritsenko and V. V. Nikulin, *Automorphic forms and Lorentzian Kac–Moody algebras.* I, Internat. J. Math. **9** (1998), no. 2, 153–199; II, 201–275.
9. ———, *The arithmetic mirror symmetry and Calabi–Yau manifolds*, Comm. Math. Phys. **210** (2000), no. 1, 1–11.
10. ———, *On the classification of Lorentzian Kac–Moody algebras*, Uspekhi Mat. Nauk **57** (2002), no. 5(347), 79–138 (Russian); English transl., Russian Math. Surveys **57** (2002), no. 5, 921–979.
11. F. Lannér, *On complexes with transitive groups of automorphisms*, Comm. Sém., Math. Univ. Lund **11** (1950).
12. D. D. Long, C. Maclachlan, and A. W. Reid, *Arithmetic Fuchsian groups of genus zero*, Pure Appl. Math. Q. **2** (2006), no. 2, 1–31.
13. V. V. Nikulin, *On the arithmetic groups generated by reflections in Lobačevskĭ spaces*, Izv. Akad. Nauk SSSR Ser. Mat. **44** (1980), no. 3, 637–669 (Russian); English transl., Math. USSR-Izv. **16** (1981), no. 3, 573–601.
14. ———, *On the classification of arithmetic groups generated by reflections in Lobachevskiĭ spaces*, Izv. Akad. Nauk SSSR Ser. Mat. **45** (1981), no. 1, 113–142 (Russian); English transl., Math. USSR-Izv. **18** (1982), no. 1, 99–123.
15. ———, *Discrete reflection groups in Lobachevsky spaces and algebraic surfaces*, Proceedings of the International Congress of Mathematicians, Vol. 1, 2 (Berkeley, CA, 1986) (A. M. Gleason, ed.), Amer. Math. Soc., Providence, RI, 1987, pp. 654–671.
16. ———, *The remark on arithmetic groups generated by reflections in Lobachevsky spaces*, Technical Report MPI/91, Max-Planck-Institut für Mathematik, Bonn, 1991.
17. ———, *Finiteness of the number of arithmetic groups generated by reflections in Lobachevsky spaces*, Izv. Math. **71** (2007), no. 1, 53–56, available at arXiv:math.AG/0609256.
18. ———, *On ground fields of arithmetic hyperbolic reflection groups*, available at arXiv:0708.3991.

19. ———, *On ground fields of arithmetic hyperbolic reflection groups*. II, available at arXiv: 0710.0162.
20. ———, *On ground fields of arithmetic hyperbolic reflection groups*. III, available at arXiv: 0710.2340.
21. K. Takeuchi, *A characterization of arithmetic Fuchsian groups*, J. Math. Soc. Japan **27** (1975), no. 4, 600–612.
22. ———, *Arithmetic triangle groups*, J. Math. Soc. Japan **29** (1977), no. 1, 91–106.
23. ———, *Commensurability classes of arithmetic triangle groups*, J. Fac. Sci. Univ. Tokyo Sect. IA Math. **24** (1977), no. 1, 201–212.
24. ———, *Arithmetic Fuchsian groups with signature* $(1; e)$, J. Math. Soc. Japan **35** (1983), no. 3, 381–407.
25. M.-F. Vignéras, *Quelques remarques sur la conjecture* $\lambda_1 \geq \frac{1}{4}$, Seminar on Number Theory (Paris, 1981–82) (M.-J. Bertin, ed.), Progr. Math., vol. 38, Birkhäuser, Boston, MA, 1983, pp. 321–343.
26. È. B. Vinberg, *Discrete groups generated by reflections in Lobačevskiĭ spaces*, Mat. Sb. (N.S.) **72(114)** (1967), 471–488 (Russian); English transl., Math. USSR-Sb. **1** (1967), 429–444.
27. ———, *The nonexistence of crystallographic reflection groups in Lobachevskiĭ spaces of large dimension*, Funktsional. Anal. i Prilozhen. **15** (1981), no. 2, 67–68 (Russian); English transl., Funct. Anal. Appl. **15** (1981), 216–217.
28. ———, *Absence of crystallographic reflection groups in Lobachevskiĭ spaces of large dimension*, Trudy Moskov. Mat. Obšč. **47** (1984), 68–102 (Russian); English transl., Trans. Moscow Math. Soc. **47** (1985), 75–112.
29. ———, *Discrete reflection groups in Lobachevsky spaces*, Proceedings of the International Congress of Mathematicians. Vol. 1, 2 (Warsaw, 1983) (Z. Ciesielski and C. Olech, eds.), PWN, Warsaw, 1984, pp. 593–601.
30. ———, *Hyperbolic groups of reflections*, Uspekhi Mat. Nauk **40** (1985), no. 1(124), 29–66 (Russian); English transl., Russian. Math. Surveys **40** (1985), no. 1, 31–75.
31. I. M. Vinogradov, *An introduction to the theory of numbers*, translated by H. Popova, Pergamon Press, London & New York, 1955.
32. P. Zograf, *A spectral proof of Rademacher's conjecture for congruence subgroups of the modular group*, J. Reine Angew. Math. **414** (1991), 113–116.

Department of Mathematical Sciences, University of Liverpool, Mathematics & Oceanography Building, Peach Street, Liverpool, L69 7ZL, England, UK

E-mail address: vnikulin@liv.ac.uk; vvnikulin@list.ru

Centre de Recherches Mathématiques
CRM Proceedings and Lecture Notes
Volume **47**, 2009

Moonshine-type Functions and the CRM Correspondence

Simon P. Norton

ABSTRACT. We describe some of the sets of modular functions associated with Moonshine, and the relations between them, and complete a listing of one of these sets.

1. Introductory remarks

We start by defining various concepts associated with Moonshine-type functions with a view to describing a range of sets of such functions, and the relations between these sets. Throughout this paper a, b, c, d, m, n denote integers, with the restrictions $m \geq -1$ and $n \geq 1$, and r, s denote rationals.

The *modular group* Γ is the set of all operations $z \mapsto (az+b)/(cz+d)$ where $ad - bc = 1$, and z is a complex number in the completed upper half plane $\mathbb{H}^+$, i.e., either $\mathrm{Im}(z) > 0$ or $z \in \mathbb{Q} \cup \{\infty\}$. (Note: ∞ is sometimes called $\mathrm{i}\infty$ because in the topology below it is normally approached along or parallel to the imaginary axis.) We do not distinguish between the operations corresponding to (a, b, c, d) and $(-a, -b, -c, -d)$, so Γ is isomorphic to $\mathrm{PSL}(2, \mathbb{Z})$.

More generally, we call a subgroup G of $\mathrm{PGL}(2, \mathbb{Z})$ *a modular group* if it is *commensurable* with Γ, i.e., $G \cap \Gamma$ has finite index in both G and Γ. Any modular group must contain a non-trivial operation $z \mapsto z + r$, which we call a *translation*, and the least positive r for which this is in G is called the *cusp width* of G at ∞. More generally, the cusp width of G at s is the cusp width at ∞ of any group obtained by conjugating G by an element of Γ that takes s to ∞.

There is a natural topology on $\mathbb{H}^+$, linked to the standard hyperbolic geometry on the upper half plane. For any modular group G, this defines a compact topology on the quotient space $\mathbb{H}^+/G$. We define the *genus* of G as the genus of this space. If G has genus zero, then the space is homeomorphic to a sphere such as $\mathbb{C}^+ = \mathbb{C} \cup \{\infty\}$, and there will exist an analytic function from $\mathbb{H}^+/G$ to $\mathbb{C}^+$ which is invertible and takes finite values when $\mathrm{Im}(z) > 0$. The corresponding analytic function from $\mathbb{H}^+$ to $\mathbb{C}^+$, which is invariant under G, is called a *Hauptmodul*, and G is called its *fixing group*.

If the fixing group of a Hauptmodul has cusp width r at ∞ then the Hauptmodul can be written as a Laurent series in $q^{1/r}$, where we write $q = e^{2\pi \mathrm{i} z}$ (here

---

2000 *Mathematics Subject Classification.* 11F03, 11F06, 20D08.
This is the final form of the paper.

and throughout the paper). The series will converge for all z such that $\mathrm{Im}(z) > 0$. Its values at s and ∞ are obtained by approaching along the line $s + ix$ with x tending to 0 and ∞ respectively (where for ∞ we can use any rational s, or even an irrational number) and taking the limit, with the value being set to ∞ if there is no finite limit.

The coefficients of $q^{-n/r}$ for all $n > 1$ must be zero, otherwise the action of the function on $\mathbb{H}^+/G$ would have a multiple pole at ∞ so could not be bijective, thus contradicting the definition of a Hauptmodul. It is clear that for any modular group G we can choose a unique function, which we call *normalized* or *in normal form*, such that the coefficients of $q^{-1/r}$ and q^0 are 1 and 0 respectively.

We may further say that the group G and the associated normalized Hauptmodul are *regular* if $r = 1$. If G is not regular, we can *regularize* it by conjugating by the map $z \mapsto z/r$.

Two Hauptmoduls in normal form are called *translates* if one of their fixing groups can be conjugated to the other by a translation. Note that the translation does *not* in general conjugate one function to the other — it is necessary to multiply the conjugate of one function by a root of unity to get the other. To be specific, if the groups are regular, then the effect of the translation $z \mapsto z+s$ is to multiply the coefficient of q^m by $e^{2\pi is(m+1)}$, so the root of unity involved in $e^{2\pi is}$. Nevertheless, it should cause no confusion if we refer to one Hauptmodul as being obtained by applying a translation to the other.

The translations define an equivalence relation among Hauptmoduls, and we use the term *translation class* to denote an equivalence class within whatever set of Hauptmoduls we are considering at the time.

We define the *period* of a regular Hauptmodul as the largest n such that the Hauptmodul is taken to itself by the translation $z \mapsto z + 1/n$. Equivalently, the period is the largest n such that the coefficients of all q^m are zero unless $m \equiv -1 \pmod{n}$.

It follows immediately from commensurability that the coefficients of any normalized Hauptmodul must be algebraic numbers. We call a function *rational* if its coefficients are rational, and a translation class rational if it contains a rational function.

2. The J-function and the CRM correspondence

The normalized Hauptmodul of Γ is the function $J(z) = q^{-1} + 196884q + 21493760q^2 + \cdots$, obtained from the classical j-function $j(z) = J(z) + 744$ by removing the constant term. The starting point of Moonshine was the observation by McKay and Thompson that the coefficients of J are "natural" character degrees of the Fischer–Griess Monster group $\mathbb{M}$. For example, the coefficient of q is the degree of the direct sum of the trivial character and the smallest non-trivial character, which has degree 196883.

Other "moonshine" functions were obtained by generalizing this observation. We systematize this by defining three "languages":

(1) *Congruence groups.* $\Gamma(n)$ is defined as the subgroup of Γ consisting of all operations with $a \equiv d \equiv 1$, $b \equiv c \equiv 0 \pmod{n}$. A *congruence group* is a modular group which contains some $\Gamma(n)$. We consider the Hauptmoduls of certain sets of congruence groups.

(2) *Replication.* If $f(z)$ has Laurent expansion $q^{-1} + h_1 q + h_2 q^2 \ldots$, there is a unique polynomial P_n, known as the nth *Faber polynomial*, such that $P_n\big(f(z)\big)$ has Laurent expansion $q^{-n}/n + h_{n,1}q + h_{n,2}q^2 \ldots$ for some $h_{n,a}$. In the case where the function is rational, if $h_{a,b} = h_{c,d}$ whenever $ab = cd$ and $(a,b) = (c,d)$ (where (a,b) denotes the greatest common divisor of a and b), then the function f is said to be *replicable*. Below we generalize this to the case where the coefficients are cyclotomic.

(3) *The Monster.* The coefficients of the J-function are character degrees of the Monster. If we consider a fixed conjugacy class of the Monster and take the values of the corresponding characters on that class, we get the *Classical Moonshine* functions. The word "classical" is used to distinguish these functions from the *Generalized Moonshine* functions, which can be obtained by the further transformations first described in [11] and reproduced below.

As we shall see later, many of the Moonshine properties were discovered by a process of cross-fertilization between these three languages that is reminiscent of the development of vocabularies by cross-fertilization between real languages (which is why we use the word "language").

We call the link between these three languages the *CRM correspondence*, both because of the initials of the headings above and to commemorate the McKay conference, held at the Centre de recherches mathématiques in Montréal, in the proceedings of which this paper is being published.

We now proceed to describe various sets of modular functions in each language. We name these sets by the initial letter of the relevant language followed by a number, which will generally get larger as the set gets smaller. We give the references where the sets are described, and discuss the status of the enumeration problems. We then go on to describe some of the relations between the sets, again with the aid of our references. (Note that [14] is not actually referenced anywhere other than here, but it is included in our references because of its general relevance to the subject matter of this paper.)

3. Congruence groups

It is known (e.g., [15, Section 6.2]) that the coefficients of the normalized Hauptmodul of any genus zero congruence group are cyclotomic numbers. We believe that they are in fact always cyclotomic *integers*, though as far as we know this hasn't been proved. For the groups in which we are most interested, we do have a proof, but only by enumeration of cases as described in this paper. These are the groups of *Moonshine type*, which are regular congruence groups of genus zero containing some $\Gamma_0(n)$, this being the subgroup of Γ consisting of those transformations $z \mapsto (az+b)/(cz+d)$ for which $n \mid c$. We denote the corresponding set of normalized Hauptmoduls (which we also describe as being of Moonshine type) by C1.

The enumeration of these groups was recently completed by Cummins [6]. There are 6486 of them in all, but the set is closed under the translations $\{z \mapsto z + n/24\}$, and the number of translation classes under this group is considerably smaller. In Tables 2 and 3 we exhibit a representative of each translation class of Hauptmoduls other than those containing functions appearing in [10] — the rest can be obtained from these by choosing an n and multiplying the coefficient of each q^m by $e^{(m+1)n\pi i/12}$.

We now describe some subsets. Some interestingly "bad" groups and functions are the *ghosts*. Three of these functions were first mentioned in [5], where they were said to be associated with "ghost elements" of the Monster. Ghosts can be defined as those groups G, or the corresponding Hauptmoduls, for which there is an s such that there is no transformation $z \mapsto (az+b)/(cz+d)$ which takes s to ∞ and G to a group containing some $\Gamma_0(n)$. This appears to be equivalent to saying that the value of the function on s, which if finite is always a cyclotomic integer, has a Galois group over $\mathbb{Q}$ with exponent > 2. Of the functions in C1 relatively few are ghosts and they can be easily enumerated; they are clearly closed under translations and in Table 4 we show a representative of each translation class in C1. The set of Hauptmoduls obtained by removing the ghosts from C1 we call C2.

A related set consists of the Hauptmoduls (not necessarily regular) which do not satisfy the "ghost" condition and for which all cusp widths are integral. It follows from the definition that this set is closed under conjugation by Γ. When regularized, and possibly translated, they lie in C2. We therefore do not expect the enumeration of this set of Hauptmoduls, which we call C3, to cause too much difficulty.

It was observed by Conway, McKay and Sebbar [4] that the subset of C2 corresponding to those groups which have a normal subgroup of type $\Gamma_0(n)$ over which the quotient has exponent at most 2 consists exactly of the Classical Moonshine functions. We call this subset C4.

See [13] for a condition on the group G for the coefficients of the associated Hauptmodul to be rational.

4. Replication

In [5] the concept of *replication* was defined. This enables one to pass from a function to its *replication powers*. The definition was clarified in [12], where a function was defined as *replicable* if the replication powers exist, and *completely replicable* if the replication powers are themselves replicable and replicate correctly (i.e., the n'th power of the nth power is the nn'th).

The definition of replicability is simplest for rational functions. However there is a natural way to generalize it to functions with cyclotomic coefficients which ensures that, like C1, the set of replicable functions is invariant under the translations $\{z \mapsto z + n/24\}$: in an equation $h_{a,b} = h_{c,d}$ we apply to the right-hand side the Galois automorphism that replaces roots of unity by their kth powers, where k is the smallest integer obtainable by dividing ac by a square. (These roots of unity must have order prime to k.)

For the function $q^{-1} + \lambda q$, it can be seen that $h_{a,b} = \delta_{a,b}\lambda^a/a$ where δ is the Kronecker delta function. It follows immediately that such functions are replicable whenever the concept is defined. However, these functions are considered trivial and called *trivialities* or *modular fictions* (the latter because they behave in some ways like modular functions but aren't).

We call the set of non-trivial replicable functions R1 or R2 according as the coefficients are assumed to be cyclotomic or rational integers. (The reason for the integrality condition is partly to correspond with the known behavior of functions in C1 and — defined below — M1, and partly because the result in the next paragraph doesn't work without this assumption.)

A "near enumeration" has been obtained for R2. To be precise, a list of functions has been obtained which is believed to be complete, and it has been shown [8] that for any other rational replicable function (which is not a triviality) and any prime $p \leq 19$ there is a function in this list all of whose coefficients are congruent to those of our function mod p.

Because of ambiguities in the algebraic conjugations involved, there does not appear to be any completely natural and general definition of replication power for irrational functions—though in many cases (including the trivialities) the ambiguities vanish in practice. We therefore limit a precise definition of complete replicability to rational functions, but one may refer loosely to an irrational function as being completely replicable if the definition of the replication powers involves no ambiguity and these powers satisfy the conditions of complete replicability. Rational completely replicable functions are enumerated in [1, 10], and after removing the trivialities we call the set of such functions R3.

We note, incidentally, that not all the irrational functions in C1 are replicable. Some are *semi-replicable* in the sense that, while $h_{a,b}$ and $h_{c,d}$ with appropriate a, b, c, d are algebraically conjugate, this may be by a different conjugation to the one referred to above. Other functions in C1 are not even semi-replicable, but they are all related to semi-replicable functions by some translation $z \mapsto z + n/24$ (where in fact we can take $n = \pm 8$).

5. The Monster

There are two basic sets of functions associated with the Monster "language", as we mentioned before, though one of them has two variations. The Generalized Moonshine Conjecture (see [11]) says that there is a set of functions, $f(g, h, z)$, associated with pairs (g, h) of commuting elements in the Monster, which satisfy the following conditions:

(1) $f(g^a h^c, g^b h^d, z) = \varepsilon f\big(g, h, (az+b)/(cz+d)\big)$ where ε is a root of unity.

(2) If g is fixed, then for every r and h the coefficient of q^r in $f(g, h, z)$ is, up to multiplication by a root of unity, the value of a proper character of a covering group of $C_{\mathbb{M}}(g)$ on any lift of h. (Note that this implies that the coefficients are all cyclotomic integers.)

(3) Conjugating g and h by the same element of the Monster leaves the function unchanged.

(4) $f(g, h, z)$ is either a constant or a Hauptmodul (see below).

Work of Dong, Li and Mason [9] has shown that a set of functions can be constructed satisfying the first three conditions, except that the first is weakened so that ε may be any complex number. Also [2] the fourth condition holds when $\langle g, h\rangle$ is cyclic.

Let us say that h is a *Fricke element* of the Monster when the value at 0 of the Classical Moonshine function corresponding to h is infinite. Then the function $f(g, h, z)$ is in normal form exactly when g is a Fricke element. Furthermore it is regular only when $g = 1$, in which case the function is the Classical Moonshine function corresponding to h.

It can be seen from the above properties that $f(g, h, a/c)$ is infinite exactly when $g^a h^c$ is a Fricke element. We therefore divide the set of functions $f(g, h, z)$ into the following three cases:

(1) g is Fricke, so $f(g,h,\infty)$ is infinite. In this case, as we said, $f(g,h,z)$ is a Hauptmodul in normal form.

(2) g is not Fricke, but there exist coprime a and c such that $g^a h^c$ is. In this case $f(g,h,z)$ is a Hauptmodul, but not in normal form. There will of course be a unique normalized Hauptmodul associated with the relevant fixing group, and this will be a bilinear transform of $f(g,h,z)$.

(3) There are no coprime a and c for which $g^a h^c$ is Fricke. (Note that this implies that g is non-Fricke, as we can take $a = 1$ and $c = 0$.) In this case $f(g,h,z)$ is a constant function of z.

We gather together all the functions in the first case plus the normalized Hauptmoduls in the second case, and call the resulting set M1. The set of regularized versions of these functions we call M2. The set of Classical Moonshine functions we call M3.

While M3 was first enumerated in [5], we have been unable to do the same for M1 or M2 because not all character tables of groups of type $C_{\mathbb{M}}(g)$ are known. At present, we have a complete enumeration of all the $f(g,h,z)$ where g belongs to one of about two-thirds of the conjugacy classes of $\mathbb{M}$.

6. Relations between the function sets

We start by tracing the historical development of the discovery of the Moonshine properties and showing the kind of cross-fertilization between the three languages that was used.

It was observed by McKay and Thompson that the coefficients of the J-function were character degrees of the Monster. Evaluating the relevant characters on general elements of the Monster gave rise to the set M3. Known recurrence relations on the coefficients of J generalized to character identities which gave rise to the replication formulas for the functions in M3, and this led naturally to the problems of classifying completely replicable and replicable functions, thus giving the sets R3, R2 and R1. Also the fixing groups of these functions were studied, and a definition of Moonshine type group emerged, leading to the set C1. It was then noted that many of these functions appeared in Moonshine-like systems for subgroups of $\mathbb{M}$. These Moonshine-like systems were coordinated into the system of Generalized Moonshine, leading to the set M2. More recently came the definition of ghost, leading to the set C2; the characterization of Classical Moonshine functions in [4], leading to the set C4; and an appreciation that regularizing fixing groups was not always natural, leading to the sets M1 and C3.

Going through the function sets we find the following containments.

(1) C4 $\subset$ C2 $\subset$ C1 (definition).

(2) R3 $\subset$ R2 $\subset$ R1 (definition).

(3) M3 $\subset$ M2, M1 (definition).

(4) Rationals in C1 = R2 ($\subset$, [7]; $\supset$, verified subject to completeness of enumeration of R2).

(5) R1 $\subset$ C1 (conjectured).

(6) C4 = M3 (proved [4]).

(7) Functions in M2 can be translated into C2 (a formulation of the Generalized Moonshine Conjecture).

(8) M1 $\subset$ C3 (a slightly stronger formulation of the Generalized Moonshine Conjecture).

(9) M3 $\subset$ R3 (conjectured [5] from character theory, later verified).

We also note that the sets M2 and C2 are definitely *not* equal, even up to translations. Two examples of functions of Moonshine type known not to occur in Generalized Moonshine are $f(2z)^{1/2}$ where f is the Classical Moonshine function corresponding to an element of order 29 or 41. These are replicable, in fact they are rational and completely replicable, but they cannot occur in Generalized Moonshine because the Monster has no elements of order 58 or 82. (In [10] the functions are called `58a` and `82a` respectively.) It would be nice to have an exact group theoretic characterization of M2 similar to that of M3 in [4].

7. A listing of Moonshine type functions

From the above it follows that, subject to certain conjectures, all the functions we are interested in, when regularized, lie in C1. Therefore Cummins' recent enumeration of all the relevant fixing groups [6] is of particular interest. Of the associated functions, those that are rational and completely replicable are enumerated in [10]. In that paper we defined the *dash operator*, which translates a function with period n by $z \mapsto z + 1/2n$ unless $8 \mid n$, in which case it fixes the function. R2 is closed under the dash operator. If the dash of a function in R3 is also in R3 then it's shown in [10, Table 1]. Otherwise we show our name for the dash in Table 1 of this paper. (Note a couple of errors in [10, Table 1]: `(32e,64a)` should be omitted and `(46A,92A)` added.)

The rest of R2 − R3 divides into pairs linked by the dash operator; we show these pairs in Table 2, giving our names together with the first 25 terms of the q-expansion of the first-named function.

In Table 3 we list subsets of the irrational translation classes of Moonshine type. In some cases we show the first 25 coefficients of the q-expansion of the first-named function, from which the others can be deduced by applying the translation $z \mapsto z + 1/nn'$ successively, where n is the period of the function shown and n' is the number of function names shown. (Note that this implies that all the functions are different, and that this translation takes the last-named function back to the first.) The remaining cases are algebraic conjugates of the ones where we show the function explicitly: in these cases the q-expansion of the first-named function can be obtained from the preceding q-expansion by applying the Galois conjugation shown (in ATLAS notation [3], where $**$, $*n$ and $**n$ replace a root of unity (of order prime to n) by its inverse, nth power and $-n$th power respectively. The other functions in these translation classes are obtained by applying the appropriate translations; it should be noted that the relevant Galois conjugates are not necessarily in the same order as in the preceding translation class. We also show the Cummins symbol used in Table 3 of [6] to denote the translation class of the relevant fixing group (which consists of a number followed by an upper case letter — we omit the superscript 0 which denotes the genus because we are only considering groups of genus zero).

These functions were computed and validated by a number of methods, especially the use of containments among the various groups to deduce relations between the corresponding functions. However, many of the implied statements (e.g., about replication orders) have not actually been proved.

In Table 4 we list the names of representatives of all translation classes that contain semi-replicables or ghosts. The latter may be either rational or irrational but the former can only be irrational. We also show sets of representatives of translation classes of functions whose fixing groups are conjugate in $PGL(2,\mathbb{Q})^+$ (i.e., the group of rational bilinear transformations with positive determinant).

The named functions divide as follows:

(1) R3 has 4 functions that are their own dashes (`32a`, `32e`, `64a` and `96a`).

(2) R3 has 49 pairs of dashes ([10, Table 1]).

(3) R3 has 223 functions whose dashes are outside R3 (Table 1 of this paper), giving a total of 325 functions in R3.

(4) R2 − R3 has 34 pairs of dashes (Table 1), giving a total of 616 functions in R2.

(5) There are 86 irrational translation classes, or 61 up to algebraic conjugation, with a total of 240 named functions, or 113 up to algebraic conjugation (Table 3). The total number of translation classes is 396, or 371 up to algebraic conjugation.

(6) The corresponding numbers for semi-replicables are 12, 10, 24, 12, 12, 10 and the numbers for ghosts are 3, 2, 12, 3, 12, 11 (Table 4).

As we would like to see function names standardized, it is appropriate to describe the guidelines used in our naming system.

For Classical Moonshine functions we use the standard ATLAS [3] names for the associated conjugacy class in the Monster (or the earlier occurring class if there are two of them), as was done in [10] (and [5]).

For other functions in R2 we again use the same names as in [10]: `Nx` where `N` is the *replication order*, i.e., the least positive integer such that the `N`th replication power of the function is the J-function, and `x` is a lower case letter. This type of name is also used for those irrational functions which we regard as being completely replicable.

For functions that are not completely replicable we use names of type `N~x`, where again `N` is the replication order and `x` a lower case letter. (There is never any ambiguity in assigning a replication order to an irrational function.) Where there are more than 26 functions with a given replication order, consistent with the corresponding convention for groups in [3], for the 27th and subsequent functions we use the symbols `N~a1`, `N~b1`,...).

The ATLAS notation assigns letters to conjugacy classes of the same order in descending order of centralizer, and in [10] we adopted the same principle for functions in R3 − M3, where the centralizer is replaced by the Frobenian with respect to R3 (a concept defined in that paper). We use the same convention for functions in R2 − R3, where the Frobenian is defined with respect to R2 rather than R3. We don't use the Frobenians for irrational functions (because though still defined they are harder to compute); instead we just continue the sequence of letters, generally assigning earlier letters to the functions that grow quicker or whose replication powers exist and have earlier letters.

Which functions in a translation class do we name? We omit the following:

(1) Functions which have translates where the coefficients lie in an algebraic number field of smaller degree over $\mathbb{Q}$, or one of the same degree but occurring in a smaller cyclotomic field.

(2) Functions that are neither replicable nor semi-replicable.

(3) Functions whose q-expansions look noticeably more complicated than others in the same translation class. (This is of course a subjective criterion.)

We have considerably more information about the functions than we are showing here, but to include it all would considerably lengthen this paper. And much of this information has not yet been systematized to cover all functions of Moonshine type. It is hoped to do this and publish the information (probably on the Web), as an "Atlas of Moonshine," in due course.

Tables 1 and 4 use the notation above. Tables 2 and 3 are shown as a printout in GAP readable format, where the notation for a given function is obtained by enclosing the relevant symbol in parentheses then in quotation marks. (The parentheses are added to facilitate certain programs.) The source files can be obtained electronically by emailing the author. Note that for the ATLAS irrationalities i, zn, bn, in, rn, shown with the corresponding upper case letters in Table 3, the GAP notation is `E(4)`, `E(n)`, `EB(n)`, `EI(n)`, `ER(n)` respectively, so the appropriate assignments have to be set up if the file is to be readable in the form shown.

TABLE 1. Dashes of completely replicable functions

(1A,4~a)	(2A,4~b)	(2a,8~a)	(3A,12~a)	(3B,12~b)
(3C,12~c)	(4B,8~b)	(4a,8~c)	(5A,20~a)	(5B,20~b)
(5a,20~f)	(6A,12~d)	(6B,12~e)	(6D,12~f)	(6a,24~a)
(6b,24~b)	(6c,24~c)	(6d,12~g)	(7A,28~a)	(7B,28~b)
(8A,8~d)	(8C,16~a)	(9A,36~a)	(9B,36~c)	(9a,36~b)
(9b,36~f)	(9c,36~g)	(9d,36~p)	(10A,20~c)	(10C,20~d)
(10D,20~e)	(10a,40~a)	(10b,20~h)	(10c,40~g)	(11A,44~a)
(12C,24~f)	(12F,24~g)	(12G,24~h)	(12H,12~h)	(12a,24~d)
(12b,24~e)	(12e,24~l)	(12f,24~n)	(13A,52~a)	(13B,52~b)
(14A,28~c)	(14C,28~d)	(14a,56~a)	(14b,56~b)	(14c,56~c)
(15A,60~a)	(15B,60~b)	(15C,60~d)	(15D,60~f)	(15a,60~h)
(15b,60~h)	(16A,16~d)	(16C,16~e)	(17A,68~a)	(18A,36~h)
(18B,36~e)	(18E,36~k)	(18a,36~i)	(18b,72~b)	(18c,72~a)
(18d,36~j)	(18f,72~d)	(18g,72~e)	(18h,36~n)	(18i,72~f)
(18j,36~r)	(19A,76~a)	(20B,40~c)	(20E,40~f)	(20F,20~g)
(20a,40~b)	(20b,40~d)	(20c,40~e)	(20e,40~j)	(21A,84~a)
(21B,84~b)	(21C,84~c)	(21D,84~d)	(22A,44~b)	(22a,88~a)
(23A,92~a)	(24B,24~m)	(24C,24~o)	(24F,48~f)	(24G,48~c)
(24H,24~q)	(24I,24~r)	(25A,100~a)	(25a,100~b)	(26A,52~c)
(26B,52~d)	(26a,104~a)	(27A,108~b)	(27a,108~c)	(27b,108~a)
(27c,108~d)	(27d,108~e)	(27e,108~h)	(28A,56~d)	(28D,56~f)
(28a,56~e)	(29A,116~a)	(30A,60~e)	(30B,60~c)	(30D,60~g)
(30E,60~i)	(30F,60~l)	(30a,120~a)	(30b,60~j)	(30c,120~b)
(30d,120~c)	(30e,60~n)	(30f,120~k)	(31A,124~a)	(32A,32~e)
(32B,32~d)	(32b,32~c)	(33A,132~a)	(33B,132~b)	(34A,68~b)
(34a,136~a)	(35A,140~a)	(35B,140~c)	(35a,140~e)	(36C,72~c)
(36D,36~q)	(36a,72~g)	(36c,72~m)	(36d,72~h)	(36g,72~n)
(36h,72~l)	(36i,72~s)	(38A,76~b)	(38a,152~a)	(39A,156~a)
(39B,156~b)	(39C,156~c)	(40A,80~a)	(40C,40~k)	(40e,80~f)
(41A,164~a)	(42A,84~e)	(42B,84~f)	(42D,84~g)	(42a,168~a)
(42b,168~b)	(42c,84~h)	(42d,168~e)	(44a,88~b)	(44c,88~c)
(45A,180~a)	(45a,180~b)	(45b,180~c)	(45c,180~g)	(46C,92~b)
(47A,188~a)	(48A,48~i)	(48h,48~n)	(49a,196~a)	(50A,100~c)
(50a,100~d)	(51A,204~a)	(52A,104~c)	(52B,104~d)	(52a,104~b)
(54A,108~g)	(54a,108~i)	(54b,108~j)	(54c,216~a)	(54d,216~b)
(55A,220~a)	(56A,56~g)	(56B,112~c)	(57A,228~a)	(58a,232~a)

TABLE 1. (continued)

(59A,236~a)	(60A,120~d)	(60D,60~m)	(60E,120~i)	(60F,120~j)
(60b,120~f)	(60c,120~l)	(60d,120~n)	(60e,120~m)	(62A,124~b)
(63a,252~a)	(66A,132~c)	(66B,132~d)	(66a,264~a)	(68A,136~b)
(69A,276~a)	(70A,140~b)	(70B,140~d)	(70a,280~a)	(71A,284~a)
(72e,144~e)	(76a,152~b)	(78A,156~d)	(78B,156~e)	(82a,328~a)
(84A,168~c)	(84B,168~f)	(84a,168~d)	(87A,348~a)	(90a,180~d)
(90b,180~e)	(93A,372~a)	(94A,188~b)	(95A,380~a)	(102a,408~a)
(104A,208~a)	(105A,420~a)	(110A,220~b)	(117a,468~a)	(119A,476~a)
(126a,504~a)	(132a,264~b)	(140a,280~b)		

TABLE 2. Pairs of replicable functions

```
dash:= [
 ["(6~a)","(24~i)",[1,0,-6,0,-18,0,16,0,-66,0,-108,0,98,0,-300,0,
  -486,0,364,0,-1128,0,-1656,0,1221]],
 ["(12~i)","(24~p)",[1,0,-2,0,6,0,8,0,-2,0,12,0,18,0,-20,0,42,0,52,
  0,-24,0,72,0,101]],
 ["(12~j)","(24~s)",[1,0,2,0,-2,0,0,0,-2,0,4,0,2,0,4,0,-6,0,-4,0,
  -8,0,8,0,5]],
 ["(16~b)","(16~c)",[1,0,8,0,6,0,-48,0,15,0,168,0,26,0,-496,0,51,0,
  1296,0,102,0,-3072,0,172]],
 ["(18~a)","(36~d)",[1,0,15,-14,-21,78,-62,-132,399,-322,-426,1332,
  -964,-1524,4278,-3072,-4059,11454,-7802,-11148,29892,-20284,
  -26268,70488,-46341]],
 ["(24~j)","(24~k)",[1,0,6,-4,5,-20,40,-44,46,-84,146,-184,226,
  -348,508,-608,747,-1108,1572,-1892,2264,-3096,4204,-5136,6245]],
 ["(24~t)","(48~o)",[1,0,0,0,2,0,0,0,-2,0,0,0,2,0,0,0,2,0,0,0,0,0,
  0,0,5]],
 ["(30~a)","(120~e)",[1,0,-1,0,2,0,1,0,-1,0,2,0,3,0,0,0,4,0,4,0,-3,
  0,4,0,6]],
 ["(32~a)","(32~b)",[1,0,0,0,4,0,0,0,-5,0,0,0,-4,0,0,0,11,0,0,0,20,
  0,0,0,-20]],
 ["(36~l)","(36~m)",[1,0,-1,6,-5,-6,18,-12,-17,42,-26,-36,92,-60,
  -74,192,-123,-150,374,-228,-284,684,-412,-504,1211]],
 ["(36~o)","(72~i)",[1,0,-3,0,3,0,2,0,-9,0,6,0,12,0,-30,0,21,0,22,
  0,-60,0,36,0,59]],
 ["(36~s)","(72~t)",[1,0,1,0,0,0,-1,0,1,0,0,0,0,0,1,0,0,0,1,0,0,0,
  0,0,-1]],
 ["(40~h)","(40~i)",[1,0,1,-2,6,-6,3,-8,16,-14,13,-26,40,-40,42,
  -64,89,-94,98,-138,190,-206,220,-296,392]],
 ["(48~a)","(48~b)",[1,0,-1,0,-3,0,-3,0,6,0,-3,0,-10,0,-1,0,15,0,0,
  0,0,-24,0,-3,0,37]],
 ["(48~d)","(48~e)",[1,0,3,0,1,0,-3,0,-2,0,9,0,-2,0,-9,0,-1,0,24,0,
  0,0,-27,0,5]],
 ["(48~g)","(48~h)",[1,0,-2,0,1,0,0,0,2,0,-6,0,6,0,-4,0,7,0,-12,0,
  12,0,-12,0,21]],
 ["(48~j)","(48~k)",[1,0,2,-2,1,-2,4,-6,10,-10,10,-12,14,-22,28,
  -32,39,-42,52,-66,76,-92,104,-120,149]],
 ["(48~l)","(48~m)",[1,0,2,0,0,0,0,0,0,0,0,0,2,0,-4,0,0,0,0,0,0,0,
  0,0,1]],
```

TABLE 2. (continued)

```
["(54~a)","(108~f)",[1,0,0,-2,0,0,1,0,0,2,0,0,-4,0,0,0,0,0,7,0,0,
 -10,0,0,3]],
["(63~a)","(252~b)",[1,0,-1,2,-2,-1,5,-3,-4,8,-5,-6,16,-8,-11,23,
 -15,-16,39,-21,-26,58,-35,-39,92]],
["(72~j)","(72~k)",[1,0,3,2,-1,-2,-2,4,7,6,2,-4,4,12,22,16,-3,-10,
 -6,28,44,36,4,-24,11]],
["(72~o)","(72~r)",[1,0,1,0,-1,0,-2,0,-1,0,2,0,4,0,2,0,-3,0,-6,0,
 -4,0,4,0,11]],
["(72~p)","(72~q)",[1,0,0,-1,2,-2,1,-2,4,-3,2,-4,7,-6,4,-8,12,-10,
 9,-14,20,-18,16,-24,32]],
["(80~b)","(80~c)",[1,0,-2,0,1,0,2,0,0,0,-2,0,1,0,4,0,1,0,-4,0,2,
 0,8,0,2]],
["(80~d)","(80~e)",[1,0,-1,0,0,0,-1,0,2,0,-1,0,0,0,-2,0,3,0,-2,0,
 2,0,-4,0,6]],
["(90~a)","(180~f)",[1,0,0,1,-1,-2,3,-2,-1,3,-1,-3,6,-4,-2,8,-4,
 -6,13,-8,-8,16,-8,-12,24]],
["(96~a)","(96~b)",[1,0,0,0,-1,0,0,0,0,0,0,0,0,0,0,0,1,0,0,0,-2,0,
 0,0,1]],
["(108~k)","(216~c)",[1,0,0,0,0,0,-1,0,0,0,0,0,0,0,0,0,0,0,1,0,0,
 0,0,0,-1]],
["(112~a)","(112~b)",[1,0,1,0,-1,0,1,0,1,0,0,0,-2,0,1,0,2,0,1,0,
 -3,0,1,0,4]],
["(120~g)","(120~h)",[1,0,1,1,0,0,0,1,1,1,1,1,1,2,3,2,2,2,2,3,4,4,
 4,4,5]],
["(126~a)","(252~c)",[1,0,1,0,0,1,1,1,0,0,1,2,2,2,1,1,1,2,3,3,2,2,
 3,5,6]],
["(144~a)","(144~b)",[1,0,-1,0,0,0,0,0,0,0,0,0,-1,0,-1,0,0,0,0,0,
 0,0,0,0,1]],
["(144~c)","(144~d)",[1,0,1,0,1,0,0,0,-1,0,0,0,0,0,2,0,1,0,0,0,0,
 0,0,0,3]],
["(160~a)","(160~b)",[1,0,0,0,-1,0,0,0,0,0,0,0,1,0,0,0,1,0,0,0,0,
 0,0,0,0]]];
```

TABLE 3. Irrational functions

```
irrats:= [
 [["(4~c)","(4~f)","(4~e)","(4~d)"],"16C",
  [1,0,-12*I,32*(I+1),66,0,-232*I,384*(I+1),639,0,-1596*I,
   2496*(I+1),3774,0,-8328*I,12032*(I+1),17283,0,-34520*I,
   48288*(I+1),66882,0,-125568*I,170112*(I+1),229244]],
 [["(6~b)","(6~c)"],"36H",
  [1,0,0,8*I3,0,0,28,0,0,64*I3,0,0,134,0,0,288*I3,0,0,568,0,0,
   1024*I3,0,0,1809]],
 [["(7a)","(28~e)"],"49B",
  [1,0,3*B7,-2*B7,9,-6*B7,4,12,0,26*B7,-15*B7,48,-25*B7,12,60,0,
   105*B7,-54*B7,190,-102*B7,57,200,0,348*B7,-184*B7]],
 [["(7b)","(28~f)"],"49B","**"],
 [["(11a)","(44~c)"],"121A",
  [1,0,-B11,2,-B11,2*B11,B11,0,3,-2,B11,6,0,-4*B11,9,-4*B11,6*B11,
   2*B11,1,6,-3,2*B11,18,0,-13*B11]],
 [["(11b)","(44~d)"],"121A","**"],
```

TABLE 3. (continued)

```
[["(12~k)","(12~n)","(12~m)","(12~l)"],"48C",
 [1,0,-3*I,-4*(I+1),3,0,-7*I,-12*(I+1),18,0,-21*I,-24*(I+1),30,0,
  -57*I,-64*(I+1),75,0,-104*I,-132*(I+1),156,0,-207*I,-240*(I+1),
  293]],
[["(12~o)","(12~r)","(12~q)","(12~p)"],"144B",
 [1,0,0,2*(I+1),0,0,-4*I,0,0,0,0,0,6,0,0,8*(I+1),0,0,-8*I,0,0,0,0,
  0,17]],
[["(14d)","(28~g)"],"98A",
 [1,0,-B7,2*B7,1,-2*B7,4,4,0,-2*B7,5*B7,0,-B7,4,12,0,-7*B7,14*B7,6,
  -10*B7,17,24,0,-12*B7,24*B7]],
[["(14e)","(28~h)"],"98A","**"],
[["(14~a)","(56~j)"],"196A",
 [1,0,B7+2,0,-1,0,2,0,-2*B7,0,-B7-2,0,B7+2,0,2,0,3*B7+6,0,-4,0,5,0,
  -4*B7,0,-I7-3]],
[["(14~b)","(56~k)"],"196A","**"],
[["(15~a)","(60~o)"],"225A",
 [1,0,0,B15+1,0,0,-B15+1,0,0,-B15-1,0,0,1,0,0,4,0,0,3*B15+3,0,0,
  -I15+3,0,0,-B15-1]],
[["(15~b)","(60~p)"],"225A","**"],
[["(16~f)","(16~i)","(16~h)","(16~g)"],"16D",
 [1,0,-16*I,16*(I+1),2,64*(I-1),-224*I,192*(I+1),-1,512*(I-1),
  -1616*I,1248*(I+1),-2,2816*(I-1),-8288*I,6016*(I+1),3,
  12288*(I-1),-34592*I,24144*(I+1),2,45952*(I-1),-125440*I,
  85056*(I+1),-4]],
[["(18~b)","(72~u)"],"324B",
 [1,0,I3,0,0,0,1,0,-I3,0,0,0,2,0,-I3,0,0,0,1,0,2*I3,0,0,0,3]],
[["(18~c)","(72~v)"],"324B","**"],
[["(18~d)","(18~e)"],"36I",
 [1,0,0,-I3,0,0,1,0,0,I3,0,0,-1,0,0,0,0,0,1,0,0,-2*I3,0,0,0]],
[["(18~f)","(18~g)"],"108F",
 [1,0,0,2*I3,0,0,1,0,0,4*I3,0,0,8,0,0,6*I3,0,0,10,0,0,16*I3,0,0,
  18]],
[["(20~i)","(20~l)","(20~k)","(20~j)"],"80A",
 [1,0,-2*I,2*(I+1),1,0,-2*I,4*(I+1),4,0,-6*I,6*(I+1),9,0,-8*I,
  12*(I+1),13,0,-20*I,18*(I+1),22,0,-28*I,32*(I+1),34]],
[["(21a)","(84~i)"],"441A",
 [1,0,0,B7,0,0,1,0,0,-B7,0,0,-B7,0,0,0,0,0,1,0,0,2,0,0,2*B7]],
[["(21b)","(84~j)"],"441A","**"],
[["(25b)","(100~m)"],"125A",
 [1,0,-B5+1,0,0,R5+1,0,-2*B5,0,0,3*B5+4,0,-R5+4,0,0,2*R5+4,0,
  -3*R5+5,0,0,5*B5+10,0,-7*B5+5,0,0]],
[["(25c)","(100~n)"],"125A","*2"],
[["(25~a)","(100~e)"],"25D",
 [1,0,5*B5+4,-10,5,-5*R5+21,-25*B5-25,30*R5+6,-85,50,-65*B5+110,
  -75*R5-75,335*B5+209,-460,230,-160*R5+664,-675*B5-675,
  715*R5+165,-1860,930,-1175*B5+1925,-1225*R5-1225,2530*R5+606,
  -6420,3080]],
[["(25~b)","(100~f)"],"25D","*2"],
[["(28b)","(56~h)"],"392A",
 [1,0,B7,0,1,0,0,0,0,0,-B7,0,-B7,0,0,0,B7,0,2,0,1,0,0,0,0]],
[["(28c)","(56~i)"],"392A","**"],
```

TABLE 3. (continued)

```
[["(32f)","(32i)","(32g)","(32h)"],"512C",
 [1,0,I2,0,0,0,0,0,1,0,-I2,0,0,0,0,0,1,0,0,0,0,0,0,0,2]],
[["(32j)","(32m)","(32l)","(32k)"],"256K",
 [1,0,-2*Z8,0,0,0,0,0,-1,0,-2*Z8,0,0,0,0,0,-1,0,-4*Z8,0,0,0,0,0,
  0]],
[["(32~f)","(32~i)","(32~h)","(32~g)"],"32G",
 [1,0,-6*I,4*(I+1),-2,8*(I-1),-20*I,16*(I+1),7,32*(I-1),-94*I,
  56*(I+1),-14,96*(I-1),-228*I,160*(I+1),27,256*(I-1),-684*I,
  404*(I+1),-50,624*(I-1),-1472*I,944*(I+1),84]],
[["(32~j)","(32~l)"],"256F",
 [1,0,2*(I2+1),0,-4*(I2+1),0,-8,0,-9,0,-14*(I2+1),0,20*(I2+1),0,24,
  0,39,0,52*(I2+1),0,-68*(I2+1),0,-96,0,-120]],
[["(32~k)","(32~m)"],"256F","**"],
[["(32~n)","(32~q)","(32~o)","(32~p)"],"256H",
 [1,0,2*I2,0,-4,0,0,0,-9,0,-14*I2,0,20,0,0,0,39,0,52*I2,0,-68,0,0,
  0,-120]],
[["(32~r)","(32~u)","(32~s)","(32~t)"],"256I",
 [1,0,2*I2,0,0,0,0,0,3,0,-6*I2,0,0,0,0,0,3,0,12*I2,0,0,0,0,0,4]],
[["(32~v)","(32~y)","(32~x)","(32~w)"],"256E",
 [1,0,2*(I+1),0,0,0,0,0,3,0,-6*(I+1),0,0,0,0,0,3,0,12*(I+1),0,0,0,
  0,0,4]],
[["(32~z)","(32~c1)","(32~b1)","(32~a1)"],"512B",
 [1,0,I+1,0,0,0,0,0,1,0,-I-1,0,0,0,0,0,1,0,0,0,0,0,0,0,2]],
[["(32~d1)","(32~g1)","(32~f1)","(32~e1)"],"32F",
 [1,0,-2*I,-4*(I+1),14,8*(I-1),4*I,-16*(I+1),39,32*(I-1),-10*I,
  -56*(I+1),162,96*(I-1),20*I,-160*(I+1),379,256*(I-1),-36*I,
  -404*(I+1),1054,624*(I-1),64*I,-944*(I+1),2228]],
[["(32~h1)","(32~m1)","(32~l1)","(32~i1)"],"256J",
 [1,0,4*(I+1)-2*Z8,0,8*I2,0,16*(I-1),0,-32*R2-1,0,-60*(I+1)-2*Z8,0,
  -104*I2,0,-176*(I-1),0,288*R2-1,0,456*(I+1)-4*Z8,0,712*I2,0,
  1088*(I-1),0,-1632*R2]],
[["(32~j1)","(32~o1)","(32~n1)","(32~k1)"],"256J","*5"],
[["(36~t)","(36~u)"],"36G",
 [1,0,-4,-3*I3,4,0,9,12*I3,-32,-21*I3,28,0,47,60*I3,-152,-96*I3,
  120,0,185,228*I3,-560,-342*I3,416,0,608]],
[["(36~v)","(36~y)","(36~x)","(36~w)"],"432A",
 [1,0,0,-I-1,0,0,-I,0,0,0,0,0,0,0,0,-I-1,0,0,-2*I,0,0,0,0,0,2]],
[["(42~a)","(42~b)"],"252A",
 [1,0,0,I3,0,0,0,0,0,I3,0,0,1,0,0,I3,0,0,1,0,0,2*I3,0,0,3]],
[["(44~e)","(44~h)","(44~g)","(44~f)"],"176A",
 [1,0,-I,-I-1,0,0,-I,-I-1,1,0,-I,-I-1,1,0,-I,-2*(I+1),2,0,-2*I,
  -2*(I+1),2,0,-3*I,-3*(I+1),4]],
[["(48~p)","(48~s)","(48~r)","(48~q)"],"48E",
 [1,0,-4*I,-2*(I+1),-1,-2*(I-1),-8*I,-6*(I+1),2,-10*(I-1),-20*I,
  -12*(I+1),-2,-22*(I-1),-56*I,-32*(I+1),3,-42*(I-1),-104*I,
  -66*(I+1),-4,-92*(I-1),-208*I,-120*(I+1),5]],
[["(48~t)","(48~w)","(48~v)","(48~u)"],"48D",
 [1,0,-I,-2*(I+1),5,4*(I-1),I,-6*(I+1),14,8*(I-1),I,-12*(I+1),34,
  20*(I-1),-5*I,-32*(I+1),75,48*(I-1),4*I,-66*(I+1),152,88*(I-1),
  5*I,-120*(I+1),293]],
```

TABLE 3. (continued)

```
[["(49~a)","(196~b)"],"49C",
 [1,0,B7-1,-3*B7-4,2,-I7+3,-3,12,7*B7,11*B7-4,-19*B7-23,6,
 [-3*I7+16,-16,53,28*B7,21*I7-42,-37*I7-57,29,-27*B7+48,-55,186,
  98*B7,72*I7-132,-241*B7-298]],
[["(49~b)","(196~c)"],"49C","**"],
[["(50b)","(200~c)"],"500A",
 [1,0,B5+1,0,0,0,0,0,0,0,-B5,0,1,0,0,0,0,0,0,0,-B5,0,B5+1,0,0]],
[["(50c)","(200~d)"],"500A","*2"],
[["(50~a)","(100~g)"],"50D",
 [1,0,B5+4,2,-3,-6*B5,3*B5+3,2*R5+10,11,-2,-13*B5+6,7*R5+7,7*B5+41,
 [28,-18,-24*R5+32,29*B5+29,13*R5+83,76,-26,-95*B5+29,41*R5+41,
  26*R5+214,164,-88]],
[["(50~b)","(100~h)"],"50D","*2"],
[["(50~c)","(100~i)"],"50E",
 [1,0,R5+1,-3,2,-R5+3,-2*R5-1,4*R5,-9,3,-3*R5+15,-6*R5-8,12*R5+5,
  -32,17,-8*R5+32,-16*R5-13,31*R5+3,-69,29,-20*R5+89,-38*R5-44,
  72*R5+24,-176,82]],
[["(50~d)","(100~j)"],"50E","*2"],
[["(54~b)","(108~l)"],"54F",
 [1,0,I3,1,-I3,I3-3,4,-2*I3-3,2*I3,5,-4*I3-3,6*I3-6,14,-7*I3-9,
  8*I3-9,18,-9-12*I3,13*I3-15,37,-20*I3-24,26*I3-18,50,-32*I3-27,
  36*I3-48,93]],
[["(54~c)","(108~m)"],"54F","**"],
[["(64~a)","(64~d)","(64~b)","(64~c)"],"256G",
 [1,0,I2,0,-2,0,0,0,-1,0,-3*I2,0,2,0,0,0,7,0,6*I2,0,-10,0,0,0,-8]],
[["(64~e)","(64~f)"],"1024C",
 [1,0,0,0,I2,0,0,0,1,0,0,0,-I2,0,0,0,-1,0,0,0,-I2,0,0,0,0]],
[["(64~g)","(64~h)","(64~i)","(64~j)"],"128D",
 [1,0,I-1,0,0,0,-2*(I+1),0,3,0,I-1,0,0,0,-2*(I+1),0,7,0,6*(I-1),0,
  0,0,-8*(I+1),0,16]],
[["(64~k)","(64~l)"],"1024B",
 [1,0,0,0,I2,0,0,0,-1,0,0,0,I2,0,0,0,-3,0,0,0,-3*I2,0,0,0,2]],
[["(64~m)","(64~o)"],"1024A",
 [1,0,I2,0,I2,0,2,0,-1,0,-I2,0,I2,0,2,0,-3,0,-2*I2,0,-3*I2,0,-4,0,
  2]],
[["(64~n)","(64~p)"],"1024A","**"],
[["(64~q)","(64~r)","(64~c1)","(64~z)","(64~u)","(64~v)",
  "(64~y)","(64~d1)"],"64D",
 [1,0,-2*Z8^3+I-1,-2*(Z8^3+I),-2*I2,2*(R2+I-1),-2*(I+1),0,4*R2+7,
  -4*(I2-I-1),-10*Z8^3+5*(I-1),-12*(Z8^3+I),-6*I2,8*(R2+I-1),
  -10*(I+1),0,12*R2+27,-16*(I2-I-1),-36*Z8^3+18*(I-1),-42*(Z8^3+I),
  -26*I2,28*(R2+I-1),-32*(I+1),0,44*R2+84]],
[["(64~s)","(64~x)","(64~e1)","(64~f1)","(64~w)","(64~t)",
  "(64~a1)","(64~b1)"],"64D","*5"],
[["(72~w)","(72~y)"],"72J",
 [1,0,Z3-1,2*Z3,Z3+4,-2,4,4*Z3^2,Z3^2,6*Z3,-I3+5,-4,10,12*Z3^2,
  I3-11,16*Z3,3*Z3+24,-10,24,28*Z3^2,-2*I3-10,36*Z3,-2*I3+34,-24,
  53]],
[["(72~x)","(72~z)"],"72J","**"],
```

TABLE 3. (continued)

```
[["(72~a1)","(72~b1)"],"72K",
 [1,0,-2,-I3,2,0,1,2*I3,-4,-3*I3,2,0,7,6*I3,-16,-8*I3,12,0,9,14*I3,
  -28,-18*I3,16,0,32]],
[["(75~a)","(300~a)"],"75B",
 [1,0,-B5+1,-1,2,R5+3,R5+1,-3*B5,-1,5,B5+8,3*B5+3,-7*B5+2,-4,8,
  2*R5+16,9*B5+9,-8*R5+12,-6,21,5*R5+33,8*R5+8,-14*R5+21,-12,38]]
[["(75~b)","(300~b)"],"75B","*2"],
[["(80~g)","(80~j)","(80~i)","(80~h)"],"80B",
 [1,0,-I,I+1,2,-I+1,I,2*(I+1),4,-3*(I-1),-I,3*(I+1),8,-4*(I-1),2*I,
  6*(I+1),13,-7*(I-1),-2*I,9*(I+1),22,-13*(I-1),0,16*(I+1),36]],
[["(96~c)","(96~e)"],"768A",
 [1,0,-I2-1,0,-I2-1,0,1,0,0,0,I2+1,0,2*(I2+1),0,-3,0,3,0,-2*(I2+1),
  0,-2*(I2+1),0,3,0,-3]],
[["(96~d)","(96~f)"],"768A","**"],
[["(96~g)","(96~j)","(96~h)","(96~i)"],"768B",
 [1,0,-I2,0,-1,0,0,0,0,0,I2,0,2,0,0,0,3,0,-2*I2,0,-2,0,0,0,-3]],
[["(96~k)","(96~n)","(96~l)","(96~m)"],"2304C",
 [1,0,-I2,0,0,0,0,0,0,0,0,0,0,0,0,0,0,0,0,0,0,0,0,0,1]],
[["(96~o)","(96~r)","(96~q)","(96~p)"],"2304B",
 [1,0,-I-1,0,0,0,0,0,0,0,0,0,0,0,0,0,0,0,0,0,0,0,0,0,1]],
[["(96~s)","(96~v)","(96~u)","(96~t)"],"96H",
 [1,0,0,I+1,1,-I+1,-2*I,I+1,4,-I+1,2*I,2*(I+1),4,-3*(I-1),0,
  4*(I+1),9,-5*(I-1),0,5*(I+1),10,-6*(I-1),-2*I,8*(I+1),21]],
[["(96~w)","(96~z)","(96~y)","(96~x)","96G",
 [1,0,-2*I,-I-1,-1,-I+1,-2*I,-I-1,0,-I+1,-4*I,-2*(I+1),0,-3*(I-1),
 [-4*I,-4*(I+1),1,-5*(I-1),-12*I,-5*(I+1),-2,-6*(I-1),-14*I,
  -8*(I+1),5]],
[["(98~a)","(196~d)"],"98B",
 [1,0,-B7-1,B7,2,2,1,4,-B7,-B7,3*B7+1,2,I7+4,0,9,-4*B7,-3*I7-2,
  4*I7-2,9,B7+12,1,22,-6*B7,-4*I7,15*B7+6]],
[["(98~b)","(196~e)"],"98B","**"],
[["(100~k)","(200~a)"],"200B",
 [1,0,-B5,0,1,0,B5+1,0,-1,0,B5+2,0,-B5+1,0,2,0,B5+1,0,0,0,B5+5,0,
  -4*B5,0,4]],
[["(100~l)","(200~b)"],"200B","*2"],
[["(100~o)","(100~p)"],"100B",
 [1,0,0,R5,-2,0,1,0,-1,-R5,6,-2*R5,-3,4*R5,-7,0,3,0,-1,-3*R5,17,
  -6*R5,-8,12*R5,-22]],
[["(108~n)","(108~o)"],"108E",
 [1,0,-1,I3,1,0,0,-I3,-2,I3,1,0,2,-2*I3,-5,2*I3,3,0,5,-4*I3,-8,
  6*I3,5,0,5]],
[["(144~f)","(144~i)","(144~h)","(144~g)"],"144A",
 [1,0,-I,I+1,-1,I-1,-2*I,0,-1,-I+1,-2*I,0,4,2*(I-1),-2*I,4*(I+1),
  -3,3*(I-1),-2*I,0,-4,-2*(I-1),-4*I,0,11]],
[["(150~a)","(300~c)"],"150B",
 [1,0,B5+1,-1,0,0,0,B5,-1,1,-B5,-B5-1,B5+2,-2,0,2,-B5-1,2*B5,-2,1,
  -R5+3,-R5-1,2*R5+1,-4,2]],
[["(150~b)","(300~d)"],"150B","*2"]]];
```

TABLE 4. Translation classes with miscellaneous properties.

Semi-replicables: `6~b 15~a/b 18~d 18~f 36~t 42~a 72~w/x 72~a1 100~o 108~n`
Rational ghosts: `25a 27a 27c 32c 32e 49a 50a 54a 54b`
Irrational ghosts: `32j 32~h1/j1`
Functions with conjugate groups: `(12H,12c) (12~j,36~l) (16C,16b) (18A,18E) (20F,20d) (24D,24H) (36D,36f) (48a,48~j) (50A,50~c/d) (54A,54~b/c) (60D,60a) (18~d,36~t) (25~a/b) (50~a/b) (64~q/s) (75~a/b) (100~k/l) (150~a/b)`

References

1. D. Alexander, C. Cummins, J. McKay, and C. Simons, *Completely replicable functions*, Groups, Combinatorics & Geometry (Durham, 1990) (M. Liebeck and J. Saxl, eds.), London Math. Soc. Lecture Note Ser., vol. 165, Cambridge Univ. Press, Cambridge, 1992, pp. 87–98.
2. R. E. Borcherds, *Monstrous moonshine and monstrous Lie superalgebras*, Invent. Math. **109** (1992), no. 2, 405–444.
3. J. H. Conway, R. T. Curtis, S. P. Norton, R. A. Parker, and R. A. Wilson, *Atlas of finite groups: Maximal subgroups and ordinary characters for simple groups*, with computational assistance from J. G. Thackray, Oxford University Press, Eynsham, 1985.
4. J. H. Conway, J. McKay, and A. Sebbar, *On the discrete groups of Moonshine*, Proc. Amer. Math. Soc. **132** (2004), no. 8, 2233–2240.
5. J. H. Conway and S. P. Norton, *Monstrous moonshine*, Bull. London Math. Soc. **11** (1979), no. 3, 308–339.
6. C. J. Cummins, *Congruence subgroups of groups commensurable with* $\mathrm{PSL}(2,\mathbb{Z})$ *of genus* 0 *and* 1, Experiment. Math. **13** (2004), no. 3, 361–382.
7. C. J. Cummins and S. P. Norton, *Rational Hauptmoduls are replicable*, Canad. J. Math. **47** (1995), no. 6, 1201–1218.
8. ———, unpublished work.
9. C. Dong, H. Li, and G. Mason, *Modular-invariance of trace functions in orbifold theory and generalized Moonshine*, Comm. Math. Phys. **214** (2000), no. 1, 1–56.
10. D. Ford, J. McKay, and S. Norton, *More on replicable functions*, Comm. Algebra **22** (1994), no. 13, 5175–5193.
11. G. Mason, *Finite groups and modular functions*, with an appendix by S. P. Norton, The Arcata Conference on Representations of Finite Groups (Arcata, CA, 1986) (P. Fong, ed.), Proc. Sympos. Pure Math., vol. 47, Part 1, Amer. Math. Soc., Providence, RI, 1987, pp. 181–210.
12. S. P. Norton, *More on moonshine*, Computational Group Theory (Durham, 1982) (M. D. Atkinson, ed.), Academic Press, London, 1984, pp. 185–193.
13. ———, *Non-Monstrous Moonshine*, Groups, Difference Sets, and the Monster (Columbus, OH, 1993) (K. T. Arasu, J. F. Dillon, K. Harada, S. Sehgal, and R. Solomon, eds.), Ohio State Univ. Math. Res. Inst. Publ., vol. 4, de Gruyter, Berlin, 1996, pp. 433–441.
14. ———, *From Moonshine to the Monster*, Proceedings on Moonshine and Related Topics (Montréal, QC, 1999) (J. McKay and A. Sebbar, eds.), CRM Proc. Lecture Notes, vol. 30, Amer. Math. Soc., Providence, RI, 2001, pp. 163–171.
15. G. Shimura, *Introduction to the arithmetic theory of automorphic functions*, Publ. Math. Soc. Japan, vol. 11, Princeton Univ. Press, Princeton, NJ, 1994.

DEPARTMENT OF PURE MATHEMATICS AND MATHEMATICAL STATISTICS, UNIVERSITY OF CAMBRIDGE, WILBERFORCE ROAD, CAMBRIDGE CB3 0WB, UK

E-mail address: s.norton@dpmms.cam.ac.uk

Centre de Recherches Mathématiques
CRM Proceedings and Lecture Notes
Volume **47**, 2009

Monodromy Evolving Deformations and Halphen's Equation

Yousuke Ohyama

1. Introduction

In this note we study examples of monodromy evolving deformations, which contain Halphen's quadratic equation [6].

A quadratic equation

$$\frac{\mathrm{d}x_i}{\mathrm{d}t} = \sum_{j,k=1}^{n} c^i_{jk} x_j x_k \quad (i = 1, 2, \dots, n) \tag{1}$$

is studied in many fields. Here c^i_{jk} is a complex constant and $c^i_{jk} = c^i_{kj}$. In 1960s Markus studies stability around the origin of a quadratic equation [7]. He uses a non-associative algebra $V = \bigoplus_{j=1}^n \mathbb{C}a_j$ associated with (1):

$$a_j \cdot a_k = \sum_{i=1}^{n} c^i_{jk} a_i.$$

Markus gave a classification of quadratic systems when $n = 2$ by means of the non-associative algebra.

Quadratic equations also appear in mathematical physics. It is well known that Halphen's equation [5] is a reduction from the Bianchi IX cosmological models or the self-dual Yang–Mills equation [3].

Halphen's first equation

$$X' + Y' = 2XY, \quad Y' + Z' = 2YZ, \quad Z' + X' = 2ZX, \tag{2}$$

can be solved by theta constants [5, 8]. If we set $y = 2(X + Y + Z)$, y satisfies Chazy's equation [1]

$$y''' = 2yy'' - 3(y')^2. \tag{3}$$

Halphen also studies his second equation

$$\begin{aligned} u_1' &= a_1 u_1^2 + (\lambda - a_1)(u_1u_2 + u_3u_1 - u_2u_3), \\ u_2' &= a_2 u_2^2 + (\lambda - a_2)(u_2u_3 + u_1u_2 - u_3u_1), \\ u_3' &= a_3 u_3^2 + (\lambda - a_3)(u_3u_1 + u_2u_3 - u_1u_2), \end{aligned} \tag{4}$$

2000 *Mathematics Subject Classification.* Primary 34M55; Secondary 17A99.
This is the final form of the paper.

which can be solved by the Gauss hypergeometric function [6]. Halphen's first equation (2) is a special case of (5). We will review Halphen's equations and Chazy's equation in Section 2.

We take new variables x_1, x_2, x_3, which are linear sums of u_1, u_2, u_3 (see (6) in Section 2). Then (4) is equivalent to the following system [9]

$$(5)\qquad \begin{aligned} x_1' &= x_1^2 + a(x_1-x_2)^2 + b(x_2-x_3)^2 + c(x_3-x_1)^2,\\ x_2' &= x_2^2 + a(x_1-x_2)^2 + b(x_2-x_3)^2 + c(x_3-x_1)^2,\\ x_3' &= x_3^2 + a(x_1-x_2)^2 + b(x_2-x_3)^2 + c(x_3-x_1)^2. \end{aligned}$$

We use (5) instead of Halphen's original equation (4) in this paper.

For Halphen's equation, the associated non-associative algebra has a unit. Conversely, a quadratic equation with rank three can be solved by (confluent) hypergeometric functions, if and only if the associated non-associative algebra has a unit [10]. The author was taught Markus' work by John McKay.

In [2], Chakravarty and Ablowitz showed that a quadratic equation with fifth order, coming from the self-dual Bianchi IX system can be represented by *monodromy evolving deformations.* Their equation contains (2) as a special case. The aim of this paper is to represent (5) as monodromy evolving deformations. Since Halphen's equation does not have the Painlevé property, it is never represented by monodromy preserving deformations.

In [3, 4], they obtained the Lax pair of Halphen's first equation and Chazy's equation, which are special cases of Halphen's second equation. And our Lax pair is different from their results even in the case of Halphen's first equation, since ours are monodromy evolving deformations but [3, 4] gave the Lax pair as a reduction of the self-dual Yang–Mills equation

Both [2] and the author treat special cases of monodromy evolving deformations. We do not have general theory of monodromy evolving deformation. But it is important to study monodromy evolving deformations to consider nonlinear equations which does not have the Painlevé property, such as Chazy's equation and Halphen's equations.

2. Halphen's equation

In this section we review Halphen's equation. Halphen's first equation (2) can be solved by theta constants. See [8].

$$\begin{aligned} X &= 2\frac{\partial}{\partial t}\log\left(\theta_2\left(0, \frac{at+b}{ct+d}\right)(ct+d)^{-1/2}\right)\\ Y &= 2\frac{\partial}{\partial t}\log\left(\theta_3\left(0, \frac{at+b}{ct+d}\right)(ct+d)^{-1/2}\right),\\ Z &= 2\frac{\partial}{\partial t}\log\left(\theta_4\left(0, \frac{at+b}{ct+d}\right)(ct+d)^{-1/2}\right). \end{aligned}$$

Here $\theta_j(z,\tau)$ is Jacobi's theta function. When $j = 2,3,4$, $\theta_j(z,\tau)$ is an even function of z. Since generic solutions of (2) have a natural boundary, (2) does not have the Painlevé property.

If we set $y = 2(X+Y+Z)$, y satisfies Chazy's equation (3), which is solved by

$$y(t) = 4\frac{\partial}{\partial t}\log\left(\theta_1'\left(0, \frac{at+b}{ct+d}\right)(ct+d)^{-3/2}\right).$$

Halphen's first equation and Chazy's equation are special cases of Halphen's second equation (4), which is solved by hypergeometric functions. We take a linear transformation

$$
\begin{aligned}
x_1 &= \frac{2a_1 + a_2 + a_3 - 2\lambda}{2} u_1 - \frac{a_1 + a_3 - 2\lambda}{2} u_2 - \frac{a_1 + a_2 - 2\lambda}{2} u_3, \\
x_2 &= -\frac{a_2 + a_3 - 2\lambda}{2} u_1 + \frac{a_1 + 2a_2 + a_3 - 2\lambda}{2} u_2 - \frac{a_1 + a_2 - 2\lambda}{2} u_3, \\
x_3 &= -\frac{a_2 + a_3 - 2\lambda}{2} u_1 - \frac{a_1 + a_3 - 2\lambda}{2} u_2 + \frac{a_1 + a_2 + 2a_3 - 2\lambda}{2} u_3.
\end{aligned} \tag{6}
$$

Then Halphen's second equation (4) is equivalent to our system (5) for

$$
\begin{aligned}
a &= \frac{-2\lambda^2 + 2(a_1 + a_2 + a_3)\lambda + a_1 a_2 + a_2 a_3 + a_3 a_1 + a_3^2}{4(a_1 + a_2 + a_3 - 2\lambda)^2}, \\
b &= \frac{-2\lambda^2 + 2(a_1 + a_2 + a_3)\lambda + a_1 a_2 + a_2 a_3 + a_3 a_1 + a_1^2}{4(a_1 + a_2 + a_3 - 2\lambda)^2}, \\
c &= \frac{-2\lambda^2 + 2(a_1 + a_2 + a_3)\lambda + a_1 a_2 + a_2 a_3 + a_3 a_1 + a_2^2}{4(a_1 + a_2 + a_3 - 2\lambda)^2}.
\end{aligned}
$$

Our system (5) can be solved by hypergeometric functions. We take a Fuchsian equation

$$
\frac{\mathrm{d}^2 y}{\mathrm{d}z^2} = \left(\frac{a+b}{z^2} + \frac{c+b}{(z-1)^2} - \frac{2b}{z(z-1)} \right) y.
$$

Let t be a ratio of two solutions of the above equation. We set

$$
x_1 = \frac{\mathrm{d}}{\mathrm{d}t} \log y, \quad x_2 = \frac{\mathrm{d}}{\mathrm{d}t} \log \frac{y}{z}, \quad x_3 = \frac{\mathrm{d}}{\mathrm{d}t} \log \frac{y}{z-1}.
$$

Then x_1, x_2 and x_3 satisfy (5).

Chazy's equation can be solved by the hypergeometric equation

$$
x(1-x)\frac{\mathrm{d}^2 u}{\mathrm{d}x^2} + \left(\frac{1}{2} - \frac{7}{6}x \right) \frac{\mathrm{d}u}{\mathrm{d}x} - \frac{1}{144} u = 0, \tag{7}
$$

by $y = 6\mathrm{d}/\mathrm{d}t \log u$ (See [1, p. 321]). The hypergeometric equation (7) gives Halphen's second equation (5) for

$$
a = -\frac{31}{288}, \quad b = -\frac{23}{288}, \quad c = -\frac{41}{288}.
$$

In case $a = b = c = -\frac{1}{8}$, (5) is equivalent to (2) by the linear transform $2X = x_2 + x_3$, $2Y = x_3 + x_1$, $2Z = x_1 + x_2$ [9].

For (5), the associated algebra has the following multiplication table:

$$
\begin{aligned}
a_1 \cdot a_1 &= a_1 + (c+a)(a_1 + a_2 + a_3), & a_1 \cdot a_2 &= -a(a_1 + a_2 + a_3), \\
a_2 \cdot a_2 &= a_2 + (a+b)(a_1 + a_2 + a_3), & a_2 \cdot a_3 &= -b(a_1 + a_2 + a_3), \\
a_3 \cdot a_3 &= a_3 + (b+c)(a_1 + a_2 + a_3), & a_3 \cdot a_1 &= -c(a_1 + a_2 + a_3).
\end{aligned}
$$

And this algebra has a unit $a_1 + a_2 + a_3$.

In [10], we showed that if and only if a non-associative and commutative algebra of rank three has a unit, the corresponding quadratic nonlinear equation can be solved by (confluent) hypergeometric functions. (There exist some exceptional cases which can be solved by elementary functions.) If the quadratic equation can be solved by hypergeometric functions, it is Halphen's second equation (5). A relation between Halphen-type equations and non-associative algebras was suggested to the author by John McKay.

3. Monodromy evolving equations

We show that Halphen's second equation is represented by monodromy evolving deformations. We set

$$Q(x) = x^2 + a(x_1 - x_2)^2 + b(x_2 - x_3)^2 + c(x_3 - x_1)^2.$$

Halphen's second equation is

$$x_j' = Q(x_j), \quad j = 1, 2, 3.$$

We set

$$P(x) = (x - x_1)(x - x_2)(x - x_3)$$

and consider the following 2×2 linear system.

$$\frac{\partial Y}{\partial x} = \frac{\mu + Cx + D}{P} Y, \tag{8}$$

$$\frac{\partial Y}{\partial t} = (\nu + C)Y - Q(x)\frac{\partial Y}{\partial x}. \tag{9}$$

Here μ is a constant parameter and

$$C = \big(k(x_1 - x_3) + l(x_2 - x_3)\big)S, \quad D = \big(kx_2(x_3 - x_1) + lx_1(x_3 - x_2)\big)S.$$

Here k, l are any constants and S is any traceless constant matrix. We assume

$$\frac{\partial \nu}{\partial x} = -\frac{x + x_1 + x_2 + x_3}{P}\mu.$$

Since ν is not a rational function of x, (9) does not give an isomonodromic deformation of (8).

Theorem 1. *The compatibility condition of* (8) *and* (9) *gives the Halphen's second equation.*

We can prove the theorem above directly. Therefore (8) and (9) are a Lax pair of Halphen's second equation.

Around each singular point $x = x_j$ the solution of (8) is of the form

$$Y(x) = Y_j(x)(x - x_j)^{L_j}, \tag{10}$$

where $Y_j(x)$ is holomorphic and invertible near $x = x_j$. The local exponent matrix L_j is independent of x and the local monodromy of $Y_j(x)$ around $x = x_j$ is adjoint to $e^{2\pi i L_j}$. The deformation does not preserve monodromy data and

$$\frac{dL_j}{dt} = \frac{2x_j + x_k + x_l}{\prod_{r \neq j}(x_j - x_r)}\mu, \tag{11}$$

where $\{j, k, l\} = \{1, 2, 3\}$ as a set. And the singular points deform as

$$\frac{dx_j}{dt} = Q(x_j), \tag{12}$$

which is nothing but Halphen's second equation.

We can eliminate the variables μ and ν in (8) and (9) by the rescaling $Y = fZ$ for a scalar function $f = f(x,t)$ [2]. f satisfies the linear equations

$$\frac{\partial f}{\partial x} = \frac{\mu}{P} f, \tag{13}$$

$$\frac{\partial f}{\partial t} = \nu f - Q(x)\frac{\partial f}{\partial x}. \tag{14}$$

The integrability condition for f is

$$\frac{\partial P}{\partial t} + Q\frac{\partial P}{\partial x} - P\frac{\partial Q}{\partial x} - (x + x_1 + x_2 + x_3)P = 0. \tag{15}$$

We remark that (13) – (14) is nothing but (8) – (9) for $k = l = 0$. The condition (15) for the scalar system (13) – (14) yields Halphen's second equation (5).

Z satisfies

$$\frac{\partial Z}{\partial x} = \frac{Cx + D}{P} Z, \quad \frac{\partial Z}{\partial t} = CY - Q(x)\frac{\partial Z}{\partial x}.$$

The integrability condition for Z is the sixth Painlevé equation. In our case, we take the Riccati solution of the sixth Painlevé equation, which reduce to the hypergeometric equation, since the residue matrix of $(Cx + D)/P$ at infinity is zero. But this hypergeometric equation is different from the hypergeometric functions which comes from (5).

4. Conclusion

Halphen's equation is a special case of monodromy evolving equations. It is similar to the Riccati solutions of the Painlevé equations. Both are very special cases and can be represented by hypergeometric solutions. We need the integrability condition for a scalar system to describe Halphen's equation and the 2×2 linear system is not necessary to derive Halphen's equation.

It is a future problem to study generic solutions of monodromy evolving equations. Generic solutions of monodromy evolving equations are believed to be highly transcendental functions like the Painlevé transcendents but we do not know how to show it is.

In this paper we treated Halphen's second equation, which corresponds to the Painlevé sixth equation. We will study confluent cases in the future. We need monodromy evolving equations to study nonlinear equations without the Painlevé property.

References

1. J. Chazy, *Sur les équations différentielles du troisième ordre et d'ordre supérieur dont l'intégrale générale a ses points critiques fixes*, Acta Math. **34** (1911), no. 1, 317 – 385 (French). MR 1555070
2. S. Chakravarty and M. J. Ablowitz, *Integrability, monodromy evolving deformations, and self-dual Bianchi* IX *systems*, Phys. Rev. Lett. **76** (1996), no. 6, 857 – 860.
3. S. Chakravarty, M. J. Ablowitz, and P. A. Clarkson, *Reductions of self-dual Yang-Mills fields and classical systems*, Phys. Rev. Lett. **65** (1990), no. 9, 1085 – 1087.
4. S. Chakravarty, M. J. Ablowitz, and L. A. Takhtajan, *Self-dual Yang-Mills equation and new special functions in integrable systems*, Nonlinear Evolution Equations and Dynamical Systems (Baia Verde, 1991) (M. Boiti, L. Martina, and F. Pempinelli, eds.), World Sci. Publ., River Edge, NJ, 1992, pp. 3 – 11.
5. G. Halphen, *Sur un système d'équations différentielles*, C. R. Acad. Sci. Paris **92** (1881), 1101 – 1103.

6. ———, *Sur certains systèmes d'équations différentielles*, C. R. Acad. Sci. Paris **92** (1881), 1404–1406.
7. L. Markus, *Quadratic differential equations and non-associative algebras*, Contributions to the Theory of Nonlinear Oscillations, Vol. V (L. Cesari, J. LaSalle, and S. Lefschetz, eds.), Ann. of Math. Stud., vol. 45, Princeton Univ. Press, Princeton, NJ, 1960, pp. 185–213.
8. Y. Ohyama, *Differential relations of theta functions*, Osaka J. Math. **32** (1995), no. 2, 431–450.
9. ———, *Systems of nonlinear differential equations related to second order linear equations*, Osaka J. Math. **33** (1996), no. 4, 927–949.
10. ———, *Hypergeometric functions and non-associative algebras*, Proceedings on Moonshine and Related Topics (Montréal, QC, 1999) (J. McKay and A. Sebbar, eds.), CRM Proc. Lecture Notes, vol. 30, Amer. Math. Soc., Providence, RI, 2001, pp. 173–184.

Graduate School of Information Science and Technology, Osaka University, Toyonaka, Osaka 560-0043, Japan

E-mail address: ohyama@math.sci.osaka-u.ac.jp

Centre de Recherches Mathématiques
CRM Proceedings and Lecture Notes
Volume **47**, 2009

Integral Solutions of Apéry-like Recurrence Equations

Don Zagier

ABSTRACT. In [4], Beukers studies the differential equation

$$(*) \qquad \big((t^3+at^2+bt)F'(t)\big)' + (t-\lambda)F(t) = 0,$$

where a, b and λ are rational parameters, and asks for which values of these parameters this equation has a solution in $\mathbb{Z}[\![t]\!]$, the motivating example being the Apéry sequence with $a = 11$, $b = -1$, $\lambda = -3$. We describe a search over a suitably chosen domain of 100 million triples (a, b, λ). In this domain there are 36 triples yielding integral solutions of (*). These can be further subdivided into members of four infinite classes, two of which are degenerate special cases of the other two, and seven sporadic solutions. Of these solutions, twelve, including all the sporadic ones, have parametrizations of Beukers type in terms of modular forms and functions. These solutions are related to elliptic curves over $\mathbb{P}^1$ with four singular fibres.

1. Beukers's recurrence equation

Beukers [4] considers the differential equation

$$(1) \qquad \big(tP(t)F'(t)\big)' + (t-\lambda)F(t) = 0$$

with $P(t)$ a quadratic polynomial, which we can take to have the form $t^2 + at + b$. This equation has a unique solution which is regular at the origin with $F(0) = 1$, given by $F(t) = \sum_{n=0}^{\infty} u_n t^n$ with $u_0 = 1$ and

$$(2) \qquad b(n+1)^2 u_{n+1} + (an^2 + an - \lambda)u_n + n^2 u_{n-1} = 0 \quad (n \geq 0)$$

so that $u_1 = \lambda/b$, $u_2 = (\lambda^2 - 2a\lambda + b)/4b^2$, etc. We are interested in finding values of (a, b, λ) for which $F(Dt) \in \mathbb{Z}[\![t]\!]$ or $D^n u_n \in \mathbb{Z}$ $(\forall n \geq 0)$ for some $D \in \mathbb{N}$. After a rescaling $(a, b, \lambda) \mapsto (a/D, b/D^2, \lambda/D)$, $u_n \mapsto D^n u_n$ we can assume $D = 1$ or $u_n \in \mathbb{Z}$, and in future we shall assume after such a rescaling that the u_n are all integral, that there is no $D > 1$ such that $D^n \mid u_n$ for all n, and that $u_1 \geq 0$. Beukers further assumes that $b = -1$, but this scaling can only be made over $\mathbb{C}$ and destroys the desired integrality property of the u_n, and in fact there seem to be almost no cases of integral solutions with $b = -1$. (See below.)

To search for rational values of a, b and λ leading to $u_n \in \mathbb{Z}$, we observe that equation (2) for $n \in \{0, 1, 2\}$ gives three linear equations for (a, b, λ) in terms of

2000 *Mathematics Subject Classification.* 11F03, 11G05, 34.
This is the final form of the paper.

(u_1, u_2, u_3) which can be solved uniquely if the corresponding determinant is non-zero. We can therefore search over a domain of $(u_1, u_2, u_3) \in \mathbb{Z}^3$, compute (a, b, λ) for each choice of these three initial values, and then see whether the further u_n (up to some pre-assigned search limit) are also integral. This search was performed in the range $0 \le u_1 \le 30$, $|u_2| \le 100$, $|u_3| \le 2000$ (time on a Sun workstation: 52 hours) and yielded 19 solutions, listed below. If we restrict to the case $b = -1$ considered by Beukers, then we need only search over $(u_1, u_2) \in \mathbb{Z}^2$, since these values already determine a, λ, and the further u_n. A search in the domain $0 < u_1 \le 100$, $|u_2| \le 3000$ yielded only a single case $(a, \lambda) = (11, -3)$ (the Apéry numbers) with integral values of u_n.

In all 19 cases found in the range searched, the numbers $A = -a/b$, $B = 1/b$ and $\Lambda = \lambda/b$ were integral. In terms of the new parameters A and B, the recurrence relation (2) becomes

$$(n+1)^2 u_{n+1} - An(n+1)u_n + Bn^2 u_{n-1} = \lambda u_n \quad (n \ge 0) \tag{3}$$

where we have changed the name of the eigenvalue from Λ back to λ. (The corresponding changes in the differential equation (1) are to take $P(t) = Bt^2 - At + 1$ and replace the factor $t - \lambda$ by $Bt - \lambda$.) The observation just made then becomes

Conjecture. *If* (3) *has a solution with* $u_0 = 1$ *and* $u_n \in \mathbb{Z}$ *for all* n, *then* A, B *and* λ *are integral.*

Notice that it is equivalent to conjecture that just A is integral, since (3) for $n = 0$ and $n = 1$ then gives $\lambda = u_1 \in \mathbb{Z}$, $B = (2A + \lambda)u_1 - 4u_2 \in \mathbb{Z}$. Also, it seems reasonable to guess that the Conjecture is true even if we do not suppose that $u_0 = 1$.

2. Numerical data

Assuming the above conjecture, we can do a new and much more rapid search over integral values of A, u_1 and u_2. As explained above, we can suppose that $u_1 \ge 0$. The search can also be speeded up by noticing that the integrality of u_3 is equivalent to a congruence for A modulo 1, 3 or 9 (namely: A is arbitrary if u_1 and u_2 are both divisible by 3, $A \equiv -u_1/3 \pmod 3$ if u_1 is divisible by 3 but u_2 is not, and $A \equiv 4u_1 + u_1u_2(2 + 3u_2 + 2u_1^2) \pmod 9$ if u_1 is not divisible by 3) and the integrality of u_4 is equivalent to the simple congruence $u_1 \equiv u_2 \pmod 2$. Inserting these two congruences into the triple loop results in a reduction of the search domain by a factor 7/54. We performed the search for the 100 million triples in the domain $|A| \le 250$, $0 \le u_1 \le 100$, $|u_2| \le 1000$ (time on the Sun workstation: 19 hours), finding altogether 36 solutions, listed in Table 1 below. The integrality of the u_n was checked up to the limit $n = 25$ (it was subsequently verified that in each of the 36 "solutions" satisfying this criterion the u_n are indeed integral for all n) and this was by no means unnecessarily large, since the maximum value of $n_0 := \min\{n \mid u_n \notin \mathbb{Z}\}$ in this range was in fact 23. This value occurred exactly once, for the triple $(a, u_1, u_2) = (229, 4, -660)$. The other values of n_0 which occurred were 5, 7, 8, 9, 11, 13, 16, 17 and 19, the last four occurring for 5926, 45, 344 and 17 triples, respectively. (Note that n_0 is automatically at least 5 since we have done the search in such a way that $u_0, \ldots, u_4$ are always integral.) The fact that n_0 is always a prime power is a consequence of a result of Beukers [4, Proposition 3.3], according to which the smallest n (if any) for which a given prime p occurs in the denominator of u_n is always a power of p.

TABLE 1.

index	A	B	λ	u_0	u_1	u_2	u_3	u_4	u_5	u_6
1	−1	0	0	1	0	0	0	0	0	0
2	0	−16	0	1	0	4	0	36	0	400
3	2	1	1	1	1	1	1	1	1	1
4	−1	0	2	1	2	0	0	0	0	0
5	7	−8	2	1	2	10	56	346	2252	15184
6	2	1	3	1	3	5	7	9	11	13
7	9	27	3	1	3	9	21	9	−297	−2421
8	10	9	3	1	3	15	93	639	4653	35169
9	11	−1	3	1	3	19	147	1251	11253	104959
10	12	32	4	1	4	20	112	676	4304	28496
11	16	0	4	1	4	36	400	4900	63504	853776
12	−1	0	6	1	6	6	0	0	0	0
13	17	72	6	1	6	42	312	2394	18756	149136
14	27	0	6	1	6	90	1680	34650	756756	17153136
15	2	1	7	1	7	19	37	61	91	127
16	−27	0	12	1	12	−126	2100	−40950	864864	−19171152
17	−16	0	12	1	12	−60	560	−6300	77616	−1009008
18	−1	0	12	1	12	30	20	0	0	0
19	32	256	12	1	12	164	2352	34596	516912	7806224
20	64	0	12	1	12	420	18480	900900	46558512	2498640144
21	2	1	13	1	13	55	147	309	561	923
22	−64	0	20	1	20	−540	21840	−1021020	51459408	−2715913200
23	−1	0	20	1	20	90	140	70	0	0
24	2	1	21	1	21	131	471	1251	2751	5321
25	54	729	21	1	21	495	12171	305919	7794171	200412801
26	32	256	28	1	28	580	10992	199524	3530352	61417616
27	−27	0	30	1	30	−180	2640	−48510	989604	−21441420
28	−1	0	30	1	30	210	560	630	252	0
29	2	1	31	1	31	271	1281	4251	11253	25493
30	−1	0	42	1	42	420	1680	3150	2772	924
31	2	1	43	1	43	505	3067	12559	39733	104959
32	−1	0	56	1	56	756	4200	11550	16632	12012
33	2	1	57	1	57	869	6637	33111	124223	380731
34	−16	0	60	1	60	420	−1680	13860	−144144	1681680
35	−64	0	84	1	84	−924	30800	−1316700	62990928	−3212537328
36	−27	0	84	1	84	630	−5460	81900	−1493856	30126096

3. Analysis of the data

Looking at Table 1, we observe that the solutions found can be divided into five classes.

Terminating solutions. The sequences numbered #1, #4, #12, #18, #23, #28, #30, and #32 in Table 1 have $u_n = 0$ for all sufficiently large n (although in the last two cases the table does not extend to the first vanishing value). By inspection, we discover that these values are given by the formula

$$(4)\qquad (A,B,\lambda)=(-1,0,d^2+d),\quad u_n=\binom{d}{n}\binom{d+n}{n},$$

where d is a non-negative integer. The corresponding function $F(t)=\sum u_n t^n$ is given by

$$(5)\qquad F(t)=F(-d,1+d;1;t)=P_d(1-2t).$$

Here $F(a,b;c;t)$ denotes the hypergeometric function and $P_d(X)$ the dth Legendre polynomial.

Polynomial solutions. The solutions numbered #3, #6 and #15 in the table are also easily recognized: they are given by $u_n=1$, $u_n=2n+1$, and $u_n=3n^2+3n+1$, respectively. Looking for further polynomial solutions, we find that the sequences #21, #24, #29, #31 and #33 are also of this type, and that these 8 solutions are the cases $0\le d\le 7$ of an infinite family given by

$$(A,B,\lambda)=(2,1,d^2+d+1),\quad u_n=H_d(n),$$

where $H_d(n)$ is a polynomial of degree d of the form

$$H_d(n)=\frac{(2d)!}{d!^3}n^d+\cdots+\left(\sum_{k=1}^{d}\frac{2}{k}\right)n+1.$$

Looking at the first few polynomials $H_d(x)$, we find that they satisfy $H_d(-x-1)=(-1)^dH_d(x)$ and $H_{2r}\left(-\frac{1}{2}\right)=2^{-4r}\binom{2r}{r}^2$, but beyond this they are not easy to recognize. However, the fact that u_n is a polynomial of degree d in n means that the corresponding solution $F_d(t)=\sum u_n t^n$ of the differential equation (1) is a rational function with denominator $(1-t)^{d+1}$, and looking at the first few values of the polynomial $(1-t)^{d+1}F_d(t)$ we immediately find that its coefficients are the squares of the binomial coefficients $\binom{d}{n}$, i.e., these solutions are given by

$$(6)\qquad F(t)=\sum_{n=0}^{\infty}H_d(n)t^n=\frac{F(-d,-d;1;t)}{(1-t)^{d+1}}=\frac{1}{1-t}P_d\left(\frac{1+t}{1-t}\right),$$

where again $P_d(X)$ denotes the dth Legendre polynomial. Notice that for these solutions the discriminant of the polynomial $P(t)=Bt^2-At+1$ in (1) is 0, a degenerate case which Beukers excluded from consideration.

Hypergeometric solutions. We next find that Table 1 contains four solutions where u_n has a simple multiplicative expression as a multinomial coefficient. These are of two types. The first corresponds to $A=\lambda=0$, so that the recursion (3) relates u_{n+1} and u_{n-1}. Up to normalization, there is only one solution of this kind, namely the sequence #2 in the table,

$$\text{Sequence \#2:}\qquad (A,B,\lambda)=(0,-16,0),\qquad u_n=\begin{cases}\binom{2r}{r}^2 & \text{if } n=2r,\\ 0 & \text{if } n \text{ is odd.}\end{cases}$$

The other type occurs when $B=0$, so that the recursion (3) relates only u_{n+1} and u_n. There are 10 sequences of this type in Table 1, with the numbers #11, #14,

#16, #17, #20, #22, #27, and #34 – #36. Some of these are easy to recognize by inspection, e.g.:

$$\text{Sequence \#11:}\qquad (A,B,\lambda)=(16,0,4),\qquad u_n=\binom{2n}{n}^2=\frac{(2n)!^2}{n!^4},$$

$$\text{Sequence \#14:}\qquad (A,B,\lambda)=(27,0,6),\qquad u_n=\binom{3n}{n}\binom{2n}{n}=\frac{(3n)!}{n!^3},$$

$$\text{Sequence \#20:}\qquad (A,B,\lambda)=(64,0,12),\qquad u_n=\binom{4n}{2n}\binom{2n}{n}=\frac{(4n)!}{(2n)!\,n!^2},$$

and there are similar formulas in the other cases, e.g., the values of u_n for the two sequences #16 and #17 are given by

$$(-1)^{n-1}\frac{(3n+1)!}{(3n-1)n!^3}\quad\text{and}\quad(-1)^{n-1}\frac{(2n)!(2n+1)!}{(2n-1)n!^4},$$

respectively. The general case is easily seen to be given by the same formula as in (4), except that d is now an arbitrary rational number rather than a non-negative integer and we must make a corresponding rescaling $A\to DA$, $\lambda\to D\lambda$, $u_n\to D^n u_n$ for some positive integer D in order to achieve integrality. (The smallest choice of D is $M^2\prod_{p|M}p$, where M is the denominator of d.) The corresponding generating function is given by the same formula as in (5), except that the variable t must be replaced by Dt and the function $P_d(X)$ is now the Legendre function of order d, which is no longer a polynomial when d is not an integer. The index d for the above-listed cases #11, #14, #16, #17, #20, #22, #27, #34, #35 and #36 are $-\frac{1}{2}$, $-\frac{1}{3}$, $\frac{1}{3}$, $\frac{1}{2}$, $-\frac{1}{4}$, $\frac{1}{4}$, $\frac{2}{3}$, $\frac{3}{2}$, $\frac{3}{4}$ and $\frac{4}{3}$, respectively. (One has a choice here between d and $-1-d$ as parameters, and we have chosen $d\ge-\frac{1}{2}$.)

Legendrian solutions. Next, we see that (apart from the polynomial solutions considered above) there are three triples in the table for which $A^2=4B$, namely, those with labels #19, #25 and #26. Here there is no evident pattern for the numbers u_n, but after a bit of thought we realize that these are (up to a normalization $t\mapsto Dt$, $u_n\mapsto D^n u_n$ with suitable $D\in\mathbb{N}$) given by the same formula as in (6), but where now d is allowed to be a rational number rather than an integer and $P_d(X)$ is again the Legendre function of index d, which for non-integral d is not a polynomial in X. (The coefficient $H_d(n)$ of t^n in $F_d(t)$ is also no longer a polynomial in n for fixed d, though it is a polynomial in d for fixed n.) This time the smallest choice of D for given $d\in\mathbb{Q}$ making the coefficients $u_n=D^nH_d(n)$ integral for all n is given by $M^2\prod_{p|M}p^*$, where M is the denominator of d and $p^*=4$ if $p=2$, $p^*=p$ if p is odd.

Sporadic solutions. Finally, Table 1 contains six solutions which do not fall into any of the four infinite families above, and these are the only really interesting ones. (They are also the only ones Beukers considered, since he assumed that the polynomial P had non-vanishing quadratic term, constant term, and discriminant.) For convenience we list these 6 solutions again, with labels **A** – **F**. We also add #2 to this list, with label **G**, since although it is hypergeometric it does not belong to the 1-parameter family of hypergeometric solutions discussed above.

TABLE 2.

new label	index	A	B	λ	u_0	u_1	u_2	u_3	u_4	u_5	u_6
A	#5	7	−8	2	1	2	10	56	346	2252	15184
B	#7	9	27	3	1	3	9	21	9	−297	−2421
C	#8	10	9	3	1	3	15	93	639	4653	35169
D	#9	11	−1	3	1	3	19	147	1251	11253	104959
E	#10	12	32	4	1	4	20	112	676	4304	28496
F	#13	17	72	6	1	6	42	312	2394	18756	149136
G	#2	0	−16	0	1	0	4	0	36	0	400

Since all of these are contained in the domain $0 < A \le 17$, $0 < u_1 \le 6$, $0 < u_2 \le 42$, which comprises less than 0.005% of the domain of our search, one could hazard a rather optimistic

Conjecture. *Up to normalizations, the only cases where the recursion* (3) *has an integral solution are the hypergeometric family*

$$\mathrm{HG}_d: \qquad (A,B,\lambda) = (-1,0,d^2+d) \quad (d \in \mathbb{Q}, d \ge -\tfrac{1}{2})$$

(*which for $d \in \mathbb{Z}_{\ge 0}$ is terminating*), *the Legendre family*

$$\mathrm{Leg}_d: \qquad (A,B,\lambda) = (2,1,d^2+d+1) \quad (d \in \mathbb{Q}, d \ge -\tfrac{1}{2})$$

(*which for $d \in \mathbb{Z}_{\ge 0}$ is polynomial*), *and the seven sporadic solutions* **A**–**G**.

4. Binomial coefficient sums

The sequence "**D**" in Table 2 just given is immediately recognized to be the Apéry numbers, which are given by a well-known expression as a sum of products of binomial coefficients, so it is reasonable to look for similar expressions for the other solutions. Apart from relatively uninteresting expressions of this form for certain hypergeometric solutions such as $(3n)!/n!^3 = \sum_{k=0}^{2n}(-1)^{n-k}\binom{2n}{k}^3$, and the evident binomial coefficient sum expression for $H_d(n)$ obtained by equating the coefficients of t^n on both sides of (6), we have the following binomial coefficient representations, covering all but one of the sporadic solutions (G is trivial):

Sequence **A**: $$u_n = \sum_{k=0}^{n}\binom{n}{k}^3 = \sum_{k=0}^{n}\binom{n}{k}^2\binom{2k}{n},$$

Sequence **B**: $$u_n = \sum_{k=0}^{[n/3]}(-1)^k 3^{n-3k}\binom{n}{3k}\binom{3k}{k}\binom{2k}{k},$$

Sequence **C**: $$u_n = \sum_{k=0}^{n}\binom{n}{k}^2\binom{2k}{k},$$

Sequence **D**: $$u_n = \sum_{k=0}^{n}\binom{n}{k}^2\binom{n+k}{k} = \sum_{k=0}^{n}(-1)^{n-k}\binom{n}{k}\binom{n+k}{k}^2,$$

Sequence **E**: $$u_n = \sum_{k=0}^{[n/2]}4^{n-2k}\binom{n}{2k}\binom{2k}{k}^2.$$

The formulas for the sequences **B** and **E** say that the corresponding generating functions $\sum u_n t^n$ have hypergeometric representations:

$$F_{\mathbf{B}}(t) = \frac{1}{1-3t}F\left(\frac{1}{3},\frac{2}{3};1;\left(\frac{-3t}{1-3t}\right)^3\right), \quad F_{\mathbf{E}}(t) = \frac{1}{1-4t}F\left(\frac{1}{2},\frac{1}{2};1;\left(\frac{4t}{1-4t}\right)^2\right).$$

We have listed here only expressions for the u_n as simple sums of products of binomial coefficients. One can always find expressions as multiple sums, e.g., by the method described at the end of Section 7, a simple example (courtesy of the referee) being the formula $u_n = \sum_{0\le j\le i\le n}(-1)^i 8^{n-i}\binom{n}{i}\binom{i}{j}^3$ in case **F**.

5. Modular properties

Thanks to Beukers [3], one knows that the Apéry sequence has a modular interpretation, i.e., that for this sequence there is a modular function $t(z)$ such that the function $F\big(t(z)\big) = \sum u_n t(z)^n$ is a modular form of weight 1. We can therefore ask whether a similar property holds for the other solutions in our list. The case when $\{u_n\}$ terminates or is a polynomial in n is uninteresting since then $F\big(t(z)\big)$ is a rational function of $t(z)$ and hence is a modular function for any choice of $t(z)$. However, all the other examples (hypergeometric, Legendrian or sporadic) are interesting, since it is not a priori evident when they can be parametrized by modular forms. It turns out that there is an algorithm to investigate this question. We explain this below. Applying it to each of the 20 non-trivial sequences in the table, we find that *all* of the sporadic solutions are modular, and that there are also four hypergeometric and two Legendrian sequences which give modular forms. (The specific parametrizations are given below.) This suggests the following

Conjecture. *Any integral solution of the differential equation* (1), *where* $P(t)$ *is a non-degenerate quadratic polynomial, has a modular parametrization.*

Whether or not this is true, the natural question arises whether it is possible to give a complete classification of the modular solutions of the differential equation (1). This probably can be done, since the restricted nature of the singularities of the differential equation should put such strong restrictions on the modular side of the picture (e.g., the corresponding modular curve should have genus 0 and a small number of cusps and elliptic fixed points) as to make a complete listing possible. The most optimistic guess would be that the 12 modular solutions already found are a complete list. (This is supported by the case that they are contained in the domain $0 \le A \le 54$, $0 \le u_1 \le 21$, $0 < u_2 \le 500$, which is less than 0.5% of the domain of our search.) If this and the above conjecture are true, then our list of 6 sporadic solutions is indeed complete.

How to recognize modularity. A general fact about modular forms is the following: if $f(z)$ is an arbitrary modular form of positive weight k and $t(z)$ a modular function (i.e., a meromorphic modular form of weight 0), then the power series $F(t)$ obtained by expressing $f(z)$ locally as a power series in $t(z)$ always satisfies a linear differential equation of order $k+1$ with algebraic (or, if $t(z)$ is a Hauptmodul, even polynomial) coefficients. A discussion of this phenomenon in the general case, and an algorithm to find the corresponding modular parametrization, if one exists, of a given linear differential equation, is explained in [5, Chapter 1, Section 5]. Here we are interested only in $k = 1$, since the differential equation (1) has second order (the case $k = 0$ is uninteresting, corresponding to the "terminating" and "polynomial" cases above when (1) is not the differential equation of lowest order satisfied by f), and the easiest proof is simply to write down the differential equation. Suppose that $t(z)$ is a modular function and $f(z)$ a modular form of weight 1. Then the function $t' := (2\pi \mathrm{i})^{-1}\,\mathrm{d}t/\mathrm{d}z$ is a (meromorphic) modular form of weight 2 and the function $2f'^2 - ff''$ is a modular form of weight 6, so we can write

$$\text{(7)} \qquad \frac{t'(z)}{f(z)^2} = \alpha\bigl(t(z)\bigr), \qquad \frac{2f'(z)^2 - f(z)f''(z)}{t'(z)f(z)^4} = \beta\bigl(t(z)\bigr)$$

where $\alpha(t)$ and $\beta(t)$ are algebraic (or even, if $t(z)$ is a Hauptmodul, rational) functions of t. From the calculation

$$\frac{1}{t'(z)}\frac{\mathrm{d}}{\mathrm{d}z}\left(\frac{\alpha(t(z))}{t'(z)}\frac{\mathrm{d}F\bigl(t(z)\bigr)}{\mathrm{d}z}\right) + \beta\bigl(t(z)\bigr)F\bigl(t(z)\bigr) = \frac{1}{t'}\left(\frac{1}{f^2}f'\right)' + \frac{2f'^2 - ff''}{t'f^4}f = 0$$

we then see that $F(t)$ is a solution of the linear second order differential equation with algebraic coefficients $(\alpha F')' + \beta F = 0$. Comparing this with (1), we find that in our case the function $\alpha(t)$, up to a constant factor which we can normalize to be 1, is equal to $tP(t) = t(1 - At + Bt^2)$. We can then integrate the equation $t'(z) = \alpha\bigl(t(z)\bigr)f(z)^2 = \alpha\bigl(t(z)\bigr)F\bigl(t(z)\bigr)^2$ to get $z \int \mathrm{d}t/\alpha(t)F(t)^2$ or, in terms of the standard expansion parameter $q = \mathrm{e}^{2\pi \mathrm{i}z}$ of modular functions and forms at infinity,

$$\text{(8)} \quad q = t\exp\left(\int_0^t \left(\frac{1}{\alpha(s)F(s)^2} - \frac{1}{s}\right)\mathrm{d}s\right) = t\exp\left(\int_0^t \left(\frac{F(s)^{-2}}{1 - As + Bs^2} - 1\right)\frac{\mathrm{d}s}{s}\right).$$

This is a power series with leading term t (recall that $F(t) = u_0 + u_1 t + \cdots$ has leading coefficient 1 by our choice of normalization) and hence can be inverted to compute $t = t(z)$ as a power series in q, after which the function $f(z) = F\bigl(t(z)\bigr)$ can also be computed. We then only have to look at these power series to see whether they are in fact the q-expansions of a modular function and a weight 1 modular form, respectively. This can often be done by inspection (in many of the cases below, $t(z)$ is a simple product of eta-functions and $f(z)$ an Eisenstein series of weight 1, which are easily recognized from their q-expansions); when this fails, one can test numerically whether the function $t(z)$ is related algebraically on the classical j-invariant $j(z)$. Once one has found a relation by looking at the first few Fourier coefficients of the candidate modular forms, the validity of the asserted differential equation can be checked algorithmically, since any equality between modular forms can be verified by verifying the equality of a specific number of terms of their q-expansions. In this way, the modular parametrizations given below were found. That the other solutions (excluding the trivial ones where the sequence $\{u_n\}$ is either polynomial or terminating and $F(t)$ is a rational function) are *not* modular is harder to verify algorithmically, but in each case the numerical calculation indicates that their q-expansions have very large denominators involving infinitely many prime numbers, and this is not a possible behavior for modular forms.

Description of all known modular solutions of (1). As already mentioned, we found altogether 12 cases in Table 1 which had modular parametrizations — the six sporadic ones, four hypergeometric ones, and two Legendrian ones. We now discuss each of these individually.

TABLE 3.

Case	α	β	$t(z)$	$1-\alpha t(z)$	$1-\beta t(z)$	$f(z)$	$f(z)$
A	-1	8	$\frac{1^3 6^9}{2^3 3^9}$	$\frac{2^5 6^1}{1^1 3^5}$	$\frac{1^8 6^4}{2^4 3^8}$	$\frac{2^1 3^6}{1^2 6^3}$	$\frac{1}{3}\theta_3(z)+\frac{2}{3}\theta_3(2z)$
C	1	9	$\frac{1^4 6^8}{2^8 3^4}$	$\frac{1^1 3^5}{2^5 6^1}$	$\frac{1^9 6^3}{2^9 3^3}$	$\frac{2^6 3^1}{1^3 6^2}$	$\frac{1}{2}\theta_3(z)+\frac{1}{2}\theta_3(2z)$
E	4	8	$\frac{1^4 2^8 4^4}{2^{10}}$	$\frac{1^4 4^{14}}{2^{14} 8^4}$	$\frac{1^8 4^4}{2^{12}}$	$\frac{2^{10}}{1^4 4^4}$	$\theta_4(z)$
F	8	9	$\frac{1^5 3^1 4^5 6^2 12^1}{2^{14}}$	$\frac{1^8 4^8 6^{20}}{2^{20} 3^8 12^8}$	$\frac{1^9 4^9 6^6}{2^{18} 3^3 12^3}$	$\frac{2^{15} 3^2 12^2}{1^6 4^6 6^5}$	$\theta_3(z)+2\theta_3(2z)-2\theta_3(4z)$

Sporadic cases. In four of the six sporadic cases, those with labels **A**, **C**, **E** and **F**, the quadratic polynomial x^2-Ax+B splits over $\mathbb{Z}$ as $(x-\alpha)(x-\beta)$. In each of these cases, the modular functions $t(z)$, $1-\alpha t(z)$ and $1-\beta t(z)$ and the modular form $f(z)$ have representations as products of Dedekind eta-functions, and the modular form $f(z)$ can also be represented as an Eisenstein series of weight 1. These representations are given in the above table, in which a notation like $1^3 2^{-3} 3^{-9} 6^9$ is a shorthand for the eta-product $\eta(z)^3\eta(2z)^{-3}\eta(3z)^{-9}\eta(6z)^9$ and the functions θ_3 and θ_4 are the Eisenstein or theta series

$$\theta_3(z)=1+6\sum_{n=1}^{\infty}\biggl(\sum_{d\mid n}\Bigl(\frac{-3}{d}\Bigr)\biggr)q^n=\sum_{m,n\in\mathbb{Z}}q^{m^2+mn+n^2},$$

$$\theta_4(z)=1+4\sum_{n=1}^{\infty}\biggl(\sum_{d\mid n}\Bigl(\frac{-4}{d}\Bigr)\biggr)q^n=\sum_{m,n\in\mathbb{Z}}q^{m^2+n^2},$$

i.e., the modular forms of weight 1 whose Mellin transforms are the Dedekind zeta-functions of the two imaginary quadratic fields $\mathbb{Q}(\sqrt{-3})$ and $\mathbb{Q}(i)$ possessing non-trivial units. Case **B** is fairly similar: here the polynomial $x^2-Ax+B=x^2-9x+27$ does not split over $\mathbb{Z}$, but we still have the eta-product representations

Case **B**: $\quad t=\frac{1^3 4^3 18^9}{2^9 9^3 36^3},\quad 1-9t+27t^2=\frac{1^9 4^9 6^{36} 9^3 36^3}{2^{27} 3^{12} 12^{12} 18^9},\quad f=\frac{2^9 3^1 12^1}{1^3 4^3 6^3}$

and the Eisenstein series representation

$$f(z)=\tfrac{1}{2}\theta_3(z)-\tfrac{3}{2}\theta_3(3z)-\theta_3(4z)+3\theta_3(12z).$$

Finally, in the last case **D** (corresponding to the Apéry numbers), the function t still has a product representation, but no longer as a product of eta-functions, and similarly for $1-At+Bt^2$ and f:

Case **D**: $\quad t=q\prod_{n=1}^{\infty}(1-q^n)^{5(n/5)},\quad 1-11t-t^2=t\cdot\frac{1^6}{5^6},\quad f^2=t^{-1}\cdot\frac{5^5}{1^1}.$

The Eisenstein series representation of f in this case is also more complicated:

$$f(z)=1+\sum_{n=1}^{\infty}\sum_{d\mid n}\Bigl(\frac{3-\mathrm{i}}{2}\chi(d)+\frac{3+\mathrm{i}}{2}\overline{\chi}(d)\Bigr)q^n,$$

where χ is the Dirichlet character of conductor 5 with $\chi(2)=\mathrm{i}$, $\chi^2=\bigl(\frac{\cdot}{5}\bigr)$.

Hypergeometric cases. Here two cases, the exceptional case #2 when $A = \lambda = 0$ and the sequence #11 with $(A, B, \lambda) = (16, 0, 4)$, behave like the first four sporadic cases:

TABLE 4.

Case	α	β	$t(z)$	$1-\alpha t(z)$	$1-\beta t(z)$	$f(z)$	$f(z)$
#2	-4	4	$\frac{2^4 8^8}{4^{12}}$	$\frac{2^{14}8^4}{1^4 4^{14}}$	$\frac{1^4 2^2 8^4}{4^{10}}$	$\frac{4^{10}}{2^4 8^4}$	$\theta_4(2z)$
#11	0	16	$\frac{1^8 4^{16}}{2^{24}}$	—	$\frac{1^{16}4^8}{2^{24}}$	$\frac{2^{10}}{1^4 4^4}$	$\theta_4(2z)$

(Note that, amusingly enough, $f(z)$ for case #11 is the *same* as for the sporadic case $\mathbf{E}$ = #10, although of course the parameter $t(z)$ and hence also the power series $F(t)$ are different.) Another case, corresponding to sequence #14 with $(A, B, \lambda) = (27, 0, 6)$, is more similar to the sporadic case $\mathbf{D}$, since we do not have product expansions for t, $1 - At$ and f separately, but only for certain multiplicative combinations, while f has an ordinary (indeed, a particularly simple) Eisenstein series representation:

Case #14: $$\frac{t}{1-27t} = \frac{3^{12}}{1^{12}}, \quad f^3 \cdot t = \frac{3^9}{1^3}, \quad f(z) = \theta_3(z).$$

Finally, for the sequence #20 with $(A, B, \lambda) = (64, 0, 12)$, we again find eta-product representations only for certain multiplicative combinations of t, $1 - At$ and f, but this time f is not a modular form at all, but only the square-root of a modular form:

Case #20: $$\frac{t}{1-64t} = \frac{2^{24}}{1^{24}}, \quad f^4 \cdot t = \frac{2^{16}}{1^8}, \quad f(z)^2 = -E_2(z) + 2E_2(2z),$$

where $E_2(z) = 1 - 24\sum \sigma_1(n)q^n$ is the usual not-quite-modular Eisenstein series of weight 2 on the full modular group.

Legendrian cases. The last two cases which turned out to be modular are again slightly different, since now f is no longer a holomorphic modular form, or even the square-root of one, but rather is a meromorphic modular form of weight one:

Case #19: $$t = \frac{1^8 4^{16}}{2^{24}}, \quad 1 - 16t = \frac{1^{16}4^8}{2^{24}}, \quad f = \frac{2^{22}}{1^{12}4^8}, \quad f(z) = \frac{-\theta_4(z) + 2\theta_4(2z)}{1 - 16t(z)};$$

Case #25: $$t = \frac{1^{12}4^{12}6^{36}}{2^{36}3^{12}12^{12}}, \quad (1-27t)^2 f^3 = \frac{2^{27}3^3 12^3}{1^9 4^9 6^9}, \quad f(z) = \frac{-\theta_3(z) + 2\theta_3(4z)}{1 - 27t(z)}.$$

Discussion. We make a few remarks about the modular parametrizations just given. We consider only the non-degenerate case when $B \neq 0$ and $A^2 \neq 4B$, so that $x^2 - Ax + B = (x - \alpha)(x - \beta)$ with α and β distinct and non-zero. We further assume that t is a Hauptmodul, i.e., that t and f are modular forms on some group $\Gamma \subset \mathrm{SL}(2, \mathbb{R})$ of genus 0 and that $z \mapsto t(z)$ gives an isomorphism of $\mathfrak{H}/\Gamma \cup \{\text{cusps}\}$ to $\mathbb{P}^1(\mathbb{C})$. From equation (7) we see that the modular forms $f(z)$, $t(z)$ and $t'(z)$ are related by $f^2 = t'/t(1 - \alpha t)(1 - \beta t)$, so that the Γ-invariant differential form $f(z)^2\,dz$ is (or pushes forward to) the differential form $dt/t(1 - \alpha t)(1 - \beta t)$ on $\mathbb{P}^1$. This form has three simple poles, at $t = 0$, $t = 1/\alpha$ and $t = 1/\beta$, and one simple zero at $t = \infty$. In terms of the "modular" variable z, the point $t = 0$ corresponds to the cusp at infinity (since we have constructed t in such a way that it has an expansion $q + \cdots$). None of the functions $t(z)$, $1 - \alpha t(z)$ or $1 - \beta t(z)$ had a pole or zero in the upper half-plane, since then the function f^2 would have a (simple) pole at that point, which is impossible. This explains why these three modular functions always had product expansions in the cases looked at above. (This actually remains formally true even in the two sporadic cases **B** and **D** when α and β are not in $\mathbb{Z}$, but must be interpreted as the statement that the logarithmic derivatives of $1 - \alpha t$ and $1 - \beta t$ are weight 2 Eisenstein series.) Also, the fact that the genus is zero means that there are no cusp forms of weight 1 or 2, so that f (and, for that matter, also f^2) must be an Eisenstein series, as indeed we found in all the examples. It seems plausible that by continuing this analysis one could get enough information about the necessary forms of f and t to classify at least all those cases when t is a Hauptmodul and f a true modular form of weight 1, but I have not tried to do this.

6. The periods associated to the six sporadic solutions

The most famous property of the original Apéry recursion $(n+1)^2 u_{n+1} - (11n^2 + 11n + 3)u_n - n^2 u_{n-1} = 0$ is, of course, that if we start with the two linearly independent solutions $\{u_n\} = \{1, 3, 19, \dots\}$ and $\{v_n\} = \{0, 1, \frac{25}{4}, \dots\}$, then the ratio v_n/u_n tends to the limit $L = \frac{1}{5}\zeta(2)$ and this convergence is rapid enough to prove the irrationality of L. In this section we look at the corresponding questions for each of the six sporadic sequences **A**–**F**. We will see that the limit exists and can be computed in five of the six cases, but that the convergence, though always exponential, is only rapid enough to imply the irrationality of the limit in the case of the Apéry equation.

We denote by $\{u_n\}$, as up to now, the solution of (3) with $u_0 = 1$ (and hence $u_1 = \lambda$, $u_n \in \mathbb{Z}$ for all n by the fundamental property of our sequences), and by $\{v_n\}$ the sequence defined by the initial values $v_0 = 0$, $v_1 = 1$ and the requirement that (3) holds for all $n \geq 1$. We want to determine whether the sequence v_n/u_n tends to a limit L and, if so, to evaluate this limit and specify the speed of the convergence.

Let α and β as in Section 5 denote the roots of $x^2 - Ax + B = 0$, chosen (except in case **B** where $\alpha, \beta = (9 \pm 3i\sqrt{3})/2$) so that $|\beta| > |\alpha|$. Then the sequence $\{u_n\}$ has the asymptotic behavior

$$u_n = C\frac{\beta^n}{n}\left(1 - \frac{c_1}{n} + O\left(\frac{1}{n^2}\right)\right) \quad (n \to \infty), \tag{9}$$

where $c_1 = \lambda - \alpha/\beta - \alpha$, while every other element of the 2-dimensional vector space of real-valued solutions of the recursion (3) (for $n \geq n_0 \geq 1$) has asymptotics given either by the same formula (with a possibly different value of C) or else by $(\text{const.})\,\alpha^n/n\bigl(1 - c_2/n + O(1/n^2)\bigr)$, where $c_2 = \lambda - \beta/\alpha - \beta$, with the space of solutions of the latter type being 1-dimensional. (To obtain these asymptotic formulas, consider a general solution $\{u_n\}$ of (3) and make the Ansatz $u_n = (\text{const.})\,n^\varepsilon x^n\bigl(1 - cn^{-1} + O(n^{-2})\bigr)$; then the ratio $\rho_n := u_{n+1}/u_n$ equals $x\bigl(1 + \varepsilon n^{-1} + \bigl(\binom{\varepsilon}{2} + c\bigr)n^{-2} + \cdots\bigr)$, and rewriting (3) in the form $(1+n^{-1})^2\rho_n - A(1+n^{-1}) - \lambda n^{-2} + B\rho_{n-1}^{-1} = 0$ we find $x \in \{\alpha, \beta\}$, $\varepsilon = -1$ and $c = (\lambda - A + x)/(2x - A)$.) It follows that there are real constants C' and L such that $v_n - Lu_n = C'\alpha^n/n\bigl(1 - c_2/n + O(1/n^2)\bigr)$ and that the quotient v_n/u_n tends to L exponentially quickly:

$$(10) \qquad \frac{v_n}{u_n} = L + C''\left(\frac{\alpha}{\beta}\right)^n \left(1 + \frac{c_1 - c_2}{n} + \left(\frac{1}{n^2}\right)\right) \qquad (n \to \infty).$$

The values of the constants β, α, C, c_1, c_2 and L occurring in equations (9) and (10) are given for each of the five cases **A**, **C**, **D**, **E** and **F** in the table below, in which ϕ in the third row denotes the golden ratio $(1+\sqrt{5})/2$ and the numbers $\zeta(2)$, $L_{-3}(2)$ and $L_{-4}(2)$ are the values at $s = 2$ of the Riemann zeta function and of the Dirichlet L-series $L(s,\chi)$ associated to the quadratic characters $\chi(n) = (-3/n)$ and $\chi(n) = (-4/n)$, each of which is a famous number: $\zeta(2)$ equals $\pi^2/6$, $L_{-4}(2)$ is "Catalan's constant," which occurs in the evaluation of many classical definite integrals, and $L_{-3}(2)$ is, up to a factor $3^{3/2}/4$, the maximum volume of a tetrahedron in hyperbolic 3-space. The constant C in (9) can be evaluated using either the explicit binomial coefficient expressions given (except for case **F**) in Section 4 or else from the modular form expressions of Section 5. The evaluation of the limiting ratio L will be discussed at the end of this section.

TABLE 5.

Case	β	α	C	c_1	c_2	L
A	8	-1	$\dfrac{2}{\pi\sqrt{3}}$	$\dfrac{1}{3}$	$\dfrac{2}{3}$	$\dfrac{1}{4}\zeta(2) = 0.4112335\ldots$
C	9	1	$\dfrac{3\sqrt{3}}{4\pi}$	$\dfrac{1}{4}$	$\dfrac{3}{4}$	$\dfrac{1}{2}L_{-3}(2) = 0.3906512\ldots$
D	ϕ^5	$-\phi^{-5}$	$\dfrac{\phi^{5/2}}{2\pi\sqrt[4]{5}}$	$\dfrac{1}{\phi\sqrt{5}}$	$\dfrac{\phi}{\sqrt{5}}$	$\dfrac{1}{5}\zeta(2) = 0.3289868\ldots$
E	8	4	$\dfrac{2}{\pi}$	0	1	$\dfrac{1}{2}L_{-4}(2) = 0.4579827\ldots$
F	9	8	$\dfrac{3\sqrt{3}}{\pi}$	-2	3	$\dfrac{5}{8}L_{-3}(2) = 0.4883140\ldots$

In each of the five cases of Table 5, formula (10) gives an interesting series of rational approximations for one of the numbers $\zeta(2)$, $L_{-3}(2)$ or $L_{-4}(2)$, but, as already mentioned, except in case **D** these do not converge quickly enough to yield the irrationality of the limit. Indeed, from the table we see that in cases **E** and **F** the quantity $v_n - Lu_n$ blows up exponentially like $4^n/n$ or $8^n/n$, respectively, and even in cases **A** and **C**, where $|\alpha| = 1$ and hence $v_n - Lu_n$ tends to zero like $O(1/n)$, this is not enough to give the irrationality of L because v_n itself has a denominator which blows up like e^{2n}. The speed of convergence is best in these two cases, with $v_n/u_n - L$ being of the order of $\left(\frac{1}{8}\right)^n$ and $\left(\frac{1}{9}\right)^n$, respectively, while the convergence in cases **E** (Catalan's constant) and **F** is only like $\left(\frac{1}{2}\right)^n$ and $\left(\frac{8}{9}\right)^n$, respectively. The approximations lead in each of the five cases to a simple infinite continued fraction expansion of the number L, the formulas in cases **C** and **E** being

$$L_{-3}(2) = \cfrac{2}{P(0) - \cfrac{9\cdot 1^4}{P(1) - \cfrac{9\cdot 2^4}{P(2) - \cfrac{9\cdot 3^4}{\ddots}}}}, \quad L_{-4}(2) = \cfrac{1/2}{Q(0) - \cfrac{2\cdot 1^4}{Q(1) - \cfrac{2\cdot 2^4}{Q(2) - \cfrac{2\cdot 3^4}{\ddots}}}}$$

with $P(n) = 10n^2+10n+3$, $Q(n) = 3n^2+3n+1$. Continued fraction expansions of a similar type, though with more complicated rational functions, have been given recently by Zudilin [12].

To find the value of L listed in the above table, we follow the method devised by Beukers [3] in 1987 to obtain modular interpretations of Apéry's irrationality proofs for $\zeta(2)$ and $\zeta(3)$, using the modular descriptions of the differential equations associated to the recursion (3) which were given in Section 5. Consider for instance case **C**. We saw that the integer solution $\{u_n\}$ of (3) in this case is given by the generating function expansion $f(z) = \sum u_n t(z)^n$, where $t(z) = \eta(z)^4\eta(6z)^8/\eta(2z)^8\eta(3z)^4$ is the "Hauptmodul" for $\Gamma_0(6)$ and $f(z)$ is the theta series (or Eisenstein series) of weight 1 for $\Gamma_0(6)$ given in Section 5. The generating function $h(z) = \sum v_n t(z)^n$ for the second solution of the recurrence (3) has a more complicated modular interpretation as $f(z)\tilde{g}(z)$, where $\tilde{g}(z) = q - \frac{5}{4}q^2 + q^3 - 11q^4/16 + \cdots$ ($q = \mathrm{e}^{2\pi i z}$) is the "Eichler integral" $\sum b(n)q^n/n^2$ associated to the weight 3 Eisenstein series $g(z) = \sum b(n)q^n = q - 5q^2 + 9q^3 - 11q^4 + \cdots$ with $b(n) = \sum_{d|n}(-1)^{d-1}\big(-3/(n/d)\big)d^2$. (To prove this, one simply verifies that the differential equation satisfied by $f\tilde{g}$ as a function of $t(z)$ coincides with the one corresponding to the recursion defining the v_n.) The group $\Gamma_0(6)$ has four cusps, at $z = 0$, $\frac{1}{2}$, $\frac{1}{3}$ and ∞, the corresponding values of the modular function $t(z)$ being $\frac{1}{9}$, ∞, 1 and 0, respectively. In particular, the asymptotics of the u_n and the v_n are determined by the singularities of the functions $f(z)$ and $h(z)$ as $z \to 0$ (this explains once again why both of these sequences grow roughly like 9^n), and hence the limiting value L of v_n/u_n is the limiting value as $z \to 0$ of the ratio $h(z)/f(z) = \tilde{g}(z)$. (Note that, since $\sum(v_n - Lu_n)t^n$ is convergent at $t = 1/\beta$, while both $\sum v_n t^n$ and $\sum u_n t^n$ have logarithmic singularities there, the value of L is automatically equal to the limit of the ratio of the two latter series as $t \to 1/\beta$.) Writing the formula defining $\tilde{g}(z)$ as $\sum_{m\ge 1}(-3/m)/m^2\, q^m/(1+q^m)$, one sees that its limit for $z \to 0$ or $q \to 1$ equals $\frac{1}{2}L_{-3}(2)$, as claimed. The proof in all the other cases is similar, the function $\tilde{g}(z)$ in case **E**, for instance, being the Eichler integral $\sum_{m\ge 1}(-4/m)/m^2\, q^m/(1+q^m)$ of a weight 3 Eisenstein series on $\Gamma_0(8)$.

Finally, we say a few words about case **B**, which has been omitted up to now. Here the two roots $\alpha = (9+3\sqrt{-3})/2$ and $\beta = \bar{\alpha}$ of $x^2 - Ax + B = 0$ have the same absolute value, so all solutions of (3) have the same growth $O(27^{n/2}/n)$, and the numbers v_n/u_n do not tend to a limit as $n \to \infty$. Instead, the asymptotic expansions of $\{u_n\}$ and $\{v_n\}$ as $n \to \infty$ are given by

$$u_n = \frac{U}{n}\alpha^n\Big(1 - \frac{\gamma_1}{n} + \cdots\Big) + \frac{\overline{U}}{n}\bar{\alpha}^n\Big(1 - \frac{\overline{\gamma}_1}{n} + \cdots\Big),$$

$$v_n = \frac{V}{n}\alpha^n\Big(1 - \frac{\gamma_1}{n} + \cdots\Big) + \frac{\overline{V}}{n}\bar{\alpha}^n\Big(1 - \frac{\overline{\gamma}_1}{n} + \cdots\Big),$$

where $U = (9\sqrt{3})/(2\pi\alpha)$, $\gamma_1 = \alpha/9$ and $V/U = \frac{1}{2}L_{-3}(2) - 4i/(9\sqrt{3})\zeta(2)$. (The ratio V/U, which is the analogue of the limit L in the previous examples, is obtained just as before: the numbers $1/\alpha$ and $1/\bar{\alpha}$ are the values of the Hauptmodul $t(z)$ given in Section 5 at the cusps $z = \pm\frac{1}{6}$ and the function $\tilde{g}(z) = h(z)/f(z)$ has the form $\tilde{g}(z) = -\sum_{n=1}^{\infty}(-3/n)n^{-2}(-q)^n/(1-(-q)^n)$, which for $z = \frac{1}{6}$, $-q = e^{-2\pi i/3}$ has the value given.) Since $\alpha/\sqrt{27} = e^{\pi i/6}$, the quotients v_n/u_n do tend to a limit $\frac{1}{2}L_{-3}(2) + 4/(9\sqrt{3})\zeta(2)\tan\pi(j-1)/6$ as n tends to infinity in any fixed residue class $n \equiv j \not\equiv 4 \pmod 6$, but the convergence is very slow, with error term of the order of $1/n$ instead of exponentially small as before, so that from the point of view of rational approximation this case is of less interest than the other five.

7. The modular cases as Picard – Fuchs differential equations

The fact that the non-degenerate solutions of our problem have parametrizations by modular forms means that they have an algebraic-geometric interpretation as the Picard – Fuchs differential equations of certain families of elliptic curves. In fact, we can be much more precise. In 1982, Beauville [2] classified completely the stable families of elliptic curves over $\mathbb{P}^1$ having exactly four singular fibres (a stable family of elliptic curves over $\mathbb{P}^1$ is a family in which the singular fibres have only double points; the minimal number of singular fibres is then 4) and found that there were exactly 6. For each family he gave a defining equation, a description of the singular fibres, and the subgroup of $SL_2(\mathbb{Z})$ describing the family. We reproduce the table from [2] in Table 6

Comparing the number six of cases found by Beauville with the number six of sporadic cases **A** – **F** in Section 5 makes it tempting to suppose that there is a 1:1 correspondence between these, with the Picard – Fuchs equations of Beauville's six families of elliptic curves being precisely the equation (1) corresponding to **A** – **F** in some order. But this is not the case, for three reasons. First of all, two of Beauville's families (V and VI) are isogenous to two of the others (II and I, respectively) and give the *same* Picard – Fuchs equations. This reduces the first "six" above to "four." Secondly, our hypergeometric (and hence by definition non-sporadic) solutions #2, #11 and #14, whose modularity was noted in Section 5, are also Picard – Fuchs equations of families on Beauville's list. This increases the second "six" above to "nine." And thirdly, each of Beauville's surfaces can give rise to several different Picard – Fuchs equations, depending on how one chooses the Hauptmodul $t(z)$. To get a differential equation of the form (1), we must require that the singular fibres correspond to the four values 0, t_1, t_2 and ∞, where t_1 and t_2 are the roots of $P(t)$, since these are the singular points of the differential equation (1). So we must put one fibre at $t = 0$ and one at ∞, giving 12 choices (the remaining freedom

TABLE 6.

Case	Equation of the family	Group
I	$X^3 + Y^3 + Z^3 + tXYZ = 0$	$\Gamma(3)$
II	$X(X^2 + Z^2 + 2ZY) + tZ(X^2 - Y^2) = 0$	$\Gamma_1(4) \cap \Gamma(2)$
III	$X(X-Z)(Y-Z) + tZY(X-Y) = 0$	$\Gamma_1(5)$
IV	$(X+Y)(Y+Z)(Z+X) + tXYZ = 0$	$\Gamma_1(6)$
V	$(X+Y)(XY - Z^2) + tXYZ = 0$	$\Gamma_0(8) \cap \Gamma_1(4)$
VI	$X^2Y + Y^2Z + Z^2X + tXYZ = 0$	$\Gamma_0(9) \cap \Gamma_1(3)$

TABLE 7.

Beauville family:	IV	I, VI	III	II,V
Picard – Fuchs equation:	**A**, **C**, **F**	**B**, #14	**D**	**E**, **G**, #11

of interchanging or rescaling t_1 and t_2 has no effect on the parameters A, B, λ of this paper). Of course, these do not necessarily all give rise to equations of the form (1) with rational coefficients, nor (since the elliptic surfaces have non-trivial automorphisms) need they all be different. The actual correspondence turns out to be as in Table 7.

One way to see this is to compute the j-invariant of each of Beauville's family as a rational function of t, invert this equation to compute t as an explicit modular function, and compare with the list of t's which we found before. The j-invariant in turn is most conveniently found by rewriting Beauville's equations in Weierstrass form $y^2 = f(x)$ (f a polynomial of degree 3). For the reader's convenience, we give the transformation needed in each of the 6 cases to do this, and the resulting formula for the j-invariant, in Table 8.

The necessary calculations are explained and carried out in detail in the recent paper [10] by Helena Verrill, in which the reader can also find a lot of other information about these elliptic surfaces and their corresponding Picard – Fuchs equations. A further discussion of some of the above differential equations and their relation to Beauville's classification and to the earlier classification by Schmickler-Hirzebruch [7] of elliptic pencils with exactly three singular fibers is given in [8], to which the reader is referred for further details. I mention briefly a few further aspects which were pointed out to me by Jan Stienstra. First of all, when different differential equations are associated to the same family of elliptic curves, then the corresponding modular functions have to be the same (possibly after a Möbius transformation $z \mapsto M(z)$, $M \in \mathrm{GL}_2(\mathbb{Z})$, of the z-variable) up to a fractional linear transformation. For instance, from the above table one sees that **A** and **C** correspond to the same fibration, and from the eta-product expansions given in Section 5 one finds that $t_{\mathbf{C}}(z) = t_{\mathbf{A}}(z)/\big(1 + t_{\mathbf{A}}(z)\big)$. Secondly, just as one has isogenies between different families of elliptic curves in Beauville's list, in certain cases one has

TABLE 8.

	Transformation to Weierstrass form	j-invariant
I	$X, Y = t(x \pm y), Z = 2(1 + 3x)$	$\dfrac{-t^3(t^3 - 216)^3}{(t^3 + 27)^3}$
II	$X = xt, Y = x + y, Z = t$	$\dfrac{2^8(t^4 - t^2 + 1)^3}{t^4(t^2 - 1)^2}$
III	$X = 1 + \dfrac{(y + tx)}{(1 - x)}, Y = 2x, Z = 2$	$\dfrac{(t^4 - 12t^3 + 14t^2 + 12t + 1)^3}{t^5(t^2 - 11t - 1)}$
IV	$X, Y = 1 + tx \pm y, Z = 2x(1 + tx)$	$\dfrac{(t + 2)^3(t^3 + 6t^2 - 12t + 8)^3}{t^3(t - 1)^2(t + 8)}$
V	$X, Y = 1 + tx \pm y, Z = 2x(1 + tx)$	$\dfrac{(t^4 + 16t^2 + 16)^3}{t^2(t^2 + 16)}$
VI	$X = 2x, Y = y - xt - 1, Z = 2x^2$	$\dfrac{-t^3(t^3 + 24)^3}{t^3 + 27}$

further transformations induced by pull-backs under $t \mapsto t^N$. Since these transformations change the nature and number of the singularities, one can get examples of Apéry-like differential equations coming from the pencils with three exceptional fibers classified in [7] as well as from the ones with four exceptional fibers classified in [2]. This is discussed in detail in [8]. Finally, and most interesting, in the modular cases one can always get explicit formulas for the coefficients u_n as sums of multinomial coefficients (although these will in general be multiple sums rather than simple sums like the ones listed in Section 4) from the equation of the corresponding pencil: if the equation of this pencil is written as $t = F(X,Y,Z)$, where $F(X,Y,Z)$ is a Laurent polynomial in X, Y and Z of degree 0 (as is the case for all of the families in Beauville's list as reproduced above, after making a linear change of variables in cases II and III), then the coefficient u_n of t^n in the corresponding power series is equal simply to the constant term of $F(X,Y,Z)^n$. For instance, the constant term in the nth power of $-(X^3+Y^3+Z^3)/XYZ$, corresponding to the "t" in Beauville's family I, is equal (up to sign) to $n!/(n/3)!^3$ if $3 \mid n$ and to 0 otherwise, and we recover the hypergeometric solution #14, and similarly the constant terms of the powers of $(X+Y)(Y+Z)(Z+X)/XYZ$, corresponding to the family IV, are the numbers 1, 2, 10, 56, ... of sequence #5 in our list.

8. Final remarks

The numerical experiments described in the first sections of this paper were carried out in 1997–98, inspired by the beautiful lecture given by Beukers on his work [4] at the conference in honor of Schinzel in Zakopane in 1997. A first version of the paper was written in 2000 and has been circulating as an informal preprint ever since, with few changes except for the addition of Section 6 on the periods associated to Apéry-like recursions (which was written in answer to a question posed by A. Connes during my course at the Collège de France in 2001) and some additions to the discussion of the connection with Beauville's classification in Section 7. The main reason for not publishing it earlier was that I thought that more explanations of the connections to geometry should be added, but did not understand these well enough. In the meantime, several other related papers, both on the geometric and on the purely differential equations/modular forms side have appeared (in some cases quoting the informal preprint version of this paper), and part of what is presented here is perhaps obsolete. I have nevertheless kept the entire text since this paper in any case contains no real theorems but is to be seen more as an informal discussion of various experimental and theoretical aspects of the three-way connection between algebraic geometry, linear differential equations, and the theory of modular forms. For further material on this subject, the reader is referred in particular to the three papers [1, 8, 10], all of whose authors — Gert Almkvist, Wadim Zudilin, Jan Stienstra and Helena Verrill, I would also like to thank here for many useful discussions and for their patience.

I should perhaps also mention explicitly that (of course) not all of the examples occurring in this paper are new. In particular, the equations **D**, **A** and **C** (= #9, #5 and #8) are discussed in the paper [9] by Stienstra and Beukers, where two of them are ascribed to Apéry and Cusick but in at least one case had already been found in the 19th century (J. Franel), as I have been informed by Zudilin. I have made no attempt at all to find the earliest reference for each differential equation. In a different direction, it should also be mentioned that the list of modular cases given in Section 5 is not complete. No further sporadic Apéry-like equations have been discovered, and the conjecture formulated in Section 3 still stands, but the infinite families of equations which we called "hypergeometric" and "Legendrian" both contain further modular cases which I overlooked because their parameters were beyond the bounds of my numerical search. Specifically, the hypergeometric case $(A, B, \lambda) = \bigl(-1, 0, d(d+1)\bigr)$ ($d \in \mathbb{Q}$, $d \geq -\frac{1}{2}$) gives a modular function for the parameter value $d = -\frac{1}{6}$ as well as for the three values $d = -\frac{1}{2}, -\frac{1}{3}$ and $-\frac{1}{4}$ (= #11, #14 and #20, respectively) which we discussed, the corresponding coefficients u_n being up to a scaling factor the multinomial coefficients $(6n)!/n!(2n)!(3n)!$, and the Legendre case $(A, B, \lambda) = (2, 1, d(d+1)+1)$ ($d \in \mathbb{Q}$, $d \geq -\frac{1}{2}$) is modular not only for the two cases $d = -\frac{1}{2}$ and $d = -\frac{1}{3}$ (= #19 and #25) which we gave, but also for $d = -\frac{1}{4}$ and $d = -\frac{1}{6}$, found by van Enckevort and van Straten. All of these examples, of course, are contained in the much more extensive tables found in the paper [1], which also gives relevant references and a discussion of how equations of the type discussed here can be used to construct (fourth order) Calabi–Yau differential equations by means of Hadamard products. Finally, I mention two recent preprints on closely related subjects. In [11], Y. Yang gives an analysis of the periods associated to Apéry-like differential equations similar to the one given in §6 here. And in [6], Vasily Golyshev describes a class of interesting higher order generalizations of the differential equation (1), which he calls equations of type DN. Here N is the order of the differential equation, so that our (1) is an example of his D2. For D2 equations he finds a connection with the quantum cohomology of Del Pezzo surfaces, and for D3 equations he gives both a connection with the quantum cohomology of certain Fano threefolds and a relation to modular forms similar to our Section 5.

References

1. G. Almkvist and W. Zudilin, *Differential equations, mirror maps and zeta values*, Mirror Symmetry. V (Banff, AB, 2003) (N. Yui, S.-T. Yau, and J. D. Lewis, eds.), AMS/IP Stud. Adv. Math., vol. 38, Amer. Math. Soc., Providence, RI, 2006, pp. 481–515.
2. A. Beauville, *Les familles stables de courbes elliptiques sur* $\mathbf{P}_1$ *admettant quatre fibres singulières*, C. R. Acad. Sci. Paris Sér. I Math. **294** (1982), no. 19, 657–660 (French, with English summary).
3. F. Beukers, *Irrationality proofs using modular forms*, Journées arithmétiques de Besançon (Besançon, 1985), 1987, pp. 271–283,.
4. ———, *On Dwork's accessory parameter problem*, Math. Z. **241** (2002), no. 2, 425–444.
5. J. H. Bruinier, G. van der Geer, G. Harder, and D. Zagier, *The* 1-2-3 *of modular forms*, Universitext, Springer, Berlin, 2008.
6. V. Golyshev, *Classification problems and mirror duality*, available at `arXiv:math/0510287`.
7. U. Schmickler-Hirzebruch, *Elliptische Flächen über* $\mathrm{P}_1\mathbf{C}$ *mit drei Ausnahmefasern und die hypergeometrische Differentialgleichung*, Schriftenreihe Math. Inst. Univ. Munster 2. Ser., vol. 33, Univ. Münster Math. Inst., Münster, 1985.
8. J. Stienstra, *Mahler measure variations, Eisenstein series and instanton expansions*, Mirror Symmetry. V (Banff, AB, 2003) (N. Yui, S.-T. Yau, and J. D. Lewis, eds.), AMS/IP Stud. Adv. Math., vol. 38, Amer. Math. Soc., Providence, RI, 2006, pp. 139–150.
9. J. Stienstra and F. Beukers, *On the Picard–Fuchs equation and the formal Brauer group of certain elliptic* K3*-surfaces*, Math. Ann. **271** (1985), no. 2, 269–304.
10. H. A. Verrill, *Picard–Fuchs equations of some families of elliptic curves*, Proceedings on Moonshine and Related Topics (Montréal, QC, 1999) (J MacKay and A. Sebbar, eds.), CRM Proc. Lecture Notes, vol. 30, Amer. Math. Soc., Providence, RI, 2001, pp. 253–268.
11. Y. Yang, *Apéry limits and special values of L-functions*, available at `arXiv:0709:1968v1`.

12. W. Zudilin, *An Apéry-like difference equation for Catalan's constant*, Electron. J. Combin. **10** (2003), R14.

MAX-PLANCK-INSTITUT FÜR MATHEMATIK, VIVATSGASSE 7, 53111 BONN, GERMANY
E-mail address: `don@mpim-bonn.mpg.de`